Future Challenges in the Framework of Integrated and Sustainable Environmental Planning: Case Studies and Innovative Proposals

Future Challenges in the Framework of Integrated and Sustainable Environmental Planning: Case Studies and Innovative Proposals

Editors

Salvador García-Ayllón Veintimilla

Antonio Tomás Espín

MDPI • Basel • Beijing • Wuhan • Barcelona • Belgrade • Manchester • Tokyo • Cluj • Tianjin

Editors
Salvador García-Ayllón Veintimilla
Technical University of Cartagena
Spain

Antonio Tomás Espín
Technical University of Cartagena
Spain

Editorial Office
MDPI
St. Alban-Anlage 66
4052 Basel, Switzerland

This is a reprint of articles from the Special Issue published online in the open access journal *International Journal of Environmental Research and Public Health* (ISSN 1660-4601) (available at: https://www.mdpi.com/journal/ijerph/special_issues/environmental_planning).

For citation purposes, cite each article independently as indicated on the article page online and as indicated below:

LastName, A.A.; LastName, B.B.; LastName, C.C. Article Title. *Journal Name* **Year**, *Volume Number*, Page Range.

ISBN 978-3-0365-7380-9 (Hbk)
ISBN 978-3-0365-7381-6 (PDF)

Cover image courtesy of Salvador García-Ayllón Veintimilla

Contents

About the Editors

Salvador García-Ayllón Veintimilla

Salvador García-Ayllón Veintimilla is an Architect and Civil Engineer. He has a PhD in Urban Planning and Territorial Management from the Technical University of Valencia. Since 2010, he has been an Associated Professor responsible for various courses related to territorial policy and environmental and infrastructure planning at the Faculty Civil Engineering of the UPCT. He has been a visiting professor at numerous universities around the world, including California Berkeley (USA), Alcalá de Henares (Spain), CEPES (Mexico), and Cyprus University of Technology. He is currently an invited professor at the Massachusetts Institute of Technology (MIT) through a Fulbright grant from the United States government. In the scientific field, he is a researcher in charge of the R&D group of Territorial Policy, Environmental and Infrastructure Planning (http://potepapi.upct.es) and the Director of the Biyectiva GIS Business Professorship of the UPCT. He is the author of more than 50 publications in international research journals and 15 books, being part of the 2% most cited authors in the world according to the list published by Stanford University.

Antonio Tomás Espín

Antonio Tomás Espín has a PhD in Civil Engineering. He is an Associated Professor of the Department of Civil and Mining Engineering of the Polytechnic University of Cartagena, of which he has been director, with more than 20 years of experience in teaching and research. In addition, he has been dean of the College of Engineers, Roads, Canals and Ports of the Region of Murcia. Currently his lines of research are specialized in issues related to sustainability and construction engineering, being the author of numerous publications in international scientific journals.

Preface to "Future Challenges in the Framework of Integrated and Sustainable Environmental Planning: Case Studies and Innovative Proposals"

The environmental planning of territory has become an indispensable tool in today's society. The complexity and dynamism of the processes of pollution and the alteration of territory are phenomena that require new, more comprehensive environmental policies and integrated and integrative approaches. This changing context forces administrations and researchers in this field to develop increasingly sophisticated and demanding frameworks. However, these actions must be carried out considering the increasingly multidisciplinary nature of these phenomena and assuming the principle of reality that forces them to be socially and economically sustainable. This Special Issue includes contributions involving innovative approaches or relevant case studies regarding environmental policy and management in topics such as sustainability frameworks linked to the optimization of land and resource consumption; socioecological approaches to solve environmental problems; protected area policies and management; risk analysis derived from the human anthropization of natural resources; regulatory applications with the aim of integrating environmental knowledge (e.g., EU INSPIRE Directive); or the spatial analysis of diffuse pollution phenomena, among others. Innovative methodologies, frameworks, or significant results from relevant case studies related to all these topics were welcomed, and similar articles were also considered for publication if they fit within the scope of this Special Issue.

Salvador García-Ayllón Veintimilla and Antonio Tomás Espín
Editors

International Journal of
Environmental Research and Public Health

Article

Sustainable Environmental Strategies for Shrinking Cities Based on Processing Successful Case Studies Facing Decline Using a Decision-Support System

Francisco Sergio Campos-Sánchez *, Rafael Reinoso-Bellido and Francisco Javier Abarca-Álvarez

Department of Urban and Spatial Planning, University of Granada, Granada 18071, Spain; rafaelreinoso@ugr.es (R.R.-B.); fcoabarca@ugr.es (F.J.A.-A.)

* Correspondence: scampos@ugr.es

Received: 30 August 2019; Accepted: 1 October 2019; Published: 3 October 2019

Abstract: Since the middle of the last century post-industrial cities around the world have been losing population and shrinking due to the decline of their structural growth models, showing important socioeconomic transformations. This is a negative phenomenon but one that cities can benefit from. The aim of this work is to verify what type of measures against urban decline would be most suitable if applied to a specific case study. To do this, international cases of shrinking cities where successful measures were already carried out facing decline: (i) are collected, (ii) are classified based on several influencing criteria, and (iii) are grouped under similar alternatives against the decline. Measures and criteria focused on achieving sustainability are emphasized. Alternatives are then prioritised using an Analytic Hierarchy Process designed at several hierarchical levels. The results are discussed based on the construction of sustainable future scenarios according to the optimal alternatives regarding the case study, improving the model validity. The work evidences that environmental and low-cost measures encouraging the economy and increasing the quality of life, regardless of the city size-population range where they were performed, may be the most replicable. Future research lines on the integration of the method together with other decision-support systems and techniques are provided.

Keywords: shrinking cities; environmental planning; territorial sustainability; analytic hierarchy process; decision-support systems

---

1. Introduction

The Shrinking Cities International Research Network (SCiRN)—a worldwide research consortium of scholars and experts from various institutions (30 members from 14 countries) pursuing research on shrinking cities in a global context to advance international understanding about population decrease and urban decline as well as causes, manifestations, and effectiveness of policies and planning initiatives so stave off decline—, agreed in 2004 that the term 'shrinking city' (i.e., declining cities) refers to a global phenomenon by which, since the middle of the last century, several hundred cities and urban areas (> 10,000 inhabitants—in 370 cities with more than 100,000 residents—) have been losing population, while undergoing major economic transformations, evidencing the decline of their structural growth models. This is mainly due to the lack of competitiveness of their manufacturing industrial production model, to which other facts contribute, e.g., suburbanization, war, natural disasters, ageing population, low fertility, or socialist systems breakdown [1,2]. Pallagst et al. introduced multidimensionality as an additional attribute of this phenomenon, adding social decline and environmental impacts to economic problems [3]. Today, many cities and regions are still trying to address this problem using different recovery strategies.

The urban decline varies according to territories, continents and particular casuistry. In Europe this problem was/is traditionally due to low fertility, emigration, the real estate boom, housing abandonment or de-industrialization, which has led to the testing of recovery strategies of diverse amplitude, profile and cost. Among others, some examples of these are those focusing on restoring market balance through housing demolitions (e.g., Germany in the 1990s), urban regeneration (e.g., in U.K.), or attracting foreign investment and increasing territorial competitiveness (e.g., in post-socialist countries) [4]. In Asia the reasons for shrinkage had/have more to do with the population ageing, exodus from rural areas to large metropolises, property speculation or economic inequality (e.g., in Japan, and in the near future in China). Since the 1970s, in North America these problems are based on suburban expansion and deterioration without regional planning perspectives, or on national economic restructuring. In Latin America they had/have to do with socio-economic inequality, emigration or the population displacement to large metropolitan regions, for example [3,5,6].

1.1. The Spanish Case

The shrinkage phenomenon was not as important in Spain as in other territories within the European context, due partly to Spain's late incorporation into the industrial process and the characteristics of its economy, which is more focused on services and tourism [7]. However, the causes and effects of shrinkage are similar to those of other countries. From the 1970s onwards, the end of industrial-mining economic and productive activity in highly specialised and resourceful areas went into decline, losing population and threatening their economic and social structures [8].

Among the few initiatives facing the decline in Spain, those developed in the Spanish (industrialised) north stand out. Perhaps because of its relevance, the most paradigmatic example is probably the deep industrial transformation of the Bilbao's waterfront from the end of the last century, involving important urban interventions (e.g., Bilbao Ria 2000 project) that would carry the name of "Guggenheim effect" [9]. Also relevant are the cases of some significant Spanish mining and industrial cities, such as (i) the partial recovery of Ponferrada, thanks to industrial investments; and (ii) to a lesser extent, Puertollano, due to its renewable energy factories and its trend towards green tourism and the reduction of CO_2; or the socio-economic regression process of some mining cities of Asturias (e.g., Mieres, Langreo) until today [10].

Similar processes (adjustment, reindustrialization, outsourcing, and urban transformation) took place in all of them, but with different results [10]. One of the most noteworthy cases was the post-industrial (after steel and iron industry), functional and landscape transformation of Avilés by urban tourism. The measures focused on the use of brownfields, environmental decontamination, regeneration of the port, historic centre improvement, and tourism policy, with important projects such as the building of the Niemeyer's Cultural Centre, which was not without management and financial problems [11]. In many of these cases, urban transformations involved parallel restoration and reuse of industrial heritage [12].

Currently, the shrinkage in the northern Spain and other areas of this country are still active. There is evidence of this, such as the increase in company closures and unemployed number, e.g. the steel companies Alcoa Company in La Coruña and Avilés, and Megasa Company in Ferrol, the naval one in Sestao, or the wind turbine plant of Vestas Company in León. Structural reasons of a productive and technological nature, lack of competitiveness and rising electricity prices are argued. To this is added the current transition of the European Union (EU) towards an economy free of greenhouse gases and the compliance with the Paris Agreement against the climate change, which lead to the closure of all non-competitive thermal and coal production plants by 2030. This has an impact on those European territories where coal still supports employment, such as Spain. Specifically, EU's decision 787 ordering the immediate closure of all non-competitive mining activity by the end of 2018, currently affects several mining companies and cities in northern Spain, such as the case of Hunosa Company in Mieres, reported in the press.

1.2. Starting Point

The working hypothesis is that specific solutions against urban decline successfully developed in certain territories could be as suitable for others, depending on the similarity and potential of each context. The lessons learned from this type of successful experiences would be useful as recovery strategic models for other cities that have not yet taken the appropriate measures or in which the measures developed have not had the expected success. This would be as long as the appropriate comparative framework was provided based on the context of both the successful urban decline experiences and the specific case study.

This would be done by answering the following research questions: Does decision-making help to explore the suitability of multiple effective alternatives against the shrinkage? What types of initiatives carried out more or less effectively in decline cities would be most likely to be replicated successfully in other similar case studies? What factors would be the most important to manage taking into account contemporary aspects such as investment priority of sustainability? The challenge is to answer these questions by analysing case studies using a hierarchical decision model, which could serve as a guide for planning professionals facing urban decline problems.

The aim of this work is to explore successful city cases against urban decline using a decision support system to prioritize lessons learned in terms of sustainability, what is done regarding a specific case study. This exploration is carried out in a comparative and integrated way at a macro level. The subsequent discussion of the results according to the potential of the case study to replicate the optimal measures against decline details the model applied. In addition, the integration suitability of the analysis method developed together with other analysis techniques is emphasized. The work is relevant as a useful method to select initially those measures facing urban decline that would best suit a specific case study, fitting as a guiding instrument within sustainable planning.

2. Research Background

2.1. Individual Case Studies VS Integrated Comparative Studies

Given this shrinking cities global scenario, since the beginning of this century, many programs and meetings were held, especially to expose the various cases of decline and to make this problem known to the international community. Some of the best known may be the Shrinking Cities project in the 2000s [13], the Shrinking Cities International Research Network (SCiRN) in 2004 supported by the Institute of Urban and Regional Development at the University of California (Berkeley), the COST-Action CIRES programme, or the Shrink Smart project. All of them have brought together academics, professionals and experts who researched the paths and results obtained in different cities.

However, it seems that this diversity of initiatives often resulted in international debates based on the exposure of individual local cases without agreement on common research strategies on the phenomenon. In contrast, comparative research between case studies should better enable the influence of local conditions to be isolated and criteria to be unified. In addition to large-scale quantitative research on urban decline, comparative case studies provide valuable information on specific patterns and paths of decline [14]. Extensive international comparative studies have helped to highlight the relevance of shrinkage [8,15]. Within this group of studies, some of them, such as that of Großmann et al. (2013) support more integrated research focused on removing transnational knowledge barriers [14].

At the UC Berkeley symposium in 2007 organized by the Centre for Global Metropolitan Studies together with the Institute of Urban and Regional Development, SCiRN-sponsored research made it clear that the comparative approach between this type of cities by studying cases from a global perspective is unique and innovative in improving the quality of life in them. Nevertheless, there is still little comparative work that collects the lesson learned in each case in an integrated way [5], which could be useful to the planning and political agenda of the involved cities.

An example of this last group of works is the research by Sánchez-Moral et al. focused on Avilés (Asturias, Spain), close to the case study of this work [16]. The strategies developed in the 2000s to revitalise the economy of Avilés included measures already taken in other manufacturing regions of Europe, e.g., privatisation of public companies, attraction of foreign investment, reorientation of the economy towards innovation and creativity, or development of iconic projects, such as Niemeyer Centre.

2.2. Studies on Urban Decline within the Sustainability Framework

The urban decline is generally something negative for a city, but it can also be used as a solution to the problem, for example from an ecological point of view: space available for other post-industrial opportunities, free land once it has been decontaminated, reuse of old industrial buildings, fresh air, etc. In addition, the urban development opportunity that this available space could offer should be taken advantage of within the context of the 2030 Agenda for Sustainable Development Goals (SDG) of the United Nations (UN). It seems that it is time for planners and decision-makers who have traditionally favoured the growth and expansion of cities to think now about appropriately planning their shrinkage, but how to do that is an important question.

Studies focused on finding answers against decline that, rather than local interest, support a recovery based on the common good are particularly interesting due to their emergence, such as those: (i) that are influenced by a growing awareness of climate change [17,18]; (ii) that suggest using a resilient perspective to deal with this type of change by means of an approach that combines ecology and economy (e.g., the use of forests as nature tourism) [19], or that address the urban form (e.g., by assessing urban compactness) [20]; (iii) that rely on the social dimension (capital, social mix, creative talent) and the use of endogenous resources to drive urban economic growth [21,22]; (iv) that suggest a change of paradigm through 'smart shrinkage' in a planned way; (v) that address the complex relationships between socio-demography, infrastructure, land-use, ecosystem services and biodiversity of shrinking cities to ensure urban quality of life and healthy urban ecosystems under shrinkage conditions [2]. This is by reviewing the principles on which traditional urban policies were based, focusing primarily on growth and expansion [5,23–25]. These last works analysed sustainable recovery strategies based, for example, on the depopulation of deteriorated neighbourhoods, the ecological management of the post-industrial land, the new challenge for both land-use and biodiversity research, or the use of economic development plans that emphasized controlled urban shrinkage.

2.3. Assessment of Case Studies Using Decision-Support Systems

Therefore, an approach that assesses successful experiences against decline in a comparative and integrated way, and obtains relevant information from them, seems to be useful for decision-making on a specific shrinking case study. In this regard, it is recognized that Decision Support Systems (DSS) are effective tools for the incorporation, integration and decision support of complex problems by reducing indeterminacy and improvisation on their resolution [26]. Basically, decision-making consists of a selection process among alternative courses of action based on a set of criteria to achieve one or more aims [27]. The multi-criteria evaluation (MCE), defined as a set of techniques to support decision-making processes, is useful for this purpose. It is based on the weighting of variables that influence the decision activity, and that have to be previously classified for selection among alternatives.

Urban shrinkage is a complex problem, due to the diversity of processes and stakeholders involved, so decision support can be a useful tool for analysis. There are many studies on decision support, and also on shrinking cities, but there are a few studies that work on both issues in an integrated way, hence its novelty. Some examples of the latter could be (i) the multi-criteria evaluation that Schetke & Haase carried out to assess the socio-environmental impact of the urban shrinkage suffered by many German cities of the old GDR [28]; or (ii) the prescriptive decision model (OR/MS) developed by Johnson et al. for the management of brownfields as a neighbourhood development strategy in a small town of Bristol (MA, USA) [29]. The former studied the socio-environmental impact of micro-scale demolition as a strategy against decline in several urban areas of Leipzig, showing development differences between these areas, such as urban expansion versus redensification and redevelopment. This is due to assumption of demolition as a generic strategy facing decline in this region. However, a preliminary analysis that initially took into account different strategic options could open up new ad hoc opportunities against decline in each case, depending on its features. In addition, other interesting empirical studies on shrinking cities from a sustainable environmental perspective were cited and compiled in the research of Haase [2].

The approach to evaluate successful initiatives against decline regarding a particular case study depends on a specific aim and process in which some alternatives are selected from a group of them to obtain the optimal one. For this reason, the Analytical Hierarchy Process (AHP), developed by Saaty, seems to be a suitable DSS for application in several problems of this type [30]. By means of comparison in pairs it is possible to identify the relative importance of each criterion and decision alternative. The AHP has already been used to weight sustainable urban decline measures. For example (i) Lee and Lim's approach based on an AHP oriented assessment of urban regeneration projects in South Korea according to three sustainability goals (physic, social and economic) by comparing economic and community indicators [31]. This is a local study that excludes aspects such as environmental sustainability and project costs, which may lead to some biases; or (ii) the multi-criteria analysis according to AHP hierarchical modelling carried out by Hemphill et al. in the light of a literature review on sustainable urban regeneration interventions and policies and their evaluation indicators [32]. It is a broad study in its approach and addressed to an expert panel but with no specific cases of applicability.

3. Materials and Methods

In general, the methodology is developed by taking the following steps: (1) Collection of case studies against urban decline and successful practices that were generally carried out in them. (2) Design of a hierarchical decision support model (AHP) by classifying the above information according to criteria (influencing factors) and alternatives (initiatives adopted). (3) Application of the AHP model by comparing criteria and alternatives, and prioritizing results. Finally, the results are discussed and future lines of research are recommended.

3.1. Case Study

This work focuses on an important post-industrial mining medium-sized city in northern Spain: Mieres (Asturias). It is a highly representative case of active processes of urban decline with economic-industrial origin in this country. Mieres has a population of 38,962 inhabitants according to the Spanish Statistical Office—known in Spanish as the INE—in 2017. It was an important mining urban node during the 19th century thanks to the coal production, being also a steel and iron centre until 1970, which is evidenced by the existence of a rich industrial heritage. (e.g. coal washer El Batán, currently being exploited by the Hunosa company). The crisis in both economic sectors has led to the loss of more than 30,000 inhabitants from 1960 (when the city had a population of 70,871) to the present day. Table 1 shows a summary characterisation of the case study based on a partial strengths, weaknesses, opportunities, and threats (SWOT) analysis.

Table 1. Case study summary table.

Scope	Weaknesses/Threats	Strengths/Opportunities
Non-residential land uses	Flood plains used for old industrial uses	Obsolescence of the industrial fabric
	Old industrial uses next to residential uses	Obsolescence of the industrial fabric
	Obsolete urban fabric	Demolitions, reuse
	The commercial use shows some development potential within the urban centre, but this area is deficient	Pedestrianisation of some streets in the city centre
	Scarce tourism	Existence of a rich industrial heritage

Table 1. *Cont.*

Scope	Weaknesses/Threats	Strengths/Opportunities
Residential land uses	40% of the population employed in Mieres resides outside Mieres	-
	Obsolete urban fabric, lack of housing	Demolitions, reuse (recovery of old and central mining neighbourhoods through public funding)
	Failed new housing developments (including social housing –VPO–)	-
Heritage	Facade interventions only, abandoned or poorly built elements	Elements of historical industrial interest
	Recovery of old railway lines as greenways (approx. 200 km), but only existing around the urban area	Post-industrial land available in urban areas
Infrastructure	High-speed train (AVE) not available	The aim is to get an AVE stop on the León-Gijón line.
	Dominant transport by private vehicle, very few km of cycling routes	Municipal bus lines available, fluvial pedestrian walkways of interest available
Economy	Non-competitive traditional mining-steel activity	Alternative economic sectors under development (e.g., thermoelectric energy, solar energy, renewed steel industry, building materials, ICTs, tertiarization, services)
	Hard to attract new economic sectors despite new incentives	-
	Excess of public funding	-
Environment	Pollution (derived from thermoelectric and cement plants, among others)	Natural environment
Others	General lack of urban land	Obsolescence of the industrial fabric

Source: Prepared by the authors based on the research of Tomé [10].

3.2. Case Collection

The response of regions and cities against the decline has been and is being very diverse, depending on the particular conditions and the socio-economic and political context in each case [33]. Through a literature review, macro-level measures against urban decline that were successfully carried out in selected international cities and regions were collected and characterised. The list is not exhaustive; only the most frequently mentioned or best documented cases were collected through multiple studies and reports. The information obtained from the compilation of experiences was detailed in order to be able to compare them with a specific case study. This would allow the subsequent design of the decision-making model. Each case was characterised by the following steps (see Table 2): (i) case identification; (ii) identification of the declining socio-economic sector; (iii) source identification; (iv) macro-level description of the dominant recovery initiatives from the decline developed in each case; and (v) identification of the dominant sustainable profile that supports or feeds each initiative.

Table 2. Some successful city cases against urban decline.

Case Collection	Declining Sector	Reference	Dominant Recovery Strategies	Initiative Profile
1. Avilés (SP)	Economic (steel and iron industry)	[11,16]	Urban tourism, functional and landscape transformation of brownfields, environmental adaptation, historic centre and port regeneration	Environmental
2. Baltimore/ Houston (US)	Economic (industrial)	[34]	Post-industrial public land given in exploitation to private property in exchange for investment and new vertical uses	Economic
3. Berlin (GE)	Economic (industrial)	[35]	Urban densification policies	Social
4. Bilbao (SP)	Economic (steel and iron industry)	[9]	Industrial restructuring, urban revitalisation, new urban facilities and services (metro, new urban nodes, etc.), tourism	Environmental
5. Asturian mining cities (e.g. Mieres, Langreo) (SP)	Economic (mining, steel and iron industry)	[10]	Adaptation, reindustrialization, tertiarization, urban transformation	Economic
6. Cleveland (US)	Economic (industrial)	[3,5,34]	Landscape transformation of the post-industrial footprint	Environmental
7. Detroit (US)	Social (racial, social, spatial segregation)	[13]	Urban transformation of central areas, cultural and creative revitalization of suburbs, guided immigration	Social
8. Estonia/ Central Germany (ES/GE)	Social (political, post-socialism, economic restructuring)	[22]	Governance focused on the accumulation of local social capital	Social
9. Fuxin (CH)	Economic (lack of resources)	[33]	Experimental structural economic change (settlement of technology parks and economic development areas in general)	Economic
10. Halle/Leipzig (GE)	Social (emigration due to German reunification)	[2,13,28]	Public subsidies, mass demolition operations	Economic
11. Ivanovo (RU)	Social (USSR's fall, globalization, deindustrialization	[13]	Subsistence agriculture, post-industrial practices, local social initiatives	Social
12. Lieksa (FI)	Economic (industrial –natural resources processing–)	[19]	Resilience and adaptability based on: wood industrial sector transformation, especially nature tourism, and internet and phone (call-centres) economy	Environmental
13. Manchester/ Liverpool (UK)	Economic (industrial) and social (insecurity, unemployment)	[13]	Recovery of empty buildings in central urban areas, new urban culture (music, fashion, media), public-private partnerships, call-centre development	Social
14. México DF (central city) (ME)	Social (gentrification, insecurity, emigration)	[8,36,37]	Urban renewal of the historic centre through large-scale investments (walkways, high-rise buildings, singular projects)	Economic
15. Mulhouse/ Roubaix/ Saint-Etienne (FR)	Economic (industrial –steel and iron, textile, weapons–)	[21]	Creative talent attraction and social mix to drive urban economic growth	Social
16. Newcastle (UK)	Economic (shipyards)	[33]	Transformation into a museum, arts and sciences city centre	Social

Table 2. *Cont.*

Case Collection	Declining Sector	Reference	Dominant Recovery Strategies	Initiative Profile
17. New York (US)	Economic (industrial)	[38]	Infrastructure resizing, flexible transport	Environmental
18. New York/ Chicago (US)	Economic (industrial)	[34]	Post-industrial public land transferred in exploitation to private property in exchange for investment and horizontal uses	Environmental
19. Philadelphia (US)	Economic (industrial)	[5,34]	Landscape transformation of the post-industrial footprint	Environmental
20. Pittsburgh (US)	Economic (steel and iron industry)	[3,5]	Settlement of prestigious universities and research centres	Social
21. Ponferrada (SP)	Economic (mining, industrial)	[10]	Industrial investments	Economic
22. Puertollano (SP)	Economic (mining, industrial)	[10]	Industrial adaptation to renewable energy, green tourism, CO_2 reduction	Environmental
23. São Paulo (central city) (BR)	Social (gentrification, overcrowding, inequity)	[36,39,40]	City centre renewal through the social reuse (cultural, major events) of historic buildings with public-private investments	Social
24. St. Louis (US)	Social (emigration)	[5,34]	Community and social cohesion oriented urban planning	Social
25. Ruhr Valley (GE)	Economic (industrial)	[41,42]	Environmental mitigation and ecological restoration by planting post-industrial forests	Environmental

Source: Prepared by the authors based on their literature review.

3.3. *Decision-Making Model Design*

3.3.1. Hierarchical Levels

The first level or model goal (N1) is to know which alternative is the most suitable to achieve the aim of the work. In other words, which initiative against the urban decline already tested in other cities or regions related to the specific case study (Mieres) would be more likely to succeed if applied to the latter.

For the second hierarchical level design (N2), the case collection information is classified according to criteria (influencing factors). In the literature review, the factors that have commonly characterised the chosen sample of frequent cases (n = 25) were the following: (C1) Population-size range. A distinction is made between large, medium-size and small cities. (C2) Origin of urban decline involving population loss. Three types of main causes are identified: environmental, social or economic. (C3) Profile of the dominant initiative facing decline. As argued above, from a sustainability perspective three types of dominant profiles are also identified: environmental, social and economic. (C4) Cost of the initiative (economic, political, social, resource consumption level, etc.) by nominal categorizing the cost levels as high, medium or low. In order to simplify the model, similar cases were grouped in this hierarchical level, i.e., cases that show the same attributes according to the mentioned criteria (C1–C4). Within alternative H (8, 11, 24, i.e., Estonia/Central Germany–Ivanovo–St. Louis, respectively) it was considered that case n° 8 (Estonia and central Germany's medium-sized and small cities) accumulates a large population as a whole, and could therefore be grouped together with cases n° 11 and 24.

The alternatives or third hierarchical level of the decision-making model (N3) were identified with the initiatives or groupings of them facing urban decline in each case, according to the cities or urban regions where they were carried out (n = 12).

Priorities (P) represent the relative weights of the nodes in any level hierarchy. 'Weight' can refer to importance (or preference, or likelihood) or whatever factor is being considered by decision making process (i.e., goal—N1—, criteria or influencing factors—N2—, and alternatives or initiatives—N3—). The priorities of each hierarchical level—local priorities—always add up to 1.00 (100%). Regarding our analysis, the sum of the alternative priorities at each criterion level adds to 1.00 (100%). The global priorities (i.e., the priority of each alternative at global level—GP—) are obtained by adding the values resulting from multiplying the alternative priority at each criterion level by the criterion priority. These alternative global priorities add to 1.00 (100%). Each GP represents the importance (i.e. weight) of its corresponding alternative in the analysis as a whole. Thus, the global priority of alternatives means a key factor to decide which of them seems to be the most successful facing urban decline in the case study. The decision-making model design is shown in Figure 1.

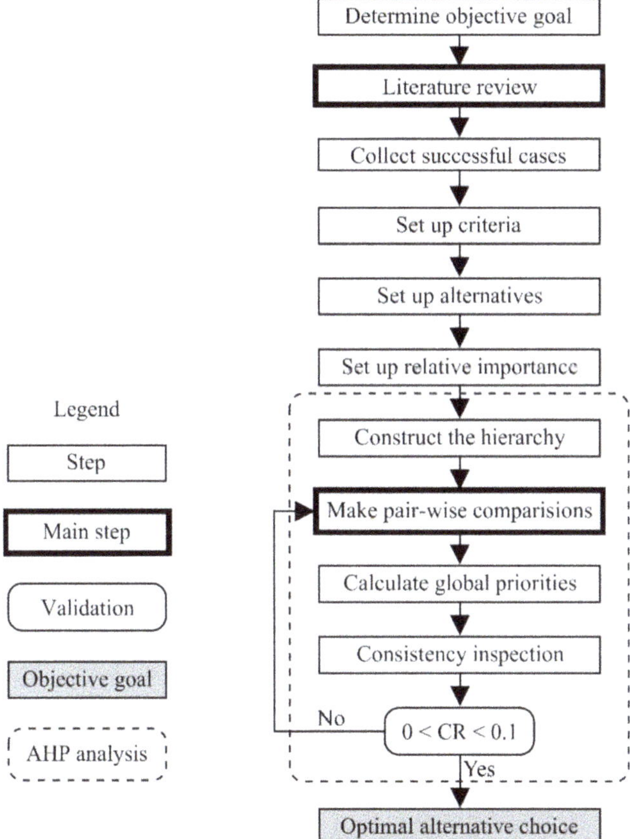

Figure 1. Decision-making model design. Source: Prepared by the authors.

3.3.2. AHP Analysis

The information obtained in the previous steps was processed using an AHP multi-criteria matrix. Pair-wise comparisons of criteria (N2) and alternatives (N3) were made according to the Saaty's numerical scale (1–9), highlighting the relative importance of some elements over others in each hierarchical level [30]. Alternatives level pair-wise comparison was carried out according to the defined criteria. The following comparative scale values are given: 1 = the same importance; 3 = weak dominance; 5 = strong dominance; 7 = demonstrated or very strong dominance; 9 = absolute dominance. In addition, the coherence of the model was checked by calculating the Consistency Index (CI), the matrix being consistent when CI < 10%. The results of the pair-wise comparisons are arranged in a matrix. The first (dominant) normalized right Eigen vector of the matrix gives the ratio scale (weighting) [30]. The Eigen value λ_{max} determines the Consistency Index: $CI = \frac{\lambda_{max}-n}{n-1}$ where n is the comparison matrix size. The AHP analysis was performed using the web tool AHP Online System (https://bpmsg.com/ahp-online-system/).

It should be noted that the design and structuring of an analytical hierarchy 'is more an art than a science' since there is not a precise expression for the identification or stratification of the elements involved in the process [43]. For this reason, an attempt was made to design the decision-making model based on logical criteria, such as: (i) dominance type in pair-wise comparisons of both criteria and alternatives; (ii) information, priorities and frequencies obtained from a literature review; or (iii) emergence and applicability level of initiative profiles. For the assignment of relative importance in each hierarchical level it is commonly used an expert panel, although there is evidence of studies supported by literature reviews regarding the study field [44,45], a methodological approach followed in this work.

The input data (attributes of successful cases facing urban decline) were collected from mostly theoretical academic studies, which are the most common in this field (see Table 2). In general, these studies do not handle large volumes of objective data produced by agreed indicators that allow empirically contrasting the strategies developed against urban decline in each case. However, they do usually include general information such as population data, goals stated, and different factors that led to involve some recovery strategies rather than others. For this reason, the AHP multi-criteria analysis was considered more suitable for assessing the different alternatives, as compared to other DSS such as decision trees (DT) from data mining computational field, despite its apparent similarity.

Briefly, AHP is a structured quantitative and qualitative method based on multiple criteria of experts' judgement that are pair-wise compared to obtain the priorities of several alternatives [46]. In contrast, DT is a quantitative technique initially more suitable for massive data analysis in which decision alternatives are divided according to probabilities and prefixed rules until the most likely one is found [47]. Both methods are useful for solving complex problems. In both cases the representation looks like a hierarchical tree, but using a different processing algorithm. A limitation of the AHP method is that subjective factors cannot be completely excluded [48]. To avoid this, an expert agreement is often taken into account, which in this work was replaced by a literature review (which in a sense is the same thing). This also makes it possible to identify objective criteria and relative importance for the criteria and for the alternatives according to each criterion. Both methods can be used together in the same analysis. For example, the AHP method would set priorities, and the DT method would show the sequential path of all possible decisions up to the most suitable alternative [49].

3.3.3. Relative Importance of Criteria and Alternatives

Given the exploratory approach of the method and in order to develop a more realistic study, the assessment of relative importance (RI) is carried out using a numerical range (1–5) of pair-wise comparison, excluding values 7 (demonstrated dominances) and 9 (absolute dominances). The values adopted at each hierarchical level are justified as follows:

(i) Criteria pair-wise comparison (C1–C4). In general, the most important thing is to strategically plan the initiative type according to the cause of the problem and the goal to be achieved in each case, and that it is also feasible. In addition, the measure scale is a factor to be taken into account, even though it is known that the same initiative can be applied in a multi-scale manner [50,51]. Therefore, at a first level of importance, an RI = 5 is provided for the initiative profile (C3), a criterion associated to the strategy against the decline developed in each case. A second level of importance is considered using RI = 3 based on decline origin (C2) and initiative cost (C4). Finally, at a third level of importance, RI = 1 is provided for the population-size range criterion of each city studied (C1), understood as a more relative decision-making factor. Regarding the results, this last choice encourages that the most suitable alternatives to the case study do not depend so much on the population-size factor as on other higher relative importance criteria.

(ii) Alternatives pair-wise comparison (1–12). (a) For population-size range (C1) and decline origin (C2) criteria, the relative importance of each alternative (i.e., initiative) is assigned according to its similarity with the case study (RI = 3 for similar cases; RI = 1 for other cases). The aim is to prioritize those initiatives that are comparatively developed in a contextual framework close to the case study, which may help to ensure the strategic suitability of the selected initiatives. (b) According to the literature review, and within the SDG framework of the UN Agenda 2030, the sustainable initiatives facing urban decline, and especially the environmental ones, seem to be the most emerging and frequently successfully applied. This type on initiatives, even those that keep an industrial use but ecologically adapted, are the ones that better preserve the environmental and natural base of a territory, being therefore the most valuable for sustainability [52,53]. On the other hand, environmental initiatives of ecological nature tend to be the lowest cost [54]. Hence, an RI = 3 is assumed for environmentally sustainable measures (being chosen RI = 1 for others) within the initiative profile criterion (C3). (c) Finally, regarding the cost factor, priority is given to the lowest cost alternative, i.e., those with the highest viability (RI = 5) as compared to those with a medium (RI = 3) or high cost (RI = 1). Each alternative is weighted by giving these RI values, obtaining a hierarchical list of initiatives against decline (A–L; n = 12) according to the case study.

4. Results

Through a literature review, general information on successful cases facing urban decline were collected and detailed in Table 2, at least in some respects, such as (a) landscape measures on post-industrial footprint in Cleveland and Philadelphia, showing an increase in close properties values [3]; (b) environmental mitigation and ecological restoration by means of the 'industrial forests' project in the Ruhr Valley [41,42]; (c) resizing infrastructure or 'flexible' transport in New York [38]; (d) densified urban 'archipelagos' proposal in Berlin [35]; (e) community oriented planning to stop population loss in St. Louis [5]; (f) incentive policies for birth rate, subsidy and capital in Western European countries since the middle of the 20th century; (g) recovery initiatives based on attracting creative social class according to the Creative Cities project [55]; or (h) central city renewal projects of large Latin American metropolitan areas (e.g., São Paulo, Buenos Aires or Mexico DF) with a view to keeping population in these areas, despite their intrinsic problems (e.g., gentrification, overcrowding, insecurity), and reaching the global city status [36,39,40], among others. In addition, some successful initiatives in the USA are based on a public-private relationship. With this, vacant land left by the earlier production system (Fordism) which demands rehabilitation, land decontamination or landscape measures, is transferred in exploitation to private property which is done in exchange for investing to recover them with new land uses (Post-Fordism). This investment may be horizontal (e.g., the implementation of parks and playgrounds in New York, or golf courses in Chicago's Downtown); or vertical by redensification interventions using mixed uses, such as in Baltimore or Houston [34].

Table 3 shows the different cases collected grouped into alternatives when the cases are similar (A–L) and classified according to the criteria adopted (C1–C4), as well as the relative importance assigned to each of them (1–5). This is the base table where the different hierarchical levels and their processing are justified, with a view to the subsequent AHP analysis. According to the literature reviewed, there is a dominance of alternatives focused on large cities (75%), with a decline origin of economic base (75%), which developed measures facing the decline of different profile (33% each case) and high cost (42%). This bias should not influence the analysis results since the sample was weighted according to the relative importance assigned at each hierarchical level. Due to its emergency, the absence of cases with a decline origin based on environmental factors is noteworthy. However, the UN and other institutions have been talking about 'climate refugees' for years, though their legal status has not yet been recognized.

Table 3. Influencing factors (criteria C1–C4), initiatives (alternatives A–L), and relative importance (1–5).

Alternatives (Cities/Regions)	C1 (1) Population			C2 (3) Decline Origin			C3 (5) Initiative Profile			C4 (3) Initiative Cost		
	LA (1)	MD (3)	SM (1)	EN (1)	SO (1)	EC (3)	EN (3)	SO (1)	EC (1)	HI (1)	ME (3)	LO (5)
A (1, 12, 22)		●				●	●				●	
B (5)		●				●			●		●	
C (4)	●					●	●			●		
D (21)		●				●			●	●		
E (6, 19)	●					●	●				●	
F (17, 18, 25)	●					●	●					●
G (3, 13, 15)	●					●		●				●
H (8, 11, 24)	●				●			●				●
I (7, 23)	●				●			●			●	
J (10, 14)	●				●				●	●		
K (16, 20)	●					●		●		●		
L (2, 9)	●					●			●	●		

Legend: LA: Large; MD: Medium-sized; SM: Small; EN: Environmental; SO: Social; EC: Economic; HI: High; ME: Medium; LO: Low; (n): relative importance (1–5). To know the cities that form each alternative (A–L) see Table 2 (case collection column). Source: Prepared by the authors based on literature review.

The results of the AHP analysis are shown in Table 4, and the AHP workflow in Figure 2. In addition, in order to clarify the decision route to the most suitable alternative, the associated sequential decision tree was represented schematically (Figure 3). In this scheme, the rest of branches were omitted to avoid a too extensive figure. According to the case study (Mieres) and the literature reviewed, the most suitable alternative is F (cases 17, 18, 25, i.e., New York–New York/Chicago–Ruhr Valley; PG = 13.9%). It is formed by large cities (see Table 2) that had problems in their productive economic sectors, and that tried to solve them using dominantly environmental and low-cost initiatives (see Table 3). On the other hand, the least suitable alternative to the case study is J (cases 10, 14, i.e., Halle/Leipzig–México DF; PG = 4.3%), which is formed by large cities showing a decline origin of social base, and that faced it using high-cost economic profile measures. Between these two extremes, the different alternatives show several combinations of the mentioned factors.

Table 4. Decision-making hierarchy.

N1	N2 (Criteria)		N3 (Alternatives)											
Goal	**Cr**	**GP%**	**A**	**B**	**C**	**D**	**E**	**F**	**G**	**H**	**I**	**J**	**K**	**L**
Suitable initiative according to the case study	C1	7.8	0.167	0.167	0.056	0.167	0.056	0.056	0.056	0.056	0.056	0.056	0.056	0.056
	C2	20.0	0.100	0.100	0.100	0.100	0.100	0.100	0.100	0.033	0.033	0.033	0.100	0.100
	C3	52.2	0.150	0.050	0.150	0.050	0.150	0.150	0.050	0.050	0.050	0.050	0.050	0.050
	C4	20.0	0.077	0.077	0.030	0.030	0.077	0.182	0.182	0.182	0.077	0.030	0.030	0.030
	GP%	**100.0**	**12.7**	**7.4**	**10.9**	**6.5**	**11.8**	**13.9**	**8.7**	**7.3**	**5.2**	**4.3**	**5.6**	**5.6**

Notes: In all cases the decision matrix is robust (CI < 2%); Cr: Criterion; GP: Global priority. Source: Prepared by the authors based on the AHP model results.

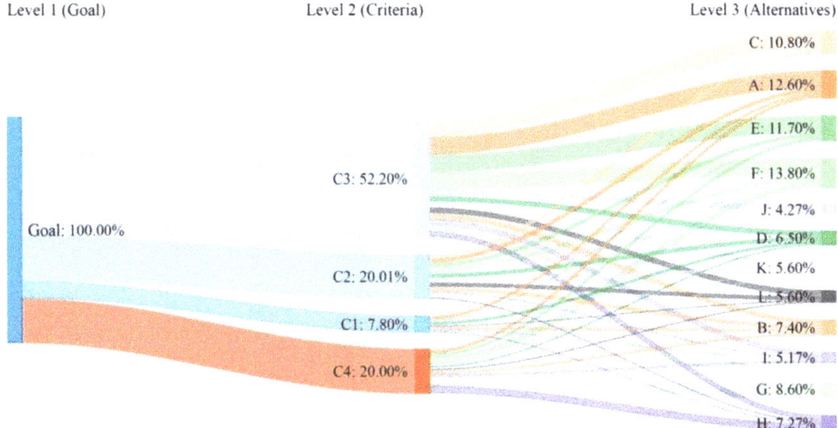

Figure 2. Priorities (%) of the elements of each hierarchical level. Notes: (1) The sum of criteria and alternatives = 100% in both cases. (2) The flow width from the criteria to the alternatives shows the weight of each criterion in each alternative. Source: Prepared by the authors (Sankey diagram).

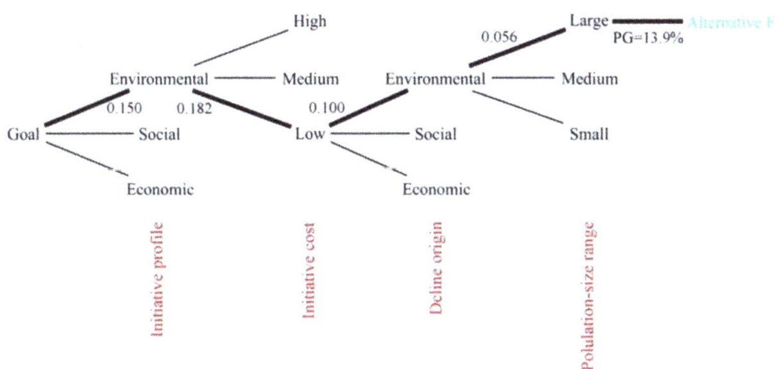

Figure 3. The decision tree structure constructed for the decision process. Notes: The criteria used in the sequential decision tree were sorted according to priorities obtained by the AHP structure. Local priorities are shown along the branches in each step. Global priority for alternative F (the one that best suits the case study) is given as a percentage. The criteria are shown in red. Due to spatial limitations only one branch of each criterion was shown. Source: Prepared by the authors.

The similarity of the cities grouped under the alternative A (1, 12, 22, i.e., Avilés–Lieksa–Puertollano; PG = 12.7%) to the case study in terms of population-size range (medium-sized) and decline origin (economic) should be noted. In addition, these cities adopted environmental correcting measures of medium cost. However, it ranks second in terms of applicability to the case study, behind alternative F which has different features.

5. Discussion

This work followed the research line of some authors such as Großmann et al., Hollander et al. and Sánchez-Moral et al. [5,14,16] and meetings (e.g., UC Berkeley–SCiRN, 2007) that call for more international and comparative studies that collect lessons learned from urban decline cases in an integrated way, which would be useful in new cases where this phenomenon is still active. In this line, in order to explore which successful cases against shrinkage, grouped under alternatives, would be more suitable to the case study (Mieres), an AHP analysis was carried out. This method is not intended as a substitute but rather as complementary to other assessment methods of already proved validity (e.g., surveys, direct observations, expert panels). Most of the cases collected in Table 2 are located in the northern hemisphere (92%), in line with Oswalt and Rieniets [6]. The complexity of the problem, due to the existence of several hierarchical levels (i.e., multiple alternatives dependent as well on several criteria of variable influence), demands the support of this useful tool for decision-making in planning [30,43,56].

Based on a literature review, in order to refine the results, several criteria of different influence were identified, prioritizing alternatives oriented towards sustainability (dominantly environmental and low-cost for the reasons argued along the text), following the steps of other works such as Hemphill et al. and Lee and Lim [31,32]. In this line, regarding the decision-making model, the penalization or even exclusion of high-cost alternatives could seem obvious without the need to develop a multi-criteria analysis. Nevertheless, as the assessment is multidimensional and therefore complex, it becomes less evident. Logically, the results showed coherent coincidences, such as the case study (Mieres) and the optimal alternative (F—New York–New York/Chicago–Ruhr Valley—) had the same decline origin (economic-industrial), which brings one closer to the other. However, there were some apparent divergences between them in the results (e.g. the population-size range factor), which were not intuited in the early methodological steps until the results were obtained. This showed how the alternative most likely to be successful if applied to the case study involves large cities. This, a priori, might seem inconsistent. However, are the largest population-size range cities those that drive the talent, creativity, resources and innovation necessary to carry out sustainable recovery initiatives more efficiently [57]. Thus, the measures successfully developed in them might be extended to lower urban categories by local adaptation.

The finding of this work does not lie in the full coincidence between the attributes of the case study and those of the alternative most likely to be applied successfully facing the urban decline of the former, but in the fact that measures against decline that involve sustainable and low-cost initiatives could be multi-scale reproduced regardless of factors such as the decline origin of the population-size range. For this reason, the optimal alternative is not A (1, 12, 22, i.e., Avilés–Lieksa–Puertollano; PG = 12.7%), apparently more similar to the case study, but F (17, 18, 25, i.e., New York–New York/Chicago–Ruhr Valley; PG = 13.9%) (see Tables 2 and 3—alternatives as groups of similar cities/regions initiatives coded by numbers are shown in Table 3. The links between numbers, i.e. cases, and cities/regions are identified in Table 2; see case collection column—). Even so, the results are not intended to be unique or exclusive but exploratory and indicative based on the arguments and steps taken.

5.1. Integration of the AHP Method Together with Data Mining Tools

Lines of future research could improve the accuracy of the analysis model by increasing, for example, the case collection number, detailing the available information, increasing the number of influencing criteria, and probably performing and additional multi-criteria analysis at the micro-level (i.e., between the case study and the optimal alternatives). These measures would significantly increase the volume of data. Therefore, the integration of AHP and data mining techniques would be useful to process the information and detail the most suitable alternatives. As an example of this, based on the study of Liu and Shih on customer lifetime and products recommendation [58], and on the study of Xi et al. on traffic accident causation [59], the integrated analysis would proceed as follows: (i) obtaining global priorities using an AHP method; (ii) profiling alternatives of similar priorities using a clustering algorithm; and (iii) using association rule learning (i.e., a rule-based machine learning method) to identify probabilities of dominant trends in criteria associations taking into account global priorities.

The last two steps would proceed as follows, considering the research of Agrawal et al. and the previous authors [58–60]. Alternatives with similar GPs according to weighted criteria would be grouped using a clustering algorithm (e.g., k-means). The factors determining the alternatives form a set of elements I. The set of possible associations of I is called D. So that the rule $X \Rightarrow Y$ is followed, where $X, Y \subseteq I$ and $X \cap Y = 0$. Condition X (antecedent) being the possible associations of criteria and subcriteria; and Y (consequent) being the similar GP alternative clusters. In order to find the most interesting associations of I, restrictions can be set based on minimum support and confidence thresholds, where:

$$supp(x) = \frac{|X|}{|D|} \text{ and } conf(X \Rightarrow Y) = \frac{supp(X \cap Y)}{supp(X)} = \frac{|X \cap Y|}{|X|}$$

This analysis would allow to know the more frequent criteria-subcriteria association in each, for example, high, medium or low global priority cluster of alternatives within a set formed by multiple cases facing urban decline. The most interesting would be the high GP associations (desired replicability) and low GP associations (avoided replicability). This would help to identify frequent successful and/or failure patterns according to specific case studies, while also removing the least influential factors and thus reducing analysis complexity, which can help decision-making against urban decline. Additionally, there is evidence of geographic information systems (GIS) as a useful tool for the objective information collection step (case collection in this work) from spatial databases. Later, this information can be weighted using an AHP analysis to prioritize decisions, such as Narimisa and Namirisa did in their work on environmental impact assessment of the introduction of a new industrial land use [61].

As noted above, the global priority determines the alternative that best suits the case study. As shown in Figure 2, this priority depends on the relevance of each alternative according to each criterion, which is shown by the width of the associated flow between criteria and alternatives. Likewise, the criteria priorities (local priorities) are different from each other. Note how the 'winner' alternative F receives the maximum possible flow (i.e., it is assessed with the highest relative importance along the pair-wise comparison in each case) from almost all criteria (see Figure 2 and Table 4). The results from the AHP method were integrated into a sequential decision tree to facilitate understanding of the decision process as for example Suner et al. did in their medical research [49].

5.2. Implementation of Alternatives Facing Urban Decline Using the Future Scenarios Tool

In Mieres, some measures were already taken under alternative A (Avilés–Lieksa–Puertollano), such as investment in renewable energies (thermoelectric, solar), but they have not yet meant definitive measures against the decline of this city today. Nevertheless, the initiatives encouraged by alternative F (New York–New York/Chicago–Ruhr Valley) have not yet been explored in this city. No other type A measures, compatible with those of type F as explained below, were taken in Mieres either. At this point, the optimal alternatives facing the decline would be clear, but how can they be captured in a specific case study?

Knowing and detailing the most suitable alternatives facing urban decline according to a specific case study is one thing, but facilitating and guiding their implementation in that case study is another. The latter can be a challenge due to high uncertainty, wide range of future states and complex interdependencies this process generates [62]. It would produce an urban transition state that should be informed to the different stakeholders to avoid social conflicts and to anticipate market potentials [63]. The construction and prospective assessment of possible urban scenarios can be useful for this transition [64–66]. To get an in-depth view from them, the approach and assessment should be multidimensional. Some key factors of this process can be (i) the evaluation indicators choice; (ii) the diversity of scenarios; (iii) the stakeholder opinions; and (iv) the type of assessment analysis. Some key evaluation indicators may include desirability, usefulness and probability [67]. This work argued for a sustainable urban transition based on the assessment of alternatives against decline. It is a type of transition widely supported by the literature on urban transition planning, which in turn involves shrinking cities [50,68]. The evaluation of the most suitable alternatives through the construction of future scenarios would make it possible to know, for example, stakeholders preferences regarding the most sustainable or less sustainable scenarios, whether the most preferred scenarios match the more or less probable ones, or whether there are differences between the evaluation of the different stakeholders.

Following the indications of Bügl et al. and Majoor, the construction of diverse future scenarios can be done based (i) on their systematic variability concerning sustainability, dominant criterion in the work approach; and (ii) on different urban areas (e.g., family, luxury, failure and transitory districts) [67,69]. To consider decision-making the structure of the scenarios should be made according to different design components of sustainable planning (environmental, social and economic), such as social environment, eco-design, building, financing, and social infrastructure. In the case study, these design components should be detailed based on measures collected by one or more of the highest global priority (GP) alternatives (such as F and A alternatives), according to Tables 2 and 4, and the basic information collected in Table 1, which can be extended in future research.

Another option for constructing future scenarios would be based, for example, on the spatial approach of Kropp and Lein [70]. For this purpose, the case site would first have to be mapped and evaluated quantitatively before and after the impact caused by the various level improvements from the highest GP alternative. Objective indicators of environmental, social and economic sustainability would be used for the assessment, such as pollution prevention, community connectivity degree, brownfield redevelopment degree, access to public transport, parking capacity, habitat restoration suitability, amount of vacant land, or heat island effect, among others. These indicators would be weighted according to their relative importance using a multi-criteria decision-making method (MCDM), such as AHP analysis. Finally, the mapping of the different scenarios would be done based on the priority given to each sustainability component through their respective indicators. This method would allow prioritizing the environmental component, connecting with the dominant profile of the highest GP alternative of this work.

Both for the design of sustainable scenarios based on the measures provided by the priority alternatives [67,69], and for the mapping and evaluation of their impact [70], the case study shows the following potentials, among others: (i) vacant urban and peri-urban post-industrial land; (ii) 6 linear km of radial riverside pedestrian paths length; (iii) 200 linear km of greenways surrounding the city (old mining railway lines); (iv) existence of expectant industrial heritage; and (v) commercial land use linked to pedestrian streets within the central urban area. For information only, these potentials suggest the following visions to be evaluated: (a) the 'industrial forests' project along the Emscher River in the Ruhr Valley occupying case study brownfields [41,42]; (b) the development of new economic activities such as forestry and logging [19]; (c) new interconnected urban green areas along with affordable land available for alternative land use options [2,5,24,25] with public-private financing and use [34], and flexibly communicated [38] that would provide the city with identity and new economic drivers [71–73].

6. Conclusions

The utility of collecting successful experiences facing urban decline and applying a DSS-AHP analysis to explore at the macro-level which would be the most suitable measures for a particular case study was evidenced. This made it possible to prioritise certain criteria and alternatives over others, which was an aim of the work. In this process, the literature review was used as a valuable complementary method to the expert panel for the relative importance assignment at different hierarchical levels. Low-cost environmental initiatives seemed to be the most sustainable and suitable to the case study, regardless of other a priori important factors such as the decline origin or the population-size range, which was a novelty.

Analysing and implementing measures against urban decline in a particular shrinking city, which have already been successfully developed in other cities, can be a challenge due to the complexity and uncertainty of the process. The integration of various decision-support systems may be useful for it. In this work, the convenience of weighting the impact of the different alternatives in the case study from a sustainable approach was evidenced using a multi-criteria assessment. Subsequently, the decision tree representation clarified the decision route until the optimal alternative facing the decline was reached. In order to improve the model, in the event of increasing information, the usefulness of clustering algorithms and associations rules learning was discussed with a view to identify patters of frequent factors based on global priorities, which may optimise the decision-making process. Finally, the value of constructing future urban scenarios as an additional decision-making tool for implementing the desired alternatives in the case study was also discussed. In this process, having detailed information on the profile of the suitable alternatives and on the case study attributes can help to decide and characterise both the different scenarios and their design components.

According to the analysis carried out, the most suitable alternative facing urban decline in the case study (GP = 13.9%) involves infrastructural flexibility (as was previously done in New York), public-private investment and exploitation for new urban uses in vacant land (as in New York/Chicago), and environmental mitigation and ecological restoration through planting and use of forests (as in Ruhr Valley). This group of measures linked economic development with sustainable planning, which is desirable given the attributes of the case study. However, according to the assessment done and the baseline information data, the case study showed significant limitations (GP = 4.3%) for the development of measures based on massive public subsidies and systematic building demolitions (as in Halle/Leipzig), and urban renewal of central urban areas supported by large-scale interventions (as in Mexico DF). These are high-cost measures that cannot be assumed by a medium population-size range and a compact central urban fabric that is not attractive enough to capture large amounts of capital. In addition, they would not take advantage of the reuse potential of certain historic urban fabric, post-industrial land uses, and heritage elements found in the city. Intermediate impact measures (GP = 7.3%) of low-cost social origin and profile such as the use of a governance focused on the social capital accumulation (as in Estonia/Central Germany), local social initiatives and the development of subsistence agriculture (as in Ivanovo), and a planning leading to social cohesion (as in St. Louis) may improve the quality of life but miss the environmental potential of the case study as the main driver of development (vacant land, partial network of walkable green areas between the main city and dispersed populations, etc.). Furthermore, it would force synergies of social nature that traditionally were not developed in a particular form to mean a contemporary incentive.

Author Contributions: Conceptualization, F.S.C.-S.; methodology, F.S.C.-S.; software, F.S.C.-S.; validation, F.S.C.-S., R.R.-B. and F.J.A.-A.; formal analysis, F.S.C.-S.; investigation, F.S.C.-S.; writing—original draft preparation, F.S.C.-S.; writing—review and editing, F.S.C.-S., R.R.-B. and F.J.A.-A; visualization, F.S.C.-S.; supervision, F.S.C.-S., R.R.-B. and F.J.A.-A.

Funding: This research received no external funding.

Conflicts of Interest: The authors declare no conflict of interest.

References

1. Wiechmann, T. *Between Spectacular Projects and Pragmatic Deconstruction. The Future of Shrinking Cities: Problems, Patterns and Strategies of Urban Transformation in a Global Context*; Institute of Urban and Regional Development: Berkeley, CA, USA, 2007.
2. Haase, D. Shrinking cities, biodiversity and ecosystem services. In *Urbanization, Biodiversity and Ecosystem Services: Challenges and Opportunities. A Global Assessment*; Elmqvist, T., Fragkias, M., Goodness, J., Güneralp, B., Marcotullio, P.J., McDonald, R.I., Parnell, S., Schewenius, M., Sendstad, M., Seto, K.C., et al., Eds.; Springer: Berlin, Germany, 2013; pp. 253–274.
3. Pallagst, K.; Aber, J.; Audirac, I.; Cunningham-Sabot, E.; Fol, S.; Martinez-Fernandez, C.; Moraes, S.; Mulligan, H.; Vargas-Hernandez, J.; Wiechmann, T.; et al. *The Future of Shrinking Cities: Problems, Patterns and Strategies of Urban Transformation in a Global Context*; IURD: Berkeley, CA, USA, 2009.
4. de Sousa, A.S.; Cottineau, C.; Dietersforfer, L.; Fernández Águeda, B.; Gonul, D.; Hoemke, M.; Jaroszewska, E.; Lella, I.; Mykhnenko, V.; Prada Trigo, J.; et al. *Mapping Urban Shrinkage in Europe*; Final Report, EU-COST Action; TSDTU-ROP: Dortmund, Germany, 2011.
5. Hollander, J.B.; Pallagst, K.; Schwarz, T.; Popper, F.J. Planning shrinking cities. *Prog. Plan.* **2009**, *72*, 223–232.

6. Oswalt, P.; Rieniets, T. *Atlas of Shrinking Cities*; Hatje Cantz: Ostfildern, Germany, 2006.
7. Riva, M. Salamanca, ¿consolidación de un sistema metropolitano o shrinking city? *Territorios en formación* **2015**, *8*, 105–122. [CrossRef]
8. Martinez-Fernandez, C.; Martinez-Fernandez, C.; Kubo, N.; Noya, A.; Weyman, T. *Demographic Change and Local Development: Shrinkage, Regeneration and Social Dynamics*; OECD: Paris, France, 2012.
9. Vegara, A.; De las Rivas, J.L. *Territorios Inteligentes*; Fundación Metrópoli: Madrid, Spain, 2004.
10. Tomé, S. Langreo, Mieres, Ponferrada, Puertollano: Cambios funcionales y morfológicos en ciudades minero industriales. *Scr. Nova* **2010**, *14*. [CrossRef]
11. Benito, P. Renovación urbana, herencia industrial y turismo: Un proceso con elementos de éxito en Avilés (Asturias). *Boletín de la Asociación de Geógrafos Españoles* **2016**, *72*, 205–304.
12. Capel, H. La rehabilitación y el uso del patrimonio histórico industrial. *Documents d'Anàlisi Geogràfica* **1996**, *29*, 21–50.
13. Oswalt, P. *Shrinking Cities—International Research*; Hatje Cantz: Ostfildern, Germany, 2005.
14. Großmann, K.; Bontje, M.; Haase, A.; Mykhnenko, V. Shrinking cities: Notes for the further research agenda. *Cities* **2013**, *35*, 221–225. [CrossRef]
15. Kabisch, N.; Haase, D. Diversifying European agglomerations: Evidence or urban population trends for the 21st century. *Popul. Space Place* **2011**, *17*, 236–253. [CrossRef]
16. Sánchez-Moral, S. Resurgent cities: Local strategies and institutional networks to counteract shrinkage in Avilés (Spain). *Eur. Plan. Stud.* **2015**, *23*, 33–52. [CrossRef]
17. Aber, J. The creative imperative in a postindustrial economy to foster a more sustainable development in shrinking cities. In *The Future of Shrinking Cities: Problems, Patterns and Strategies of Urban Transformation in a Global Context*; Pallagst, K., Aber, J., Audirac, I., Cunningham-Sabot, E., Fol, S., Martinez-Fernandez, C., Moraes, S., Mulligan, H., Vargas-Hernandez, J., Wiechmann, T., et al., Eds.; IURD, UCA: Berkeley, CA, USA, 2009; pp. 111–121.
18. Mulligan, H. Environmental policy action: Comparative importance in differing categories of shrinking city. In *The Future of Shrinking Cities: Problems, Patterns and Strategies of Urban Transformation in a Global Context*; Pallagst, K., Aber, J., Audirac, I., Cunningham-Sabot, E., Fol, S., Martinez-Fernandez, C., Moraes, S., Mulligan, H., Vargas-Hernandez, J., Wiechmann, T., et al., Eds.; IURD, UCA: Berkeley, CA, USA, 2009; pp. 101–111.
19. Kotilainen, J.; Eisto, I.; Vatanen, E. Uncovering mechanisms for resilience: Strategies to counter shrinkage in a peripheral city in Finland. *Eur. Plan. Stud.* **2015**, *23*, 53–68. [CrossRef]
20. Nelson, A.C.; Petheram, S.; Ewing, R.; Stoker, P.; Hamidi, S. Compact development as a factor in income resilience among shrinking counties in the United States. Statistical analysis with policy implications. In *Shrinking Cities. A Global Perspective*; Richardson, H.W., Nam, C.W., Eds.; Routledge: London, UK; New York, NY, USA, 2014; pp. 301–310.
21. Miot, Y. Residential attractiveness as a public policy goal for declining industrial cities: Housing renewal strategies in Mulhouse, Roubaix and Saint-Etienne (France). *Eur. Plan. Stud.* **2013**, *23*, 104–125. [CrossRef]
22. Leetmaa, K.; Kriszan, A.; Nuga, M.; Burdack, J. Strategies to cope with shrinkage in the lower end of the urban hierarchy in Estonia and central Germany. *Eur. Plan. Stud.* **2013**, *23*, 147–165. [CrossRef]
23. Elzerman, K.; Bontje, M. Urban shrinkage in Parkstad Limburg. *Eur. Plan. Stud.* **2015**, *23*, 87–103. [CrossRef]
24. Hollander, J.B.; Németh, J. The bounds of smart decline: A foundational theory for planning shrinking cities. *Hous. Policy Debate* **2011**, *21*, 349–367. [CrossRef]
25. Pallagst, K.; Wiechmann, T. Shrinking smart? Städtische Schrumpfungsprozesse in den USA. In *Jahrbuch StadtRegion 2004/05 Schwerpunkt Schrumpfende Städte*; Gestring, N., Glasauer, H., Hannemann, C., Petrowsky, W., Pohlan, J., von Christine Hannemann, M.B., Wixforth, J., Benke, C., Pallagst, K., Wiechmann, T., et al., Eds.; VS Verlag fuer Sozialwissenschaften: Wiesbaden, Germany, 2005; pp. 105–127.
26. Ayedi, B. The design of spatial decision support systems in urban and regional planning. In *Decision Support Systems in Urban Planning*; Timmermans, H., Ed.; Routledge: London, UK; New York, NY, USA, 1998.
27. Simon, H.A. *The Ford Distinguished Lectures: Vol. 3. The New Science of Management Decision*; Harper & Brothers: New York, NY, USA, 1960.
28. Schetke, S.; Haase, D. Multi-criteria assessment of socio-environmental aspects in shrinking cities. Experiences from eastern Germany. *Environ. Impact Assess. Rev.* **2008**, *28*, 483–503. [CrossRef]

29. Johnson, M.P.; Hollander, J.; Hallulli, A. Maintain, demolish, re-purpose: Policy design for vacant land management using decision models. *Cities* **2014**, *40*, 151–162. [CrossRef]
30. Saaty, T.L. *The Analytical Hierarchy Process: Planning, Priority, Resource Allocation*; RWS Publications: Pittsburgh, PA, USA, 1980.
31. Lee, J.H.; Lim, S. An analytic hierarchy process (AHP) approach for sustainable assessment of economy-based and community-based urban regeneration: The case of South Korea. *Sustainability* **2018**, *10*, 4456.
32. Hemphill, L.; McGreal, S.; Berry, J. An aggregated weighting system for evaluating sustainable urban regeneration. *J. Prop. Res.* **2002**, *19*, 353–373. [CrossRef]
33. Richardson, H.W.; Nam, C.W. *Shrinking Cities. A Global Perspective*; Routledge: London, UK; New York, NY, USA, 2014.
34. Rybczynski, W.; Linneman, P.D. How to save our shrinking cities. *Public Interest* **1999**, *135*, 30–44.
35. Cepl, J. Oswald Mathias Unger's urban archipelago for shrinking Berlin. In *Shrinking Cities. Vol. 2: Interventions*; Oswalt, P., Ed.; Hatje Cantz Verlag: Ostfildern, Germany, 2006.
36. Torres, H.; Alves, H.; Aparecida De Oliveira, M. São Paulo peri-urban dynamics: Some social causes and environmental consequences. *Environ. Urban.* **2007**, *19*, 207–219. [CrossRef]
37. Gilbert, A. *The Mega-City in Latin America*; The United Nations University Press: Tokyo, Japan, 1996.
38. Böhm, F. On-demand infrastructure for shrinking regions. In *Shrinking Cities. Vol. 2: Interventions*; Oswalt, P., Ed.; Hatje Cantz Verlag: Ostfildern, Germany, 2006.
39. Isunza, V.G. Política de vivienda y movilidad residencial en la Ciudad de México. *Estudios Demográficos y Urbanos del Colegio de México* **2010**, *25*, 277–316. [CrossRef]
40. Nadal, L. City center revitalization: Tapping São Paulo's global potential. *Sustain. Transp.* **2006**, *18*, 16–21.
41. Dettmar, J. Forests for shrinking cities? The project 'Industrial forests of the Ruhr'. In *Wild Urban Woodlands*; Kowarik, I., Körner, S., Eds.; Springer: Berlin/Heidelberg, Germany, 2005; pp. 263–276.
42. Dettmar, J. Naturally determined urban development. In *Shrinking Cities. Vol. 2: Interventions*; Oswalt, P., Ed.; Hatje Cantz Verlag: Ostfildern, Germany, 2006.
43. Saaty, T.L. *Decision-Making for Leaders—The AHP for Decisions in a Complex World*; RWS Publications: Pittsburgh, PA, USA, 1995.
44. Campos-Sánchez, F.S. Spanish medium-sized cities of the XXI century. Review and bibliometric analysis of prevailing topics and approaches. *Cuadernos Geográficos* **2017**, *56*, 217–241.
45. Valenzuela, L.M.; Talavera, R. Entornos de movilidad peatonal: Una revisión de enfoques, factores y condicionantes. *Eure* **2015**, *41*, 5–27. [CrossRef]
46. Saaty, T.L. Decision making with the analytic hierarchy process. *Int. J. Serv. Sci.* **2008**, *1*, 83–98. [CrossRef]
47. Witten, I.H.; Frank, E.; Hall, M.A.; Pal, C.J. *Mining Practical Machine Learning Tools and Techniques. Data Mining*, 4th ed.; Elsevier: Amsterdam, The Netherlands, 2011; Volume 277.
48. Whitaker, R. Criticisms of the analytic hierarchical process: Why they often make no sense. *Math. Comput. Model.* **2007**, *46*, 948–961. [CrossRef]
49. Suner, A.; ÇElikoğLu, C.C.; Dicle, O.; Sökmen, S. Sequential decision tree using the analytic hierarchy process for decision support in rectal cancer. *Artif. Intell. Med.* **2012**, *56*, 59–68. [CrossRef]
50. Campos-Sánchez, F.S.; Abarca-Álvarez, F.J.; Domingues, A. Sostenibilidad, planificación y desarrollo urbano. En busca de una integración crítica mediante el estudio de casos recientes. *ACE Archit. City Environ.* **2018**, *12*, 39–72.
51. Shen, L.Y.; Ochoa, J.J.; Shah, M.N.; Zhang, X. The application of urban sustainability indicators. A comparison between various practices. *Habitat Int.* **2011**, *35*, 17–29. [CrossRef]
52. Leigh, N.G.; Hoelzel, N.Z. Smart growth's blind side. Sustainable cities need productive urban industrial land. *J Am. Plan. Assoc.* **2012**, *78*, 87–103. [CrossRef]
53. Pike, A.; Rodríguez-Pose, A.; Tomaney, J. What kind of local and regional development and for whom? *Reg. Stud.* **2007**, *41*, 1253–1269. [CrossRef]
54. Rave, T. Environmental impacts of shrinking cities. In *Shrinking Cities. A Global Perspective*; Richardson, H.W., Nam, C.W., Eds.; Routledge: London, UK; New York, NY, USA, 2014; pp. 311–325.
55. Florida, R. *The Rise of the Creative Class*; Basic Books: New York, NY, USA, 2002.
56. Curtis, C.; Scheurer, J. Planning for sustainable accessibility. Developing tools to aid discussion and decision-making. *Prog. Plan.* **2010**, *7*, 53–106. [CrossRef]

57. Portney, K.E. *Taking Sustainable Cities Seriously: Economic Development, the Environment, and Quality of Life in Americans Cities*; MIT Press: Cambridge, MA, USA, 2013.
58. Liu, D.R.; Shih, Y.Y. Integrating AHP and data mining for product recommendation based on customer lifetime value. *Inf. Manag.* **2005**, *42*, 387–400. [CrossRef]
59. Xi, J.; Zhao, Z.; Li, W.; Wang, Q. A traffic accident causation analysis method based on AHP-Apriori. *Procedia Eng.* **2016**, *137*, 680–687. [CrossRef]
60. Agrawal, R.; Imieliński, T.; Swami, A. Mining association rules between sets of items in large databases. In Proceedings of the SIGMOD '93 Proceedings of the 1993 ACM SIGMOD International Conference of Management Data, Washington, DC, USA, 25–28 May 1993; pp. 207–216.
61. Narimisa, M.R.; Narimisa, M.R. The research of environmental impact assessment for oil refineries in Iran based on AHP and GIS. *Int. J. Hum. Cult. Stud.* **2016**, *1*, 1396–1416.
62. Priemus, H. Mega-projects: Dealing with pitfalls. *Eur. Plan. Stud.* **2010**, *18*, 1023–1039. [CrossRef]
63. Haase, D.; Haase, A.; Kabisch, S.; Bischoff, P. Guidelines for the 'perfect inner city': Discussing the appropriateness of monitoring approaches for reurbanization. *Eur. Plan. Stud.* **2008**, *16*, 1075–1100. [CrossRef]
64. Larsen, K.; Gunnarsson-Östling, U. Climate changes scenarios and citizen-participation: Mitigation and adaptation perspectives in constructing sustainable futures. *Habitat Int.* **2009**, *33*, 260–266. [CrossRef]
65. Scholz, R.W.; Tietje, O. *Embedded Case Study Methods: Integrating Quantitative and Qualitative Knowledge*; Sage: Thousand Oaks, CA, USA, 2002.
66. Wiek, A.; Binder, C.; Scholz, R.W. Functions of scenarios in transition processes. *Futures* **2006**, *38*, 740–766. [CrossRef]
67. Bügl, R.; Stauffacher, M.; Kriese, U.; Pollheimer, D.L.; Scholz, R.W. Identifying stakeholders' views on sustainable urban transition: Desirability, utility and probability assessments of scenarios. *Eur. Plan. Stud.* **2012**, *20*, 1667–1687. [CrossRef]
68. Evans, J.; Jones, P. Rethinking sustainable urban regeneration: Ambiguity, creativity, and the shared territory. *Environ. Plan. A* **2008**, *40*, 1416–1434. [CrossRef]
69. Majoor, S.J.H. The disconnected innovation of new urbanity in Zuidas Amsterdam, Ørestad Copenhagen and Forum Barcelona. *Eur. Plan. Stud.* **2009**, *17*, 1379–1403. [CrossRef]
70. Kropp, W.W.; Lein, J.K. Assessing the geographic expression of urban sustainability: A scenario based approach incorporating spatial multicriteria decision analysis. *Sustainability* **2012**, *4*, 2348–2365. [CrossRef]
71. Aguado, I.; Barrutia, J.M.; Echebarria, C. El anillo verde de Vitoria-Gasteiz. Una práctica exitosa para un planeamiento urbano sostenible. *Boletín de la Asociación de Geógrafos Españoles* **2013**, *61*, 401–404. [CrossRef]
72. Giseke, U. Urbane Freiräume in der schrumpfenden Stadt. *Der Architekt* **2002**, *8*, 44–46.
73. Sieverts, T. *Zwischenstadt—Inzwischen Stadt Entdecken, begreifen verändern*; Müller + Busmann: Wuppertal, Germany, 2005.

International Journal of
Environmental Research and Public Health

Article

Dissipation Theory-Based Ecological Protection and Restoration Scheme Construction for Reclamation Projects and Adjacent Marine Ecosystems

Faming Huang [1,*], Yanhong Lin [1,2], Rongrong Zhao [1], Xuan Qin [3], Qiuming Chen [1] and Jie Lin [1]

1 Third Institute of Oceanography, Ministry of Natural Resources, Xiamen 361005, China; linyanhong@tio.org.cn (Y.L.); zhaorongrong@tio.org.cn(R.Z.); chenqiuming@tio.org.cn(Q.C.); linjie@tio.org.cn(J.L.)
2 College of Environment and Ecology, Xiamen University, Xiamen 361005, China; linyanhong@tio.org.cn
3 Xiamen Branch of Tianjin Urban Planning and Design Institute, Xiamen 3611005, China; lydia710_qin@hotmail.com
* Correspondence: huangfaming@tio.org.cn; Tel.: +86-592-219-5001

Received: 26 September 2019; Accepted: 4 November 2019; Published: 5 November 2019

Abstract: According to the 2017 results of the Special Inspector of Sea Reclamation, a substantial number of idle reclamation zones existed in 11 provinces (cities) along the coast of China. To improve the protection level of coastal wetlands and strictly control reclamation activities, it is necessary to carry out ecological restoration of reclamation projects and adjacent marine ecosystems. The characteristics of Guanghai Bay and its reclamation project are typical in China's coastal areas, making it an optimal representative site for this study. The dissipative structure and entropy theory was used to analyze ecological problems and environmental threats. The analytic hierarchy process was applied to determine the order of the negative entropy flow importance. The entropy increase and decrease mechanism was used to determine an ecological protection and restoration scheme for the reclamation, including the reclamation of wetland resource restoration, shoreline landscape restoration, environmental pollution control, and marine biological resource restoration. Finally, based on system logic, a typical ecological restoration system was constructed east of Guanghai Bay, with the mangrove wetland area as the model in the north and the artificial sandbeach recreation area as the focus in the south.

Keywords: ecological restoration; dissipative structure; entropy change; analytic hierarchy process

1. Introduction

Coastal wetlands (including beaches, estuaries, shallow seas, mangroves, and coral reefs) are precious wetland resources with significant ecological functions, which provide important habitats, breeding sites, and migration stations for birds [1]. To provide sufficient land space for the socio-economic development, China's coastal cities have carried out long-term and large-scale reclamation activities from 2002 to 2018. Due to this, its coastal wetlands have substantially decreased, and the natural coastline has sharply reduced, damaging marine and terrestrial ecosystems [2]. However, according to the survey results of the Special Inspector of Sea Reclamation in 2017 [3], all provinces (cities) in China's coastal areas face the problem of sea-reclamation projects remaining largely idle [4–11]. To improve the protection level of coastal wetlands and control reclamation activities, it is necessary to ecologically assess the reclamation projects as soon as possible and propose reasonable and feasible ecological restoration measures for reclamation projects and their adjacent marine ecosystems, thereby minimizing the impacts on ocean hydrodynamics and biodiversity [12–14].

Reclamation projects and their adjacent marine ecosystems possess typical features of a dissipative structure, which include (1) an open system, (2) nonequilibrium, (3) nonlinear interaction, and (4) fluctuation. A dissipative structure is a kind of dynamic ordered structure state, and change processes can be demonstrated based on the entropy change of the dissipative structural system [15]. Entropy is a state parameter that measures the degree of order of the system. A higher entropy value indicates a more disordered system and a lower value indicates a more ordered system [16,17]. In recent years, a combination of the dissipative structure theory and entropy change method has been developed to evaluate and improve various ecosystems. Zhang et al. (2006) developed an indicator system based on the dissipative structure theory and a model based on information entropy to estimate various flows in an urban ecosystem [18]. Wu et al. (2013) proposed the maintenance and orderly development of island ecological–environmental systems by increasing negative entropy flow [15]. Ludovisi et al. (2014) studied the directional role of several entropy-based indicators (e.g., structural information, specific entropy production, and the eco-exergy index) in the ecological succession of eutrophication [17]. Di and Han (2014) used the information entropy method to judge the sustainable development capability of marine ecosystems at the national level [19]. Wang et al. (2018) constructed the "Marine Ecosystem Evolution Index System" based on the information entropy theory [20]. Xu et al. (2019) proposed the modified entropy weight of the AHP model, which can be used in the construction of regional information of ecological environments [21]. Zhou (2019) analyzed the relationship of ecological compensation based on the principle of maximum entropy and designed the calculation method of ecological compensation standard for adjacent administrative districts [22]. These studies showed that the combination of the dissipative structure theory and entropy change method for studying the evolution and sustainable development of urban ecosystems has been extremely common. However, this combination has rarely been applied for the construction of ecological restoration schemes for reclamation zones and adjacent marine ecosystems.

Therefore, the objectives of this study are to analyze ecological problems and environmental threats faced by typical reclamation projects and their adjacent marine ecosystems, based on the theory of dissipative structure and the entropy theory, and then to utilize the mechanism of entropy increase and decrease to determine ecological protection and restoration plans. In addition, the characteristics of Guanghai Bay and the reclamation project in it are typical at China's coastal areas, making it a representative site for analyzing the ecological problems and environmental threats arising from reclamation based on the dissipative structure theory and entropy change. Finally, a restoration scheme for wetland resource restoration, shoreline landscape restoration, environmental pollution control, and marine biological resource restoration was proposed, and a typical ecological restoration system considering the mangrove wetland area in the north and the artificial sandbeach recreation area in the south was constructed. The results of this paper can be used as a reference for formulating ecological protection and restoration schemes of other reclamation projects and their adjacent marine ecosystems.

2. Materials and Methods

2.1. Study Area

Guanghai Bay is located at the south sea area of Jiangmen City, Guangdong Province (Figure 1). The east, north, and west sides are connected to the open inland hinterland, while facing the sea in the south. Guanghai Bay is a typical semi-enclosed bay. Currently, there is only one reclamation project being conducted in the Bay, which has formed a land area.

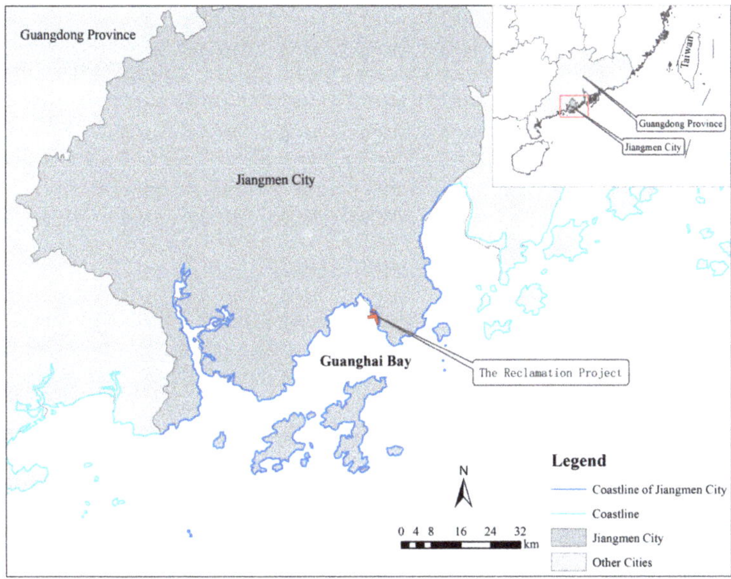

Figure 1. Location of Guanghai Bay in Guangdong Province, China.

The east side of Guanghai Bay and the land behind it are currently planned to be reformed as Guanghai harbor industrial park, whose harbor industries include marine engineering equipment manufacturing and science and education research and development. The planned land area is approximately 21.2 km^2, of which approximately 10.4 km^2 needs to be obtained via reclamation activities. Therefore, the Guanghaiwan reclamation project was launched in Guanghai Bay to solve the problem of land shortage in Guanghaiwan Industrial Park. Currently, only approximately 6.6 km^2 of these reclamation projects has been completed (divided into areas C, D, and E from the north to the south). As the Chinese government has formulated the strictest reclamation policy from 2018, subsequent reclamation plans have been shelved. The scale of the 6.6 km^2 of reclamation and adjacent marine ecosystem were consider as the study area, and case studies were conducted based on the assumption that subsequent reclamation projects would not be implemented.

2.2. Research Methods

2.2.1. Dissipative Structure and Establishment of Entropy Model

Considering the four characteristics of a dissipative structure, reclamation projects and their adjacent marine ecosystems are a typical dissipative structure. First, they are open systems [23], where animals, plants, and microorganisms are constantly exchanged with the surrounding environment for material and energy, and they are significantly affected by environmental factors. Second, there are spatial differences, functional differences, and four seasons of transformation in the system, which results in the system being far from the equilibrium state [24]. Third, the dissipative system is a nonlinear dynamic process. The reclamation project and its adjacent ecosystem display a positive and negative feedback mechanism, and the system is evolved and updated according to nonlinear laws. Fourth, as a typical complex ecosystem, the reclamation project and its adjacent ecosystem are likely to suffer from random and uncertain fluctuations, such as human interference and natural evolution. This causes the system to deviate from its normal state, thereby forming fluctuations, and to rely on its self-organizing abilities to adjust the structure and functions, finally forming a new dissipative structure.

In 1948, Shannon proposed the concept of "information entropy", which is used to represent the degree of disorder of the system and interpret the evolution direction of the system [15,25,26]. A higher entropy value (S) indicates a more disordered system and a lower value indicates a more ordered system [16,17,27]. The entropy change (ΔS) of an open system consists of two parts: the entropy increase ($\Delta S\text{i}$) caused by the internal irreversible process of the system (its value is always greater than zero) and the entropy reduction ($\Delta S\text{e}$) caused by the exchange of matter or energy with the outside world (its value may be positive, negative, or zero). That is, the total entropy of the open system is the sum of the entropy increase and decrease, and the total entropy (ΔS) of the open system is determined as follows [28]:

$$\Delta S = \Delta S\text{i} + \Delta S\text{e} \tag{1}$$

According to Equation (1), the positive and negative values of the total entropy change (ΔS) of the system are dependent on the scale and magnitude of the increase and decrease in the entropy. That is, the system can only progress toward ordering when $\Delta S\text{e} < 0$ and $|\Delta S\text{e}| > \Delta S\text{i}$. At this time, $\Delta S = \Delta S\text{i} + \Delta S\text{e} < 0$.

A nonlinear open system that is far from the equilibrium state can only self-organize if the system draws a sufficient negative entropy from the outside to reduce the total entropy. Therefore, to develop in a dynamic and orderly manner, the system must acquire a large amount of negative entropy by exchanging matter and energy with the outside world.

As it is a dissipative-structure system, the entropy values inside reclamation projects and their adjacent marine ecosystems change constantly during the development process. As illustrated in Figure 2, when entropy reduction dominates, the system efficiency will gradually increase, and the system will change from disordered to ordered (segment AB). As time increases, the system entropy increases continuously. When the entropy increases and plays the dominant role, the system efficiency decreases and the system will change from ordered to disordered (segment BC). Here, if the system does not introduce high-quality and effective negative entropy reduction in time, the system will eventually be eliminated (segment CD). If appropriate amounts of high-quality negative entropy reduction of matter, information, and energy, among others, are introduced in time to break the balance, the system development will regain its vitality (segment CE). The increase and decrease in the entropy of the system exists in a state of confrontation and growth [29]. Therefore, the entropy increase and decrease factors of the reclamation projects and their adjacent marine ecosystems must be analyzed under different stages and conditions.

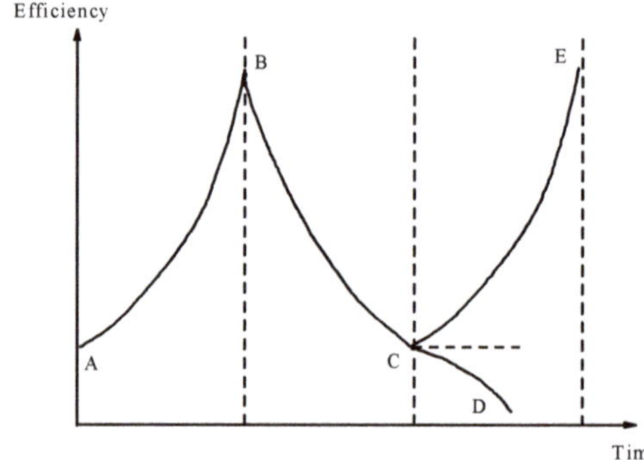

Figure 2. The entropy change of reclamation projects and their adjacent marine ecosystems with time changes [30].

2.2.2. Analytic Hierarchy Process

Analytic hierarchy process (AHP) is a hierarchical weighted decision analysis method proposed by Professor T.L. Satty of the University of Pittsburgh [31–33]. It decomposes the relevant elements of a decision-making problem into goals, criteria, and programs, and accordingly conducts qualitative and quantitative analyses. The application of AHP involves four main steps:

(1) Establishment of a hierarchical structure model:

First, the objectives of the decision, factors considered (evaluation criteria) and decision objects (action plans), are divided into the top layer (target layer-AW), middle layer (guidelines layer-BW), and bottom layer (measures layer-CW), according to their mutual relationship.

(2) Construction of a comparison discriminant matrix:

Once the hierarchy is established, the evaluator or experts score the factors according to their experience, starting from the first criterion level. The importance of the different factors of each layer relative to other factors is gradually weighted, generally through a pair-wise comparison method. The comparative relationship of each element is summarized in a comparison judgment matrix A, as follows:

$$\mathrm{A} = \begin{bmatrix} a_{11} & a_{12} & \cdots & a_{1n} \\ a_{21} & a_{22} & \cdots & a_{2n} \\ \vdots & \vdots & \vdots & \vdots \\ a_{n1} & a_{n2} & \cdots & a_{nn} \end{bmatrix} \tag{2}$$

where a_{ij} generally takes a positive integer from 1 to 9 (known as the scale) and its reciprocal; that is, if factor i is compared with factor j to obtain a_{ij}, factor j is compared with factor i by using $1/a_{ij}$ (where i and $j = 1, 2, \cdots, n$). The rule for establishing the value of a_{ij} is indicated in Table 1 [34]. Then, the hierarchical single-sorted weight vector (W) and maximum eigenvalue ($\lambda_{\max}$) can be calculated using matrix A.

(3) Consistency test:

Table 1. Analytical method with two-two comparison scale [23].

Element	Scaling	Value Rule (a Factor in the Above Layer is the Criterion, and at the Current Level, Factor *i* is Compared with Factor *j*)
a_{ij}	1	Equally important
	3	*i* is slightly more important than *j*
	5	*i* is more important than *j*
	7	*i* is more important than *j*
	9	*i* is extremely more important than *j*
	2, 4, 6, 8	Comparison between the importance of the two factors *i* and *j* is in the middle of the above results
a_{ji}	Reciprocal	Comparison between the importance of factors *i* and *j* is the reciprocal of the comparison between their importance

When constructing a comparison judgment matrix, the judgment does not require complete consistency; however, the direction should be substantially uniform. Therefore, a consistency test is required for each level of single-criterion ordering.

Let A be an n-order positive cross-reverse matrix, where the consistency standard (CI) is defined as

$$CI = \frac{\lambda_{\max} - \mathrm{n}}{\mathrm{n} - 1}, \tag{3}$$

where $\lambda_{\max}$ is the maximum eigenvalue of matrix A and CI is the quantitative standard for measuring the inconsistency degree.

When the maximum eigenvalue of matrix A is slightly larger than n, A is said to exhibit satisfactory consistency. The following evaluation method was adopted in this study: a fixed n is used to construct a positive reciprocal matrix A = $(a_{ij})_n$ randomly, where a_{ij} ranges from 1, 2, 3, ... , 9, 1/2, 1/3, ... , 1/9 out of the total of 17 numbers. Such a positive cross-reverse matrix A is the most inconsistent. The maximum characteristic λ_{max} of the above-mentioned random judgment matrix is calculated 1000 times, and the given random consistency index (*RI*) value is presented in Table 2.

Table 2. Average random consistency indicator, *RI*.

n	1	2	3	4	5	6	7	8	9
RI	0	0	0.58	0.94	1.12	1.24	1.32	1.41	1.45

In Table 2, for n = 1 and 2, RI = 0 because the first- and second-order judgment matrices are always consistent. When $n \geq 3$, the consistency ratio (*CR*) is determined as follows:

$$CR = CI/RI. \tag{4}$$

When $CR < 0.1$, the consistency of the comparison judgment matrix is acceptable; otherwise, the judgment matrix should be appropriately corrected.

3. Results

3.1. Main Form of Entropy Increase of Reclamation Project and Adjacent Marine Ecological Environment

3.1.1. Entropy Increases in Resources: Reduced Wetland Area and Loss of Biological Resources

The reclamation project is approximately 6.6 km^2, and it is located within the 0 m of water depth (Figure 3), which belongs to the coastal wetland (within the 6-m of water depth) in a broad sense. Development and utilization methods, including the extension toward the sea level, cutting off the bend to make it straight, and other simple types, which are adopted in this reclamation project, resulted in a series of problems such as shortening of the natural shoreline, a loss of wetland resources, and a reduction in marine biodiversity. The occupation of wetlands transforms the original sea area into land and changes the natural attributes and ecological environment of the original coast, intertidal zone, and sea area. This destroys the ecological service functions of wetlands, with wetland organisms losing their living and reproductive space. Moreover, with the destruction of marine biological resources, their output is reduced, leading to the loss of wetland habitats; this has a negative impact on the population composition and the spatio-temporal distribution of benthic animals and birds. During the process of construction behaviors, such as riprap, blasting to squeeze out silt, push-fill overflow in the land area, and permanent occupation of the seabed subsoil through sea reclamation, has resulted in the loss of marine resources. According to a July 2017 survey of bio-ecological and fishery resources in the Guanghai Bay conducted by the Ocean Monitoring and Testing Center of the Ocean University of China, this reclamation project resulted in a loss of approximately 10 t of plankton, approximately 17 t of swimming creatures, 1.2×10^8 grains of fish eggs, 9.0×10^7 tails of larvae, approximately 320 t of benthic organisms, approximately 800 t of intertidal organisms, 5.4 t of fish, and 1.8 t of crustaceans.

3.1.2. Entropy Increases in Environment: Soil Erosion, Near-Shore Pollution, and Reduced Environmental Capacity

The reclamation materials in this reclamation project contained high soil content and formed a land area over more than 10 years (Figure 4a,b). At present, the soil is still directly exposed or only grows sporadic pioneer plants, mostly herbaceous and vine plants (Figure 4c). The soil in the reclamation area is soft and does not comprise a large-scale vegetal cover. Under the conditions of hydraulics, gravity, and wind erosion, and particularly high-intensity rainstorm conditions, soil erosion occurs easily. Soil erosion destroys land resources and causes a large amount of sediment to flow into rivers. Increased sediment inflow into rivers can easily lead to siltation of the surrounding rivers, such as the Dama and Xiaoma Rivers, and could easily cause flooding disasters, resulting in a decrease in the quality of the ecological environment and pollution of the surrounding water bodies.

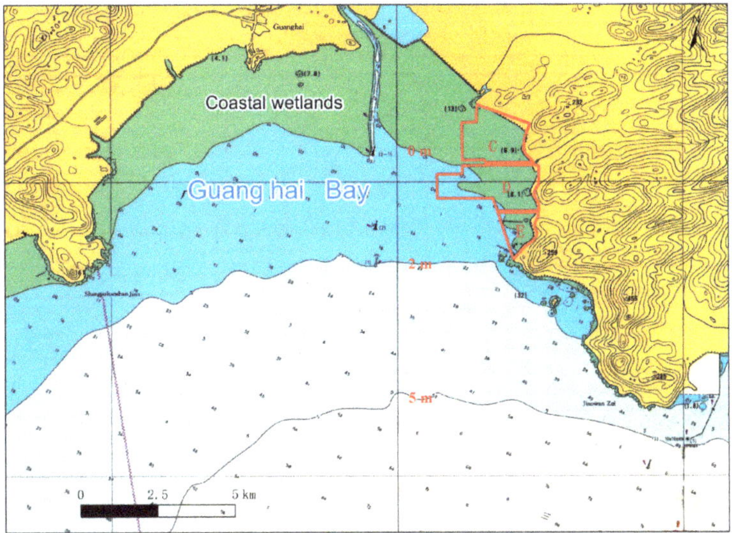

Figure 3. Schematic occupancy map of coastal wetlands through engineering construction.

Simultaneously, the sea reclamation has intensified the near-shore environmental pollution of Guanghai Bay, and the permanent change in the natural attributes of the marine ecosystem due to sea reclamation has led to the disappearance of the environmental capacity of this part of the ocean. According to the results of a marine environmental quality survey conducted in July 2017, the loss in the environmental capacity of chemical oxygen demand caused by sea reclamation in the reclamation area was 2.77 t/a, while the capacity loss of total nitrogen (TN) and total phosphorus (TP) caused by the occupied sea area was 7.99 t/a.

3.1.3. Entropy Increases in Landscape: Poor Public-Service Function and Landscape Effect

The rigid flat design and single hard revetment form had led to poor public-service function, and landscape effect. The layout of the reclamation, particularly in area D in the reclamation area, is stiff, forming a plurality of near 90° revetment corners (Figure 4d). The rivers, including the Dama and Xiaoma rivers between the reclamation areas, are straight, which reduces the diversity of the ecological environment of the river water and leads to the degradation of the river ecosystem [35]. This, in turn, would lead to the entrainment of pollutants in upstream waters that directly enter the reclamation areas, and the organisms that could take refuge in the slow-flow area of the bay during floods would not survive. The erosion of the tidal current on such a revetment is more severe than that of a meandering revetment, thus affecting the revetment stability. The revetments of the reclamation area are illustrated in Figure 4e–g and are made up of hard stone blocks; the hard sea reclamation revetment form cuts off biological exchange between the water body and land. Apart from a small area of mangroves that has been planted in front of the revetment, coastal plants are scarce. The revetment has not been implemented with a reasonable and attractive hydrophilic design. The inert materials were simply stacked and treated to form a large piece of bare soil, and the landscape effect on the revetment is poor. The resource advantages of the coastal wetlands are not effectively reflected, with lack of space sharing, public hydrophilic zones, and public service functions [36,37].

Figure 4. (**a**) (**b**) Soil in the reclamation area; (**c**) Plants in the reclamation area; (**d**) Reclamation layout; (**e**) Sloping shoreline in area C; (**f**) Sloping shoreline in area D; and (**g**) Sloping shoreline in area E.

3.2. Relative Importance of Calculating Negative Entropy Flow by Analytic Hierarchy Process (AHP)

The maintaining of the dynamic and orderly development of the system does not negatively affect the status quo but constantly breaks the balance, with external matter and energy continually being absorbed to dissipate the internal low quality and high entropy to maintain an orderly structure. The analysis of the reclamation project and the entropy increase of its adjacent marine ecological environment showed that the reclamation projects and their adjacent marine ecosystems rely solely on their self-organization capabilities for structure and function owing to problems with resources, environment, and landscape ecology. It is difficult for the adjustment to form a new dissipative structure, and a negative entropy input outside the system is required. Combining the requirements for ecological protection and restoration of reclamation projects in the "Technical Guidelines for the Preparation of Ecological Protection and Restoration Schemes for Reclamation Projects (Trial)," the restoration of marine biological resources, coastal wetland restoration, shoreline restoration, and pollution prevention measures may be adopted to introduce negative entropy flow.

To evaluate the importance of the proposed restoration measures, a hierarchical structure model was constructed using the AHP (Figure 5), and its weight was calculated.

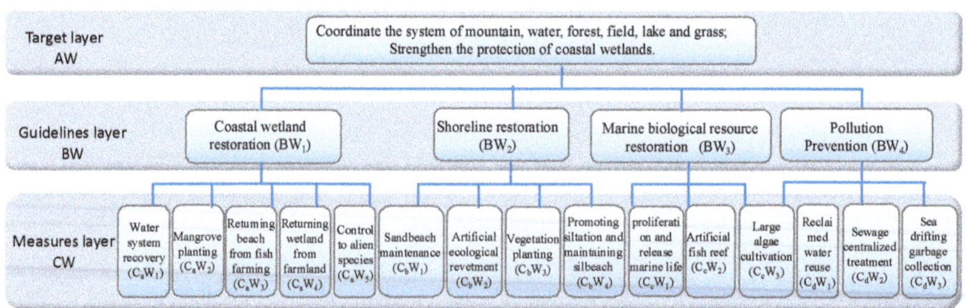

Figure 5. Hierarchical model of the reclamation projects and adjacent marine ecosystems. AW: target layer, BW: guidelines layer, CW: measures layer.

Thereafter, a comparative discriminant matrix was constructed, and after consultation, the importance of the different factors of each layer relative to the upper layer factor was gradually weighted. The comparison discriminant matrix A is formulated as follows:

$$AW = \begin{bmatrix} BW_{11} & BW_{12} & BW_{13} & BW_{14} \\ BW_{21} & BW_{22} & BW_{23} & BW_{24} \\ BW_{31} & BW_{32} & BW_{33} & BW_{34} \\ BW_{41} & BW_{42} & BW_{43} & BW_{44} \end{bmatrix} = \begin{bmatrix} 1 & 2 & 5 & 3 \\ 1/2 & 1 & 3 & 2 \\ 1/5 & 1/3 & 1 & 1/2 \\ 1/3 & 1/2 & 2 & 1 \end{bmatrix} \quad (5)$$

$$BW_1 = \begin{bmatrix} C_aW_{11} & C_aW_{12} & C_aW_{13} & C_aW_{14} & C_aW_{15} \\ C_aW_{21} & C_aW_{22} & C_aW_{23} & C_aW_{24} & C_aW_{25} \\ C_aW_{31} & C_aW_{32} & C_aW_{33} & C_aW_{34} & C_aW_{35} \\ C_aW_{41} & C_aW_{42} & C_aW_{43} & C_aW_{44} & C_aW_{45} \\ C_aW_{51} & C_aW_{52} & C_aW_{53} & C_aW_{54} & C_aW_{55} \end{bmatrix} = \begin{bmatrix} 1 & 2 & 3 & 6 & 9 \\ 1/2 & 1 & 2 & 5 & 7 \\ 1/3 & 1/2 & 1 & 2 & 5 \\ 1/6 & 1/5 & 1/2 & 1 & 2 \\ 1/9 & 1/7 & 1/5 & 1/2 & 1 \end{bmatrix} \quad (6)$$

$$BW_2 = \begin{bmatrix} C_bW_{11} & C_bW_{11} & C_bW_{11} & C_bW_{11} \\ C_bW_{21} & C_bW_{22} & C_bW_{23} & C_bW_{24} \\ C_bW_{31} & C_bW_{32} & C_bW_{33} & C_bW_{34} \\ C_bW_{41} & C_bW_{42} & C_bW_{43} & C_bW_{44} \end{bmatrix} = \begin{bmatrix} 1 & 2 & 3 & 5 \\ 1/2 & 1 & 2 & 4 \\ 1/3 & 1/2 & 1 & 2 \\ 1/5 & 1/4 & 1/2 & 1 \end{bmatrix} \quad (7)$$

$$BW_3 = \begin{bmatrix} C_cW_{11} & C_cW_{12} & C_cW_{13} \\ C_cW_{21} & C_cW_{22} & C_cW_{23} \\ C_cW_{31} & C_cW_{32} & C_cW_{33} \end{bmatrix} = \begin{bmatrix} 1 & 2 & 3 \\ 1/2 & 1 & 2 \\ 1/3 & 1/2 & 1 \end{bmatrix} \quad (8)$$

$$BW_4 = \begin{bmatrix} C_dW_{11} & C_dW_{12} & C_dW_{13} \\ C_dW_{21} & C_dW_{22} & C_dW_{23} \\ C_dW_{31} & C_dW_{32} & C_dW_{33} \end{bmatrix} = \begin{bmatrix} 1 & 2 & 3 \\ 1/2 & 1 & 2 \\ 1/3 & 1/2 & 1 \end{bmatrix} \quad (9)$$

where BW_1 is coastal wetland restoration, BW_2 is shoreline restoration, BW_3 is marine biological resource restoration and BW_4 is pollution prevention.

In the third step, the weight vector was calculated and tested once (Table 3).

Table 3. Weight vector calculations and one-time test results.

Matrix	*n*	Hierarchical Single-Sorted Weight Vector (*W*)	Maximum Eigenvalue (λmax)	Average Random Consistency Indicator (*RI*)	Consistency Indicator (*CI*)	Consistency Ratio (*CR*)	Acceptable Consistency
AW	4	(0.4824, 0.2718, 0.0883, 0.1575)	4.015	1.12	0.005	0.004	Yes
BW_1	5	(0.4461, 0.2864, 0.1567, 0.0716, 0.0392)	5.050	1.12	0.042	0.034	Yes
BW_2	4	(0.4758, 0.2884, 0.1544, 0.0813)	4.021	0.94	0.007	0.007	Yes
BW_3	3	(0.5390, 0.2973, 0.1638)	3.009	0.58	0.004	0.008	Yes
BW_4	3	(0.5390, 0.2973, 0.1638)	3.009	0.58	0.004	0.008	Yes

AW: target layer, BW: guidelines layer.

Table 3 indicates that the consistency of the comparison judgment matrix is acceptable. The ecological restoration measures to be taken are sorted according to importance. From the perspective of the guidelines, the following order is considered: coastal wetland restoration (BW_1) > shoreline restoration (BW_2) > pollution prevention (BW_4) > marine biological resource restoration (BW_3).

From the perspective of measures, the importance is as follows:

(1) BW_1: water system recovery > mangrove planting > returning beach from fish farming > returning wetland from farmland > control to alien species;

(2) BW_2: sandbeach conservation > artificial ecological revetment > vegetation planting > promoting siltation and maintaining siltbeach;
(3) BW_3: proliferation and release marine life > artificial fish reef > large algae cultivation;
(4) BW_4: reclaimed water reuse > sewage centralized treatment > sea-drifting garbage collection.

3.3. Main Ecological Restoration Scheme for Reclamation Projects and Adjacent Marine Ecosystems

According to the entropy change model and the degree of importance calculated via the analytic hierarchy process, ecological restoration schemes should be proposed in the following order: ecological restoration of the wetland system, ecological seawall construction, pollution prevention, and restoration of marine biological resources.

3.3.1. Negative Entropy Flow of Resources: Ecological Restoration of Wetland Systems

The restoration of water system and mangrove plantation with the highest importance value were selected as restoration schemes of the wetland space (Figure 6a).

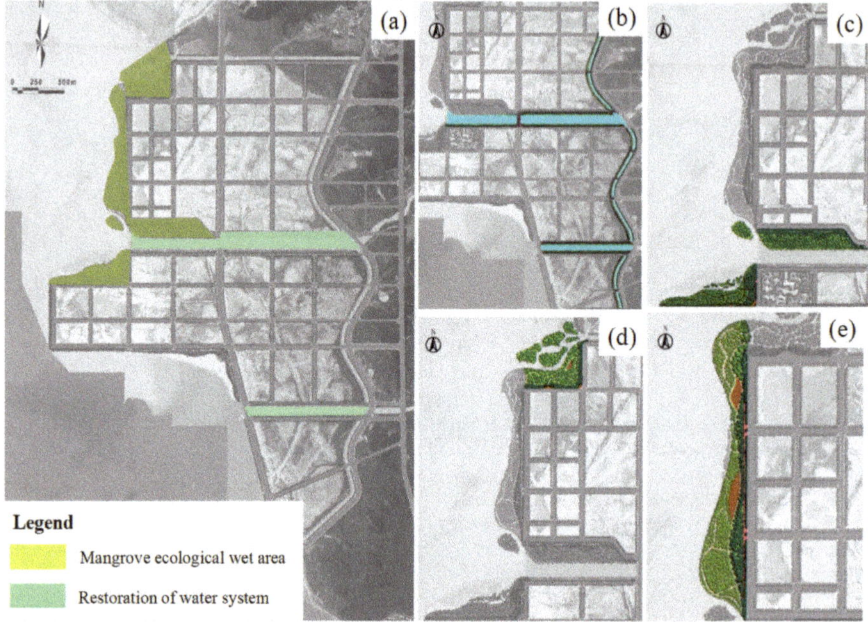

Figure 6. (**a**) The restoration schemes of wetland space; (**b**) Restoration of water system; (**c**) Mangrove ecological wetland area outside the Dama River; (**d**) Mangrove ecological wetland area at north side of area C and (**e**) Mangrove ecological wetland area at west side of area C.

(1) Restoration of the water system:

Widen the width of the Dama and Xiaoma Rivers to ensure flood discharge and drainage. Green spaces, including a green gallery, green ring, and green heart, should be constructed at various areas, including sheltering belts on both sides of the coastal river channel (Figure 6b), to build a complete and coherent green space system and increase landscape, leisure, entertainment, and ecological functions.

(2) Mangrove ecological wetland area:

The entrance to the sea outside the Dama River and the north side of area C were selected as areas for mangrove planting (Figure 6c–e) because the water at these locations is relatively open, which facilitates the silt deposition. Furthermore, a small number of mangroves with eggplant and white bone soil were successfully planted at the north side of area C. The total area of the mangrove ecological restoration is approximately 270,000 m^2.

In addition, there is still silt formed by riprap during revetment construction at the west side of area C. Therefore, mangroves can be planted on the outside, semi mangroves and herbaceous salt tolerant plants can be planted on the inside, and vegetation can be used to slow down wave erosion. This would improve the biodiversity in the coastal wetland and restore the habitats of foraging seabirds and beach organisms at these areas.

3.3.2. Negative Entropy Flow of Landscape: Ecological Seawall Construction to Improve Landscapes

The sandbeach conservation and artificial ecological revetment (including vegetation planting) with the highest importance values were selected as measures for ecological shoreline restoration (Figure 7).

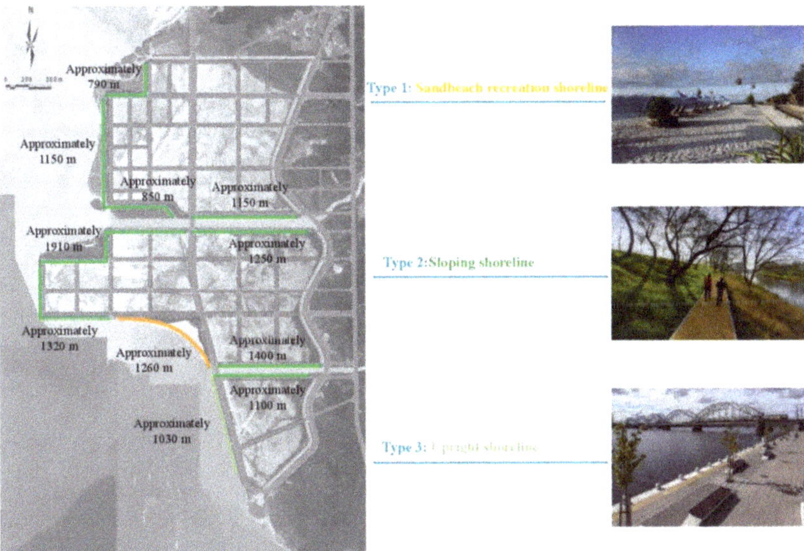

Figure 7. The ecological-shoreline-repair scheme.

(1) Sandbeach shoreline:

Artificial sandbeach restoration on the south side of area D would protect and enhance the connection between the coastal city and the coastal space, providing space for public activities. The shoreline length of the restored sandbeach is approximately 1260 m, forming a sandbeach shoulder width of about 40–60 m and a sandbeach area of approximately 120,000 m^2. It is estimated that the amount of sand to be backfilled is approximately 250,000 m^3. The south end of the sandbeach is provided with a sand retaining (diversion) dike with a length of approximately 150 m.

(2) Artificial ecological revetment and vegetation planting

The shoreline have the form of slope revetment and vertical revetment. Based on the existing embankment structure in the reclamation area, the former is arranged in the northwest, west, and south of area C, as well as the northwest, west, and south of area D; the latter is mainly arranged in the west of area E (Figure 7).

Sloping shoreline: Vines are proposed to be planted above the masonry revetment to allow them to cover the revetment. In the gap of the masonry face, the revetment surface would be filled with fillers with a high efficiency of water absorption and nutrition, in which herbs or vines could be planted to form a plant surface covering the masonry face.

Upright shoreline: The greening center will mainly be the space enclosed by the inner side of the embankment and road. This offers anti-wave, windproof, and landscape effects, and will exhibit sound ecological stability. This is a complex coastal ecological protection system with a high ecological service function.

3.3.3. Environmental Negative Entropy Flow: Pollution Prevention

The reclaimed water reuse and sewage centralized treatment with the highest importance value were selected as measures of sewage discharge and control.

Industrial wastewater and domestic sewage recycling should be promoted in the reclamation region, which should be combined with the construction of artificial ecological wetland and water-system. If sewage and wastewater must be discharged to the sea, they should be treated using the highest standard treatment method according to the water quality requirements of the marine functional area and total pollutant control requirements. Moreover, centralized drainage, offshore drainage, and ecological discharge should be adopted as much as possible. Ecological discharge encourages the discharge of sewage after its ecological treatment and fully exerts the repurification effect of ecological projects such as constructed wetlands.

3.3.4. Bio-Ecological Negative Entropy Flow: Restoration of Marine Living Resources

The proliferation and release of marine life with the highest importance value were selected as measures of the restoration of marine living resources.

According to the local main biological species, the most damaged species and the species with the most obvious effects of proliferation and release, the main proliferation and release species are Black snappers, Yellow fin snappers and Penaeus monodon. Most importantly, the amount of proliferation and release of marine life must not be less than the loss amount.

4. Discussion

Due to the increasingly intensive sea use activities in the coastal zone, ecological protection and restoration work has been gradually taken seriously by regions and governments around the world to address resource loss and functional decline in many coastal ecosystems [38]. However, the current ecological protection and restoration works are mostly limited to a single level of technical restoration [39], such as only mangrove planting [40] or only coral reef restoration [41], but the ecosystem needs to be considered as a whole [42]. The single level of restoration hinders people from using the system concept to realize overall protection and restoration. However, it's necessary that all factors of the restoration project should be connected in series to form an independent, interconnected, and mutually dependent whole, according to the different ecological restoration objects, varying damage degrees, and different stages. In addition, the ecosystem structure must be reconstructed or repaired. Social, economic, environmental, and other factors are combined with the superposition effect of point, line, and surface repairs [41]. Therefore, in this study, based on the current water system and natural environment resource background conditions, we considered the ranking of the importance of the negative entropy reduction calculated through AHP, and reorganized the ecological pattern of the Guanghai Bay reclamation project and its adjacent marine ecosystem. Through the integration management concepts of the mountain, water, forest, field, lake and grassland systems, a typical ecological restoration system was constructed at the east of Guanghai Bay, with the mangrove wetland area as the model in the north and the artificial sandbeach recreation area as the focus in the south (Figure 8).

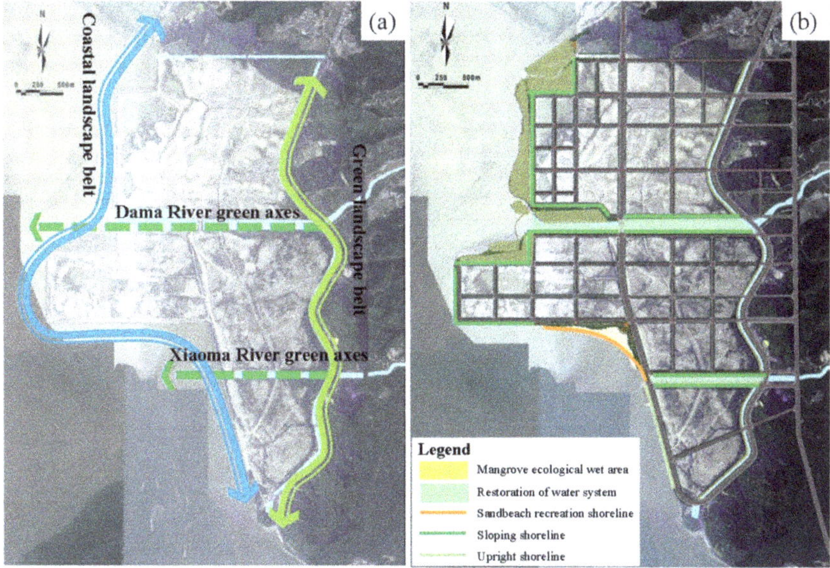

Figure 8. Restoration space pattern: (**a**) landscape structure and (**b**) ecological restoration.

Entropy increase caused by problems in resources, ecology, environment, and landscape results in the disordered development of reclamation projects and their adjacent marine ecosystems. By introducing external negative entropy flow, the overall entropy reduction of the system can be realized. The reclamation projects and their adjacent marine ecosystems are significantly different from the average value under the action of entropy increase and decrease, thereby promoting their orderly and dynamic development (Figure 9) [16]. However, this study only qualitatively analyzed the entropy model of the reclamation projects and their adjacent marine ecosystems and supplemented the relative importance ranking of the negative entropy flow using AHP. The construction of a quantitative analysis model for the entropy would be considered in the future research. The change in the value of entropy can further clarify the evolution mechanism of the reclamation projects and their adjacent marine ecosystems to develop a more targeted ecological protection and restoration program.

Figure 9. Reclamation projects and adjacent marine ecosystems development mechanism.

5. Conclusions

In this study, the dissipative structure and entropy theory was used to analyze ecological problems and environmental threats faced in the Guanghai Bay and the restoration. The problems associated with resources, ecology, environment, and landscape are addressed. The AHP was applied to determine the order of negative entropy flow importance. Finally, the entropy increase and decrease mechanism was used to determine the ecological protection and restoration scheme for reclamation, including the reclamation of wetland resource restoration, shoreline landscape restoration, environmental pollution control, and marine biological resource restoration. The conclusions provide a reference for formulating ecological protection and restoration schemes for reclamation projects and their adjacent marine ecosystems in other regions, as well as for promoting the construction of a marine ecological civilization in coastal areas.

Author Contributions: Conceptualization, F.H.; Data curation, F.H., R.Z. and X.Q.; Formal analysis, Y.L.; Funding acquisition, F.H.; Investigation, Y.L. and R.Z.; Methodology, F.H. and Q.C.; Project administration, Y.L.; Software, X.Q. and Q.C.; Validation, R.Z.; Writing—Original draft, Y.L.; Writing—Review & editing, F.H. and J.L.

Funding: This study was supported by the Ministry of Natural Resources, People's Republic of China (grant no. HD01-190501). The authors express their sincerest thanks to experts who provided suggestions for this research

Conflicts of Interest: The authors declare no conflict of interest.

References

1. Calder, R.S.D.; Shi, C.J.; Mason, S.A.; Olander, L.P.; Borsuk, M.E. Forecasting ecosystem services to guide coastal wetland rehabilitation decisions. *Ecosyst. Serv.* **2019**, *39*, 101007. (In Chinese) [CrossRef]
2. Yang, W.; Jin, Y.W.; Sun, T.; Yang, Z.F.; Cai, Y.P.; Yi, Y.J. Trade-offs among ecosystem services in coastal wetlands under the effects of reclamation activities. *Ecol. Indic.* **2018**, *92*, 354–366. [CrossRef]
3. Ministry of Natural Resources. The First National Marine Inspector Was Officially Launched. Available online: http://gtfwj.dl.gov.cn/info/43800_251170.vm (accessed on 8 January 2019). (In Chinese)
4. Ministry of Natural Resources. The National Marine Inspectorate Group Reported to Liaoning on the Special Inspection of Reclamation. Available online: http://www.mnr.gov.cn/zt/hy/2017wthzxdc/xwzx/201801/t20180114_2102459.html (accessed on 8 January 2019). (In Chinese)
5. Ministry of Natural Resources. The National Marine Inspectorate Group Provides Regular Inspectors and Special Inspections for Reclamation in Hebei. Available online: http://www.mnr.gov.cn/zt/hy/2017wthzxdc/xwzx/201801/t20180116_2102462.html (accessed on 8 January 2019). (In Chinese)
6. Ministry of Natural Resources. The National Marine Inspectorate Group Sent Feedback to Tianjin on the Special Inspection of the Reclamation. Available online: http://www.mnr.gov.cn/dt/zb/2018/2018bhsd/beijingziliao/201807/t20180709_2183957.html (accessed on 8 January 2019). (In Chinese)
7. Ministry of Natural Resources. National Ocean Inspector Group: There Are Four Major Problems in Shandong Ocean Work. Available online: http://www.mnr.gov.cn/dt/mtsy/201807/t20180705_2328845.html (accessed on 8 January 2019). (In Chinese)
8. Ministry of Natural Resources. The National Marine Inspectorate Group Reports to Jiangsu on the Special Inspection of Reclamation. Available online: http://www.mnr.gov.cn/zt/hy/2017wthzxdc/xwzx/201801/t20180114_2102458.html (accessed on 8 January 2019). (In Chinese)
9. Ministry of Natural Resources. The National Marine Inspectorate Group Sent Feedback to the Special Inspection of the Reclamation in Zhejiang. Available online: http://www.mnr.gov.cn/dt/zb/2018/2018bhsd/beijingziliao/201807/t20180705_2183956.html (accessed on 8 January 2019). (In Chinese)
10. Ministry of Natural Resources. The National Marine Inspectorate Group Reports to Guangdong on Routine Inspections and Special Inspections on Reclamation. Available online: http://www.mnr.gov.cn/dt/zb/2018/2018bhsd/beijingziliao/201807/t20180703_2183953.html (accessed on 8 January 2019). (In Chinese)
11. Ministry of Natural Resources. National Ocean Inspector Group. The Four Major Issues of Guangxi Reclamation Control Cannot Be Ignored. Available online: http://www.mnr.gov.cn/zt/hy/2017wthzxdc/mtbd/201801/t20180116_2102486.html (accessed on 8 January 2019). (In Chinese)
12. State Council. *Notice of the State Council on Strengthening the Protection of Coastal Wetlands and Strictly Controlling the Reclamation*; State Council: Beijing, China, 2018. (In Chinese)
13. Ministry of Natural Resources. Notice on Further Addressing the Relevant Requirements for the Remaining Issues Concerning the Reclamation of the Reclamation (Natural Resources [2018] No. 7). Available online: http://www.gov.cn/xinwen/2019-01/08/content_5355787.htm (accessed on 11 January 2019). (In Chinese)
14. Ministry of Natural Resources. *Technical Guidelines for the Preparation of Ecological Protection and Restoration Program for Reclamation Projects (Trial)*; Ministry of Natural Resources: Beijing, China, 2018. (In Chinese)

15. Wu, J.C. *Island Ecological Restoration and Environmental Protection*; China Ocean Press: Bejing, China, 2013. (In Chinese)
16. Jenssen, M. Ecological potentials of biodiversity modelled from information entropies: Plant species diversity of North-Central European forests as an example. *Ecol. Inform.* **2007**, *2*, 328–336. [CrossRef]
17. Ludovisi, A. Effectiveness of entropy-based functions in the analysis of ecosystem state and development. *Ecol. Indic.* **2014**, *36*, 617–623. [CrossRef]
18. Zhang, Y.; Yang, Z.; Li, W. Analyses of urban ecosystem based on information entropy. *Ecol. Model.* **2006**, *197*, 1–12. [CrossRef]
19. Di, G.B.; Han, Y.Q. Analysis of the sustainable development capacity of China's marine ecosystem from the perspective of entropy. *Geogr. Sci.* **2014**, *34*, 664–671. (In Chinese)
20. Wang, Y.M.; Wang, X.; Zhang, S.; Ding, J.X. Analysis of sustainable development of regional human-sea complex ecosystem based on information entropy. *Soil Water Conserv. Res.* **2018**, *25*, 332–338. (In Chinese)
21. Xu, S.; Xu, D.; Liu, L. Construction of regional informatization ecological environment based on the entropy weight modified AHP hierarchy model. *Sustain. Comput.* **2019**, *22*, 26–31. [CrossRef]
22. Zhou, Y.J.; Zhou, J.X.; Liu, H.L.; Xia, M. Study on eco-compensation standard for adjacent administrative districts based on the maximum entropy production. *J. Clean Prod.* **2019**, *221*, 644–655. [CrossRef]
23. Beronvera, F.J.; Olascoaga, M.J.; Haller, G.; Farazmand, M.; Triñanes, J.; Wang, Y. Dissipative inertial transport patterns near coherent Lagrangian eddies in the ocean. *Chaos* **2015**, *25*, 087412. [CrossRef] [PubMed]
24. Goldbeter, A. Dissipative structures and biological rhythms. *Chaos* **2017**, *27*, 104612. [CrossRef] [PubMed]
25. Shannon, C.E. A mathematical theory of communication. *Bell Syst. Tech. J.* **1948**, *27*, 379–423. [CrossRef]
26. Jürgen, K. A mathematical theory of communication: Meaning, information, and topology. *Complexity* **2011**, *16*, 10–26.
27. Aoki, I. Ecological pyramid of dissipation function and entropy production in aquatic ecosystems. *Ecol. Complex.* **2006**, *3*, 104–108. [CrossRef]
28. Miotto, M.; Monacelli, L. Entropy evaluation sheds light on ecosystem complexity. *Phys. Rev.* **2018**, *98*, 042402. [CrossRef]
29. Zhao, J.C.; Ji, G.X.; Tian, Y.; Chen, Y.L.; Wang, Z. Environmental vulnerability assessment for mainland China based on entropy method. *Ecol. Indic.* **2018**, *91*, 410–422. [CrossRef]
30. He, L.; Zhou, G.H. Analysis into development of pastoral complex based on dissipative structure theory. *Acta Agric. Zhejiangensis* **2019**, *31*, 1388–1398.
31. Saaty, T.L.; Vargas, L.G. Models, methods, concepts & applications of the analytic hierarchy process. *International* **2017**, *7*, 159–172.
32. Saaty, T.L. Fundamentals of the Analytic Hierarchy Process. In *The Analytic Hierarchy Process in Natural Resource and Environmental Decision Making*; Springer: Dordrecht, The Netherlands, 2001; Volume 3.
33. Satty, T.L. *Analytic Hierarchy Process. Encyclopedia of Biostatistics*; John Wiley & Sons, Ltd.: Hoboken, NJ, USA, 2005.
34. Liang, Z.X.; Li, W. Evaluation of ecological environment tourism resource based on AHP method and countermeasure research of Yanhu district Yuncheng City. *Sichuan Environ.* **2017**, *36*, 91–96. (In Chinese)
35. Fan, C.H. Discussion on the feasibility of various forms of fluvial cross section in northen cities. *China Water Transp.* **2014**, *14*, 288–289. (In Chinese)
36. Zhang, X.F.; Wang, Y.L.; Li, Z.G.; Li, W.F. Relationship between soil bareness and landscape pattern in the Loess Plateau: A case study on the tower region of the Yan'an City. *Quat. Sci.* **2004**, *24*, 709–715. (In Chinese)
37. Xing, Y.; Wang, B.; Bian, Q.Q. Research on Urban Waterfront Spatial Planning from the Perspective of Public Demand: A Case Study of Haihe Bund Area in Tianjin Binhai New Area. In Proceedings of the Annual Conference on Urban. Planning in China, Guiyang, China, 19 September 2015. (In Chinese).
38. Sun, S.; Wang, B. Study on ecological restoration technology of Yanxi River wetland in Beijing. *Energy Procedia* **2018**, *153*, 330–333. [CrossRef]
39. Mao, X.F.; Wei, X.Y.; Jin, X.; Tao, Y.Q.; Zhang, Z.F.; Wang, W.Y. Monitoring urban wetlands restoration in Qinghai Plateau: Integrated performance from ecological characters, ecological processes to ecosystem services. *Ecol. Indic.* **2019**, *101*, 623–631. [CrossRef]
40. Das, S. Ecological Restoration and Livelihood: Contribution of Planted Mangroves as Nursery and Habitat for Artisanal and Commercial Fishery. *World Dev.* **2017**, *94*, 492–502. [CrossRef]

41. Hein, M.Y.; Birtles, A.; Willis, B.L.; Gardiner, N.; Beeden, R.; Marshall, N.A. Coral restoration: Socio-ecological perspectives of benefits and limitations. *Biol. Conserv.* **2019**, *229*, 14–25. [CrossRef]
42. Pueyo-Ros, J.; Garcia, X.; Ribas, A.; Fraguell, R.M. Ecological restoration of a coastal wetland at a mass tourism destination. Will the recreational value increase or decrease. *Ecol. Econ.* **2018**, *148*, 1–14. [CrossRef]

International Journal of
Environmental Research and Public Health

Article

Dynamic Analysis of the Coupling Coordination Relationship between Urbanization and Water Resource Security and Its Obstacle Factor

Kaize Zhang [1,2], Juqin Shen [1], Ran He [1,*], Bihang Fan [3,*] and Han Han [1]

1 Business School, Hohai University, Nanjing 211100, China; kzzhang@hhu.edu.cn (K.Z.); jqshen@hhu.edu.cn (J.S.); 170212070002@hhu.edu.cn (H.H.)
2 Department of Ecosystem Science and Management, The Pennsylvania State University, State College, PA 16802, USA
3 State Key Laboratory of Hydraulics and Mountain River Engineering, Sichuan University, Chengdu 610065, China
* Correspondence: heran88081@163.com (R.H.); buf143@psu.edu (B.F.)

Received: 1 November 2019; Accepted: 26 November 2019; Published: 28 November 2019

Abstract: Water resource security is an important condition for socio-economic development. Recently, the process of urbanization brings increasing pressures on water resources. Thus, a good understanding of harmonious development of urbanization and water resource security (WRS) systems is necessary. This paper examined the coordination state between urbanization and WRS and its obstacle factors in Beijing city, utilizing the improved coupling coordination degree (ICCD) model, obstacle degree model, and indicator data from 2008 to 2017. Results indicated that: (1) The coupling coordination degree between WRS and urbanization displayed an overall upward tendency during the 2008–2017 period; the coupling coordination state has changed from an imbalanced state into a good coordination state, experiencing from a high-speed development stage (2008–2010), through a steady growth stage (2010–2014), towards a low-speed growth (2014–2017). (2) In urbanization system, both the social and spatial urbanizations have the greatest obstruction to the development of urbanization-WRS system. The subsystems of pressure and state are the domain obstacle subsystems in WRS system. These results can provide important support for urban planning and water resource protection in the future, and hold great significance for urban sustainable development.

Keywords: urbanization; water resource security; improved coupling coordination model; obstacle degree model; Beijing

1. Introduction

Water is not only the natural resources to maintain the human existence, but also the material basis to guarantee the socio-economic development. Water resources directly or indirectly provide resources guarantee for social development [1]. As the important strategic socio-economic resource for the development of urbanization, the rational and efficient utilization of water resources will affect the implementation of sustainable urban development strategies [2,3]. Urbanization system includes complex relationships of dependence between socio-economic activities and resources. Water resource security and urban development systems are intricately linked [4]. Due to the rapid development of urbanization, urban population has increased dramatically, and the scale of urbanization has been expanding, resulting in increasing pressures on urban water resources [5]. Recently, water scarcity caused by urbanization has brought a series of socio-economic issues. According to statistics, about 2.6 billion people in the world lack access to safe water resource [6]. Clarifying the relationship between urbanization and water resource security (WRS) is not only the basis for effective water resources management, but is also required to achieve a sustainable urban development.

The concept of WRS was first developed in the late 20th century. In a broad sense, WRS refers to a state in which water resources can ensure the healthy and stable existence and development of human beings [7,8]. The term WRS indicates the risks to water supply and demand, and to social and economic development. In our research, WRS means that the function and state of the water environment system on which humans rely should be in a healthy, complete, and stable, and can provide ecological services for human survival and socio-economic development. Previous studies have appreciated that urbanization has a strong interaction with water resources [9,10]. On the one hand, the urbanization process relies heavily on water resource consumption. Water resources are required to ensure the healthy operation of industry, agriculture, and other economic activities [11]. On the other hand, water shortage and water pollution caused by socio-economic development can induce economic losses and threaten the sustainable development of human society [12]. At present, there is evidence that problems related to water scarcity and water pollution are increasing, mainly due to overconsumption of water resources and sewage discharge. It appears that there is a significant unbalance between urbanization and WRS. The need to balance between these two aspects is clear, and is necessary to ensure a sustainable development.

Academic study on the effects of urbanization on the environment have mainly focused on water resources [13], land use [14], and energy consumption [15]. These studies examined the impacts of urban scale, population density, and transportation network on the eco-environment. An increasing number of scholars are concerned about ensuring a harmonious relationship between economic development and the environment, as well as the reasonable utilization of resources [16]. At present, the concept of coupling has been widely used in the fields of ecology and climate change. It mainly means that two or more subjects interact reciprocally in various forms [17]. Environmental assessments have gradually evolved from a single environmental analysis to the exploration of the harmonious relationship among economic development and environment [18,19]. For example, Han et al., [20] analyzed the coupling state between urbanization and water ecosystem from the spatial–temporal perspective. Al-Mulali et al., [21] found an inverted U-shaped coupling relationship between ecological environment and urbanization in both middle- and high-income countries. Yao et al., [22] assessed the spatiotemporal changes in coordinating state of new urbanization and ecological-environmental stress.

Recently, one of the biggest challenges facing society is how to maintain natural resources while promoting social development and quality of human life [23]. Sustainable development theory holds that economic growth should be based on strict population control, improvement population quality, sustainable resources use and environmental protection. This theory regards the coordination state between the resources and urbanization as a key indicator to judge whether the urban development is sustainable [24]. Society should examine the coordination between urbanization development and resources to protect resources, especially water resources. However, there are few studies on the harmonious relationship between WRS and urbanization. Therefore, there is an urgent need to accurately assess the coupling degree of these two systems, as this is crucial for the realization of sustainable urban development. In addition, about the analysis of urbanization, most scholars pay attention to the urban land expansion [25], such as land use conflicts, land use synergies and tradeoffs, and land-use/land-cover change [26]. In fact, the urbanization process includes not only the land use, but also demographic change, economic growth and social development [20], which is the combination process of these four aspects. The coupling coordination model can integrate the four aspects of urbanization with WRS, which contributes to the analysis of sustainable urban development. Thus, it is important to have an adequate understanding of the coordination between urbanization and WRS to support for new-urban planning.

To fill this gap, we attempt to introduce the improved coupling coordination degree (ICCD) model and an obstacle degree model to analyze the coordination relationship between WRS and urbanization and its obstacle factors in Beijing city. We firstly establish an indicator system including the urbanization system and the WRS system. Then, we use the ICCD model to examine the coordination state between urbanization-WRS in Beijing city from 2008 to 2017. Finally, we introduce the obstacle degree model to determine the dominant obstacle indicators. This study will provide new insights into the water resources management and new-urban development in the future.

2. Materials and Methods

2.1. Study Area

Beijing city, the capital of China, is located in the north of China. It covers an area of 1.64 million square kilometers, and is divided into 16 major districts (see Figure 1). Beijing lies between 39°26′–41°03′ N and 115°25′–117°30′ E, in a warm, temperate, with moderate warmth and four seasons. Beijing, as one of the fastest growing urbanization areas in China, has a high socio-economic level; however, the gap between urbanization and water resource is very large. As the center of politics, economy, and culture in China, in 2016 the urbanization rate of Beijing reached about 86.5%; the total amount of water resources availability was 3.506 billion cubic meters, while the water resource amount per capita was only 161 billion cubic meters, i.e., about 1/15 of the national average [27,28]. Water resource security in Beijing is a quite prominent problem, especially water supply and demand, and seriously influences the ecological health and the socio-economic development.

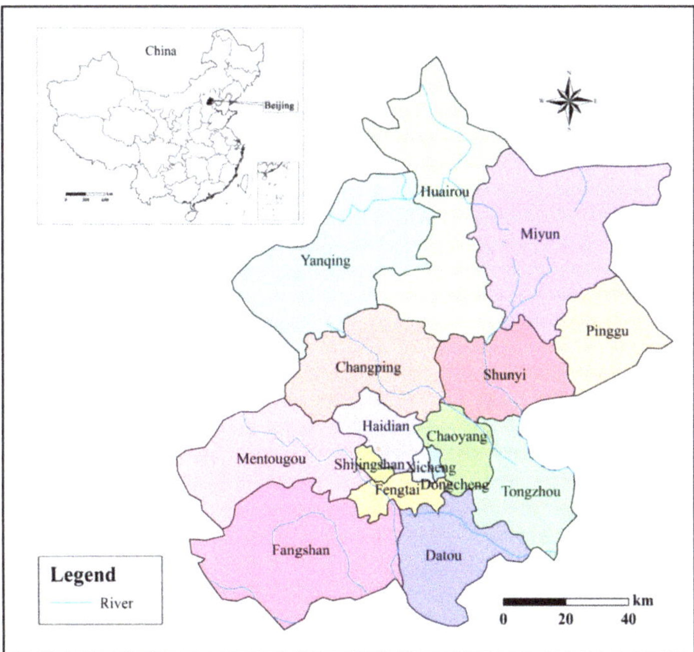

Figure 1. Location of Beijing city.

2.2. Methods

2.2.1. Research Framework

This paper tries to explore the coordination relationship between WRS and urbanization by using a systematic framework (see Figure 2). For this purpose, (1) we establish a comprehensive indicator system, including the urbanization and WRS systems; (2) the data of forward and reverse indicators are preprocessed into dimensionless values; (3) the ICCD model and obstacle degree model are established; and (4) the coupling coordination state between the urbanization-WRS system is analyzed.

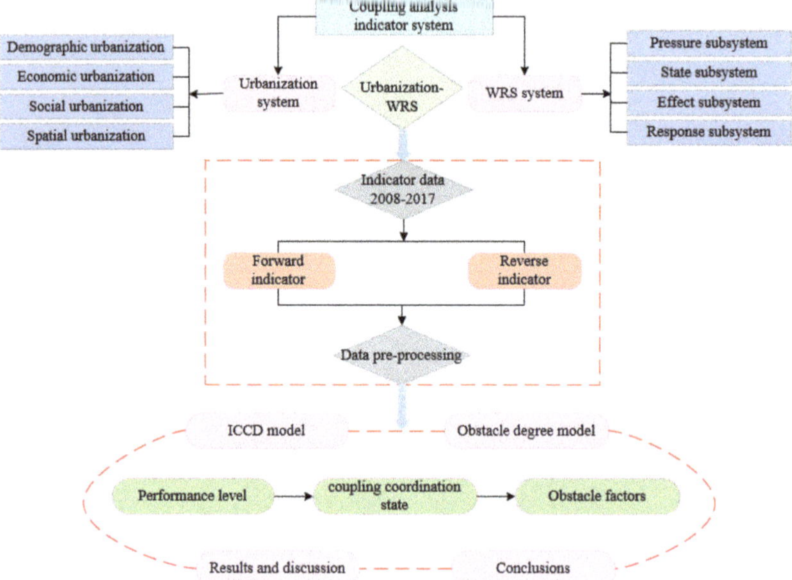

Figure 2. The research framework.

2.2.2. Construction of Indicator System

Urbanization is a way to describe the process of continuous concentration of population into urban areas. This is a process of continuous development and change, in which the type of urbanization changes from "traditional" to "new-type", the social economy is constantly developing, population quality is constantly improving, and urban development moves toward modernization [29]. Urbanization has an important impact on several aspects of society, population, and economic life [30]. Recently, numerous scholars have carried out research on China's urbanization development from four aspects of demographic urbanization, economic urbanization, social urbanization, and spatial urbanization [31,32]. In this study, we followed this classification to establish the primary indicators for the urbanization system. In terms of the specific secondary indicators, we selected urban population density, non-agricultural population rate, and population growth rate to represent demographic urbanization. These indicators are commonly used as secondary indicators for demographic urbanization [33]. Economic urbanization is the geographical concentration of urban economic activities in the process of urbanization. Thus, the indicators of per capita financial revenue, per capital GDP, and per capita investment in fixed assets can be employed to evaluate economic urbanization. The extent of social urbanization is quite wide; therefore, we chose three representative indicators (i.e., per capita education funds, number of doctors per 10^4 people, and number of public transports per million people) to represent it, based on existing research [34]. Spatial urbanization mainly shows the changes in land use and fixed assets investment. Dwelling area per capita, urban road area per capita, and fixed asset investment growth rate were selected as indictors of spatial urbanization. Thus, a total of 12 representative indicators were determined for urbanization.

The construction of the WRS indicator system followed the Pressure-State-Effect-Response (PSER) framework, which is widely used in the field of environment and sustainable development [35,36]. The PSER framework includes the pressure subsystem, the state subsystem, the effect subsystem, and the response subsystem, as well as their interactions and constraints. The PSER framework can clearly show the causal relationship between these four WRS subsystems. In terms of secondary indicators, we determined the indicator system of WRS through 4 primary indicators and 14 secondary representative indicators based on the status quo of the water resources and the water environment in Beijing [20,37].

In general, the urbanization system and WRS system couple with each other and have complex interactions. We selected the 12 representative urbanization indicators and 14 WRS indicators to reflect the relationship between the two systems. The structure model of urbanization-WRS system in Beijing is shown in Figure 3 [38]. The indicator systems for urbanization and WRS are shown in Table 1.

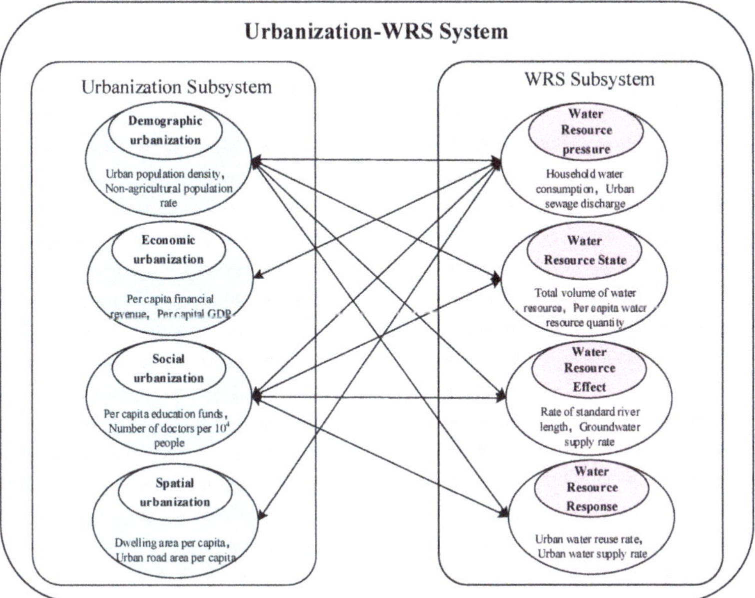

Figure 3. Urbanization-WRS system connotation structure model.

Table 1. Urbanization-WRS indicator system.

System	Subsystem	Indicators	Unit	Type
Urbanization	Demographic urbanization	C_1 Urban population density	Person/km^2	+
		C_2 Non-agricultural population rate	%	+
		C_3 Population growth rate	‰	+
	Economic urbanization	C_4 Per capita financial revenue	10^4 Yuan/person	+
		C_5 Per capital GDP	10^4 Yuan/person	+
		C_6 Per capita investment in fixed assets	10^4 Yuan/person	+
	Social urbanization	C_7 Per capita education funds	Yuan/person	+
		C_8 Number of doctors per 10^4 people	/	+
		C_9 Number of public transports per million people	/	+
	Spatial urbanization	C_{10} Dwelling area per capita	m^2/person	+
		C_{11} Urban road area per capita	km^2/person	+
		C_{12} Fixed asset investment growth rate	%	+

Table 1. *Cont.*

System	Subsystem	Indicators	Unit	Type
WRS	Pressure	D_1 Water consumption for industrial production	10^9 m^3	—
		D_2 Water consumption of agricultural irrigation	10^9 m^3	—
		D_3 Household water consumption	10^9 m^3	—
		D_4 Urban sewage discharge	10^9 m^3	—
	State	D_5 Total volume of water resource	10^9 m^3	+
		D_6 Per capita water resource quantity	m^3/person	+
	Effect	D_7 Water producing coefficient	%	+
		D_8 COD emissions	10^4 tons	—
		D_9 Rate of standard river length	%	+
		D_{10} Groundwater supply rate	%	+
	Response	D_{11} Rate of water resources management investment to GDP	%	+
		D_{12} Rate of industrial wastewater up to discharge standard (RIWDS)	%	+
		D_{13} Urban water reuse rate	%	+
		D_{14} Urban water supply rate	%	+

2.2.3. The Variation Coefficient Method

(1) Pre-processing of the indicators.

The WRS indicator system includes both forward and reverse indicators. For the forward indicators (i.e., the state subsystem indicators D5–D6), a larger indicator value means a better performance level. For the reverse indicators (i.e., the pressure subsystem indicators D1–D4), a bigger indicator value means a worse performance level. Thus, it is necessary to convert the indicators to dimensionless values to avoid the effect of the characteristics and the scope of the selected indicator. The following normalization formulas were used:

$$\text{Forward indicator}: \ f_{ij} = (x_{ij} - \min\{x_i\})/(\max\{x_i\} - \min\{x_i\}), \tag{1}$$

$$\text{Reverse indicator}: \ f_{ij} = (\max\{x_i\} - x_{ij})/(\max\{x_i\} - \min\{x_i\}), \tag{2}$$

where x_{ij} is the original index value and f_{ij} is the standardized value of x_{ij}.

(2) Determination of indicators' weight.

The urbanization and WRS systems cover 12 and 14 indicators, respectively. A way to distinguish the importance of these indicators in each system is to assign them different weights. The variation coefficient method is based on the information contained in the index data to evaluate the importance of difference indexes, which is a method of objectively calculating weights [39,40]. The variation coefficient method believes that when the variation degree of the indicator is greater, the ability of the indicator to distinguish different evaluated objects should then be stronger, and the weight should be larger [41]. The variation coefficient method has been widely introduced to environmental quality assessments [42]. The variation coefficient weight was calculated with the following steps:

$$\overline{x_i} = \frac{1}{n}\sum_{j=1}^{n} x_{ij} \tag{3}$$

$$\sigma_i = \sqrt{\sum_{j=1}^{n} (x_{ij} - \overline{x_i})^2/n} \tag{4}$$

$$d_i = \sigma_i/\overline{x_i} \tag{5}$$

$$w_i = d_i/\sum_{i=1}^{m} d_i \ (0 < w_i < 1, \sum_{i=1}^{m} w_i = 1) \tag{6}$$

where $\overline{x_i}$ refers to the average value of x_i during the period studied; σ_i represents the standard deviation of x_i; d_i is the variation coefficient of x_i, it represents the variation degree of x_i; and w_i is the weight of the indicator x_i.

2.2.4. The ICCD Model

The system composed of both WRS and urbanization can be seen as a coupling system, in which WRS and urbanization influence and restrict reciprocally [43]. The degree of interaction among systems is measured by coupling coordination degree [44]. The contribution coefficients describe the influence degree of each single system to the whole coupling system. The ICCD model is shown below [45]:

(1) Determination of the performance levels.

The performance levels of the two systems were calculated in the following way:

$$U_j = \sum_{i=1}^{m} w_i^u \cdot f_i^u \tag{7}$$

$$W_j = \sum_{i=1}^{m} w_i^w \cdot f_i^w \tag{8}$$

where U_j is the urbanization system's performance level; W_j refers to the WRS system's performance level; $f_i{}^u$ and $w_i{}^u$ are the standardized value and the weight of the indicator x_i in urbanization system, respectively; and $f_i{}^w$ and $w_i{}^w$ are the standardized value and the weight of indicator x_i in WRS system, respectively.

(2) Determination of the improved contribution coefficients.

In the ICCD model, it needs to determine the contribution coefficients of urbanization system and WRS system. The traditional CCD model usually utilizes subjective assignment methods to determine the contribution coefficients. The two contribution coefficients are arbitrarily assigned the value of 0.5 by some scholars. The way determines the contribution coefficients by assigning the fixed value (0.5) in a subjective way, which highlights the subjective judgment of the decisionmakers [46,47]. Thus, we introduced an ICCD model to determine the contribution coefficients [48] in the following way:

$$a^* = W/(U+W) \tag{9}$$

$$b^* = U/(U+W) \tag{10}$$

where a^* is improved contribution coefficient of urbanization system; b^* is the improved contribution coefficient of WRS system.

(3) Calculation of the coupling coordination degree D.

The D was determined in the following way:

$$C = \left[(U * W)/\left(\frac{U+W}{2}\right)^2\right]^{1/k} \tag{11}$$

$$T = a^*U + b^*W \tag{12}$$

$$D = \sqrt{C \cdot T} \tag{13}$$

where C is the degree of coupling; k is the adjustment coefficient, it is usually assigned the value of 2. T is the overall performance level; and D refers to the coupling coordination degree between these two systems.

2.2.5. Obstacle Degree Model

To effectively improve the coordination status of urbanization and WRS, it is necessary to explore the main obstacle factors that have a negative impact on the harmonious relationship between urbanization-WRS system. The main step of obstacle degree model is proposed as follows [49]:

$$Q_i = w_i \cdot (1 - f_{ij}) / \sum_{i=1}^{m} w_i (1 - f_{ij}) \tag{14}$$

where Q_i represents the obstacle degree, it means the influence degree of each subsystem or indicator on the two systems. $1 - f_{ij}$ refers to the deviation degree of the indicator, it represents the difference between the actual indicator value and the optimal target value.

2.2.6. Classification Standard of Coupling Coordination Degree

Previous studies have shown that the states of coupling coordination between urbanization and WRS are divided into five grades, namely serious imbalance, imbalance, basic coordination, coordination, and good coordination [21,32]. Accordingly, this study adopted this classification. The D was used to describe the coordination status between two interacting systems, and the five grades were defined as follows:

(1) Serious imbalance state: when D^* assumes a value within the range $0 \leq D^* \leq 0.25$. In this case, the nexus between urbanization and WRS is very poor.
(2) Imbalance state: when D^* assumes a value within the range $0.25 < D^* \leq 0.45$. In this case, the interaction between urbanization and WRS is weak.
(3) Basic coordination state: when D^* assumes a value within the range $0.45 < D^* \leq 0.65$. In this case, the link between urbanization and WRS begins to reinforce.
(4) Coordination state: when D^* assumes a value within the range $0.65 < D^* \leq 0.75$. In this case, the relationship between urbanization and WRS is coordinated.
(5) Good coordination state: when D^* assumes a value within the range $0.75 < D^* \leq 1$. In this case, the coordination between urbanization and WRS is very good.

2.3. Data Source

The urbanization system data during the period 2008–2017 were collected from the Beijing Statistical Yearbook [27] and the China Urban Statistical Yearbook [50]. The annual data on the WRS system were collected from the Beijing Water Resource Bulletin [28] and the Beijing Statistic Yearbook on Environment [51].

3. Results

3.1. Performance Level of Two Systems

The overall performance level of urbanization displayed an upward trend during the period 2008–2017, as shown in Figure 4. Figure 4 shows that, except for 2010 and 2015, the overall performance level of urbanization has been increasing steadily in the other years, from 0.13 in 2008 to 0.79 in 2017. It is noteworthy that the value in 2017 is five times larger than that in 2008. In terms of the subsystems' performance level, the performance values of the four urbanization subsystems showed different degrees of increase for each year. Spatial urbanization displayed a clear fluctuation, with its performance value increasing from 0.11 in 2008 to 0.22 in 2009, before declining to 0.06 in 2012, and finally reaching 0.13 in 2017. The other three subsystems showed a steady upward trend during the period studied, with social urbanization recording the highest growth trend, from 0.02 in 2008 to 0.256 in 2017.

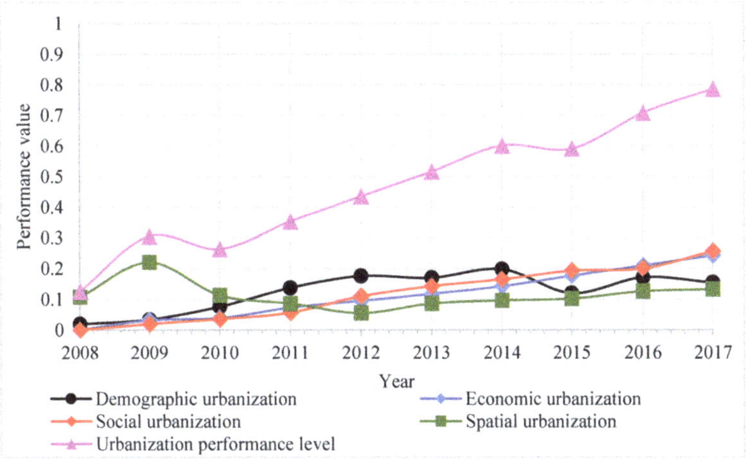

Figure 4. Performance level trends in the urbanization system from 2008 to 2017.

Figure 5 shows the trends of the performance level in the WRS system from 2008 to 2017. The overall performance level of WRS followed an S-shaped growth curve, which initially decreased, and subsequently clearly increased from 2008 to 2012. Thereafter, the performance value in the WRS system largely increased from 2014 to 2017. The fluctuation of the performance level in the WRS system was mainly due to the performance level of both the state and the effect subsystems, which declined from 2008 to 2010, but raised continuously after 2010. From 2012 onwards, the two subsystems showed a downward trend of volatility. It is worth noting that the changing trend of the state subsystem was similar to that of the overall performance level. The fluctuation of both the state and the effect subsystems caused the S-shaped growth curve of the overall trend. The performance level of the pressure and response subsystems gradually increased with a significant growth trend from 2014 to 2017.

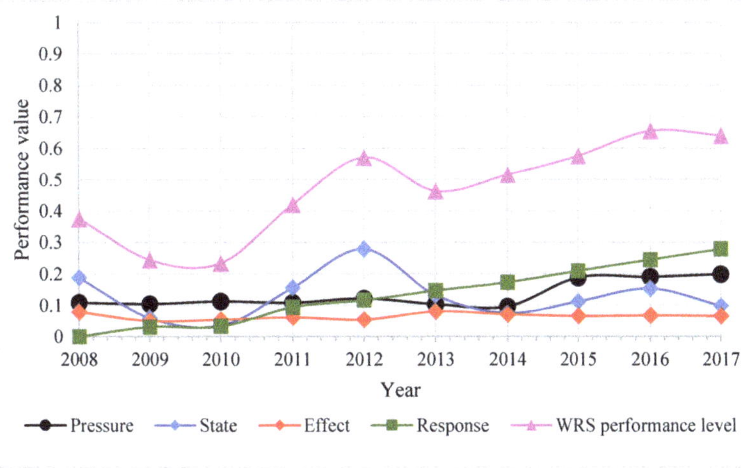

Figure 5. Performance level trends in the WRS system from 2008 to 2017.

3.2. Coupling Coordination State

Figure 6 reveals the results of the coupling coordination degree D between urbanization and WRS in Beijing from 2008 to 2017. As shown in Figure 6, the D between urbanization and WRS followed an overall rising tendency during the period studied, from 0.354 in 2008 to 0.796 in 2017. The results of D mean that the coupling coordination state between the urbanization-WRS system has improved. The state of the coupling coordination between the two systems shows a dynamic evolution, passing from an imbalance state to a good coordination state during the 2008–2017 period. Specifically, by comparing the standard D^* with the actual D, we find that the two systems in Beijing city was in an imbalance state in 2008. During the 2009–2012 period, the coordination between the two system in Beijing maintained in a basic coordination state. In the following four years, D values between urbanization-WRS in Beijing were basically in a good coordination state.

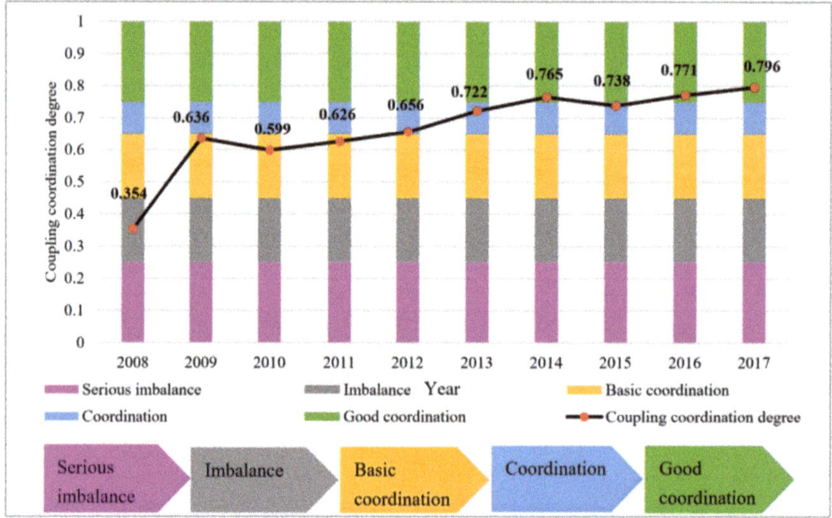

Figure 6. Coupling coordination degree trends from 2008 to 2017.

3.3. Dominant Obstacle Factors

Using the obstacle degree model, the obstacle degrees of the subsystem in the urbanization and WRS systems were obtained; they are shown in Table 2. As shown in Table 2, in the urbanization system, the subsystem with the greatest obstacle degree was social urbanization during 2008 to 2010; the subsystem of spatial urbanization had the largest obstacle degree from 2011 to 2017. In comparison, the demographic urbanization subsystem had the lowest obstacle degree during the 2008–2017 period. This means that in Beijing city, both the social and spatial urbanizations were the domain obstacle subsystems, while the demographic urbanization was the least important subsystem. In the WRS system, the pressure subsystem had the greatest obstacle degree in the first six years during the surveyed period and the state subsystem had the biggest obstacle in terms of degree in the remaining four years, whereas the effect subsystem had the smallest during the period 2008–2017. These findings suggest that, both the pressure and state subsystems were considered to have the greatest hindrance to the development of urbanization-WRS system. On the contrary, the effect subsystem had the smallest impact on the coordination state between urbanization-WRS system in Beijing city.

Table 2. Obstacle degree of subsystem in the two systems.

System	Urbanization				WRS			
Year	Demographic Urbanization	Economic Urbanization	Social Urbanization	Spatial Urbanization	Pressure	State	Effect	Response
2008	21.2%	27.8%	29.3%	21.7%	45.4%	15.9%	6.9%	31.8%
2009	11%	30.3%	34.1%	24.5%	33.5%	30.1%	9.6%	26.8%
2010	17.5%	27.6%	29.9%	25%	32.8%	25.4%	9.0%	32.8%
2011	10.4%	26.2%	30.8%	32.6%	34.3%	22.5%	10.6%	32.6%
2012	5.1%	26.1%	26.0%	42.8%	43.3%	16.0%	1.8%	38.9%
2013	7.1%	25.8%	23.5%	43.6%	37.9%	28.8%	7.9%	25.4%
2014	1.7%	25.1%	22.9%	50.3%	36.1%	36.2%	8.9%	18.8%
2015	15.6%	16.1%	20.7%	47.7%	28.1%	41.0%	13.5%	17.4%
2016	11.1%	11.3%	19.2%	58.4%	34.0%	38.6%	11.3%	16%
2017	0.0%	23.3%	0.1%	76.6%	30.1%	52.4%	1.4%	16%

In urbanization system, the top five indicators that hinder the coordination development of urbanization-WRS system are shown in Table 3. From 2008 to 2010, the indicators of *C12* fixed asset investment growth rate, *C9* number of public transports per million people, and *C10* dwelling area per capita were ranked in the top three in the urbanization system. After 2010, the indicator of *C11* urban road area per capita replaced the *C10* dwelling area per capita became one of the top three indicators. In terms of the WRS system, the indicators of *D1* water consumption for industrial production, *D2* water consumption of agricultural irrigation and *D14* urban water supply rate ranked as the top three indicators from 2008 to 2017. This means that the three indicators had a profound effect on the coordination state between urbanization and WRS.

Table 3. The dominant obstacle indicators in the two systems.

	Urbanization					WRS				
Indicator Order	1	2	3	4	5	1	2	3	4	5
2008	C12	C9	C10	C8	C4	D1	D14	D2	D8	D11
2009	C9	C12	C10	C8	C4	D1	D14	D2	D8	D5
2010	C9	C10	C12	C8	C4	D1	D14	D2	D8	D5
2011	C11	C9	C12	C8	C7	D1	D14	D2	D12	D9
2012	C11	C12	C9	C10	C4	D1	D14	D9	D2	D12
2013	C11	C12	C9	C4	C8	D1	D14	D2	D6	D5
2014	C12	C11	C4	C8	C10	D1	D2	D14	D7	D5
2015	C12	C11	C9	C3	C3	D14	D2	D1	D10	D5
2016	C12	C11	C9	C3	C7	D6	D14	D1	D10	D5
2017	C12	C11	C3	C2	C1	D14	D1	D2	D10	D4

4. Discussion

4.1. Analysis of Performance Level

The results of the urbanization system's performance level suggest that in the past decade, the performance level of urbanization system in Beijing city has improved. During the four subsystems in urbanization, social urbanization is considered to make the greatest contribution for the whole performance level. In the past 10 years, Beijing city was undergoing a period of urbanization and industrialization, relying on its location advantage and policy support, thus achieving a fast increase in the social urbanization level. In relation to WRS, the overall performance levels showed a fluctuating upward trend. The performance levels of the pressure and response subsystems were constantly optimized. This is mainly due to the fact that in 2016, Beijing was listed as the second demonstration "Sponge City". The concept of "Sponge City" is new in urban water management; it aims to effectively divert rain and sewage, store rainwater, and purify sewage. In Beijing city, the response measures adopted in the context of the "Sponge City" released the pressures on water resources, thereby contributing to the improvement of the overall performance levels in the WRS system.

4.2. Analysis of Coupling Coordination State

We examined the coordination state between the WRS and urbanization systems by analyzing the performance gap between the urbanization and the WRS systems. The trend comparison of urbanization performance level and WRS performance level is shown in Figure 7. We discussed this dynamic evolution from the following three stages (2008–2010, 2010–2014, and 2014–2017).

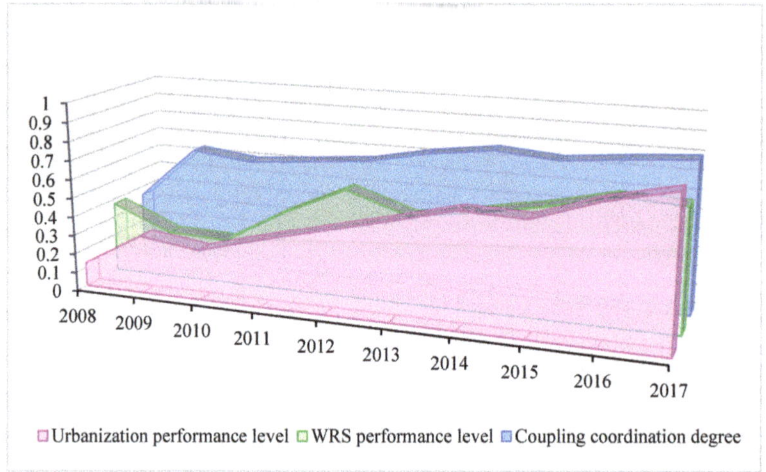

Figure 7. Trend comparison of urbanization performance level and WRS performance level.

(1) 2008–2010: The D between urbanization and WRS was low, and grew rapidly. The nexus between urbanization and WRS was initially weak, and gradually began to strengthen. In 2008, the performance level of urbanization was poor, while the performance level of WRS was considerably higher. As a result, an unbalanced state occurs between the two systems in Beijing. Subsequently, as the gap between urbanization performance level and WRS performance level decreases gradually, while the D among urbanization and WRS grew more rapidly, increasing from 0.354 in 2008 to 0.636 in 2009. Therefore, the coupling coordination state was optimized. During the period from 2009 to 2010, the impact of urbanization on WRS strengthened, and the performance level of WRS was at its lowest point. As a result, the D showed a slight decline. Fortunately, this decline did not cause changes in the coupling coordination state. Both in 2009 and 2010, the urbanization-WRS system was in a basic coordination state.

(2) 2010–2014: The D between urbanization and WRS raised steadily. During this period, urbanization in Beijing entered a stage of rapid development, and the WRS state improved due to the implementation of the South-to-North Water Diversion Project in 2012. The levels of urbanization and WRS increased simultaneously. The gap in the performance levels between the two systems was reduced, signaling their simultaneous development. In this period, the D between the two systems increased from 0.599 in 2010 to 0.765 in 2014. Therefore, the coupling coordination state improved greatly, displaying a cross-level change. The coupling coordination state moved from a basic coordination state in 2010–2011, to a coordination state in 2012–2013, and eventually to a good coordination state in 2014. This finding demonstrates that these processes were gradually coordinated during this period.

(3) 2014–2017: The D between urbanization and WRS slowly increased. After the implementation of the "new-type urbanization" plan in 2014, Beijing has experienced a rapid urbanization process.

The "new-type urbanization" is a new concept of urban development, based on the tenets of resource conservation, eco-friendliness, and sustainable development. During this period, a series of new-type urbanization measures were taken, such as the construction of the "Sponge City", the protection of the eco-environment, and the renovation of the underground pipeline network, which contributed to the improved quality urbanization and water environment. Not only the urbanization level, but also water resource quality was constantly enhanced. The urbanization and WRS systems were in a good coordination state during the period 2016–2017. Urbanization went in a coordinated development phase with water environmental protection. The coupling coordination state among WRS and urbanization improved slowly.

4.3. Analysis of Obstacle Factors

In the urbanization system, both social and spatial urbanizations had the greatest influence on coordination development of urbanization-WRS in Beijing city, while demographic urbanization had the smallest impact. Thus, social and spatial urbanizations subsystems need to be considered by policy-makers wishing to improve the coordination development of the two systems. In parallel, the pressure and state subsystems played a dominant role in the WRS system, while the effect subsystem had a minimal effect. The research results are consistent with the current situation of WRS in Beijing. They suggest that, the development of socio-economy has brought tremendous pressure on water resources in Beijing over the past ten years. The state of WRS hinder the development of two systems. Given these circumstances, it is necessary to take more reasonable water resource conservation measures to release the stress on water resources.

In relation to the secondary indicators, the statistical results of obstacle factors show that fixed asset investment growth rate, number of public transports per million people, dwelling area per capita and urban road area per capita were the key indicators restricting the healthy development of urban and water resources. These findings show that, during the 2008–2017 period, the urbanization population has increased rapidly, and Beijing's infrastructure is insufficient to meet the needs of the increased urban population. Water consumption for industrial production, water consumption of agricultural irrigation and urban water supply rate were the three key indicators that had the highest influence in the health development of two systems. Beijing is located in the north of China, in an area with a dry climate; this has led in the past to serious shortages of water resources. Since the reform and opening-up of the economy, Beijing is experiencing a process of industrialization and urbanization, and the industrial and agricultural development has led to an increase in industrial and agricultural water use which pose a significant risk to water resources security.

5. Conclusions

This study explored the coordination relationship between urbanization and WRS in Beijing by using the ICCD model and obstacle degree model. The main contributions of this research are as follows. Firstly, the indicator systems of urbanization and WRS have been established. These indicators can provide criteria to measure changes in the coupling degree in both systems. Secondly, this study revealed the dynamic trends present in the coupling coordination state between urbanization and WRS, using the ICCD model. Thirdly, the indicators that restrict the coordination development of urbanization and WRS systems are determined by using the obstacle degree method, which can provide a solid knowledge base for policy-makers to adjust water resources use plan and urbanization development policies.

Over the past ten years, Beijing has experienced a rapid urbanization process, which put a lot of pressure on water resources security. Overall, the state of the coupling coordination between the two systems has improved, passing from an imbalance state towards a good coordination state. However, the coupling coordination degree between the two systems is still low. The social urbanization, spatial urbanization, pressure and state subsystems are the main obstacle subsystems.

According to the research results, we propose the following suggestions and prospects: (1) policy-makers should adjust the urban development strategy by changing the development mode of spatial urbanization. The "chain effect" of medical, education, and infrastructure development on social urbanization can be improved through a rational allocation of resources. Moreover, more attention should be paid to the environmental factors when formulating the new-urbanization development strategy; (2) In order to guarantee the sustainable development of urbanization and water resources, the government should pay more attention on relieving the pressures on water resources caused by the rapid urban development. The local government needs to increase the investment to optimize the industrial structure, promote the adoption of innovative water-saving technologies, and develop clean energy to reduce sewage discharge. In addition, it is necessary to propose a water consumption strategic plan to improve water utilization effect and reduce water resource consumption of industrial output.

The coordination relationship between urbanization and WRS is a very complicated system. Due to data source limitations, this study focused only on the investigation of one city, i.e., Beijing. Future research should focus on the methodology to perform a comprehensive analysis of China's urban agglomerations, such as the Beijing-Tianjin-Hebei urban group. A comparative analysis should be performed of different regions in the urban group through the use of spatial gradients and temporal scales, to have a win-win effect.

Author Contributions: Conceptualization, K.Z.; Methodology, K.Z., H.H. and J.S.; Validation, H.H.; Formal Analysis, R.H.; Visualization, K.Z. and B.F.

Funding: This research was funded by the National Natural Science Foundation of China (No.71801130), Foundation for Postgraduate Research & Practice Innovation Program of Jiangsu Province (No. SJKY19_0409) and Fundamental Research Funds for the Central Universities of China (No. B19020686X).

Conflicts of Interest: The authors declare no conflict of interest.

References

1. Dong, G.H.; Shen, J.Q.; Jia, Y.Z.; Sun, F.H. Comprehensive evaluation of water resource security: Case study from Luoyang City, China. *Water* **2018**, *10*, 1106. [CrossRef]
2. Al-Jawad, J.Y.; Alsaffar, H.M.; Bertram, D.; Kalin, R.M. A comprehensive optimum integrated water resources management approach for multidisciplinary water resources management problems. *J. Environ. Manag.* **2019**, *239*, 211–224. [CrossRef] [PubMed]
3. Fezzi, C.; Harwood, A.R.; Lovett, A.A. The environmental impact of climate change adaptation on land use and water quality. *Nat. Clim. Chang.* **2015**, *5*, 255–260. [CrossRef]
4. Song, X.M.; Kong, F.Z.; Zhan, C.S. Assessment of water resources carrying capacity in Tianjin City of China. *Water Resour. Manag.* **2011**, *25*, 857–873. [CrossRef]
5. Giordano, M.; Shah, T. From IWRM back to integrated water resources management. *Int. J. Water Resour. Dev.* **2014**, *30*, 364–376. [CrossRef]
6. Zeng, Z.; Liu, J.; Koeneman, P.H.; Zarate, E.; Hoekstra, A.Y. Assessing water footprint at river basin level: A case study for the Heihe River Basin in northwest China. *Hydrol. Earth Syst. Sci.* **2012**, *16*, 2771–2781. [CrossRef]
7. Zhang, J.Y.; Wang, L.C. Assessment of water resource security in Chongqing City of China: What has been done and what remains to be done? *Nat. Hazards* **2015**, *75*, 2751–2772. [CrossRef]
8. Xia, J.; Zhu, Y.Z. The measurement of water resources security: A study and challenge on water resources carrying capacity. *J. Nat. Resour.* **2002**, *17*, 262–269.
9. Mcdonald, R.I.; Green, P.; Balk, D. Urban growth, climate change, and freshwater availability. *Proc. Natl. Acad. Sci. USA* **2011**, *108*, 6312–6317. [CrossRef]
10. Srinivasan, V.; Seto, K.C.; Emerson, R.; Gorelick, S.M. The impact of urbanization on water vulnerability: A coupled human–environment system approach for Chennai, India. *Glob. Environ. Chang.* **2013**, *23*, 229–239. [CrossRef]
11. Rasul, G.; Sharma, B. The nexus approach to water–energy–food security: An option for adaptation to climate change. *Clim. Policy* **2016**, *16*, 682–702. [CrossRef]

12. Han, H.; Li, H.M.; Zhang, K.Z. Urban water ecosystem health evaluation based on the improved fuzzy matter-element extension assessment model: Case study from Zhengzhou City, China. *Math. Probl. Eng.* **2019**. [CrossRef]
13. Bao, C.; Fang, C.L. Water resources constraint force on urbanization in water deficient regions: A case study of the Hexi Corridor, arid area of NW China. *Ecol. Econ.* **2007**, *62*, 508–517. [CrossRef]
14. Fu, P.; Weng, Q. A time series analysis of urbanization induced land use and land cover change and its impact on land surface temperature with Landsat imagery. *Remote Sens. Environ.* **2016**, *175*, 205–214. [CrossRef]
15. Wang, S.J.; Liu, X.P.; Zhou, C.S.; Hu, J.C.; Ou, J.P. Examining the impacts of socioeconomic factors, urban form, and transportation networks on CO_2 emissions in China's megacities. *Appl. Energy* **2017**, *185*, 189–200. [CrossRef]
16. Wan, G.; Wang, C. Unprecedented urbanisation in Asia and its impacts on the environment. *Aust. Econ. Rev.* **2014**, *47*, 378–385. [CrossRef]
17. Wang, S.J.; Fang, C.L.; Wang, Y. Quantitative investigation of the interactive coupling relationship between urbanization and eco-environment. *Acta Ecol. Sin.* **2015**, *35*, 1–14. (In Chinese) [CrossRef]
18. He, J.Q.; Wang, S.J.; Liu, Y.Y.; Ma, H.T.; Liu, Q.Q. Examining the relationship between urbanization and the eco-environment using a coupling analysis: Case study of Shanghai, China. *Ecol. Indic.* **2017**, *77*, 185–193. [CrossRef]
19. Fang, C.; Wang, Z.; Xu, G. Spatial-temporal characteristics of $PM_{2.5}$ in China: A city-level perspective analysis. *J. Geogr. Sci.* **2016**, *26*, 1519–1532. [CrossRef]
20. Han, H.; Li, H.; Zhang, K. Spatial-temporal coupling analysis of the coordination between urbanization and water ecosystem in the Yangtze River Economic Belt. *Int. J. Environ. Res. Public Health* **2019**, *16*, 3757. [CrossRef]
21. Al-Mulali, U.; Weng, W.C.; Sheau-Ting, L. Investigating the environmental Kuznets curve (EKC) hypothesis by utilizing the ecological footprint as an indicator of environmental degradation. *Ecol. Indic.* **2015**, *48*, 315–323. [CrossRef]
22. Yao, L.; Li, X.L.; Li, Q.; Wang, J.K. Temporal and spatial changes in coupling and coordinating degree of new urbanization and ecological-environmental stress in China. *Sustainability* **2019**, *11*, 1171. [CrossRef]
23. Jordan, S.J.; Hayes, S.E.; Yoskowitz, D.; Smith, L.M.; Summers, J.K.; Russell, M.; Benson, W.H. Accounting for natural resources and environmental sustainability: Linking ecosystem services to human well-being. *Environ. Sci. Technol.* **2010**, *44*, 1530–1536. [CrossRef] [PubMed]
24. Cui, X.G.; Fang, C.L.; Liu, H.M.; Liu, X.F. Assessing sustainability of urbanization by a coordinated development index for an Urbanization-Resources-Environment complex system: A case study of Jing-Jin-Ji region, China. *Ecol. Indic.* **2019**, *96*, 383–391. [CrossRef]
25. Rimal, B.; Zhang, L.F.; Stork, N.; Sloan, S.; Rijal, S. Urban expansion occurred at the expense of agricultural lands in the Tarai region of Nepal from 1989 to 2016. *Sustainability* **2018**, *10*, 1341. [CrossRef]
26. Rimal, B.; Zhang, L.F.; Keshtkar, H.; Haack, B.N.; Rija, S.; Zhang, P. Land use/land cover dynamics and modeling of urban land expansion by the integration of cellular automata and markov chain. *ISPRS Int. J. Geo-Inf.* **2018**, *7*, 154. [CrossRef]
27. Beijing Statistics Bureau. *Beijing Statistical Yearbook*; China Statistics Press: Beijing, China, 2016.
28. Beijing Water-affair Authority. *Beijing Water Resources Bulletin*; China Statistical Press: Beijing, China, 2016.
29. Li, Y.H.; Jia, L.R.; Wu, W.H.; Yan, J.Y.; Liu, Y.S. Urbanization for rural sustainability-rethinking China's urbanization strategy. *J. Clean. Prod.* **2018**, *178*, 580–586. [CrossRef]
30. Zang, W.B.; Liu, S.; Huang, S.F.; Li, J.R. Impact of urbanization on hydrological processes under different precipitation scenarios. *Nat. Hazards* **2019**, *99*, 1233–1257. [CrossRef]
31. Zhao, Y.; Wang, S.; Zhou, C. Understanding the relation between urbanization and the eco-environment in China's Yangtze River Delta using an improved EKC model and coupling analysis. *Sci. Total Environ.* **2016**, *571*, 862–875. [CrossRef]
32. Wang, S.J.; Ma, H.; Zhao, Y.B. Exploring the relationship between urbanization and the eco-environment—A case study of Beijing–Tianjin–Hebei region. *Ecol. Indic.* **2014**, *45*, 171–183. [CrossRef]
33. Guo, Y.; Wang, H.; Nijkamp, P.; Xu, J. Space-time changes in interdependent urban-environmental systems: A policy study on the Huai River Basin in China. *Habitat Int.* **2015**, *45*, 135–146. [CrossRef]
34. Liu, N.N.; Liu, C.Z.; Xia, Y.F.; Da, B. Examining the coordination between urbanization and eco-environment using coupling and spatial analyses: A case study in China. *Ecol. Indic.* **2018**, *93*, 1163–1175. [CrossRef]

35. Liu, W.J.; Jiao, F.C.; Ren, L.J.; Xu, X.G. Coupling coordination relationship between urbanization and atmospheric environment security in Jinan City. *J. Clean. Prod.* **2018**, *204*, 1–11. [CrossRef]
36. Zhang, Z.F.; He, W.J.; An, M. Water security assessment of China's One Belt and One Road Region. *Water* **2019**, *11*, 607. [CrossRef]
37. Jia, Y.Z.; Shen, J.Q.; Wang, H.; Dong, G.H.; Sun, F.H. Evaluation of the spatiotemporal variation of sustainable utilization of water resources: Case Study from Henan Province (China). *Water* **2019**, *11*, 607. [CrossRef]
38. Coscieme, L.; Pulselli, F.M.; Niccolucci, V.; Patrizi, N.; Sutton, P.C. Accounting for 'land-grabbing' from a biocapacity viewpoint. *Sci. Total Environ.* **2016**, *539*, 551–559. [CrossRef]
39. Zhao, W.; Lin, J.; Wang, S.F.; Liu, J.L.; Chen, Z.R. Influence of human activities on groundwater environment based on coefficient variation method. *Environ. Sci.* **2010**, *31*, 1277–1283.
40. Li, M.C.; Mao, C.M. Spatial-Temporal variance of coupling relationship between population modernization and eco-environment in Beijing-Tianjin-Hebei. *Sustainability* **2019**, *11*, 991. [CrossRef]
41. Sun, Y.; Liang, X.J.; Xiao, C.L. Assessing the influence of land use on groundwater pollution based on coefficient of variation weight method: a case study of Shuangliao City. *Environ. Sci. Pollut. Res.* **2019**, *10*, 1–13. [CrossRef]
42. Ding, L.; Zhao, W.T.; Huang, Y.L.; Cheng, S.G.; Liu, C. Research on the coupling coordination relationship between urbanization and the air environment: A case study of the area of Wuhan. *Atmosphere* **2015**, *6*, 1539–1558. [CrossRef]
43. Zhao, Y.B.; Wang, S.J. The relationship between urbanization, economic growth and energy consumption in China: An econometric perspective analysis. *Sustainability* **2015**, *7*, 5609–5627. [CrossRef]
44. Ai, J.; Feng, L.; Dong, X.; Zhu, X.; Li, Y. Exploring coupling coordination between urbanization and ecosystem quality (1985–2010): A case study from Lianyungang City, China. *Front. Earth Sci.* **2016**, *10*, 527–545. [CrossRef]
45. Wang, L.; Wang, Y.; Zhang, Y. Coupling coordination degree of urban land use benefit and urbanization: A case of 27 Cites in the Bohai Rim Region. *Ecol. Econ.* **2017**, *33*, 25–28.
46. Geng, X.; Huang, M.H.; Feng, Y.Q.; Chen, D.H. Coupling analysis between economic development and eco-environment for a comprehensive industrial park. *Environ. Sci. Technol.* **2011**, *34*, 196–200.
47. Zhang, Y.J.; Jun, M.A.; Business, S.O.; University, H. Research on coordination development of ecological-environment and economy in North Jiangsu based on coupling degree model. *Environ. Sci. Technol.* **2016**, *29*, 11–15.
48. Shen, L.Y.; Huang, H.; Huang, Z.H. Improved coupling analysis on the coordination between socio-economy and carbon emission. *Ecol. Indic.* **2018**, *94*, 357–366. [CrossRef]
49. Lei, X.P.; Qiu, R.; Liu, Y. Evaluation of regional land use performance based on entropy TOPSIS model and diagnosis of its obstacle factors. *Trans. Chin. Soc. Agric. Eng.* **2016**, *32*, 243–253.
50. National Bureau of Statistics of China. *China Urban Statistical Yearbook (2008–2017)*; China Statistical Press: Beijing, China, 2017.
51. Beijing Bureau of Statistics of China, Ministry of Environmental Protection of Beijing. *Beijing Statistic Yearbook on Environment (2008–2017)*; Beijing Statistical Press: Beijing, China, 2017.

International Journal of
Environmental Research and Public Health

Article

Ecological Security Assessment Based on Ecological Footprint Approach in Hulunbeir Grassland, China

Shanshan Guo and Yinghong Wang *

School of Environment Science and Spatial Informatics, China University of Mining and Technology, Xuzhou 221116, China; guoshanshan@cumt.edu.cn
* Correspondence: 1369@cumt.edu.cn; Tel.: +86-181-1525-7389

Received: 24 September 2019; Accepted: 25 November 2019; Published: 29 November 2019

Abstract: Hulunbeir grassland, as a crucial ecological barrier and energy supply base in northwest China, suffers from a fragile ecological environment. Therefore, it is crucially important for Hulunbeir grassland to achieve the sustainable development of its social economies and ecological environments through the evaluation of its ecological security. This paper introduces the indexes of the ecological pressure index (EPI), ecological footprint diversity index (EFDI), and ecological coordination coefficient (ECC) based on the ecological footprint model. Furthermore, the Stochastic Impacts by Regression on Population, Affluence, and Technology (STIRPAT) model was applied to analyze the main driving factors of the change of the ecological footprint. The results showed that: The ecological footprint (EF) per capita of Hulunbeir grassland has nearly doubled in 11 years to 11.04 ha/cap in 2016, while the ecological capacity (EC) per capita was rather low and increased slowly, leading to a continuous increase of per capita ecological deficit (ED) (from 5.7113 ha/cap to 11.0937 ha/cap). Within this, the footprint of fossil energy land and grassland contributed the most to the total EF, and forestland and cropland played the major role in EC. The EPI increased from 0.82 in 2006 to 1.25 in 2016, leading the level of ecological security to increase from level 3 (moderately safe) to level 4 (moderately risky). The indexes of the EFDI and ECC both reached a minimum in 2014 and then began to rise, indicating that Hulunbeir steppe's ecological environment, as well as its coordination with economy, was considered to be worse in 2014 but then gradually ameliorated. The STIRPAT model indicated that the main factors driving the EF increase were per capita GDP and the proportion of secondary industry, while the decrease of unit GDP energy consumption played an effective role in curbing the continuous growth of the EF. These findings not only have realistic significance in promoting the coordinated development between economy and natural resource utilization under the constraint of fragile environment, but also provide a scientific reference for similar energy-rich ecologically fragile regions.

Keywords: ecological footprint; ecological capacity; ecological security; STIRPAT model; Hulunbeir grassland

---

1. Introduction

Ecological security means that a country or a region enriches ecological resources that could continuously meet social progress and economic development and lessens the restriction of the ecological environment on social and economic development [1]. However, with the acceleration of industrialization and urbanization, the increase of population, and a subsequent greater use of natural resources, ecological security issues, such as grassland degradation, forest decline, desert encroachment, and biodiversity loss are gradually becoming prominent. Maintaining a dynamic equilibrium between humans and nature becomes particularly hard and important in ecologically fragile areas [2]. This requires to research the relationship between human activity and environmental change, optimize the structure of resource utilization, and minimize the impact of human behavior on the ecological environment under the premise of maintaining sustainable economic growth and safeguarding human well-being [3,4].

To further understand the change of the ecological environment, many beneficial explorations in terms of concept connotation, evaluation framework, and control approach of ecological security have been conducted [5–7]. The related researches have ranged from the empirical study of environmental change and security to the interactions between them, emphasizing the harmonious coexistence of socio-economic development and ecological security [1,8]. Additionally, various approaches have been applied to evaluate the state of ecological security, such as the pressure-state-response [9], landscape ecology [10,11], material/substance flow analysis [12,13], ecosystem services [14,15], and ecological footprint [8,16,17]. Among which, the ecological footprint method is the most widely used. It is defined as a bio-productive land area that maintains human living needs while absorbing pollution caused by human activities. The primary advantages of this method are that it is easy to apply, repeatable, and simple to understand [18,19]. Scholars have used this method to study multiple scales of ecological footprints. On a large scale, Wackernagel et al. [20] researched the ecological footprints of 52 countries that include 85% of the global population and 95% of the global economic output, finding that 35 of these countries were experiencing ecological risk. Niccolucci et al. [21] studied the ecological footprints of 150 countries from 1961 to 2007 by dividing the change tendencies of ecological footprint and ecological capacity into four types: parallel, scissors differential, wedge, and downtrend. Li et al. [19] researched the five countries in arid areas of Central Asia, showing that the ecological pressure in all five countries was rising, and the overall ecological security situation changed from comparably safe before 2000 to unsafe in 2005 and then may be at risk in 2025.

However, these large-scale studies generally ignored the differences in resources, technologies, and other aspects among the various regions. In response to this gap, some scholars have recently researched the ecological footprint based on comparatively smaller scales of province, city, and county. For example, Dong et al. [22] evaluated the ecological security and natural capital utilization of Hainan Province, China, from 2005 to 2016, uncovering the main factors influencing the changes of ecological footprint by partial least squares regression model. Pan et al. [23] analyzed the dynamic changes in supply and demand of resources and revealed that China's Shanghai is suffering a high ecological footprint intensity and a poor coordination relationship between its economy and environment. Zhao et al. [24] evaluated the ecological security of Lhaze County in China's Tibetan Autonomous Region, showing that ecological environment has deteriorated from "early stages of damage status" in the 1980s to "moderately damaged status" today.

Scholars have also used different models to analysis the main factors that affect the change of the ecological footprint, such as the Stochastic Impacts by Regression on Population, Affluence, and Technology (STIRPAT) [25], partial least squares regression (PLS) [26], and log-mean divisia index (LMDI) [27]. One of the most representative methods for exploring the influencing factors is the STIRPAT model. Its primary advantage is that it allows for non-monotonic or non-proportional effects from the driving forces [28]. Currently, this model has been widely used in studies to analyze the effect of population, affluence, and technology on environment.

Affected by climate change and human activities, mining areas often overlap with fragile ecological regions [29]. Hulunbeir grassland, as an important natural ecological barrier and energy supply base in north China, provides a significant contribution to the country's economic development and ecological security. However, global climate change and continuous socio-economic development with excessive exploitation of resources have jointly induced soil and water loss, grassland desertification, degradation, and salinization [13]. Eco-environmental issues have been paid close attention by the whole society [30]. At present, the pertinent literatures in this area mainly focuses on the discussion of the variation tendency of grassland production capacities [31] and correspondingly ecological protection measures [32]. While research rarely shows how natural resources are used and ecological security changes over time, these exhibit the remarkable effects of ecological policy implementation [33].

Especially since the eleventh "Five Plan", Hulunbeir city has further strengthened ecological restoration and made efforts towards industrial transformation. Therefore, in this paper, the ecological footprint (EF) framework and the ecological capacity concept are introduced to clarify the supply and demand of the regional ecosystem. Three indicators of ecological pressure index (EPI), ecological coordination coefficient (ECC), and ecological footprint diversity index (EFDI) from 2005 to 2016 are calculated to provide a more comprehensive assessment of historical and current ecological security of mining areas in Hulunbeir grassland. Additionally, the STIRPAT model is constructed to further study the main socio-economic driving factors that lead to the emergence or even aggravation of ecological deficits and quantify the importance of each driving factor. This study can provide a scientific basis for solving the contradiction between the rapid development of the social economy and degradation of ecological environment and thereby provide a reference for ecological environment management decisions for mining areas in arid and semi-arid grasslands in China. 2. Study Area and Data Sources.

1.1. Study Area

Hulunbeir grassland is mostly distributed in Hulunbeir city (E 115°31′–126°04′, N 47°05′–53°20′), northeast of Inner Mongolia (Figure 1). Except for the transition area of forest and grassland in the eastern region, the rest are basically natural grasslands with an area of approximately 11.27×10^4 km^2, covering about 11.54% of Inner Mongolia's total grassland area. Hulunbeir grassland as a whole is a plateau landform with an altitude of 650–1000 m. It belongs to a continental arid to semi-arid climate with an annual average temperature of −3–0 °C and annual precipitation of approximately 350 mm. Hailaer River, Yimin River, and Gen River originating from the Greater Khingan Mountain are the major water supplies for grasslands. From the western foothills of the Greater Khingan Mountains to the Mongolian Plateau, there is a distribution of zonal grasslands with arid steppe, meadow steppe, and forest steppe. The perennial herbaceous community is the basic feature of the grassland ecology in this region, with about 1000 plant species. Moreover, Hulunbeir grassland is rich in coal resources, with an area of about 2.7×10^4 km^2 accounting for 31% of the total region. Since the reform and opening up of China, the coal industry has undergone a rapid development. Especially in 2006, the coal mines of Jalainur, Baorixile, and Yimin were listed in the second batch of 26 national planned coal mining areas [34], with the coal industry expanding vigorously. At present, its proved reserves of coal resources reach 29.785 billion tons, which is 1.8 times higher than the combined storage of the three provinces in Northeast China (Heilongjiang Province, Jilin Province, and Liaoning Province) [35], and more than 370 mining sites and 355 mining enterprises of various categories have been constructed. The accumulated solid waste output produced by mining is about 370 million tons, and the total area of mined-out subsidence area is approximately 42 km^2 [36]. All of these problems are concentrated in the grassland area, such as Hailar Basin, Labudalin, and Miandu River. The area of the degradation, desertification, and salinization in Hulunbeir grassland has increased from 13% in the 1960s to 21% in the 1980s, to about 30% in the 1990s and to nearly 50% in the beginning of this century [37], which seriously handicaps the sustainable development of the regional economy and environment.

1.2. Sources of Data

Hulunbuir's rapid economic growth is mainly supported by a high consumption of energy resources. There are three large mining areas listed in the second batch of the 26 nationally planned coal mining areas in 2006, which have promoted the rapid development of energy-related industries. At the same time, the new eleventh "Five Plan" emphasizes the importance of ecological environment protection and the necessary of industrial transformation to reduce dependence on energy-intensive industries. The disparity between economic development and environmental protection is becoming increasingly acute. Based on this situation, we mainly select the time series of 2006–2016 to explore the dynamic change of the ecosystem's health and natural resource utilization after the new eleventh "Five Plan". Detailed data sources and descriptions are shown in Table 1.

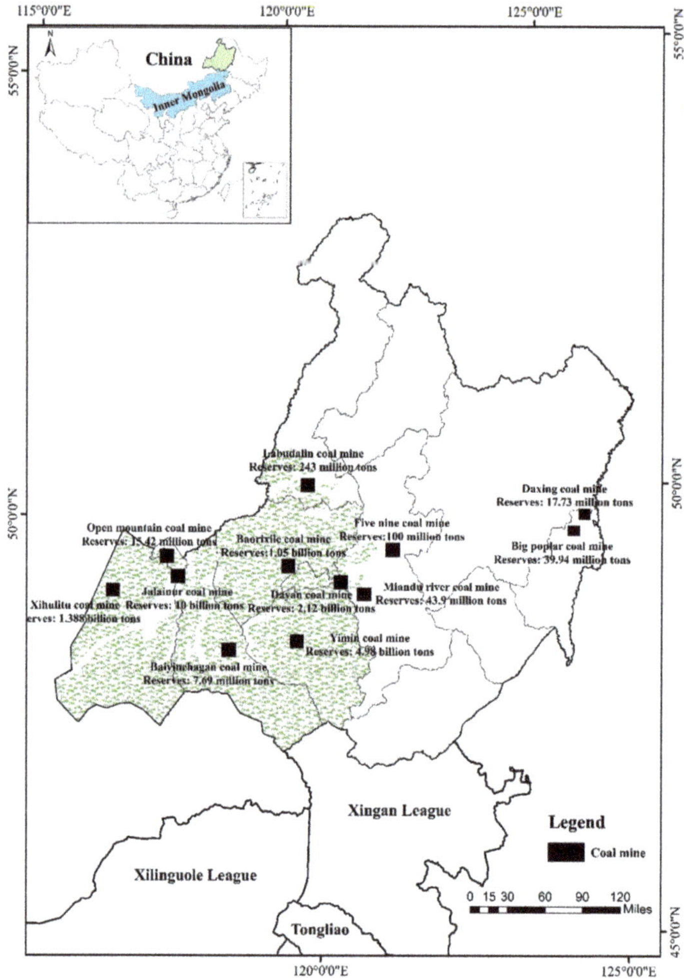

Figure 1. Location of the overlapped areas of grassland and coal resources.

Table 1. Indicators and data sources.

Items	Indicators	Data Sources
Biological account	Agricultural products: wheat, corn, rice, sorghum, potato, oil crop, vegetables, beans, wine, sugar, pork and eggs Forest products: fruits and wood Grass products: beef, lamb, poultry, milk, dairy products, sheep wool, goat wool, cashmere Aquatic products: freshwater	«Hulunbeir Statistical Yearbook» (2007–2017)
Energy account	The consumption of raw coal, crude oil, coke, gasoline, kerosene, diesel oil, fuel oil, electricity, heat	«Hulunbeir Statistical Yearbook» (2007–2017)
Land use	Land use area	Land Resources Data of the Ministry of Natural Resources (2006–2016) and «Hulunbeir Statistical Yearbook» (2007–2017)
Equivalence factor	cropland (2.8), grassland (0.5), forest land (1.1), water (0.2), fossil energy land (1.1), build-up land (2.8)	«Calculation of China's equivalence factor under ecological footprint mode based on net primary production» [38]
Yield factor	cropland (1.7), grassland (0.19), forestland (0.91), water (1), fossil energy land (0), build-up land (1.7)	«Calculating national and global ecological footprint time series: resolving conceptual challenges» [39] «Quantitative analysis of sustainability development of inner Mongolia» [40]
Population, economy, and technology	Population: year-end resident population, urbanization rate Economy: per capita GDP, proportion of secondary industry output value Technology: unit GDP energy consumption	«Hulunbeir Statistical Yearbook» (2007–2017) «Inner Mongolia statistical yearbook» (2007–2017) National economy and society developed statistical bulletin in Hulunbeir (2006–2016)

2. Methods

2.1. Construction of Ecological Security Evaluation Framework

Here, we selected four indicators of Flux (F), Pressure (P), Diversity (D), and Coordination (C) on the basis of the method of ecological safety evaluation (Figure 2). F reflects the difference between ecological footprint capacity (EC) and ecological footprint (EF). It indicates changes on an ecosystem's support to socio-economic system, thereby reflecting ESS from an absolute value change perspective. F is the equivalent of the ecological balance, which can be calculated through the classic EF model. The ecological pressure index (EPI), as a substitute for P, is the pressure level on the ecosystems generated by a given socio-economic system or a specific population scale. It is defined as the ratio of per capita EF (*ef*) to per capita EC (*ec*). EPI reflects ESS by analyzing stress suffered by ecosystem per unit of ecological capacity. Diversity (D) expresses both the amount and the distribution of the different EF components, which reflects ESS from an EF composition and structure perspective. Finally, Coordination (C) is used to measure the coordination degree between socio-economic development and ecosystems. The ecological footprint is affected by population, economy, and technology. Thus, we created multiple regression equations between the ecological footprints and variables to explore the main driving factors of the ecological footprint change.

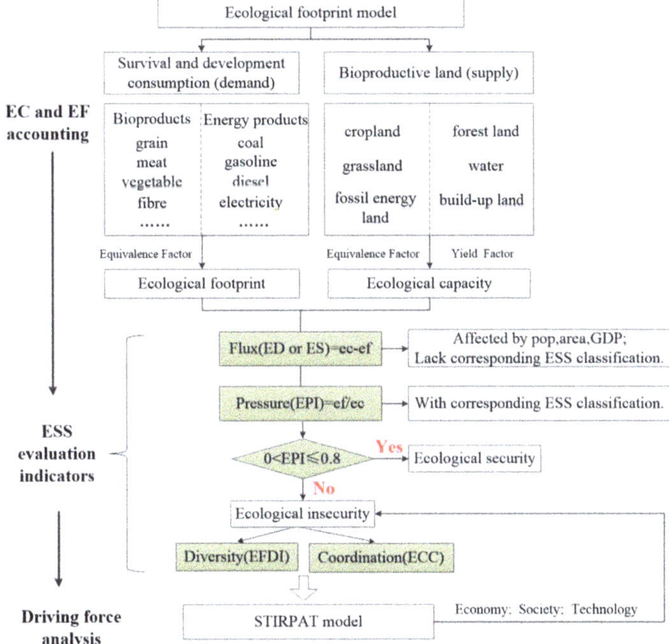

Figure 2. The framework of ecological security evaluation.

2.2. Evaluation Model of Ecological Footprint

The ecological footprint is the area of productive land and water needed to support the regional population and the land needed to absorb the waste produced by those populations [41]. This approach provides a kind of simple methodology but comprehensive way to measure direct and indirect human consumption on the regional regenerative capacity. Then, by comparing it with the biocapacity available, we can judge whether the development pattern of this region is in a sustainable state. According to the theory of ecological footprint, the biologically productive land can be divided into 6 types, i.e., cropland, grassland, forestland, water, fossil energy land, and build-up land [20], each type has a different production capacity of per unit area. The specific calculations formulas are as follows:

$$EF = N \times ef = N \times \sum_{i=1}^{n} (aa_i \times r_i) = N \times \sum_{i=1}^{n} \left(\frac{c_i}{p_i} \times r_i \right). \quad (1)$$

In the formula above: EF is the total ecological footprint (ha); N is the total population; i is the category of items consumed by a certain population (i = 1,2, ... , n); aa_i is the ecologically productive area from the ith consumption item; p_i is the average productivity of the ith item in a certain area (kg/ha); c_i is the per capita quantity of the ith item (kg/ha) affected by the productivity and trade balance amount; r_i is an equivalence factor, which describes the ratio of the productive capacity of a certain type of bioproductive land to the productive capacity of all the world's bioproductive land. The equivalence factor of each biologically productive land is shown in Table 1

$$EC = N \times ec = N \times (1 - 12\%) \sum_{j=1}^{6} \left(a_j \times r_j \times y_j \right) \quad (2)$$

In the formula above: EC is the total ecological capacity (ha); j represents the area of the biologically productive land required; a_j represents the per capita area of the biologically productive land for items of the jth category (ha); y_j is yield factor, which describes the ratio of average land productivity of a country or region to the global average productivity of the same land type. The yield factor of each biologically productive land is shown in Table 1. In addition, the area of biologically productive land should be decreased by 12% to account for biodiversity conservation [42].

2.3. Evaluation Model of Ecological Security

2.3.1. Ecological Deficit/Surplus

As the EF and EC are both measured by the area of biologically productive land, they can be compared directly. Ecological deficit/surplus (ED/ES) presents the profit and loss of supply and the demand situation of the regional ecological system [17]. When a region's EC is less than the EF in a region, an ecological deficit (ED) appears, indicating that the supply of regional ecological resources neither meet the demands of social development nor bear the corresponding environmental purification and renewal. Therefore, the region may import resources from surrounding cities or even other faraway cities to satisfy increasing local demand for natural resources and energy. Conversely, ecological surplus (ES) indicates that the supply of regional ecological resources is sufficient to meet the needs of human production. The formula for the ED/ ES is as follows:

$$ED/ES = EF - EC = N \times (ef - ec) \quad (3)$$

2.3.2. Ecological Pressure Index

The ecological pressure index (EPI) mainly reflects ecological pressure which is caused by the consumption of resources and sequestering carbon dioxide emissions, and so forth, in the industry and daily life of local residents [8], representing the pressure intensity suffered by regional ecological environment. If 0 < EPI < 1, EPI will be positive and the supply of ecological resources exceeds the demand for it, indicating that the ecological security remains in a sustainable status. If EPI = 1, the supply of the ecological resource and the demand for it are equal, indicating that the ecological security is in a critical status. Finally, if EPI > 1, people's demand for ecological resources is greater than its supply, indicating that the regional ecology is in a threatened status. Using data calculated for the EFs of 147 countries or regions, as provided in the Living Planet Report by the International Monetary Fund in 2004, Yuan [43] and Chu [17] detailed the classification standard of ecological security (Table 2). The formula is shown as follows:

$$EPI = ef/ec. \quad (4)$$

Table 2. Classification standard of ecological security.

Ecological Security Grade	Range of EPI	Characterization State	Ecological Security Alarm Level
1	<0.5	Pretty safe	No alarm
2	0.50–0.80	Safe	
3	0.81–1.00	Moderately safe	Low alarm
4	1.01–1.50	Moderately risky	Moderate alarm
5	1.51–2.00	Risky	High alarm
6	>2	Very risky	Severe alarm

2.3.3. Ecological Coordination Coefficient and Ecological Footprint Diversity Index

Ecological deficit is an absolute value and cannot reflect its relationship with resource endowment conditions. Therefore, it is necessary to introduce the concept of an ecological coordination coefficient (ECC) to compensate for this deficiency in the ecological deficit [4]. Ecological coordination coefficient represents the coordination degree between regional ecological environment and socio-economic development. The formula is as follows:

$$ECC = (ef + ec)/\sqrt{(ef)^2 + ec^2} = \left(\frac{ef}{ec} + 1\right)/\sqrt{\left(\frac{ef}{ec}\right)^2 + 1} = (EPI + 1)/\sqrt{EPI^2 + 1}. \quad (5)$$

Due to ef and ec being larger than 0, the ECC ranges from 1 to 1.414. The closer it is to 1.414, the better the coordination. Conversely, the closer it is to 1, the worse the coordination.

The ecological footprint diversity index (EFDI) reflects the abundance of different land types and the fairness of ecological footprint distribution in a region [16,44]. The more equal the ecological footprint distribution in an eco-economic system, the higher the ecological diversity is for the ecological economy of given system components. Generally speaking, in the early stage of regional development, the EFDI is relatively low. However, with the development of social economy, the diversity index gradually increases, which can promote the improvement of energy utilization efficiency. The formula is as follows:

$$EFDI = -\sum p_i \ln p_i. \quad (6)$$

In the formula above: p_i is the proportion of the ith category of land type in regional ecological footprint.

2.4. The STIRPAT Modelling Approach

Ehrich and Holdren [45] firstly proposed the IPAT (Environmental Impact, Population, Affluence, Technology) model to analyze the relationship between population, economy, technology, and environment. While the weakness of the IPAT model is that it regards population, economy, technology, and environmental issues as change relations in equal proportion, this is inconsistent with the reality. In addition, the importance degree of each driver cannot be clearly judged. To overcome these problems, Dietz and Rosa [46] proposed the Stochastic Impacts by Regression on Population, Affluence, and Technology (STIRPAT) model, making quantitative analysis of environmental problems more flexible. The generic STIRPAT model is given as:

$$I = aP^b A^c T^d \xi . \quad (7)$$

In the formula above: I represents the environmental impact; a and ξ are the coefficient and random error of the model; P, A, T represent population, affluence, and technology, respectively; b, c, and d denote the exponentials of the driving forces.

In order to measure hypotheses and assess the importance of each influencing factor, the STIRPAT model is taken as a logarithm to get a linear model [25]:

$$lnI = lna + blnP + clnA + dlnT + ln\xi \quad (8)$$

However, the accuracy of the evaluation results will be reduced because of the multicollinearity among socio-economic variables [47]. Therefore, in this study, the principal component analysis method (PCA) was used to improve the STIRPAT model. The basic principle is: (1) Standardizing the original data and then calculating the correlation coefficient matrix; (2) obtaining the eigenvalues and variance contribution rates of the correlation matrix, and determining the principal components; (3) sing principal components to replace the original variables for multiple regression *F*; (4) substituting the original variables into the principal component regression model; (5) based on the results of the principal component analysis, the logarithms of the extracted principal components were taken respectively and then restored to the original variable form in the STIRPAT model.

3. Results and Discussion

3.1. Evaluation Results of Ecological Footprint and Ecological Capacity

The *ef* of the study area increased from 5.71 ha/cap in 2006 to 11.04 ha/cap in 2016, at a range of 93.34% (Table 3). From 2006 to 2014, the average annual rate was 8.01%, while, from 2015 to 2016, the value of the *ef* showed a negative growth. This could be due to the effect of the global coal market and emerging energy, as well as the regulation of illegal and irregular mining enterprises (37 mine enterprises were closed in 2015) in Hulunbeir grassland. In terms of components of the *ef*, the change in the water *ef* and build-up land *ef* over time was not obvious, but the the *ef* of fossil energy land and cropland increased obviously, especially in 2010 to 2014, the fossil energy *ef* rose from 2.94 ha/cap to 5.29 ha/cap, which could be resulted from the boom of industrial enterprises above designated size (increased nearly 100 over those 4 years). Similarly, the *ef* of grassland and cropland increased by 1.75 ha/cap and 0.65 ha/cap respectively over these 11 years, which reflected the increase of people's consumption of milk, meat, and agricultural products and that the living standard was gradually improving. In addition, the *ef* of fossil energy land exceeded the *ef* of grassland since 2007 and became one of the most important issues in daily life in recent years. In other words, the EF in the study area has increasingly come from energy-related production's consumption from then on. On the contrary, the forest *ef* showed a declining trend in recent years, mainly because the study area implemented effective measures against the commercial logging of natural forest since 2012 and cancelled the original production index of 320,000 m^3 of natural forest commodity timber, which significantly reduced the consumption of wood.

Table 3. Changes of ecological footprint in study area from 2006 to 2016 (ha/cap).

Year	Cropland	Forestland	Grassland	Water	Fossil Energy Land	Build-Up Land	*ef*	EF (ha)
2006	1.33	0.64	1.80	0.07	1.79	0.07	5.71	1.51×10^7
2007	1.27	0.69	1.75	0.07	1.85	0.06	5.69	1.55×10^7
2008	1.63	0.64	2.00	0.07	2.50	0.06	6.90	1.88×10^7
2009	1.87	0.67	2.12	0.08	2.63	0.04	7.39	2.01×10^7
2010	2.24	0.55	2.24	0.08	2.94	0.17	8.22	2.23×10^7
2011	2.27	0.39	2.27	0.09	3.99	0.26	9.26	2.50×10^7
2012	2.65	0.42	2.41	0.09	4.23	0.13	9.93	2.52×10^7
2013	2.72	0.37	2.23	0.09	4.64	0.15	10.21	2.75×10^7
2014	2.99	0.39	2.51	0.08	5.29	0.18	11.43	2.89×10^7
2015	3.10	0.30	2.33	0.10	5.14	0.25	11.23	2.84×10^7
2016	3.09	0.19	2.45	0.10	4.95	0.26	11.04	2.79×10^7

Table 4 shows that the per capita EC of the study area experienced fluctuant growth from 6.81 ha/cap to 8.85 ha/cap throughout the study period, with an average annual rate of only 1.03%.

But compared with the overall decreasing trend in China [16], the slight increase of the *ec* in Hulunbeir grassland indicates that the implementation of ecological construction projects have played a positive role in improving the supply capacity of regional resources. Specifically, as the biological productivity of cropland and forestland was higher than that of other land use types, the *ec* of these two land use types accounted for the largest proportion (58% and 33.55%, respectively) in Hulunbeir grassland, followed by build-up land (4.64%), grassland (3.39%), and water area (0.42%). Additionally, the *ec* of cropland grew fastest, from 2.15 in 2010 to 3.83 in 2016, indicating that the reclamation intensity of cropland in the study area was strengthened and the other land utilization types (mainly grassland) had been converted into cropland. The *ec* of forestland and grassland both experienced a trend of decreasing first and then increasing, mainly because of a series of treatment measures, such as managing water pollution, enclosure, returning grazing land to grassland, and the prohibition of commercial logging.

Table 4. Changes of ecological capacity in study area from 2006 to 2016 (ha/cap).

Year	Cropland	Forestland	Grassland	Water	Build-Up Land	Biodiversity Conservation Area	*ec*	Total EC (ha)
2006	2.15	5.04	0.30	0.03	0.22	0.93	6.81	1.84×10^7
2007	2.11	4.98	0.35	0.04	0.25	0.93	6.79	1.85×10^7
2008	2.50	4.97	0.29	0.04	0.27	0.97	7.10	1.94×10^7
2009	2.84	5.09	0.28	0.04	0.41	1.04	7.62	2.07×10^7
2010	3.17	4.95	0.28	0.04	0.47	1.07	7.84	2.13×10^7
2011	2.88	4.97	0.28	0.04	0.43	1.03	7.56	2.05×10^7
2012	3.37	5.30	0.30	0.04	0.49	1.14	8.36	2.12×10^7
2013	3.15	4.98	0.28	0.04	0.46	1.07	7.84	2.11×10^7
2014	3.41	5.31	0.30	0.04	0.50	1.15	8.41	2.13×10^7
2015	3.58	5.32	0.30	0.04	0.53	1.17	8.59	2.17×10^7
2016	3.83	5.31	0.30	0.04	0.57	1.21	8.85	2.24×10^7

3.2. Evaluation Results of Ecological Security

Based on the results of the *ef* and *ec* in study area, we obtained per capita ED/ES of the six productive land types, shown in Figure 3. The *ec* was larger than the *ef* during 2006–2009, meaning that during these 4 years, the ability of regional natural resources that supported human activities still remained in a sustainable state as a whole. However, the *ef* exceeded the *ec* in 2010, and then the per capita ES gradually transformed into per capita ED with the increase of the fossil energy consumption and the grassland deterioration. This indicated that the equilibrium between demand and supply showed an unbalanced status and the unsustainable tendency was increasingly obvious in Hulunbeir grassland. Furthermore, the cropland, forestland and build-up land were in a state of ecological surplus during the past eleven years, and the bearing capacity of forest land experienced a small upward trend after the implementation of various protection policies. Build-up land and water area distributed near the boundary between deficit and surplus changed slightly. However, the fossil energy land and grassland showed an obvious deficit, and their change trend was basically consistent with the per capita ED, indicating that these two components contributed the most to the change of per capita ED.

Trends of ecological pressure index (EPI), ecological footprint diversity index (EFDI) and ecological coordination coefficient (ECC) in Hulunbeir grassland from 2006 to 2016 were shown in Figure 4. As a whole, the reverse trend of ECC and EPI was obvious. The coordination of ecological environment was good when ecological pressure was low; conversely, it was poor when ecological pressure was high. Therefore, the threshold value of EPI could be determined by the critical value of ECC, and then the ecological security level and ecological security alarm level could be determined. Specifically, EPI rose from 0.83 in 2006 to 1.36 in 2014 after which it showed a slight downward trend from 1.36 in 2014 to 1.25 in 2016. In 2010, it was greater than 1, which meant that the degree of security increased from level 3 (moderately safe) to level 4 (moderately risky) (Table 5). However, from 2014 to 2016, although ecological security was still at level 4, the EPI decreased obviously, indicating that the ecological security condition in the study area showed an evident improvement. It could be related to the "Three Horizontal and Three Vertical" industrial transformation and development strategy implemented by governments, in 2014, and grassland ecological protection measures, such as returning grazing to grassland and enclosure projects.

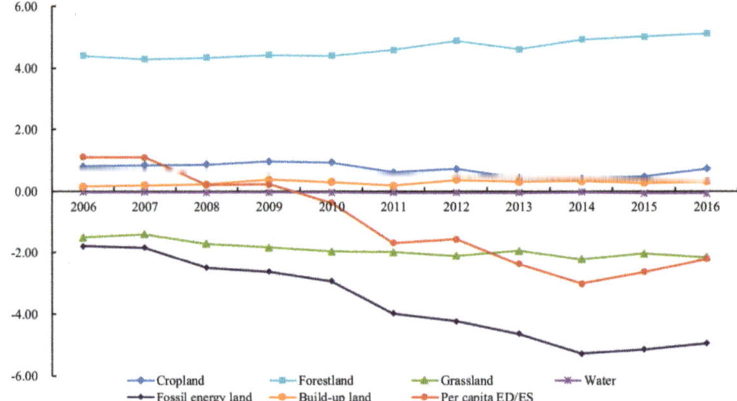

Figure 3. Changes of per capita ecological deficit/ecological surplus in Hulunbeir grassland from 2006 to 2016.

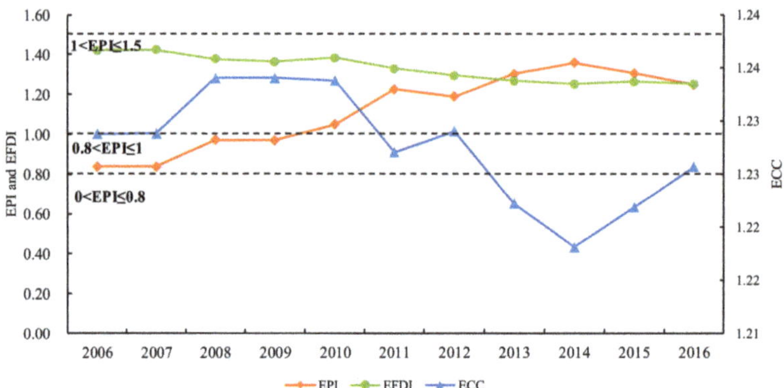

Figure 4. Evolution of ecological pressure index, ecological coordination coefficient and ecological footprint diversity index in Hulunbeir grassland from 2006 to 2016.

Table 5. Changes of ecological security in Hulunbeir grassland from 2006 to 2016.

Year	Ecological Pressure Index	Ecological Security Grade	Characterization State	Ecological Security Alarm Level
2006	0.83	3	Moderately safe	Low alarm
2007	0.84		Moderately safe	
2008	0.97		Moderately safe	
2009	0.97		Moderately safe	
2010	1.05	4	Moderately risky	Moderate alarm
2011	1.22		Moderately risky	
2012	1.19		Moderately risky	
2013	1.30		Moderately risky	
2014	1.36		Moderately risky	
2015	1.31		Moderately risky	
2016	1.25		Moderately risky	

In addition, with an increase in the ecological deficit and ecological pressure, the EPI reached its maximum in 2014, indicating that the degree of the use of natural resources, and the rate at which waste was being released, had already exceeded the system's recycling and self-purification capabilities. The ECC declined slightly over the study period, indicating the coordination between regional ecological environment and socio-economic development was gradually decreasing. In 2014, it reached the minimum, contrary to the EPI, indicating that when the ecological pressure rises to a certain point, the ecological environment will become uncoordinated and unsafe.

The EFDI showed a downward trend from 2006 in 1.42 to 1.25 in 2016. At the same time, we can see that the annual average decline rate of EFDI during the period 2006–2013 was 1.4%, the increase rate of EPI was 5.6% at the same time. However, when the annual average decline rate of EFDI gradually approached 0, the change rate of the EPI began to decline obviously. What is more, when the EFDI reached the minimum in 2014, the EPI reached the maximum. Therefore, an increase of ecosystem diversity by using different types of land resources equally and improving resource utilization efficiency could help alleviate ecological stress and enhance the development capability of the ecological system.

3.3. Identification of Driving Factors of Ecological Footprint

Climate, precipitation, soil, and other factors have to go through a long evolution process before they can affect the ecological footprint [22,48], which could be ignored for the time being. Therefore, what we explore are internal drivers of the ecological footprint primarily skewing towards the impact of regional economic and social development. On the basis of the STIRPAT model, the index of energy footprint driving forces in study area was selected from three aspects of economy, society, and technology. Economic growth requires a large amount of resources and energy from nature. Meanwhile, it will produce various emissions of pollution, leaving a deep imprint on the natural ecology and further affect the change of ecological footprint. Therefore, we selected per capita GDP (A) (represents the comprehensive situation of regional economic development) and the proportion of secondary industry (T) (highly relies on energy and raw materials and generates more wastes) to represent the ecological footprint consumption of economic growth. Social development is another important aspect that determines the change of ecological footprint, mainly reflected by population scale and consumption structure [22,49]. Therefore, the year-end resident population (P) and urbanization rate (U) were selected to explore the effect on the ecological footprint. Generally, the improvement of science and technology levels will reduce the energy consumption, that is, the unit of energy consumption will a produce greater value, and the waste rate of resources will be reduced. Hence, we adopted the unit GDP energy consumption (C) to reflect the technology development level.

Initially, the correlation matrix method was utilized to check the correlation coefficients among variables (Table 6). It is obvious that the collinearity among these independent variables was higher because many correlation coefficients were greater than 0.8, indicating that there was serious multicollinearity existing among them. This would impact the accuracy and credibility of STIRPAT model. Thus, it was necessary to adopt the principal component analysis (PCA) to eliminate this effect.

Table 6. Correlation coefficient matrix between independent variables.

Index	A	T	P	U	C
A- per capita GDP	1	0.905 **	−0.811 **	0.816 **	−988 **
Significance test		0.000	0.002	0.002	0.000
T-proportion of the second industry	0.905 **	1	−587	0.561	−0.882 **
Significance test	0.000		0.057	0.072	0.000
P- year-end resident population	−0.811 **	0.587	1	−0.955 **	0.800 **
Significance test	0.002	0.057		0.000	0.003
U- urbanization rate	0.816 **	0.561	−0.955 **	1	−0.814 **
Significance test	0.002	0.072	0.000		0.002
C- unit GDP energy consumption	−0.988 **	0.882 **	0.800 **	−0.814 **	1
Significance test	0.000	0.000	0.003	0.002	

** Correlation is significant at the 0.01 level (2-tailed).

In order to avoid the influence between variables, Kaiser–Meyer–Olkin (KMO) test of sample data should be conducted before principal component analysis (PCA). Results showed that the KMO measurement value was 0.695, greater than 0.5, the Bartlett sphericity test value was 71.849, and the sig < 0.001, passing the test, which indicated that PCA was feasible. According to the results of PCA (Table 7), when the first two principal components were extracted, the cumulative contribution rate was 97.332%, greater than 85%, indicating that it contained 97.332% of the original variable information and could replace the original variable to achieve a satisfactory effect.

Table 7. Extraction results of principle component characteristic value and contribution rate.

Component	Initial Eigenvalues			Extraction Sums of Squared Loadings			Rotate Sums of Squared Loadings		
	Total	% of Variance	Cumulative%	Total	% of Variance	Cumulative%	Total	% of Variance	Cumulative%
F_1	4.199	83.979	83.979	4.199	83.979	83.979	2.445	48.908	48.908
F_2	0.668	13.354	97.332	0.668	13.354	97.332	2.421	48.425	97.332

As showed in Table 8, the principal component F1 was mainly related to the total population, urbanization rate, and unit GDP energy consumption, and the variance percentage was 48.908%. The principal component F2 was mainly related to the per capita GDP, the proportion of the secondary industry and unit GDP energy consumption, and the variance percentage reached 48.425%. On the basis of the results of the principal component analysis, the two extracted principal components, F1 and F2, were respectively expressed by the influence factors after the logarithm, and then restored to the original variable form in the STIRPAT model, shown as follows:

$$F_1 = -0.135LnA - 0.384LnT + 0.584LnP + 0.585LnU - 0.121LnC \quad (9)$$

$$F_2 = 0.459LnA + 0.675LnT - 0.286LnP - 0.284LnU - 0.206LnC \quad (10)$$

$$LnI = 0.795 + 1.160F_2 + LnK. \quad (11)$$

Table 8. Rotational component matrix and principal component score coefficient matrix.

Category	Rotational Component Matrix		Principal Component Score Coefficient Matrix	
Indicator	Component		Component	
	F_1	F_2	F_1	F_2
LnA	0.481	0.874	−0.135	0.459
LnT	0.253	0.957	−0.384	0.675
LnP	−0.923	−0.339	0.584	−0.286
LnU	0.929	0.344	0.585	−0.284
LnC	−0.660	−0.713	−0.121	−0.206

The original time series data of the ecological footprint were converted into a natural logarithm, and the converted data were represented by LnI. With the dependent variable LnI as the control variable and the comprehensive variables F1 and F2 as the explanatory variables, ordinary least square regression was adopted to conduct regression analysis on the variables based on SPSS19.0 software. The regression results and equation test are shown in the Table 9. The R^2 and adjusted R^2 were both greater than 0.9, and the F statistic value is significant at the level of less than 0.01, indicating the overall fit was very good. According to the regression coefficients, the regression equations of dependent variable LnI and comprehensive variables F1 and F2 were obtained, shown in Equation (11).

Table 9. Analysis coefficient of principal component regression.

Component	Unstandardized Coefficients		Sig
	B	Std. Error	
F_1	0.795	0.477	0.134
F_2	1.160	0.091	0.000
R^2	0.976		
Adjusted R^2	0.970		
F-statistic	160.668		
Sig.	0.000		

After the reduction of Equation (11), the expression of Hulunbeir grassland's ecological footprint and five driving factors based on STIRPAT model was obtained:

$$I = NA^{0.425}T^{0.478}P^{0.135}U^{0.136}C^{-0.335} \quad (12)$$

Results of STIRPAT model indicated that the per capita GDP, the proportion of the output value of the secondary industry, the total population, as well as the urbanization rate in the study area are positively correlated with the ecological footprint, with force indexes 0.425, 0.478, 0.135, and 0.136 respectively. Among them, the proportion of the secondary industry had the largest effect on ecological footprint. Every 1% increase in the proportion of secondary industry would cause an increase of 0.478% in the total EF. With the rising of per capita GDP, the proportion of the secondary industry experienced a rapid increase, from 27.6% in 2006 to 44.7% in 2016. The development of the secondary industry is often accompanied by the consumption of energy resources and the discharge of wastes, putting great pressure on the local environment. The rapid growth of the secondary industry will undoubtedly lead to the increase of the regional ecological footprint. Therefore, the optimization of industrial structure is an important factor to improve the quality of the ecological environment.

In addition, the positive driving effect of the population and urbanization rate on the ecological footprint was small. Every 1% increase in the population and urbanization rate would cause an increase of 0.135% and 0.136% in the EF, respectively. According to the statistics, the urbanization rate increased slowly during 11a, from 65.72% in 2006 to 71.52% in 2016; the year-end resident population dropped from 2.7 million in 2006 to 3.4 million in 2016. This was consistent with the analysis results of the STIRPAT model, indicating that the population and city size of Hulunbeir grassland had little impact on the change of the EF. The results of this study are inconsistent with Zheng's research [50], i.e., urbanization rate and ecological capacity are the main factors affecting the ecological footprint in China, but they are similar to Yang's results [51], i.e., population and urbanization rate have little influence on the ecological footprint, but the output value of the secondary industry has a great influence in energy-rich ecologically fragile regions.

On the contrary, the technology level (unit GDP energy consumption) was the main negative driving force on the growth of the EF, and every 1% increase in technology level may induce about 0.335% decrease in the EF, supplementing Yang's [52] theoretical analysis on the inhibitory effect of technical factors on the growth of the ecological footprint to a certain extent, i.e., the improvement of science and technology provides clean technologies and production processes for industrial production, which could, theoretically, reduce resource consumption, decrease the amount of contaminant emission, and promote the utilization of renewable energy resources.

4. Conclusions

The EF framework used in this study offers an intuitive to rationally judge the relationship between regional socio-economic development and ecological capacity from a supply-and-demand perspective. Based on this theory, the indexes of the EPI, ECC, and EFDI were obtained to determine the level of ecological security and coordination between the ecosystem and economy. The main conclusions are as follows:

The per capita EF of Hulunbeir grassland nearly doubled during the past 2006–2016 years. While, the per capita EC was rather low and increased by only 29.9%, the footprint of fossil energy land contributed the most to the total EF, followed by grassland, cropland, forestland, build-up land, and water. In terms of time variation characteristics, the EF exceeded the EC in 2010, and the ecological deficit began to show and gradually expand from then on. At the same time, the EPI increased obviously, causing the degree of ecological security to rise from level 3 (moderately safe) to level 4 (moderately risky) and the alarm level changed from low alarm to moderate alarm. Additionally, the value of the EFDI and ECC dropped to different extents during 11a, but both reached a minimum in 2014 and have increased slightly since. Contrarily, the EPI reached a maximum in 2014 and then decreased slightly. These changes show that the status of ecological environment and its coordination with economy were in a worse position in 2014 but have been gradually alleviated since.

Driving force analysis shows that the per capita GDP and the proportion of secondary industry are mainly positive driving factors of EF growth. However, technological advances played an important role in curbing the growth of the EF during the study period. The increase of the population and city size had little influence on the ecological footprint. Therefore, Hulunbeir city should continue to strengthen the optimization and upgrading of the industrial structure, enhance the capacity of technological innovation and promote the use of clean energy. At the same time, ecological environment restoration should be strengthened to improve the carrying capacity of the environment.

Although the ecological footprint model has been widely used in ecological environment and sustainable development research, the parameters involved in the model, such as the equivalence factor, yield factor, and average yield, are various and the comparison scale is different (global scale, national scale, and regional scale), which may have a certain impact on the evaluation results. Therefore, it is necessary to further optimize the model to achieve a more comprehensive regional ecological assessment in future studies.

Author Contributions: Conceptualization, S.G. and Y.W.; data collection process: S.G.; methodology: S.G.; data analysis: S.G.; paper writing: S.G.; paper review: Y.W.

Funding: This research was funded by the Fundamental Research Funds for the Central Universities (2017XKZD14).

Conflicts of Interest: The authors declare no conflict of interest.

References

1. Pang, Y.S.; Wang, L. A Review of Regional Ecological Security Evaluation. *China Popul. Resour. Environ.* **2014**, *24*, 337–344. [CrossRef]
2. Teixidó-Figueras, J.; Duro, J.A. The building blocks of International Ecological Footprint inequality: A Regression-Based Decomposition. *Ecol. Econ.* **2015**, *118*, 30–39. [CrossRef]
3. Kuosmanen, T.K. Measurement and Analysis of Eco-efficiency. An economist's Perspective. *J. Ind. Ecol.* **2010**, *9*, 15–18. [CrossRef]
4. Wang, Z.; Yang, L.; Yin, J.; Zhang, B. Assessment and prediction of environmental sustainability in China based on a modified ecological footprint model. *Resour. Conserv. Recycl.* **2018**, *132*, 301–313. [CrossRef]
5. Nakajima, E.S.; Ortega, E. Carrying capacity using emergy and a new calculation of the ecological footprint. *Ecol. Indic.* **2016**, *60*, 1200–1207. [CrossRef]
6. Peng, W.; Wang, X.; Li, X.; He, C. Sustainability evaluation based on the emergy ecological footprint method: A case study of Qingdao, China, from 2004 to 2014. *Ecol. Indic.* **2018**, *85*, 1249–1261. [CrossRef]
7. Liu, D.; Chang, Q. Ecological security research progress in China. *Acta Ecol. Sin.* **2015**, *35*, 111–121. [CrossRef]
8. Li, X.; Tian, M.; Wang, H.; Wang, H.; Yu, J. Development of an ecological security evaluation method based on the ecological footprint and application to a typical steppe region in China. *Ecol. Indic.* **2014**, *39*, 153–159. [CrossRef]
9. He, G.; Yu, B.; Li, S.; Zhu, Y. Comprehensive evaluation of ecological security in mining area based on PSR-ANP-GRAY. *Environ. Technol.* **2018**, *39*, 3013–3019. [CrossRef]

10. Aspinall, R.; Pearson, D. Integrated geographical assessment of environmental condition in water catchments: Linking landscape ecology, environmental modelling and GIS. *J. Environ. Manag.* **2000**, *59*, 299–319. [CrossRef]
11. Kuchma, T.; Tarariko, O.; Syrotenko, O. Landscape Diversity Indexes Application for Agricultural Land Use Optimization. *Procedia Technol.* **2013**, *8*, 566–569. [CrossRef]
12. Huang, C.; Vause, J.; Ma, H.; Yu, C. Using material/substance flow analysis to support sustainable development assessment: A literature review and outlook. *Resour. Conserv. Recycl.* **2012**, *68*, 104–116. [CrossRef]
13. Chen, P.; Crawford-Brown, D.; Chang, C.; Ma, H. Identifying the drivers of environmental risk through a model integrating substance flow and input–output analysis. *Ecol. Econ.* **2014**, *107*, 94–103. [CrossRef]
14. Huang, Z.; Wang, F.; Cao, W. Dynamic analysis of an ecological security pattern relying on the relationship between ecosystem service supply and demand: A case study on the Xiamen-Zhangzhou-Quanzhou city cluster. *Acta Ecol. Sin.* **2018**, *38*, 4327–4340. [CrossRef]
15. Ren, Z.; Liu, Y. Exploring the Regional Ecological Security Evaluation Methods Based on Values: A Case Study in the Energy Region of Northern Shaanxi. *Geogr. Res.* **2013**, *32*, 5–15.
16. Yang, Q.; Liu, G.; Hao, Y.; Coscieme, L.; Zhang, J.; Jiang, N.; Casazza, M.; Giannetti, B.F. Quantitative analysis of the dynamic changes of ecological security in the provinces of China through emergy-ecological footprint hybrid indicators. *J. Clean. Prod.* **2018**, *184*, 678–695. [CrossRef]
17. Chu, X.; Deng, X.; Jin, G.; Wang, Z.; Li, Z. Ecological security assessment based on ecological footprint approach in Beijing-Tianjin-Hebei region, China. *Phys. Chem. Earth* **2017**, *101*, 43–51. [CrossRef]
18. Wackernagel, M.; Schulz, N.B.; Deumling, D.; Linares, A.C.; Jenkins, M.; Kapos, V.; Monfreda, C.; Loh, J.; Myers, N.; Norgaard, R.; et al. Tracking the ecological overshoot of the human economy. *Proc. Natl. Acad. Sci. USA* **2002**, *99*, 9266–9271. [CrossRef]
19. Li, J.; Chen, Y.; Xu, C.; Li, Z. Evaluation and analysis of ecological security in arid areas of Central Asia based on the emergy ecological footprint (EEF) model. *J. Clean. Prod.* **2019**, *235*, 664–677. [CrossRef]
20. Wackernagel, M.; Onisto, L.; Bello, P.; Callejas Linares, A.; Susana López Falfán, I.; Méndez García, J.; Isabel Suárez Guerrero, A.; Guadalupe Suárez Guerrero, M. National natural capital accounting with the ecological footprint concept. *Ecol. Econ.* **1999**, *29*, 375–390. [CrossRef]
21. Niccolucci, V.; Tiezzi, E.; Pulselli, F.M.; Capineri, C. Biocapacity vs Ecological Footprint of world regions: A geopolitical interpretation. *Ecol. Indic.* **2012**, *16*, 23–30. [CrossRef]
22. Dong, H.; Li, P.; Feng, Z.; Yang, Y.; You, Z.; Li, Q. Natural capital utilization on an international tourism island based on a three-dimensional ecological footprint model: A case study of Hainan Province, China. *Ecol. Indic.* **2019**, *104*, 479–488. [CrossRef]
23. Pan, H.; Zhuang, M.; Geng, Y.; Wu, F.; Dong, H. Emergy-based ecological footprint analysis for a mega-city: The dynamic changes of Shanghai. *J. Clean. Prod.* **2019**, *210*, 552–562. [CrossRef]
24. Zhao, Y.; Zou, X.; Cheng, H.; Jia, H.; Wu, Y.; Wang, G.; Zhang, C.; Gao, S. Assessing the ecological security of the Tibetan plateau: Methodology and a case study for Lhaze County. *J. Environ. Manag.* **2006**, *80*, 120–131. [CrossRef] [PubMed]
25. Bargaoui, A.S.; Liouane, N.; Nouri, F.Z. Environmental Impact Determinants: An Empirical Analysis based on the STIRPAT Model. *Procedia Soc. Behav. Sci.* **2014**, *109*, 449–458. [CrossRef]
26. Yang, Y.; Hu, D. Natural capital utilization based on a three-dimensional ecological footprint model: A case study in northern Shaanxi, China. *Ecol. Indic.* **2018**, *87*, 178–188. [CrossRef]
27. Xie, P.; Gao, S.; Sun, F. An analysis of the decoupling relationship between CO_2 emission in power industry and GDP in China based on LMDI method. *J. Clean. Prod.* **2019**, *211*, 598–606. [CrossRef]
28. York, R.; Rosa, E.A.; Dietz, T. STIRPAT, IPAT and ImPACT: Analytic tools for unpacking the driving forces of environmental impacts. *Ecol. Econ.* **2003**, *46*, 351–365. [CrossRef]
29. Giampietro, M.; Saltelli, A. Footworking in circles: Reply to Goldfinger et al. (2014) "Footprint Facts and Fallacies: A Response to Giampietro and Saltelli (2014) Footprints to nowhere". *Ecol. Indic.* **2014**, *46*, 260–263. [CrossRef]
30. Bai, Y.; Deng, X.; Zhang, Q.; Wang, Z. Measuring environmental performance of industrial sub-sectors in China: A stochastic metafrontier approach. *Phys. Chem. Earth Parts A/B/C* **2017**, *101*, 3–12. [CrossRef]
31. Sun, Y.; Yang, Y.; Zhang, Y.; Wang, Z. Assessing vegetation dynamics and their relationships with climatic variability in northern China. *Phys. Chem. Earth Parts A/B/C* **2015**, *87*, 79–86. [CrossRef]

32. Zhang, Y.; Wang, Q.; Wang, Z.; Yang, Y.; Li, J. Impact of human activities and climate change on the grassland dynamics under different regime policies in the Mongolian Plateau. *Sci. Total Environ.* **2020**, *698*, 134304. [CrossRef] [PubMed]
33. Ulucak, R.; Lin, D. Persistence of policy shocks to Ecological Footprint of the USA. *Ecol. Indic.* **2017**, *80*, 337–343. [CrossRef]
34. China Electricity Council. Eight Mining Areas in Inner Mongolia have been Included in the State Plan of Coal Mining Areas. Available online: http://www.cec.org.cn/xiangguanhangye/2010-11-27/8907.html (accessed on 11 November 2019).
35. Yang, C.M. Legal Research on Mineral Resources Exploitation and Ecological Environment Protection in Hulunbuir Grassland Area. *J. Hulunbeir Univ.* **2016**, *21*, 36–39. [CrossRef]
36. Ma, Y.; Ju, X. Geological environment problems and prevention measures of mines in Hulunbeir city. *Opencast Min. Technol.* **2012**, 85–88. [CrossRef]
37. Chang, L. Research on the Prediction Model of Hulunbeir Grassland Degradation. Master's Thesis, Northeast Forestry University, Harbin, China, 2014.
38. Liu, M.C.; Li, W.H. The Calculation of China's Equivalence Factor under Ecological Footprint Mode Based on Net Primary Production. *J. Nat. Resour.* **2009**, *24*, 1550–1559.
39. Wackernagel, M.; Monfreda, C.; Schulz, N.B.; Erb, K.; Haberl, H.; Krausmann, F. Calculating national and global ecological footprint time series: Resolving conceptual challenges. *Land Use Policy* **2004**, *21*, 271–278. [CrossRef]
40. Ren, J.J. Quantitative Analysis of Sustainability Development of Inner Mongolia Based on Ecological Footprint Model. Master's Thesis, Inner Mongolia University, Hohhot, China, 2012.
41. Sheng, Z.; Wu, C.; Hong, H.; Zhang, L. Linking the concept of ecological footprint and valuation of ecosystem services: A case study of economic growth and natural carrying capacity. *Int. J. Sust. Dev. World* **2009**, *16*, 137–142. [CrossRef]
42. Dang, X.; Liu, G.; Xue, S.; Li, P. An ecological footprint and emergy based assessment of an ecological restoration program in the Loess Hilly Region of China. *Ecol. Eng.* **2013**, *61*, 258–267. [CrossRef]
43. Yuan, W.B. Ecological Security Evaluation Based on the Ecological Footprint in Nanning City. Master's Thesis, Jilin University, Changchun, China, 2010.
44. Shannon, C.E.; Weaver, W. *The Mathematical Theory of Communication Urbana*; University of Illinois Press: Urbana, IL, USA, 1950.
45. Ehrlich, P.R.; Holdren, J.P. Impact of Population Growth. *Science* **1971**, *171*, 1212–1217. [CrossRef]
46. Dietz, T.; Rosa, E.A. Effects of population and affluence on CO_2 emissions. *Proc. Natl. Acad. Sci. USA* **1997**, *94*, 175–179. [CrossRef] [PubMed]
47. Chai, J.; Liang, T.; Lai, K.K.; Zhang, Z.G.; Wang, S. The future natural gas consumption in China: Based on the LMDI-STIRPAT-PLSR framework and scenario analysis. *Energy Policy* **2018**, *119*, 215–225. [CrossRef]
48. Hu, M.J.; Zhou, N.; Li, Z.; Qi, X. Calculation and driving factor analysis of three-dimensional ecological footprint in Nanjing city. *Geogr. Geo Inf. Sci.* **2015**, *31*, 91–95. [CrossRef]
49. Wen, L.; Shao, H. Influencing factors of the carbon dioxide emissions in China's commercial department: A non-parametric additive regression model. *Sci. Total Environ.* **2019**, *668*, 1–12. [CrossRef]
50. Zheng, D.; Liu, X.; Wang, Y.; Lv, L. Spatiotemporal evolution and driving forces of natural capital utilization in China based on three-dimensional ecological footprint. *Prog. Geogr.* **2018**, *37*, 1328–1339. [CrossRef]
51. Yang, Y.; Hu, D. Dynamic changes and driving factors of three dimensional ecological footprint in Yulin. *J. Nat. Resour.* **2018**, *33*, 1204–1217. [CrossRef]
52. Yang, C.; Zhu, Y. Analysis on Driving Force Factors of Ecological Footprint in Hunan Province from the Perspective of Green Development. *Econ. Geogr.* **2019**, *14*, 1–15.

International Journal of
Environmental Research and Public Health

Article

Quality Criteria to Evaluate Performance and Scope of 2030 Agenda in Metropolitan Areas: Case Study on Strategic Planning of Environmental Municipality Management

María de Fátima Poza-Vilches *, José Gutiérrez-Pérez and María Teresa Pozo-Llorente

Department Methods of Research, Faculty of Education, University of Granada, 10871 Granada, Spain; jguti@ugr.es (J.G.-P.); mtpozo@ugr.es (M.T.P.-L.)

* Correspondence: fatimapoza@ugr.es

Received: 7 November 2019; Accepted: 6 January 2020; Published: 8 January 2020

Abstract: The United Nations' (UN) 2030 Agenda brings new governance challenges to municipal environmental planning, both in large urban centres and in metropolitan peripheries. The opportunities of the new framework of action proposed by the United Nations (UN) and its integrative, global, and transversal nature constitute advances from the previous models of municipal management based on the Local Agenda 21. This text provides evidence to apply quality criteria and validated instruments of participatory evaluation. These instruments have been built on the foundation of evaluative research, a scientific discipline that provides rigour and validity to those decisions adopted at a municipal level. A case study focused on a metropolitan area serves as a field of experimentation for this model of the modernization of environmental management structures at a local level. Details of the instruments, agents, priority decision areas, methodologies, participation processes, and quality criteria are provided, as well as an empirically validated model for participatory municipal management based on action research processes and strategic planning that favours a shared responsibility across all social groups in the decision-making process and in the development of continuous improvement activities that are committed to sustainability. Finally, a critical comparison of weaknesses and strengths is included in light of the evidence collected.

Keywords: 2030 Agenda; strategic planning; quality criteria

---

1. Introduction

Recent advances in the field of Sustainability Sciences open the gate to emerging disciplinary areas such as Global Urban Science [1–4], built from new paradigms and ways of doing science through models that are participatory in nature. These models provide significant novelties to the ways of developing socio-environmental knowledge and justify decision-making outcomes in municipal management. They also open an inexhaustible field of exploration for the progress and modernization of municipal governance models and urban environment management [5–8]: 'science-policy interactions between urban scholars and urban practitioners have, in the wake of the Paris Agreement, Sendai Framework, Sustainable Development Goals (SDGs) and the United Nations' (UN) Habitat III-New Urban Agenda (NUA), undergone important steps towards greater integration' [9] (p. 12).

From this logic, environmental strategic planning (ESP) applied to the field of municipal management emerges as a governance instrument that provides rigour and rationality to the interventions and decisions proposed to counter environmental problems and scenarios. The principles that inspire it, and the methodologies it applies, base the actions on quality criteria endowed with instruments for evaluating achievement and compliance with standards. These instruments help to ensure decisions are made following a certain direction or to convert them following the demands and requirements of the new planning and urban governance agendas [10] such as the 2030 Agenda.

The strategic plan (2030A) launched by the United Nations (UN) in 2015 represents an integrative framework for the development of environmental governance within the framework of municipal management. This agenda inherits the spirit with which the Agendas 21 [10–12], which started at the Rio 92 Summit, were built, giving them continuity from a new and more comprehensive integrative framework that demands cross-cutting commitments around the Sustainable Development Goals, which contain 17 goals, 169 targets, and 241 indicators.

The 2030A takes up the torch for the advances and successes of Agendas 21, overcoming its limitations [13]. This is especially true in relation to the processes involving social intervention and its methodology; providing them with a timeline and the means and instruments of a strategic nature that go beyond the immediate, that contemplate the basic ingredients of environmental complexity, and that are approached from a holistic and inclusive vision. In addition, this vision cannot be achieved if it is not based on participatory and innovative action methodologies that evaluate management processes, consensual decisions and challenges.

Therefore, from this logic of addressing the 2030A implementation processes in municipal management, in this document, we show the steps to follow in the application of participatory methodologies for the development of 2030A in a municipal context using the strategic environmental evaluation (SEE) and participatory action research (PAR) approaches, involving researchers, citizens, and managers.

Academic environmental research can play a key role in informing the design, implementation, and evaluation of sustainable urban strategies at the global scale. In addition, the active involvement of various non-academic actors in the production of urban knowledge for policy, 'as well as the multitude of actors involved in urban affairs (beyond government) requires the scholarly community to look beyond academia and forge new collaborations to enhance research use into urban strategies' [9] (p. 14). Dominant research modes are not enough to guide the societal transformations necessary to achieve the 2030A. Researchers, practitioners, decision makers, funders, and civil society should work together to achieve universally accessible and mutually beneficial sustainability science [14]. New approaches to science, such as action research [15], mode two knowledge production [16], transdisciplinary research [17,18], and post-normal science [19], propose that scientists should engage in deliberative learning processes with societal actors, with a view to jointly reflect on existing development visions and create new, contextualized ones [20].

Therefore, this research aims to show how the approach to urban sustainability assessment developed from participatory action research processes can be a good methodological reference to address the design and implementation of the 2030 Agenda in contexts of local management.

Within the framework of this logic of strategic environmental evaluation addressed from participatory action research, our study starts from the following aims and research questions (Table 1).

Table 1. Aims and research questions. SDGs, Sustainable Development Goals; PAR, participatory action research; SEE, strategic environmental evaluation.

Aims of Research	Research Questions (RQs)
1. Characterize a methodological model of strategic environmental planning based on democratic evaluation (participatory research action approach): define stages, obstacles, conditions, and limitations from a practical case study. 2. Analyse and assess the contribution of this strategic planning model to the development of the 2030A in the case study analysed. 3. Provide the necessary guidelines to address the 2030A in local municipal management through a citizen leadership model. 4. Identify new challenges set by the 2030A for the strategic and sustainable management of municipalities. 5. Define and model the planning and management stages, and analyse the possibilities of transferring these to different contexts.	RQ1. What are the novelties that the 2030A framework brings to sustainable municipal management? RQ2. What stages do the new methodologies associated with collaborative, transdisciplinary, and action research models involve for the grounds of municipal decision-making? RQ3. What criteria and quality indicators should be required from processes and instruments? RQ4. What are the most significant weaknesses and strengths of this new stage of municipal planning and management? RQ5. What viability and transfer possibilities do these new management models have in implementing them in different contexts?
Aims of Case Study	**Research Questions of Case Study (CS)**
1. Characterize the socio-environmental situation of the municipality studied from the SDGs. 2. Identify and prioritize the social and environmental needs that arise in this municipality according to the SDGs. 3. Collect the opinions and perceptions of citizens regarding environmental issues. 4. Involve all social sectors and the population in general, in processes of participation and decision-making in municipal environmental management. 5. Channel communication and dissemination processes of the 2030A and SDGs. 6. Promote communication, participation, negotiation, and reflection processes to prioritize collective needs. 7. Define by consensus on environmental quality indicators for the implementation of 2030A and SDGs. 8. Define by consensus some lines of action that favor the improvement of the socio-environmental situation of the municipality under investigation. 9. Create stable citizen participation structures that facilitate municipal decision making.	RQ(CS)1. Taking as reference the SDGs, what image of the studied municipality is projected? RQ(CS)2. What participation and communication platforms have been generated as a result of implementing the PAR? RQ(CS)3. After the SEE is carried out in the municipality, what responsibilities do the agents involved in the decision making assume? RQ(CS)4. What quality criteria and indicators have been generated after the application of the SEE? RQ(CS)5. What lines of action have been generated following the participatory diagnosis carried out?

2. Framework

2.1. The Agenda 2030 as Planning and Action Frameworks

The 2030A proposes a framework for knowledge-based transformations to sustainable development that reconciles evidence and socio-political deliberations for accelerated action [20,21]: understanding systemic interactions, understanding competing development agendas, and understanding transformations in concrete contexts.

Defining the term 'strategy' in the field of municipal management can be complicated because of the same complexity derived from the delimitation of the concept of management in local administration. The idea of strategic planning can be understood as the articulation of a set of operational elements aimed at establishing processes with the capacity for social and territorial transformation; processes that, in the medium or long-term, revert the conservation of ecosystems and/or in the improvement of the quality of life of citizens. In the fields of organizational management and business, every strategy involves establishing a work scheme, and designing an organized action protocol that facilitates interventions within a solid framework, and at the same time makes it possible to control external variables and factors that can influence the process, generating a competitive advantage that allows it to successfully remain in the market [22].

From this logic, we can consider municipal management as a typology of activity for municipal organizations whose priority activity is to define goals aimed at improving the quality of life and welfare of citizens in the territory they live, from approaches based on participation and democracy.

This last aspect is perhaps the main element of agreement among authors when conceptualizing the strategic elements in the field of public management, in opposition to their application within the business environment; those aspects that make mention of the idea of involvement and consensus and the need to jointly build plans that affect those involved in one way or another and that look to the future [23].

Under this action prism, the new environmental strategic planning (ESP) models require citizen participation and leadership as instruments of change and improvement. Some examples of successful case studies that have become reference models are the following: the ESP model to address issues related to the management of closed coastal seas [24–26]; effectiveness of ESP in Thailand to achieve the legitimacy of the processes [27]; ESP and the use of indicators in water resources planning [28]; integrated models that develop coherent and policy relevant socio-ecological strategies, which are relevant to policies where appropriate decision frameworks need to be co-developed across the range of stakeholders and decision-makers in Coastal Bangladesh [29]; integrated sustainability models in management organizations [30]; identify key governance challenges that cultivating collective action, accountability, decision spaces for stakeholder interaction regarding decision-making, investment, action, and outcomes [31]. All these approaches have the potential to integrate knowledge across the natural and social sciences, and to do so in a new way; that is, a framework to overcome ontological differences between the social and natural sciences, and thereby also overcome ontological barriers in sustainability research [24,32]. The priorities and needs are identified, defined, and planned from the consensus and unique interests of the various segments of citizenship, harmonizing demands of majorities and minorities, giving voice and a vote to all sectors of the population (from childhood, youths, the elderly, ethnic minorities, and so on). In this sense, 'the generation and support of small foci of social change in the field of environmental sustainability seems an immense field of opportunities, which has, among other advantages, the ability to: demonstrate that another way of doing things is possible, overcome mental obstacles and prejudices about alternative solutions, normalize or improve the image of models considered exotic, if not, marginal, and, amplify the positive effects of actions that have a moderate implication' [33] (pp. 12–13).

Traditionally, Agendas 21 for local development (L21A) have been a clear example of these small plots of social and environmental change demanded by society today, even in spite of the discrepancies, controversies, resistance, and frequent divorces that usually accompany any sphere of citizen intervention. Within the L21A processes, strategic and participatory planning acquires true meaning as a methodology of intervention and local transformation that has no reason for existence if it is not for citizen involvement and social leadership [34]. Currently, the 2030A and the Sustainable Development Goals (SDG) are the frameworks of reference guiding municipal administrative institutions, ensuring that their management model is sustainable and incorporates sustained strategic planning.

If we take as reference the 17 SDGs and 169 targets associated with them, we could affirm that the majority are linked to local competences regulated in the laws, norms, and regulations in which the municipal management is structured. This reflection highlights how transcendental the application and adaptation of 2030A is for the City Councils to comply with the SDG.

While the 17 SDGs are not legally binding treaties, there is a political and ethical commitment that must be addressed by every Municipal Program of Action for the coming years up to 2030, being a strategic priority in the achievement of local goals and thus meeting the goals of the SDG, in terms of providing basic services and promoting endogenous, inclusive, and sustainable territorial development.

This is a great challenge for City Councils at present and a pending issue. They are, therefore, responsible for the design of a strategic plan that connects their political action program with the requirements of 2030A and the SDG, taking citizenship leadership in the decision-making process as prescriptive, as well as the establishment of multi-level articulations that favour the fulfilment of all SDGs, whether or not they are municipal management competencies (Figure 1).

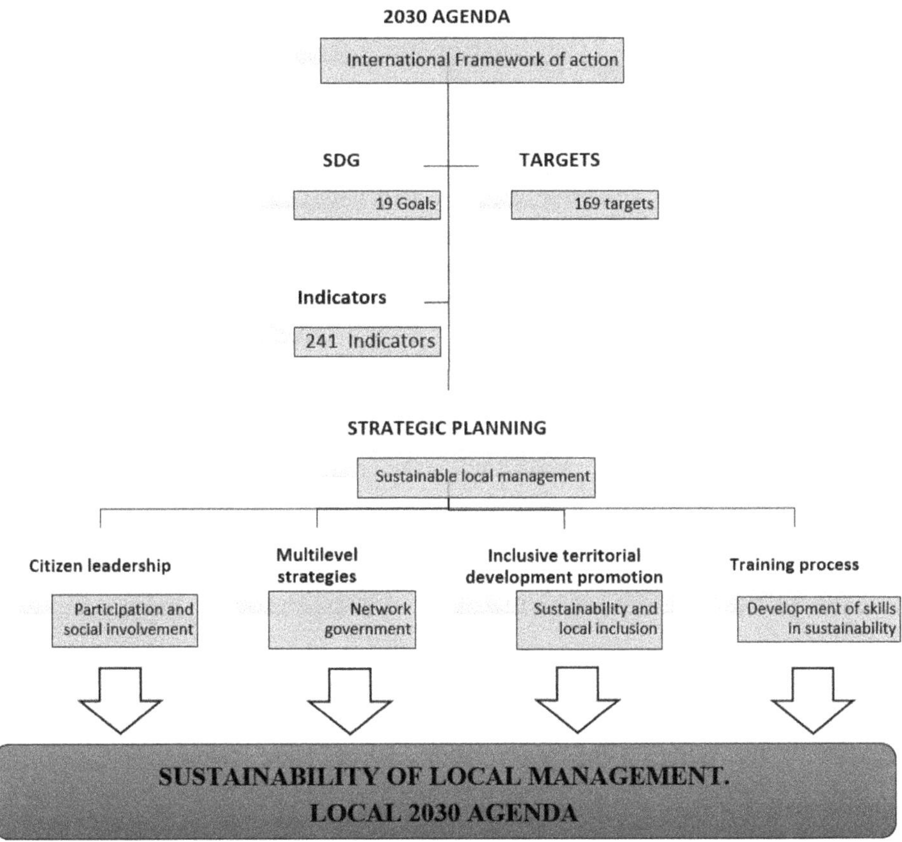

Figure 1. Principles of sustainable local management for a Local 2030 Agenda (L2030A). SDG, Sustainable Development Goal.

Along these lines, within the framework of Sustainability Science, a new emerging disciplinary field gains prominence in the form of what some authors call 'Global Urban Science', which plays an essential role in strategic planning processes in municipal management, as argued in the following section and whose characteristics, according to the Nature Sustainability Network, are summarized in three key messages [9] (p. 2):

1. A new global science is needed for the urban era: there is a need to develop an 'urban science', not as a single science, but as a cross-cutting field of engagement across multiple disciplines;
2. Urban science needs a broad range of experts and information: the urban science community will need to include a wide range of experts, including non-academic actors such as Non-Governmental Organizations (NGOs), residents, consultancies, industry, international organizations, city networks, and the scholarly edifice of academic research;
3. An urbanizing planet calls upon the sciences and policymaking to rethink and enhance their relationship across complex systems: the pathways to reform and improvement of the role of science in the future of cities goes, inevitably, through multiple sectors and scales of governance.

2.2. Strategic Planning in Local Management

The new approaches to citizen leadership point to new-generation models of democracy based on deliberative approaches, which consists of a transition from "I" to "we" through the creation of participatory will. These models are committed to the educational value of the process in deliberative management. They emphasize the different vital stages of decision making, assess the need for debate, the exchange of arguments and the value of consensus. Shared agreements help to fulfill collective commitments, to act together and to assume each one of their share of responsibility. In the face of aggregative policies, which resolve conflicts by voting a final vote ignoring the value of consensus [35,36]. This new prism of local action brings to light the importance of framing the desirable scenarios towards which we direct the change in an explicit model that provides a base and gives ideological, political, and social legitimacy to the interventions. If these values and principles govern the intervention, 'the contribution of citizens (participation) and the position of the rulers (leadership) become key factors in determining the reasons, foundations and interests of a strategic plan' [37], (p. 45). In this case, these management plans become 'Participatory Strategic Plans'; where *participation* is considered to be a tool for citizen involvement in decision-making and in the assumption of responsibilities and commitments in the construction of their future; and *leadership* is seen as the new role that governments have to assume in order to be mediators between the interests of citizens and the final decisions of those who represent those interests [35,36,38]. This movement that we advocate must overcome the citizen distrust that manifests itself in 'anti-democratic´ and demobilization actions of civil society [39]; as disenchantment against the broken promises of democracy [40] and the disaffection that exists between citizens towards political parties [41]. Today we are witnessing the 'emergence of new leaderships that can be associated with the expansion and reconfiguration of public space and that contribute to consolidating other representation links' [42] (p. 35).

Therefore, in our case, by taking strategic planning to the design and implementation of sustainable municipal management models built out of the principles of deliberative democracy, we would be talking about a networked or relational municipal government model based on participation and political leadership [43–45]. The following diagram (Figure 2) illustrates the different poles that can result from a combination of both aspects, marking as a favourable scenario for the elaboration of strategic plans in those cases in which there is high participation and marked political leadership [46].

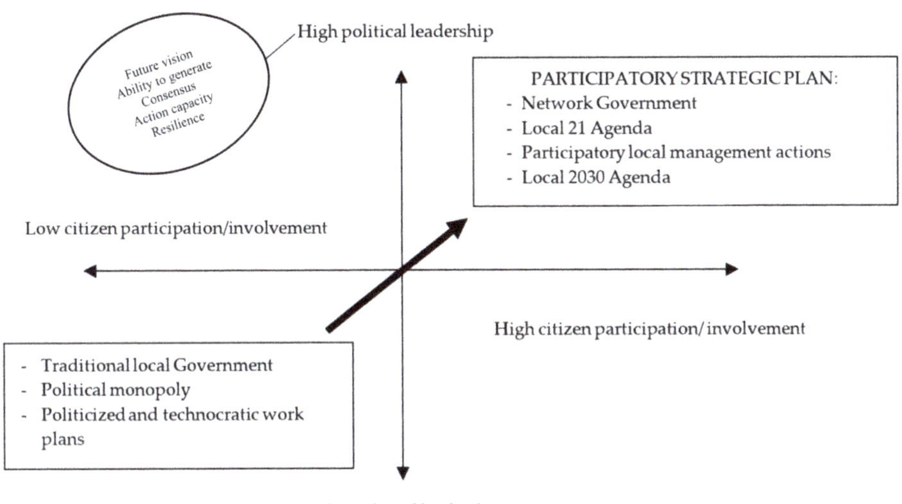

Figure 2. Network Government (authors' elaboration from the work of Font, 2001 [46]).

In general, and taking into account the four axes mentioned above, participatory strategic planning applied to the development of sustainable municipal management models is defined by four dimensions that govern its methodological process and its objectives, becoming an ideal method for the development of participatory processes at a municipal level (Figure 3).

Figure 3. Participatory strategic planning in municipal environmental management.

This intervention methodology, in the case of municipal environmental management, favours the definition of future scenarios that, in this case, translates into city and territory management models adapted to demands and the starting situation, from an approach focused on consensus and citizen dialogue in decision-making.

This is the starting point of this research and the conceptual reference in the field work carried out. In municipal management, participatory decision-making must be based on management models that involve participation structures that involve all social actors: political representatives, social representatives, business, and economic references and citizens in general. The diversity of actors who have participated in the information collection strategies used in the research (farmers, managers, women, youth, and so on) and that have covered a double aim are proof of the following: diagnose local environmental management, on the one hand, and, on the other hand, to foster intervention and improvement actions generated from debate and consensus processes. The assumption of shared responsibilities among the agents involved has been one of the achievements of this participatory because the recommendations provided by these groups in the different focus group sessions were transferred to the municipal council with a series of real improvements, as will be seen later in the 'Discussion' section.

2.3. Environmental Evaluation as a Participation and Planning Tool

From the logic of the activities in strategic planning, the strategic environmental evaluation (SEE), inspired by the proposals of the PAR (participatory action research) [47], is one of the 'most complete instruments for decision support on wide-ranging development initiatives with potential effects on the environment. The SEE has been defined as the formalized, systematic and global process for assessing the environmental impacts of a policy, plan or program, as well as its alternatives, including the preparation of a written report on the results of that evaluation and their use for the adoption of public decisions on which to account [48–50]. At the same time, it is considered to be a process to integrate the concept of sustainability from the highest levels at which decisions about development models are taken' [51] (p. 27). Historically, at international level, the SEE appears in the first National Environmental Policy Act (NEPA) (EE.UU, 1969) that required reports on the environmental consequences of federal actions. In the European Union, the development of the SEE is framed in the European Council Guidelines (85/337/EEC). Its premise was to incorporate environmental assessments at all levels of 'decision making' [50,52].

In general, the following application process (Figure 4) can be distinguished in the SEE [50,52].

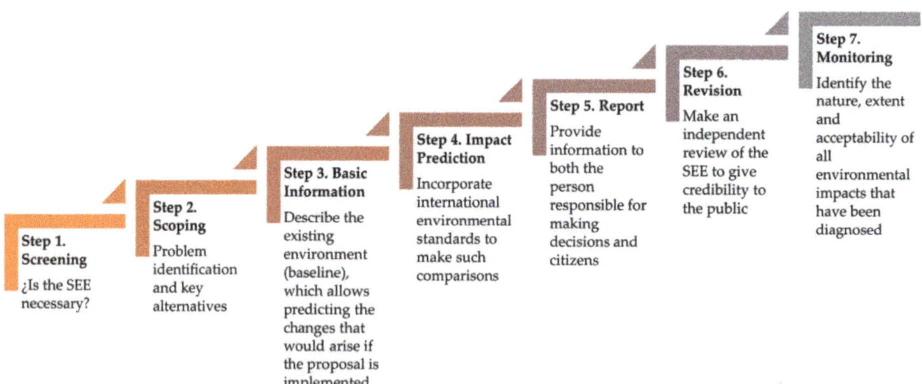

Figure 4. Strategic environmental evaluation (SEE) process.

This concept, in our field of work, gives meaning to the term strategic and action research planning. The SEE intends to serve to implement a sustainable local development process that integrates evaluation and decision-making at all stages of municipal management. Without forgetting the need to monetize the options of the environment as a source of local development and respect for natural cycles, ecosystems, spaces, and species, this task stresses the role of environmental policy as an important branch, interrelated with local actions, and not as a work area that is separate to general municipal policy, so that it helps to promote an intelligent, harmonious, and sustainable development.

Evaluation plays an essential role in this environmental planning as a scientific instrument that gives quality assurances to the decisions adopted [53]. The 2030A was developed through a largely political rather than a scientific process, the goals and targets—as well as the specific indicators developed to assess progress against these goals and targets—are formulated in a limited and somewhat inconsistent way [20]. The uniqueness of the environmental planning field requires the selection of proven evaluation models, inspired by methodologies validated in practice, built on bottom-up models [54,55] in which bottom-up participation is an essential requirement in a decision paradigm that places citizens at the heart of democratic decision-making processes, from the empowerment provided by the SEE [45] and, specifically, the participatory action research (PAR) [15,56] introduced at a time when undertaking an analysis of needs and prioritizing decisions on the actions to be undertaken strategically in the short, medium, and long-term in the municipal and global context in which they develop as historical subjects.

'Wicked' sustainability problems, defined as problems that are multi-dimensional, appear intractable, and for which there is no one clear solution, are increasing in number and intensity [57,58]. These problems differ fundamentally from technical problems that can be isolated and controlled using standard scientific methodologies. The unique characteristics of knowledge production that can address complex sustainability problems were first defined by Gibbons and Nowtony in their formulation of 'Mode 2' knowledge, defined as follows: knowledge production that is applied, integrates multiple disciplines and stakeholders, is reflexive, and offers novel ways to assess quality [59].

Experiments in science–policy collaboration at the local level are fundamental. Academia and local governments should take tangible steps towards joint investments for science–policy collaboration. 'This includes suggested practical actions such as: City-regional and metropolitan science policy mechanisms, such as urban observatories', need to be taken seriously by both universities and local governments, but with the support of national governments and the UN system. Appoint academically-grounded 'chief scientific advisors' to local government to advise on evidence use in city policymaking. Include peer review processes within the production of major private sector and city network datasets, engaging in scholarly outputs as much as reports from these analyses, including clear outlines of methodologies' [9] (p. 5).

In the case of local environmental management, the SEE will make sense as long as it is part of the decision-making process for the definition of a strategic framework for participatory and consensual intervention on the road to sustainability. The environmental assessment will be more strategic the more directly it is associated with the decision-making process. To capture and materialize the SEE, it is necessary to take into account a series of conditions that will guide the process, among which we can highlight the following [51] (Figure 5).

The SEE, therefore, is presented as a tool for participatory strategic planning in the field of local management and aims to be an instrument that favours an analysis of the impacts of planning in the territory and in the community. On the other hand, it is proposed as a work proposal to achieve the local environmental objectives that must be assumed by the local corporation as part of its management and its policy.

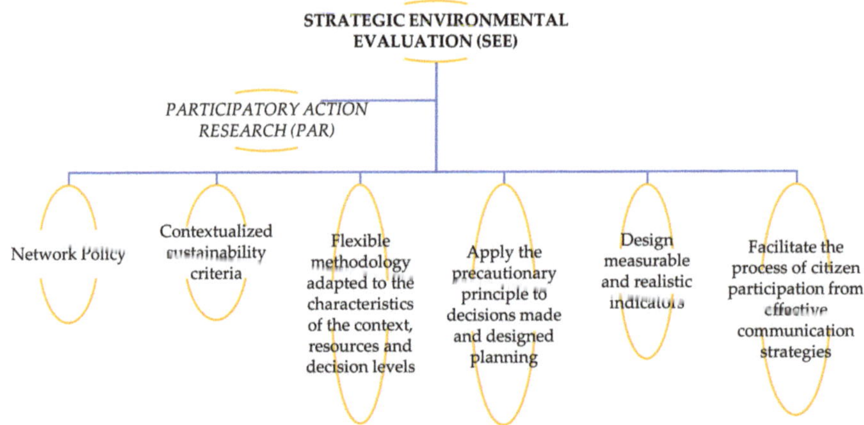

Figure 5. Strategic environmental evaluation: principles of action.

The SEE is part of the territorial planning processes as a strategy of impact assessment, compliance with environmental objectives, and monitoring of policies and design of recommendations to be incorporated into management policies in a cyclical and continuous manner, based on the participation of citizens in decision making, in the search for consensus, negotiation, and in the incorporation of alternatives to local political actions [51].

In short, these are some of the premises that must be taken into account to draw up a strategic plan: the need to develop rigorous evaluative research processes, implement PAR methodologies, apply change assessment instruments, promote tools for citizen participation and for the analysis of inclusive needs that involve all sectors of citizens, and mobilize municipal management actions from the bottom-up that democratize environmental decisions.

Following this descriptive situational analysis of the potential of strategic planning in local environmental management addressed using the SEE's approaches, a case study is presented to validate the model advocated in this research, focused on a metropolitan area.

3. Materials and Methods

3.1. Methodological Framework

Structuring a participatory strategic plan and implementing it across all its phases requires a coordinated, dynamic, and flexible process that favours citizen participation and decision-making, taking as basis the search for consensus and the prioritization of needs. Community-engaged, action-oriented research approaches involve communities that are impacted by the issues being studied. Such approaches include the overlapping traditions of participatory action research and community-based participatory research [60,61]. These processes manifest as a complex framework that requires continuous adjustments about negotiation and agreements, and re-adjustments around local management based on participatory and direct methodologies.

It is a model immersed in a structure that overcomes political-administrative management and favours transversal actions of citizen leadership across each of the stages into which it is divided. Reaching the balance between political action and social action is the essential ingredient that guarantees the success of this type of methodological structures based on teamwork, the search for consensus, the adoption of responsibilities, and decision-making in the definition and launching of the strategies.

The model that we intend to validate with this action research process, following the logic of strategic planning and through adapting the model of SEE [43,62], advocates sustainable municipal management such as the proposal shown below (Figure 6).

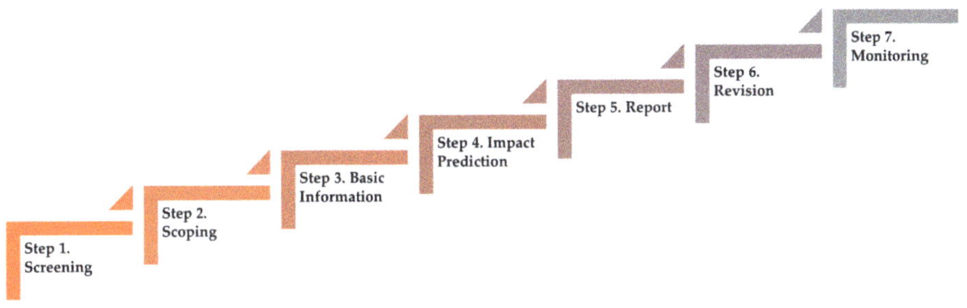

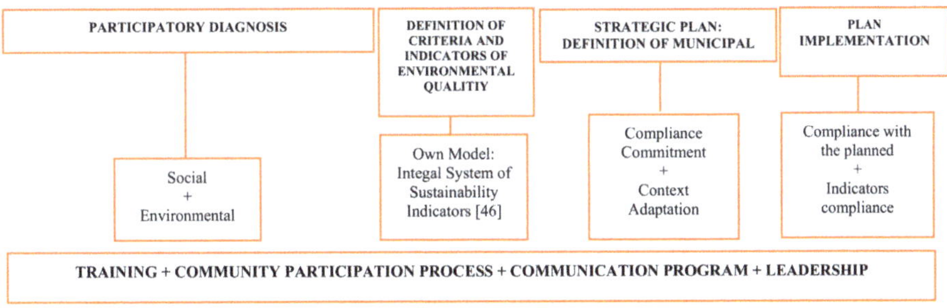

Figure 6. SEE model for participatory municipal management applied in our case study [63].

Once the model is defined, the exemplification of each of these phases is complex. The research that we include in this article brings forward the framework used at the beginning of each of these phases, so it has been indispensable to carry out complex triangulation processes both at a level addressing the techniques and key agents and informants, which have resulted in clues, suggestions, strengths, and weaknesses, in order to ultimately establish and extrapolate the results, in order to design quality criteria that we have deemed essential to defining a quality and sustainable municipal management model based on the philosophy of participatory strategic planning linked to compliance with 2030A and SDGs.

From this logic, this model incorporates an integral system of indicators for sustainability [63] from the approaches of a participatory diagnosis backed by technical indicators (territorial framework, water, energy, or mobility, among others); social indicators (population, economy and employment, heritage, or education, among others); and sustainability indicators such as the following: (1) Citizen perception of environmental and social problems. Prioritization. (2) Citizen representation in the diagnosis of local environmental management. (3) Level of commitment and responsibility of the different social, political, and technical groups in the process of socio-environmental transformation. (4) Definition and consensual design of the proposals and lines of action. (5) The decision-making process in the participatory diagnosis of local sustainable management. (6) The communication process as a training strategy.

Attending to this multidimensional indicator system, and as an example of this complex action research process, of the implementation of a strategic planning process based on the processes of individual and community reflection that have been carried out throughout the research, we exemplify this complexity with the analysis resulting from the diagnostic phase (steps of SEE—screening/scoping/basic information) of one of the sustainability indicators addressed in the investigation (step 4 of SEE—prediction of impacts) (Figure 6): 'Citizen perception of environmental and social problems. Prioritization'.

From the presentation of these results, we address a discussion related to the fulfilment of objectives. There is a methodological reflection on the process followed through the research that will enable us to validate the proposed model from five basic elements that respond to the objectives and questions proposed at the beginning of the paper (step 5 of SEE—report):

- Suitability of the information collection instruments used and of their quality.
- Strengths of the model.
- Weaknesses of the model.
- Contribution of the model for the development of sustainable local management.
- Feasibility and transfer possibilities to other contexts.

3.2. Instruments for Data Collection

To articulate this strategic framework, the instruments used for data collection are as follows (Table 2).

Table 2. Information collection strategies.

Strategy	Objectives	Description	Use (Phase)
Citizen opinion questionnaire	- Opinions and perceptions of citizens in environmental matters. - Involve the population in municipal participation processes. - Communicate the participatory environmental management process. - Promote that this local management model is known by all citizens.	Section 1: Environmental situation of the municipality. Item 1. What do you think are the most important environmental problems in your municipality? Item 2. Which of the following things do the neighbors of your town and which do you usually do? Item 3. What is the current situation of different aspects of your municipality? Section 2: Local environmental management in the municipality Item 4. What degree of responsibility do the different social groups have in the protection of the environment and in the socioeconomic improvement of the municipality? Item 5. You indicate the order of importance of the following solutions linked to the improvement of the environmental management of your municipality. Item 6. What do you think will improve the local sustainable development agenda in your town? Item 7. Which sector should be further developed with the realization of the local sustainable development agenda for your people to improve? Item 8. How far would you be willing to get involved so that improvement actions were carried out? Item criterion: Municipal environmental situation (in general)	- For all population sectors from 12 years old. - Diagnostic phase: To correctly direct the actions and strategies of action and participation.
Monitoring commission	- Promote a process of collective reflection from the monitoring of actions. - Consolidate a platform for monitoring and evaluation of the local environmental management process developed. - Involve the population in decision-making processes in local management.	A control and evaluation body of the local environmental management process has been created. There have been different control sessions to ensure compliance with the actions. Representatives of the different social groups in the municipalities studied participate in this commission. In this commission, issues relate to the following: - functions of the commission; - rights and duties of the commission; - election of commission members; - monthly commitments for the sustainable management of the municipality.	It has been carried out during the diagnostic phase owing to the relevance of this body throughout the process.

Table 2. *Cont.*

Strategy	Objectives	Description	Use (Phase)
Discussion groups/citizen participation forums	- Promote a process of collective reflection. - Establish socio-environmental and participatory management indicators. - Triangulate the information collected with the different techniques. - Involve the population in decision-making processes in local management.	- Two citizen participation forums - Four groups discussion with two sessions: (1) councillors and technicians; (2) women; 3 (farmers); (4) youth Forum 1/Session 1 Discussion group: Characterization of the town. Prioritization of the problems.Forum 2/Session 2 Discussion group: Reflection and negotiation of intervention strategies.	At the end of the diagnostic phase and beginning of phase 2 (design of indicators): citizens contribute to the consensual definition of indicators and action strategies from the results obtained in this initial diagnosis.
Letter to municipal political representatives	- Extract the perceptions and opinions of the child population about the environmental problems of their municipality. - Promote the participation of the child population in the development of this model of participatory local management. - Raise awareness among the youngest population in care and respect for the environment. - Involve the education system in municipal management processes.	Taking advantage of the Christmas season, an activity has been developed, with the elementary courses, entitled "Letter to municipal representatives" to reflect the situation of the municipality from the point of view of children.  CARTA A LOS REYES MAGOS	- Diagnostic phase: strategy linked to the participation plan, and the communication plan. Aimed at children.
SWOT TECHNIQUE‰ (Strengths—Weaknesses—Opportunities—Threats)	- Agree and negotiate problems and solutions. - Favour a process of collective reflection. - Triangulate the information collected with the different techniques and according to different population sectors. - Promote the establishment of socio-environmental indicators. - Reference for the action plan.	First phase: problems are reorganized into weaknesses, threats, strengths, and opportunities considering the internal level and external elements. Second phase: the data are crossed and the proposals and action strategies are elaborated. Immediate actions are prioritized and established.	Transversal action: diagnostic phase, criteria and indicators design phase, and action plan design phase: the research team with a heterogeneous work group formed by process participants identifies these elements to implement the action plan.

These information collection instruments have enabled a diagnosis and characterization of the municipality through the strategic environmental evaluation.

The application of these instruments has favored citizen participation and participatory decision making, which has allowed us to draw a real image of the municipality evaluated and, therefore, from the work groups created, to be able to profile in a participatory way, strategic lines for the improvement of sustainable local management.

One of the great criticisms that have been made to all previous models of environmental diagnosis (L21A, Environmental Education Strategies, National, or Local Strategic Sustainability Plans) has been ignorance or non-consideration of social aspects and fragmentation of all this information. The advantage of this model is that it provides an integrated diagnostic model that rotates in the structure from the bottom-up, as mentioned above.

3.3. Context, Sample and Agents

The study was carried out in a municipality in the metropolitan area of the city of Granada (South Spain, Europe), located 7 km from the capital, considered to be a "dormitory town" (linked to work in the capital), with approximately 20,000 inhabitants, and whose main economic sources are agriculture and the service sector (Figure 7).

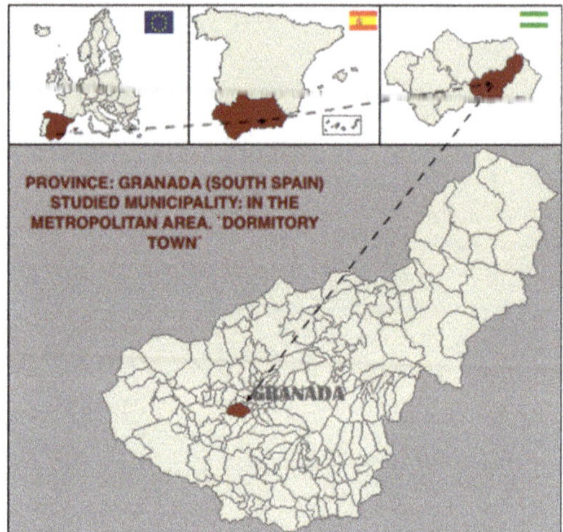

Figure 7. Territorial map of the municipality studied.

The municipality is distributed in six population centers and is governed by a conservative political group, but one that promotes the design of policies and strategic plans for local development.

The City Council has different strategic plans, such as 'municipal action plan', 'action plan for sustainable energy'; 'first local health plan', 'social tourism program'; 'economic-financial plan'; and 'programs for equality', for 'youth', among others.

On the other hand, the transparency portal, the citizen service, the consumer information office, and the youth information center, among other services, promote the development of these policies based on more participatory and bidirectional work proposals between political organizations and civil organizations.

This study generated a process of citizen reflection where all the key agents of the municipality have played a leading role as informants who have their own requirements. All citizens were involved, diversifying the sample as shown in the following tables (Table 3).

3.4. Analysis Procedure

The analyses carried out on the information collected were of a different nature depending on the technique used. We followed a mixed methodology to analyze quantitative and qualitative information with software for data analysis and treatment; that is, SPSS v.23 for the analysis of quantitative data, and Nudist Vivo v.10 for qualitative data.

With regard to the validity of the questionnaire, we highlight that the analysis of different documents related to the subject and other instruments used in previous studies and the consultation of a group of experts has allowed us to guarantee the validity of content; we have ensured the construct validity through a factorial analysis, and the criterial validity through the correlation of all the items of each of the blocks involved with the total of each of them (with the exception of itself), having, for the majority, obtained Pearson's product-moment correlation coefficients, statistically significant at alpha levels of 0.01 and, to a lesser extent, at 0.05.

Table 3. Sample Total (N = 966).

Citizen Opinion Questionnaire (Sub-Total N = 507)					
Study level	No studies	5	Current activity	Student	424
	Primary studies	51		Employee	54
	Secondary studies	432		Unemployed	14
	University studies	19		Others	15
Age	Less than 15	193	Professional activity	Student	421
	15–25	239		Services and culture	25
	26–35	24		Housewife	24
	36–45	34		Industry	14
	46–55	10		Retired	7
	56–65	3		Official	9
	More than 65	4		Others	5
Gender	Women	257		Non-official executives	2
	Men	250			
Monitoring Commission (Sub-Total N = 14)					
1 representative of farmers and livestock.			1 representative of the women's associations.		
1 representative of sports and cultural associations.			1 representative of shopkeeper associations.		
1 representative of youths			1 representative of retirees and pensioners.		
1 representative of the neighbourhood associations.			1 representative of the parents' associations.		
1 representative of environmental associations.			1 representative of teacher of the educational centres.		
1 repres. with recognized prestige in environment/university or research institute.					
3 repres. of the political groups with representation in the town council					
Discussion Groups (Subtotal N = 30)					
Concillors: 4	Technicians: 5	Farmers: 6	Women Association: 6	Youth: 9	
Forums 1 & 2 (Subtotal N = 49)					
Forum 1 (N = 23)			Forum 2 (N = 26)		

- Political groups women group: housewives and working women.
- School community: teachers.
- Representatives of associations.
- Representatives "media": radio and photography.
- Administration technicians: sociocultural animator, woman informant.
- Environmental volunteer representatives.
- Youth group.

Letter to Municipal Political Representatives (Primary education students) (Subtotal N = 366)
TOTAL SAMPLE N = 966

For the calculation of the reliability of this questionnaire, we used the internal consistency procedure. The results achieved in Cronbach's alpha per instrument and thematic blocks are satisfactory [64], ranging from 0.70 to 0.86, as shown in the following table (Table 4).

Table 4. Analysis of the reliability of the citizen opinion questionnaire.

Instrument	Value α de Cronbach	Elements/Subitems	Sections of Items
Citizen opinion questionnaire	0.79	21	Environmental problems (item 1)
	0.73	22	Things that the residents of the town do (item 2)
	0.77	8	Responsibility of social groups (item 3)
	0.88	24	Current situation of the municipality (item 4)
	0.80	9	What can improve participatory local environmental management in your town (item 7)
	0.70	8	Sector to be developed with this management model (item 8)
	0.70	5	Global claims (item 10)

Another indicator that supports this consideration is the presence of reliability coefficients of little or no gain, if not some loss, when we have eliminated, one-by-one and in various rounds, each of the items that made up each thematic block.

Regarding the qualitative information (discussion groups, citizen participation forums, letter to municipal political representatives, monitoring commission), we based our analysis on the four quality criteria that need to be considered in the analysis of qualitative information (credibility, applicability, consistency, and neutrality) [65,66]. (i) Credibility: during the analysis process, conversations were held with participants in the study to corroborate the interpretations made based on their answers. (ii) Applicability or transferability: the study was carried out in only one municipality of the province of Granada, but instruments and results obtained can be applied in other contexts with similar characteristics. (iii) Consistency: we consider that similar results would be obtained if the study was to be replicated in other municipalities because the analysis was carried out in a meticulous way from a process of triangulation of sources and techniques. (iv) Neutrality: the detailed description of the research process carried out indicated in this article shows that it was a neutral and non-biased process.

4. Results

Next, we broadly present the most relevant results achieved after an analysis of the information collected through the different instruments used in the diagnostic phase for the indicator 'Citizen perception of environmental and social issues. Prioritization', with the dual purpose of (1) defining the socio-environmental problems of the municipality from the citizen's perception; and (2) establishing lines of action that allow us to justify the validation of the model presented in this article as lines of action and as a proposal for the development of sustainable management at a local level.

4.1. Citizen Opinion Questionnaire

To identify the socio-environmental problems from the perception of citizenship, a factorial analysis (the type of factor analysis calculated is exploratory; the extraction method used was that of main components and the rotation method, varimax with Kaiser normalization, that is, eliminating components with a percentage of explained variance under 1% (<1)) of the answers given in the questionnaire was undertaken in order to identify response patterns, or whether these are related across common dimensions.

Through the analysis, we were able to identify five factors that, together, explain 48.26% of the total variance, with a first factor that explains 10.94% of it, and the rest that range between 7.96% and 10.75% of variance explained. The values achieved by the communalities are between 0.14 and 0.67, and indicate the acceptable representation that the items included in the scale have acquired.

Finally, Bartlett's sphericity test, with a value of 164.70 and a $p = 0.000$, and the KMO (Kaiser–Meyer–Olkin) sample adequacy measure, with a value of 0.814, allow us to state that the correlation matrix is not an identity matrix [67]. Therefore, there are a number of significant high inter-correlations, as the value found in the Bartlett test is significantly high [68]. This, together with the value obtained in the KMO test, a meritorious value [69], and the value obtained by the determinant of the correlation matrix (R = 0.016) indicates that the data matrix is suitable for the factor analysis (Tables 5–7).

Table 5. Kaiser–Meyer–Olkin (KMO) and Barlett test.

KMO and Barlett Test (Question:Environment Problems)		
Kaiser–Meyer–Olkin sample adequacy measure		0.814
Bartlett's sphericity test	Chi-square approximate	164.70
	gl	210
	Sig.	0.00

Table 6. Λ total and % variance.

Λ Total	2.29	2.25	2.02	1.87	1.67	
% variance	10.94	10.75	9.64	8.94	7.96	48.24%
% accumulated	10.94	21.69	31.33	40.28	48.24	

Table 7. Extraction method: analysis of main components. Rotation method: varimax normalization with Kaiser.

Matrix of Rotated Components (a) "Environmental Issues"	Components 1	2	3	4	5	Communalities
FACTOR 1. Environmental-context problem						
Lack of care and cleanliness of the environment	0.73					0.61
Pollution rivers and vegetation and forest areas	0.79					0.67
Loss of landscape and agricultural land	0.44					0.38
Discharge of illegal taste on the outskirts of the municipality	0.63					0.45
FACTOR 2. Labour problem						
Lack of stable work		0.77				0.61
Jobs that require low training and qualification		0.57				0.43
Low salaries		0.71				0.57
High number of unemployed		0.69				0.61
FACTOR 3. Executive-legislative problem						
Lack of communication between municipal political representatives: put political interests before social needs			0.63			0.60
Poor coordination between town council technicians			0.67			0.59
Urban growth			0.54			0.37
Lack of urban planning			0.63			0.52
FACTOR 4. Normative-educational problem						
Lack of green areas				0.32		0.23
Lack of awareness towards environmental problems				0.73		0.56
Lack of constant training that makes people care for and respect their environment				0.72		0.56
Weak legislation in environment that allows the guilty get through 'in good shape'				0.51		0.36
FACTOR 5. Technical-environmental problem						
Recycling waste					0.55	0.39
Existence of very loud and annoying noises					0.64	0.54
The passage of so many vehicles through the town center					0.58	0.41
Lack of bins and containers					0.44	0.43
Misuse of containers and bins					0.26	0.14

4.2. Monitoring Commission

The monitoring commission was intended to be the participatory body that guided and evaluated the process of diagnosis and implementation of the local environmental management model. This entails the presence of a relevant social representation to ensure that most of possible perspectives are present in the process, in order to approach a vision as integral and real as possible.

The following table shows the decisions made by this participation body (Table 8 and Figure 8).

Table 8. Decisions achieved by the monitoring commission.

Strategy	Decisions Achieved
Creation of a newsletter	• Writing and approval of the relevant contents about the municipality in environmental, social, and economic issues
Logo design that identifies this local management model	• Drawing competition proposed by the commission and addressed to all elementary students (first and second year of primary school) • Approval and definition of the final logo
Website	• Approval of the contents to be disseminated on the website • Design: technical team of the local corporation • The page offers the possibility for the population to participate through forums and virtual surveys on social, economic, and environmental issues and to know the actions that are being carried out in local management

Figure 8. Logo referring to the sustainable local management model of the municipality. Approved by the monitoring commission and designed by ten-year-old children, "world without rubbish".

4.3. Discussion Groups

The four discussion groups included councillors and technicians, farmers, women, and young people. The results obtained make visible the problems detected by the participants across the different socio-environmental areas according to the importance given to each one (Tables 9–12).

Table 9. Socio-environmental problems extracted from the discussion group 'councillors and technicians'.

Discussion Group: Councillors and Technicians		
Importance Level	**Area**	**Problems**
Very important	Recycling and selective collection of rubbish. Container use	• 'Lack of containers. Selective collection' • 'Uncontrolled focus of all types of waste, rubbish and all types of packaging' • 'Lack of citizen awareness in the generation of waste and deposit. Respect for the collection of equipment and debris' • 'Lack of network of clean points'
	Optimization and expansion of green areas	• 'Lack of green areas and parks' • 'Low maintenance of green areas'
	Noise	• 'Noisy' • 'Urban centre loaded with vehicles and, consequently, with smoke' • 'Noise pollution in the urban centre'
	Water Quality	• 'Poor water quality' • 'Existence of sanitation discharges to irrigation ditches' • 'Poor citizen awareness in the use of water' • 'Absence of wastewater treatment plant'
	Citizen awareness	• 'Lack of citizen awareness in environmental matters' • 'Respect for street furniture'

Table 9. *Cont.*

Discussion Group: Councillors and Technicians		
Importance Level	**Area**	**Problems**
Important	Development of the women' sector	• 'Lack of work initiatives for women' • 'Shortage of resources that favour the insertion of women into the job market'.
	Town planning	• 'Uncontrolled housing growth'
	Space adaptation	• 'Few public parking' • 'Existence of architectural barriers' • 'Existence of industry within the urban area'
	Citizen security	• 'Unsafe entrance to schools. Matching vehicles and pedestrians'
	Cleaning	• 'Unclean streets and public areas' • 'Lack of citizen awareness and respect for the cleanliness of the town'
	Population density	• 'High Population density in the urban area'
	Sector involvement	• 'Difficulty in developing actions where all sectors are involved'
	Training and employment	• 'Few resources for training and employment'
	Health	• 'Health Services Deficiency'

Table 10. Socio-environmental problems extracted from the discussion group 'farmers'.

Discussion Group: Farmers		
Importance Level	**Area**	**Problems**
Very important	Costs	• 'High labour cost with respect to the product price' • 'Land cost' • 'Products price' • 'High cost of phytosanitary products' • 'Renewal of planting products (monocultures)' • 'Expensive labour in relation to the price for which the collected product is sold'
	Water	• 'Irrigation, wastewater' • 'Wastewater' • 'Channeling of ditches, roads' • 'Water of the swamp' • 'Old ditches' • 'In winter there is plenty of water with rain and in summer it is missing'
	The product	• 'Low value of corn at this time' • 'There are no alternative fruits for this type of agricultural land' • 'Low tobacco prices' • 'Regarding the cultivation of olive trees, it is difficult because there is dry land'
	The job	• 'Aging of the sector' • 'Renewal difficulty' • 'Delay in machinery in general (methods, machines, systems...)'
	Administration support	• 'Support for farmers with tobacco companies' • 'More support for cooperatives to expedite subsidies. High administrative requirements'
Important	External variables	• 'The brick factories that harm smoke and dust' • 'Ways of the valley in very bad conditions'

Table 11. Socio-environmental problems extracted from the discussion group 'women'.

Discussion Group: Women		
Importance Level	**Area**	**Problems**
Very important	Citizen awareness	• 'Awareness' • 'Respect/Education' • 'Indifference of people' • 'Lack of citizen collaboration' • 'Lack of mutual respect between groups'
	Citizen security	• 'Surveillance service' • 'We cannot walk quietly at certain times through the streets'
	Development of the women's sector	• 'Municipal nursery'
	Recycling and selective collection of rubbish. Containers use	• 'Container service' • 'There is no good waste collection plan'
Important	Optimization and expansion of green areas	• 'Park care'
	Town planning	• 'Street arrangement'
	Cleaning	• 'You cannot walk the street or square without fear of cuts and infections' • 'Bad smells, infections, poor vision of the town'
	Economy	• 'Limited family and economic well-being' • 'Lowering of the general economy'

Table 12. Socio-environmental problems extracted from the discussion group 'youth'.

Discussion Group: Youth		
Importance Level	**Area**	**Problems**
Very important	Services	• 'Lack of leisure equipment such as swimming pools' • 'Public transport and traffic deficit' • 'Social and community services deficit'
	Citizen security	• 'Insecurity' • 'That the mayor listen to the young people, the local police are not in the places where this insecurity is suffered'
	Culture and education	• 'A space of cultural encounter' • 'Promotion of cultural in general, in the town' • 'Promotion of the culture of the town abroad' • 'More resources for the library' • 'Specific activities for young people' • 'Training in topics such as indiscipline and classroom conflicts'
Important	Optimization and expansion of green areas	• 'Lack of green areas'
	Water quality	• 'Poor water quality'

4.4. Citizen Participation Forum

The problems detected in the different areas and their importance were also expressed by the citizens participating in the first citizen participation forum (Table 13).

Table 13. Socio-environmental problems extracted from the citizen participation forum.

Citizen Participation Forum		
Importance Level	**Area**	**Problems**
Very important	Recycling: use of bins	- Lack of containers - Recycling - Separate rubbish collection plan - Use of bins - Citizen awareness
	Existence of noise	- Motor vehicle noise - Acoustic pollution - Atmospheric pollution
Important	Lack of green areas and natural environment	- Increase of green areas - Loss of natural spaces - Protection of plant species - Improvement and conditioning of existing gardens and green areas - Citizen awareness
	Education and environmental awareness	- Civic education - Environmental education for different population sectors - Lack of citizen awareness - Disrespect for the environment
Less important	Care, cleanliness, and respect for the environment	- Cleaning the environment - Street arrangement - Lack of sanitation - Dirt, aesthetic conservation of the municipality - Citizen awareness for the respect and care of the environment
	Waste	- Uncontrolled landfills
	Others	- Poor water quality - Residual collectors - Lack of public spaces - Stock of electric towers in the urban area

4.5. Letters to the Council Representatives

Children and adolescents also participated in this process through letters addressed to municipal representatives. An analysis of the 366 letters written allowed us to prioritize the needs identified by these groups, as shown in the following table (Table 14).

Table 14. Socio-environmental problems extracted from the letter to municipality political representatives.

Letter to Municipal Political Representatives				
Importance Level	**Area**	**Problems**	**No. Passages**	**%**
Very important	Services	(A) Public Services	577	40%
		1. Infrastructures and equipment	443	
		2. Quality and improvement of services	114	
		3. Social services	12	
		4. Citizen security	8	
		(B) Private Services	169	
		TOTAL	746	
	Leisure	(A) Equipment	374	21%
		(B) Activities	16	
		TOTAL	390	
Important	Environment	(A) Pollution and cleaning	124	13%
		(B) Recycling	76	
		(C) Traffic	39	
		(D) Water	17	
		TOTAL	256	
	Town planning	(A) Job	179	12%
		(B) Living place	57	
		TOTAL	236	
	Civic education	(A) Pro-social behaviors	67	8%
		(B) Pro-environmental behaviors	49	
		(C) Pro-social attitudes	42	
		TOTAL	158	
Less important	Natural environment and green areas	TOTAL	102	5%
	Employment and job stability	TOTAL	8	0.8%
	Cultural heritage	TOTAL	7	0.2%

5. Discussion

Carrying out the socio-environmental diagnosis of a municipality from the citizen's perception held by the different agents of the community requires a process of the triangulation of information, as well as synthesis and prioritization. Thus, the SWOT ((Strengths—Weaknesses—Opportunities—Threats)) analysis and the triangulation of the information collected in the diagnostic phase through the various participatory instruments used allowed us to identify the most urgent and greatest priority needs, as shown in Figure 9.

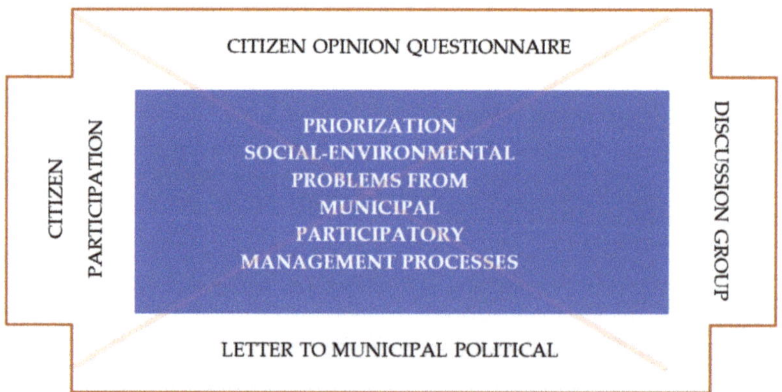

Figure 9. Triangulation process of environmental problems.

The problems addressed by the population living in this municipality through the different participation processes provided us with a generic vision of the environmental and social situations of the municipality studied, as well as its limitations, which will enable us to subsequently develop action strategies that minimize existing needs.

Within each of the areas in which the needs and problems detected in the municipalities are grouped, there are first-order problems that need to be addressed during the next stage of the implementation of this management model: the action plan.

If we take into account the principles in which 2030A is framed, making that diagnosis and addressing environmental issues in an integral way means to address them from a double perspective: on the one hand, we must analyse the objective data of the reality of the environment (physical–environmental diagnosis) and its associated problems and, on the other, understand the perception and assessment that citizens make of it (participatory diagnosis) [10,11]. From this logic, the problems derived from citizen perception and assessment linked to the SDGs are summarized into the following broad categories (Figure 10).

Finally, and in response to the objectives of the study, we can respond to them by addressing the results from a quadruple approach:

(a) *Methodological reflection on the model*: 1. The design and implementation of the participatory strategies presented in this study enabled us to collect information on the citizen's perception of the environmental and social problems existing in the municipality studied, as well as possible proposals for improvement. 2. Foster the participation of the population in decision-making related to local management in the environmental field. 3. Promote a process of personal and collective self-reflection that favours addressing any doubts and assumptions of responsibilities by the population in the process of developing a participatory local environmental management model endorsed by the principles of 2030A.

These information collection strategies enabled the PAR model because some of these participation platforms were permanently established in the municipality (i.e., monitoring commission) to make a formative evaluation of the actions and to make binding decisions regarding sustainability in the daily local management of the investigated municipality. This commission, in fact, made decisions related to issues to work in the discussion groups, and in the forum, they voted the election of the logo, established the internal regulations of the commission, and decided on the first actions to be carried out based on the report obtained in the diagnostic phase, among others. In the process of conducting a socio-environmental diagnosis for the implementation of this innovative local management model, there are lights that strengthen the process and shadows that weaken it:

(b) *Strengths of the validated management model*: 1. Allows gathering of broad perceptions of the population, facilitating the participation of all citizens in the process. 2. Facilitates initial contact with the population in order to consolidate much more complex structures of citizen participation. 3. Systematizes and structures procedures for collecting information that encourage citizen participation in municipal management. 4. Provides a formula for collective and individual reflection of the citizen in relation to their behaviours and attitudes within the global municipal structure. 5. Facilitates the knowledge of the premises of the 2030A and its application by the neighborhood. 6. Favours the involvement of all those most representative in the municipality.

(c) *Weaknesses*: 1. Difficulty in ensuring the representativeness of the entire population and that the demands described are really those that exist and not just a reflection of individual issues. 2. Political opportunism conceived as occasional strategies of a circumstantial nature that are intended to merely meet political and economic targets from specific subsidies. 3. Risk of becoming decontextualized and discontinuous actions that do not facilitate results in the medium or long-term. 4. Lack of motivation and trust in these types of structures by the population, including political groups and the municipal technical group. 5. Compliance with expectations.

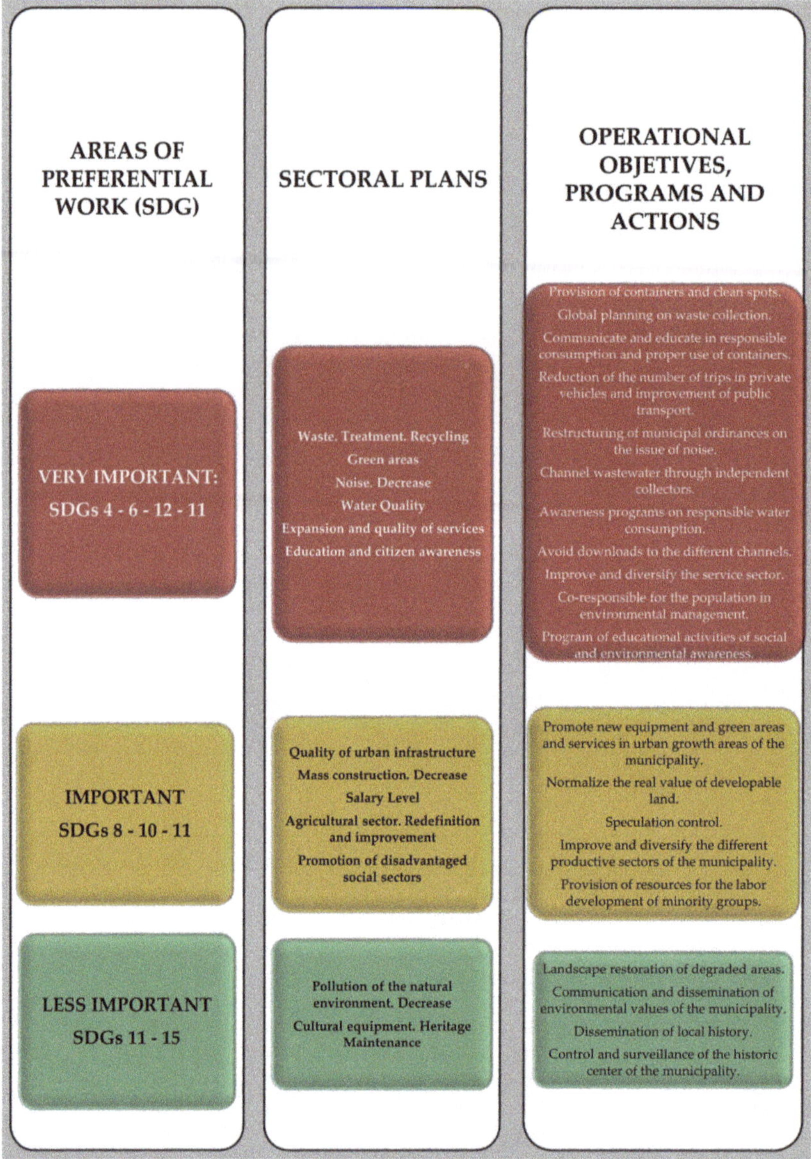

Figure 10. Perceived problems and SDGs. Triangulation (adaptation matrixed participatory process of Garcia-Ayllon, S, 2018 [24]).

(d) *Contribution to Participatory Municipal Management and feasibility of application to other contexts*: Among the contributions of the municipal management model we supported in this study, and which can serve as a reference and be suitable for application in other contexts, we highlight the following: 1. Provides information for contextualized management. 2. Gives ground to the political and local management actions carried out, which enables a relevant degree of success and effectiveness. 3. Introduces the population to innovative processes of citizen participation and local development. 4. Consolidates reflective processes and continuous training in the development of sustainable actions. 5. Promotes the involvement of representative social sectors in the municipality in the decision-making process of municipal management.

In general, therefore, we can say that this study reflected on how the SEE can be applied in sustainable local management. In the document, the first five steps in the application of the SEE model were exemplified, as a previous step to the design and application of the actions. In addition, this model is defended from the logic of the PAR.

Some of the most outstanding strengths that we can point out in light of the results detected are as follows. The monitoring commission was established as a platform for stable participation in the municipality, which has led to the generation of public initiatives (i.e., creation of a peri-urban park, mobility plan). Short-term evaluations and evaluations based on the permeability of government teams were made to the concrete proposals derived from the participatory diagnosis made with all the agents involved. The model advocated in research is a model applied and transferable to other contexts.

There are limitations became visible with this study, such as the shortage of stable citizen participation platforms. Controversies that arise when seeking consensus between political, social and economic representatives; as well as citizen and institutional obstacles to implement the recommendations that arise from participatory processes. The scope of the research was not contemplated in the medium- and long-term impact assessment (only in the short term). All these constitute future lines and research challenges.

6. Conclusions

As conclusions, we can highlight that this research has contributed with key elements for the theoretical and methodological body of the SEE and the ESP, such as the following: (1) The entire process developed in this investigation has had a formative character for the different agents involved, including the municipal government team and opposition political groups. (2) All the agents involved have experienced the dynamics of participatory work in the first person and examples of information gathering instruments that support and enrich the decision-making process taking into account the diversity of interests and opinions expressed in the process. (3) The research has allowed a matrix of criteria and indicators, with which they are working for the improvement plan. These criteria and indicators focus on analyzing the SEE application process and its associated steps, taking into account the adaptation of the case study model (according to the model described in Figure 6) and arising from this process of continuous reflection that was present in all the work. It is also nurtured and based on the conclusive data and the results of this study, taking the principles of 2030A and SDG as a reference.

Taking these research conclusions as a reference and taking into account that, by 2050, 70% of the planet's population will be concentrated in large urban centers, and by 2100, this percentage will reach 85%, great sustainability challenges are staged that involve placing cities and metropolitan areas at the heart of the issue. As has been raised throughout this work, it is a priority to respond to this problem from the decision-making process and from the leadership of the citizens. Application of the principles of 2030A represents an important challenge in addressing these challenges related to the modernization of urban management models. Decisions about human mobility, car traffic, transportation, pollution, urban planning, urban infrastructure planning, collection, treatment and waste management, lighting, tourism, water supply, garden irrigation, maintenance of green spaces, and so on require models of intelligent decision-making in which citizen participation is in the DNA of planning and management [70]. Leaning into participatory strategic action methodologies in management plans means betting on an intelligent, sustainable, and networked government model that favours the harmony between the natural and artificial, which stimulates the balance between social, environmental, economic, and political dimensions, aiming to improve the quality of life from dialogue, reflection, and citizen involvement in the decision-making of local management from a global perspective. The technological instruments at the service of the SmartCity must facilitate a creative participatory management process of decision-making that is informed, consensual, and grounded, which measures itself against taking advantage of the opportunities offered by the digitalization of a large number of processes in which the citizen can contribute and provide relevant information in real-time [71].

Author Contributions: Conceptualization, M.d.F.P.-V. and J.G.-P.; methodology, M.d.F.P.-V., J.G.-P., and M.T.P.-L.; software, M.d.F.P.-V.; validation, J.G.-P. and M.T.P.-L.; formal analysis, M.d.F.P.-V. and J.G.-P.; investigation, M.d.F.P.-V., J.G.-P., and M.T.P.-L.; resources, M.T.P.-L.; data curation, J.G.-P.; writing—original draft preparation, M.T.P.-L.; writing—review and editing, M.d.F.P.-V., J.G.-P. and M.T.P.-L.; visualization, M.d.F.P.-V.; supervision, M.d.F.P.-V.; project administration, J.G.-P.; funding acquisition, M.d.F.P.-V., J.G.-P., and M.T.P.-L. All authors have read and agreed to the published version of the manuscript.

Funding: Project: 'Sustainability in Higher Education: Evaluation of the scope of the 2030 Agenda in curriculum innovation and teacher professional development in Andalusian Universities'. B-SEJ-424-UGR18. Principal researchers: José Gutiérrez-Pérez and María de Fátima Poza-Vilches.

Conflicts of Interest: The authors declare no conflict of interest.

References

1. Acuto, M.; Parnell, S.; Seto, K.C. Building a global urban science. *Nat. Sustain.* **2018**, *1*, 2. [CrossRef]
2. Batty, M. *The New Science of Cities*; The MIT Press: Cambridge, MA, USA, 2013.
3. Bai, X.; Elmqvist, T.; Frantzeskaki, N.; McPhearson, T.; Simon, D.; Maddox, D.; Watkins, M.; Romero-Lankao, P.; Parnell, S.; Griffith, C.; et al. New Integrated Urban Knowledge for the Cities We Want. In *Urban Planet: Knowledge towards Sustainable Cities*; Bai, T., Frantzeskaki, N., Griffith, C., Maddox, D., McPhearson, T., Eds.; Cambridge University Press: Cambridge, MA, USA, 2018; pp. 462–482.
4. Wachsmuth, D.; Cohen, D.A.; Angelo, H. Expand the frontiers of urban sustainability. *Nat. News* **2016**, *536*, 391. [CrossRef] [PubMed]
5. Barnett, C.; Parnell, S. Ideas, Implementation and Indicators: Epistemologies of the post-2015 Urban Agenda. *Environ. Urban.* **2016**, *28*, 87–98. [CrossRef]
6. Elmqvist, T.; Bai, X.; Frantzeskaki, N.; Griffith, C.; Maddox, D.; McPhearson, T.; Parnell, S.; Romero-Lankao, P.; Simon, D.; Watkins, M. *The Urban Planet: Knowledge Towards Sustainable Cities*; Cambridge University Press: Cambridge, MA, USA, 2018.
7. Parnell, S. Globalization and Sustainable Development: At the Urban Crossroad. *Eur. J. Dev. Res.* **2018**, *30*, 169–171. [CrossRef]
8. Smith, M.S.; Cook, C.; Sokona, Y.; Elmqvist, T.; Fukushi, K.; Broadgate, W.; Jarzebski, M.P. Advancing sustainability science for the SDGs. *Sustain. Sci.* **2018**, *13*, 1483–1487. [CrossRef]
9. Nature Sustainability Network. Science and the Future of Cities. *Report of the International Expert Panel on Science and the Future of Cities.* 2018. Available online: http://sites.nationalacademies.org/cs/groups/depssite/documents/webpage/deps_191052.pdf (accessed on 7 January 2020).
10. Dellas, E.; Schreiber, F. Follow-up and Review of the New Urban Agenda. *Plan. Theory Pract.* **2018**, *19*, 133–137.
11. Birch, E. A Midterm Report: Will Habitat III Make a Difference to the World's Urban Development? *J. Am. Plan. Assoc.* **2016**, *82*, 398–411. [CrossRef]
12. Nilsson, M.; Chisholm, E.; Griggs, D.; Howden-Chapman, P.; McCollum, D.; Messerli, P.; Neumann, B.; Stevance, A.-S.; Visbeck, M.; Stafford-Smith, M. Mapping interactions between the sustainable development goals: Lessons learned and ways forward. *Sustain. Sci.* **2018**, *13*, 1489–1503. [CrossRef]
13. Lafortune, G.; Schmidt-Traub, G. SDG Challenges in G20 Countries. *Sustain. Dev. Goals* **2019**, *2*, 219–234.
14. Messerli, P.; Kim, E.M.; Lutz, W.; Moatti, J.-P.; Richardson, K.; Saidam, M.; Smith, D.; Eloundou-Enyegue, P.; Foli, E.; Glassman, A.; et al. Expansion of sustainability science needed for the SDGs. *Nat. Sustain.* **2019**, *2*, 892–894. [CrossRef]
15. Greenwood, D.; Levin, M. *Introduction to Action Research*; SAGE Publications: Thousand Oaks, CA, USA, 2006.
16. Nowotny, H.; Gibbons, M.; Scott, P. *Re-Thinking Science: Knowledge and the Public in an Age of Uncertainty*; Polity: Cambridge, MA, USA, 2001.
17. Hirsch Hadorn, G.; Bradley, D.; Pohl, C.; Rist, S.; Wiesmann, U. Implications of transdisciplinarity for sustainability research. *Ecol. Econ.* **2006**, *60*, 119–128. [CrossRef]
18. Adler, C.; Hadorn, G.H.; Breu, T.; Wiesmann, U.; Pohl, C. Conceptualizing the Transfer of Knowledge across Cases in Transdisciplinary Research. *Sustain. Sci.* **2018**, *13*, 179–190. [CrossRef] [PubMed]

19. Funtowicz, S.; Ravetz, J.K. Post-Normal Science. In *Companion to Environmental Studies*; Castree, N., Hulme, M., Proctor, J.D., Eds.; Routledge: Abingdon, UK, 2018; pp. 443–447.
20. Schneider, F.; Kläy, A.; Zimmermann, A.B.; Buser, T.; Ingalls, M.; Messerli, P. How can science support the 2030 Agenda for Sustainable Development? Four tasks to tackle the normative dimension of sustainability. *Sustain. Sci.* **2019**, *14*, 1593–1604. [CrossRef]
21. Schinas, M. The EU in 2030: A long-term view of Europe in a changing world: Keeping the values, changing the attitudes. *Eur. View* **2012**, *11*, 267–275. [CrossRef]
22. Destro, M. Planificación Estratégica participativa para la calidad y competitividad. **2000**. Unpublished work.
23. Charalabidis, Y.; Theocharopoulou, C. A Participative Method for Prioritizing Smart City Interventions in Medium-Sized Municipalities. *Int. J. Public Adm. Digit. Age* **2019**, *6*, 41–63. [CrossRef]
24. Garcia-Ayllon, S. The Integrated Territorial Investment (ITI) of the Mar Menor as a model for the future in the comprehensive management of enclosed coastal seas. *Ocean Coast. Manag.* **2018**, *166*, 82–97. [CrossRef]
25. García-Ayllón, S.; Tomás, A.; Ródenas, J.L. The Spatial Perspective in Post-Earthquake Evaluation to Improve Mitigation Strategies: Geostatistical Analysis of the Seismic Damage Applied to a Real Case Study. *Appl. Sci.* **2019**, *9*, 3182. [CrossRef]
26. García-Ayllón, S. New Strategies to Improve Co-Management in Enclosed Coastal Seas and Wetlands Subjected to Complex Environments: Socio-Economic Analysis Applied to an International Recovery Success Case Study after an Environmental Crisis. *Sustainability* **2019**, *11*, 1039. [CrossRef]
27. Chanchitpricha, C.; Morrison-Saunders, A.; Bond, A. Investigating the effectiveness of strategic environmental assessment in Thailand. *Impact Assess. Proj. Apprais.* **2019**, *37*, 356–368. [CrossRef]
28. Santos, R.; Coelho, P.; Antunes, P.; Barros, T. Stakeholders Perspectives on the Use of Indicators in Water Resources Planning and Related Strategic Environmental Assessment. *J. Environ. Assess. Policy Manag.* **2019**, *21*, 1950001. [CrossRef]
29. Hutton, C.W.; Nicholls, R.J.; Lázár, A.N.; Chapman, A.; Schaafsma, M.; Salehin, M. Potential Trade-Offs between the Sustainable Development Goals in Coastal Bangladesh. *Sustainability* **2018**, *10*, 1108. [CrossRef]
30. Rodríguez-Olalla, A.; Avilés-Palacios, C. Integrating Sustainability in Organisations: An Activity-Based Sustainability Model. *Sustainability* **2017**, *9*, 1072. [CrossRef]
31. Bowen, K.J.; Cradock-Henry, N.A.; Koch, F.; Patterson, J.; Häyhä, T.; Vogt, J.; Barbi, F. Implementing the "Sustainable Development Goals": Towards addressing three key governance challenges—Collective action, trade-offs, and accountability. *Curr. Opin. Environ. Sustain.* **2017**, *26*, 90–96. [CrossRef]
32. Olsson, L.; Jerneck, A. Social fields and natural systems: Integrating knowledge about society and nature. *Ecol. Soc.* **2018**, *23*, 26. [CrossRef]
33. Sintes, M. A Importancia da Educación Ambiental no Eido Local. In *Iniciativas Municipais de Educación Ambiental en Galicia*; Fernáncez, L., Iglesias, L., Eds.; SGEA-Proyecto Fénix Xunta de Galicia-CEIDA: Coruña, Spain, 2010.
34. Pozo-Llorente, M.T.; Gutiérrez-Pérez, J.; Poza-Vilches, M.F. Local Agenda 21 and Sustainable Development. In *Encyclopedia of Sustainability in Higher Education*; Leal Filho, W., Ed.; Springer: Cham, Switzerland, 2019.
35. Benson, J. Knowledge and communication in democratic politics: Markets, forums and systems. *Political Stud.* **2019**, *67*, 422–439. [CrossRef]
36. Copeland, P. A Trade-Off Analysis of the Normative Values of Deliberative Democracy. Ph.D. Thesis, University of Guelph, Department of Philosophy, Guelph, ON, Canada, 2019.
37. Jarque, M.; Brugué, J. Planes estratégicos locales y redes participativas: Entre el discurso y la práctica. In *Gobiernos Locales y Redes Participativas*; Blanco, I., Gomà, R., Eds.; Ariel: Barcelona, Spain, 2002; pp. 43–63.
38. Poza-Vilches, M.F.; Gutiérrez-Pérez, J.; López-Alcarria, A. Participation and Sustainable Development. In *Encyclopedia of Sustainability in Higher Education*; Leal Filho, W., Ed.; Springer: Cham, Switzerland, 2019.
39. Rosanvallon, P. *La Contrademocracia. La Política en la era de la Desconfianza*; Ediciones Manantial SRL: Buenos Aires, Argentina, 2007.
40. Bobbio, N. *El Futuro de la Democracia*; Fondo de Cultura Económica: México City, Mexico, 1986.
41. Cardona-Zuleta, L.M.; Roll, D. Participación, partidos y liderazgo político. Entre la desafección y la esperanza. *Forum. Rev. Dep. Ciencia Política* **2019**, *16*, 7–12.
42. Cheresky, I. *Poder Presidencial, Opinión Pública y Exclusión Social*; Ediciones Manantial SRL: Buenos Aires, Argentina, 2008.

43. Müller, U.; Lude, A.; Hancock, D.R. Leading Schools towards Sustainability. Fields of Action and Management Strategies for Principals. *Preprints* **2019**. [CrossRef]
44. Trilaksono, T.; Purusottama, A.; Misbach, I.H.; Prasetya, I.H. Leadership Change Design: A Professional Learning Community (PLC) Project in Eastern Indonesia. *Int. J. Eval. Res. Educ.* **2019**, *8*, 47–56. [CrossRef]
45. Rüeg-Stürm, J. *The New St. Gallen Management Model. Basic Categories of an Approach to Integrated Management*; Springer: London, UK, 2004. [CrossRef]
46. Font, J. Participación Ciudadana: Una Panorámica de Nuevos Mecanismos Participativos. In *Papers de la Fundació*; N. 128; Fundació Rafael Campalans: Barcelona, Spain, 2001.
47. Gutiérrez, J. El proceso de investigación cualitativa desde el enfoque interpretativo y de la investigación acción. In *Modelos de Análisis de la Investigación Educativa*; En Buendía, L., González, D., Gutiérrez, J., Pegalajar, M., Eds.; Alfar: Sevilla, Spain, 1999; pp. 7–60.
48. Thérivel, R.; Partidário, M.R. *The Practice of Strategic Environmental Assessment*; Earthscan Publications: London, UK, 1996.
49. Thérivel, R. *Strategic Environmental Assessment in Action*; Routledge: London, UK, 2012.
50. Clark, R. Making EIA count in decision-making. In *Perspectives on Strategic Environmental Assessment*; Partidário, M.R., Clark, R., Eds.; Lewis Publishers: Boca Raton, FL, USA, 2000; pp. 15–28.
51. Oñate, J.J.; Jerez, D.P.; Cardona, F.S.; Mesa, J.C.; Rodriguez, J.J. *Evaluación Ambiental Estratégica. La Evaluación Ambiental de Políticas, Planes y Programas*; Ediciones Mundi—Prensa: Madrid, Spain, 2002.
52. Wu, Y.-Y.; Ma, H.-W. Challenges for Integrating Strategic Environmental Assessment to Enhance Environmental Thinking: A Case Study of Taiwan Energy Policy. *Sustainability* **2019**, *11*, 609. [CrossRef]
53. Simon, D.; Arfvidsson, H.; Anand, G.; Bazaz, A.; Fenna, G.; Foster, K.; Jain, J.; Hansson, S.; Marix Evans, L.; Moodley, N.; et al. Developing and testing the Urban Sustainable Development Goal's targets and indicators—A five-city study. *Environ. Urban.* **2016**, *28*, 49–63. [CrossRef]
54. Pohl, C. From science to policy through transdisciplinary research. *Environ. Sci. Policy* **2008**, *11*, 46–53. [CrossRef]
55. Fraser, E.D.G.; Dougill, A.J.; Mabee, W.E.; Reed, M.; McAlpine, P. Botton-up or Top-down: Analysis of participatory processes for sustainable indicator identification as a pathway to community empowerment and sustainable environmental management. *J. Environ. Manag.* **2006**, *78*, 114–127. [CrossRef] [PubMed]
56. Knapp, C.N.; Reid, R.S.; Fernández-Giménez, M.E.; Klein, J.A.; Galvin, K.A. Placing Transdisciplinarity in Context: A Review of Approaches to Connect Scholars, Society and Action. *Sustainability* **2019**, *11*, 4899. [CrossRef]
57. Termeer, C.; Dewulf, A.; Biesbroek, R. A critical assessment of the wicked problem concept: Relevance and usefulness for policy science and practice. *Policy Soc.* **2019**, *38*, 167–179. [CrossRef]
58. Head, B.W. Forty years of wicked problems literature. *Policy Soc.* **2019**, *38*, 180–197. [CrossRef]
59. Gibbons, M. *The New Production of Knowledge: The Dynamics of Science and Research in Contemporary Societies*; SAGE: London, UK, 2010.
60. Wallerstein, N. *Community-Based Participatory Research for Health: Advancing Social and Health Equity*, 3rd ed.; John Wiley & Sons: Hoboken, NJ, USA, 2017.
61. Binet, A.; Vedette, G.; Leigh, C.; Mariana, A. Designing and Facilitating Collaborative Research Design and Data Analysis Workshops: Lessons Learned in the Healthy Neighborhoods Study. *Int. J. Environ. Res. Public Health* **2019**, *16*, 324. [CrossRef]
62. Oble, B.F.; Storey, K. Towards a structured approach to strategic environmental assessment. *J. Environ. Assess. Policy Manag.* **2001**, *3*, 483–508.
63. Gutiérrez, J.; Poza, M.F. Instrumentos de evaluación comunes en las redes municipales europeas: El necesario protagonismo de los indicadores de educación ambiental en la sostenibilidad local. In *Estratexias de Educación Ambiental: Modelos, Experiencias e Indicadores para a Sostenibilidade Local*; Iglesias, L., Ed.; Eixo Atlántico: Vigo, Spain, 2008; pp. 83–128.
64. Rodríguez, C.; Gutiérrez, J.; Pozo, T. *Fundamentos Conceptuales y Desarrollo Práctico con SPSS de las Principales Pruebas de Significación Estadística en el Ámbito Educativo*; Grupo Editorial Universitario: Granada, Spain, 2007.
65. Guba, E.G.; Lincoln, Y.S. *Effective Evaluation: Improving the Usefulness of Evaluation Results through Responsive and Naturalistic Approaches*; Jossey-Bass: San Francisco, CA, USA, 1991.

66. Poza-Vilches, F.; López-Alcarria, A.; Mazuecos-Ciarra, N. A Professional Competences' Diagnosis in Education for Sustainability: A Case Study from the Standpoint of the Education Guidance Service (EGS) in the Spanish Context. *Sustainability* **2019**, *11*, 1568. [CrossRef]
67. García, E.; Gil, J.; Rodríguez, G. *Análisis Factorial*; La Muralla: Madrid, Spain, 2000.
68. Comrey, A.L. A method for removing outliers to improve factor analytic results. *Multivar. Behav. Res.* **1985**, *20*, 273–281. [CrossRef]
69. Bisquerra, R. *Introducción Conceptual al Análisis Multivariable*; PPU: Barcelona, Spain, 1989; Volume II.
70. Shao, Q.; Weng, S.S.; Liou, J.J.H.; Lo, H.W.; Jiang, H. Developing A Sustainable Urban-Environmental Quality Evaluation System in China Based on A Hybrid Model. *Int. J. Environ. Res. Public Health* **2019**, *16*, 1434. [CrossRef] [PubMed]
71. Aldegheishem, A. Success factors of smart cities: A systematic review of literature from 2000–2018. *J. Land Use Mobil. Environ.* **2019**, *12*, 53–64.

International Journal of
Environmental Research and Public Health

Article

Making Paving Stones from Copper Mine Tailings as Aggregates

Elizabeth J. Lam [1,*], Vicente Zetola [2], Yendery Ramírez [1,3], Ítalo L. Montofré [4,5] and Franco Pereira [1]

1 Chemical Engineering Department, Universidad Católica del Norte, Antofagasta CP 1270709, Chile; yendery.ramirez@ucn.cl (Y.R.); franco.pereira@alumnos.ucn.cl (F.P.)
2 Construction Management Department, Universidad Católica del Norte, Antofagasta C.P. 1270709, Chile; vzetola@ucn.cl
3 School of Engineering Sciences, Lappeenranta-Lahti University of Technology, FI-53851 Lappeenranta, Finland
4 Mining Business School, ENM, Universidad Católica del Norte, Antofagasta CP 1270709, Chile; imontofre@ucn.cl
5 Mining and Metallurgical Engineering Department, Universidad Católica del Norte, Antofagasta CP 1270709, Chile
* Correspondence: elam@ucn.cl

Received: 17 February 2020; Accepted: 26 March 2020; Published: 3 April 2020

Abstract: Copper mining, the central axis of Chile's economic development, produces a large number of tailings, which become a potential environmental risk. This study aims to evaluate the mechanical properties resulting from the making of Portland cement mixtures with tailings as aggregates so that they can be eventually used in paving stones for building inactive tailings dams. Tailings coming from two dams at a concentration plant located in Taltal (Chile) were used. Currently, Dam 1 is inactive, while Dam 2 is active. The tailings samples obtained from both dams were granulometrically characterized by sieving. In addition, pH, humidity, Eh, and mineralogical assays (sulfides, oxides, sulfates, carbonates, phosphates, and silicates) were measured. The fines content of the tailings from Dams 1 and 2 with a sieve size of N°200 ASTM were 76.2% and 29.6%, respectively. Therefore, owing to their high percentage of fines, they cannot be as used as concrete aggregates. Aggregates must contain a maximum percentage of fines so that mortars and concrete can meet Chilean standards. In this paper, to comply with a 7% and 15% fines content lower than 0.075 mm, tailings materials were mixed with conventional aggregates containing very little fines. In addition, a reference mixture was made with only tailings aggregates with and without a superplasticizer additive. To measure the mixtures of cement, aggregates, and tailings, bending and compression strength assays were made of the specimens after a 28-day curing, according to the Chilean standard. The results of the study show that the addition of only part of the tailings to the mixture increases bending strength by 26% and compression strength by 180% compared with the reference mortar, with a fines content lower than 0.075 mm in the 7% mixture, thus allowing paving stone manufacture with tailings materials. In addition, it was possible to increase the workability of the reference mixture by using superplasticizers as additives.

Keywords: tailing bricks; copper mine tailings; mine tailing stabilization; mining environmental liabilities

1. Introduction

Chile is the world's largest copper producer, a fact that has brought about a serious environmental impacts associated with risks to the environment and the population. Among these serious risks is the presence of mining environmental liabilities, mainly consisting of tailings dams, some of them active and others inactive. Some of the latter are abandoned. Tailings storage, for decades in some cases, has been associated with a big environmental footprint, considering the space occupied and the time the storage remains. The penetration of physical and chemical substances in the ground poses huge additional challenges to stabilize the ground and avoid mining acid drainage [1].

The mining industry produces a great amount of solid waste such as mining tailings and rock waste, damaging the environment and human health. In addition, the mining industry is responsible for significant environmental damage such as atmospheric contamination by particles, toxic atmospheric emissions, and the exhaustion and contamination of surface and underground water resources [2,3]. Currently, stricter regulations and public awareness are challenging the mining industry to reduce, reuse, and recycle waste via ecological and profitable strategies.

Historically, tailings have been stored in terrestrial systems called dams and reservoirs. To optimize the final disposal conditions, new technologies have been developed to reduce their impact on the environment. These emerging technologies include tailings thickening and filtration and marine disposal, the latter technology being highly controversial and restricted in many countries [4].

Some technologies studied for mining tailings stabilization are in situ paste disposal, coverage systems, cement addition, mixing, and agglomeration. In the case of rock acid drainage treatment, permeable reactive barriers are used. The physical solidification of waste materials improves their engineering properties and chemical stabilization changes the chemical form of heavy metals, modifying their mobility [5]. The leaching of a porous medium is influenced by waste and the chemical composition of the leaching medium, waste engineering and physical properties—that is, hydraulic conductivity, porosity and particle size, and the hydraulic gradient of the waste [6]. Mining tailings of sulfide deposits usually contain high concentrations of toxic elements whose mobility represents a potential environmental risk for surrounding ecosystems [7,8].

Mine tailings may contain high concentrations of sulfides and sulfates, which may generate sulfuric acid owing to the oxidation of sulfide minerals and leaching of associated metals from rocks containing sulfides when they are exposed to air and water. Due to the salinity usually present in tailings, metals such as Cu, Fe, and Mn show high mobility [2].

Particularly, mine tailings in Chile have been historically abandoned without any proper management due to inadequate legislation. Currently, the legislation requires that tailings be physically and chemically stabilized. Mining companies are forced to have a closure plan to ensure tailings landfill stability to protect the surrounding environment and the local population health [2].

Several studies have been conducted to make building bricks from waste materials: Ahmari and Zhang [9] studied the feasibility of using copper mine tailings to produce bricks using geopolymerization technology, without using clay and shale and requiring a lower temperature for kiln firing. Geopolymers are inorganic synthetic polymers made of aluminosilicates made up of minerals with Al_2O_3 and SiO_2 content [10,11]. Among aluminosilicates are feldspar, chlorites, clay minerals, and some types of pozzolana. Geopolymers have become highly interesting as Portland cement substitutes owing to their low CO_2 emission during production, high chemical and thermal resistance, and good mechanical properties at both ambient and extreme temperatures [1], with their use being interesting in the Atacama Region (Chile); Chen et al. [12] utilized hematite tailings with clay and Class F fly ash to produce bricks requiring up to 84% tailings weight; Chou et al. [13] used Class F fly ash to replace part of the clay and shale to produce bricks by the conventional method, with up to 40% fly ash in test runs on a commercial basis, with properties exceeding ASTM specifications; Morchhale et al. [14] produced bricks with copper mine tailings and ordinary Portland cement, by compressing the mixture in a mold. These bricks have a higher compressive strength and lower water absorption when the ordinary Portland cement content increases. Roy et al. [15] used gold mill tailings to produce bricks by mixing them with ordinary Portland cement, black cotton soils, or red soils. The ordinary Portland cement–tailings bricks were cured by submerging them into water, but the soil–tailings bricks were sundried and fired at high temperatures (<950 °C). Shon et al. [16] used stockpiled circulating fluidized bed combustion ash with Type I cement, lime, Class F fly ash, or calcium chloride to manufacture compressed bricks, with a 55.2 compaction pressure, placing the specimens at 23 °C and 100% relative room humidity for one day before air curing at room temperature.

Additionally, several techniques have been used for developing mine tailings bricks, each one with its advantages and disadvantages. Several methods using waste to make bricks require baking in a high-temperature oven or using alkaline solutions for geopolymers. In many applications, geopolymer offers a performance comparable to ordinary Portland cement, with several advantages such as the rapid development of mechanical strength, high acid resistance, expansion related to the reaction without and with the low dioxide content of alkaline silicon, excellent adhesion to aggregates, immobilization of toxic and hazardous materials, and low greenhouse gas emissions [9,17–21]. Curing temperature is an essential factor affecting geopolymerization and, thus, the strength of geopolymer brick specimens. For the copper mine tailings studied by [9], the optimum curing temperature is around 90 °C. Therefore, they still show the disadvantages of high energy consumption and high greenhouse gas emissions [9].

Ahmari and Zhang [9] studied the effects of sodium hydroxide solution concentration, water content, forming pressure, and curing temperature on the physical and mechanical properties of geopolymer bricks made of copper mine tailings. Conventional brick production requires clay, shale, and high kiln firing temperatures (900–1000 °C) [9]. Clay and shale are experiencing a shortage in several parts of the word. Additionally, the quarrying operation required to obtain clay and shale is energy-intensive, releasing a high volume of waste and impacting on the landscape [9].

Chen et al. [12] made building bricks using secondary resources such as 84% hematite tailings with clay and fly ash to improve the mechanic strength of the bricks and reduce bulk density, by mixing, forming, drying, and firing (980–1030 °C for 2 h). The main disadvantage of hematite tailings for making bricks is the presence of excess iron and little silica, with compressive strength decreasing with the addition of fly ash [12].

Numerous studies reveal the significant influence of the physical (specific gravity, particle shape, size, texture, size distribution, porosity) and chemical (mineralogy, active clay content) properties of a filler on primary pavement distresses (rutting, fatigue, low-temperature cracking, aging, and moisture susceptibility) [22–26]. Choudhary et al. [22] investigated waste materials as fillers in a graded bituminous macadam mix, i.e., copper tailings, carbide lime, brick dust, rice straw ash, red mud, limestone dust, and glass powder which, when mixed with fines, had a superior stiffness and cracking resistance, while mixtures with calcium predominance presented superior adhesion and moisture resistance. Fines filler particles influence the physical (porosity, size distribution, texture, size, particle shape, specific gravity) and chemical (active clay content, mineralogy) properties of primary pavement distresses [22]. The use of wastes as fillers can limit the consumption of conventional fillers, providing an effective solution for waste disposal [22].

A cemented tailings backfill or cemented paste backfill mixture is used as a supporting structure in underground mining operations, where the main components are mine tailings and binder [27]. Xu et al. [27] analyzed the tailings effect on the solid content, the type of cement reagent, and the effect of the binder proportion on unconfined compression strength and microstructure evolution.

A higher solid content increases the compactness of mine tailings; mechanical strength is proportional to the amount of binder content, with the pore and hydrated material distribution influencing solid content, binder type, and curing time (Xu et al., 2018) [27].

Nehdi and Tariq [6] studied Class C fly ash and cement kiln sulfate-resistant cement in the process of the stabilization–solidification of polymetallic sulfidic mine tailings, analyzing the leaching properties, hydraulic conductivity, and compression strength. Sulfidic tailings waste disposal is site-specific and involves a mine tailings facility, waste characterization, and site characterization and selection [6].

According to the National Geology and Mining Service (SERNAGEOMIN, for its acronym in Spanish), there are 740 tailings, 469 of which are inactive and 170 abandoned, the latter being the greatest threat due to their physical and chemical instability. Apart from this, mining producers in some Chilean regions create tailings via small mineral processing plants, which are deposited on unsafe sites owing to faulty legislation. Historically, most accidents and disasters in tailings are connected to or caused by physical stability failure, a fact that turns into a potential risk in Chile because it is a highly seismic country. One way of reducing instability in these inactive and abandoned tailings is utilizing them as inputs to make mixtures for building materials, thus encapsulating the heavy metals present in the tailings.

To reduce the negative impact on the environment and use the secondary resource of copper mine tailings, this study provides an attractive method for recycling mine tailings into building materials and creating zero-emission tailings, which could be helpful for the production of construction bricks using green raw materials and industrial waste recycling [12,22]. Another way is to install a tailings surface coverage system, thus allowing us to decrease water infiltration and, therefore preventing the generation of acids and sulfide oxidation. This coverage system could be made with the mixtures obtained in this study, as shown in Figure 1.

Figure 1. Scheme for using paving stones. The figure on the left shows a graphic. The figure on the right shows the proposed tailings surface coverage system at Andacollo commune.

The natural decrease in mineral content in the dams makes it necessary to process increasingly greater volumes and, therefore, produce bigger amounts of waste for the same production level. So, it is necessary to find methodologies to reduce risks associated with the presence of these wastes, some of which are being found in urban areas, as is the case in Andacollo in Chile [28]. This commune has nine big waste deposits in its urban section, which could destabilize at any moment since they are exposed to natural phenomena such as heavy rains and/or earthquakes, among others. The use of different waste materials as fillers instead of conventional material in densely graded bituminous macadam mixtures such as copper tailings could contribute to reducing the amount of Portland cement required [22].

The objective of this study is to develop a Portland cement-based mixture to be used in paving stones, with the physical and mechanical characteristics needed for its use, by mixing cement, as an agglomerate, with conventional aggregates (gravel and sand), tailings aggregates and water. Aggregates must consist of copper mine tailings in order to reuse them and reduce their impact on the environment. Additionally, this process to encapsulate tailings in concrete mixtures allows us to immobilize and insolubilize heavy metals present in tailings. Tailings used as inputs allow us to reduce the tailings volumes and the risks associated with them. From an economical viewpoint, tailings can be used as input for making building materials. The manufacture of concrete paving stones allows a reduction in contamination costs and levels. These paving stones may be used for the physical stabilization of the tailings, by using them on the upper part of the tailings. They could also be used in mixtures for making blocks to support their walls or for building tailings dams.

This study contributes a methodology to develop new uses for mine tailings materials, which currently make up a potential environmental risk, by mixing tailings with conventional stone aggregates and examining the possibility of using superplasticizers as additives with tailings aggregates to improve their workability.

2. Materials and Methods

This study focuses on the use of tailings aggregates for making Portland cement mixtures for paving stones from two tailings dams at a copper sulfide concentration plant located in Taltal commune, Antofagasta Region, Chile. Figure 2 shows the location of the mining deposit (MC) and the two tailings dams (D1 and D2) from which four surface samples (two from each dam) were extracted (beach zone), weighing 60 kg each, thus totaling 240 kg of tailings. Dam D1 is inactive, while Dam D2 is active and, therefore, still receives material.

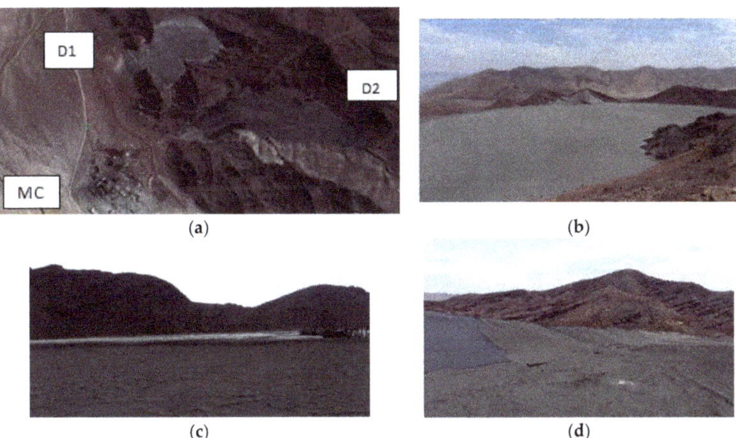

(a) (b)

(c) (d)

Figure 2. (a) Study area, tailings Dam 1 (D1) currently closed; (b) Dam 1 (close up view); (c,d) tailings Dam 2 (D2).

Assays were conducted in the Materials Assay and Research Lab (LIEMUN, for its acronym in Spanish) at Universidad Católica del Norte (UCN) since it has all the equipment required for the study—that is, a mixer, a metal plate compaction table, and a machine for testing mortar compression and bending.

2.1. Analysis Method

2.1.1. Mine Tailings Analysis

Tailings geocharacterization was conducted to obtain the main properties of the tailings and calculate the mixture design. The tailings material was used as an aggregate to make concrete mixtures.

a. Physical analysis

The percentage of tailings humidity was determined and the following assays were conducted, according to Chilean Norm (NCh), specifications: (1) determination by washing the material and passing it through mesh N° 200 (0.075 mm) in mineral aggregates—Standard NCh 1223 [29], (2) analysis by sieving the fines aggregates—standard NCh 165 [30], adapted to tailings conditions, and (3) actual density and absorption, NCh 1239 [31].

Particle size distribution was determined by sieving, according to Chilean standards [29,30], by making the following changes: the material retained in mesh N° 200, after washing, was dried at 110 ± 5 °C for 24 h. The sample was later passed through a set of sieves meeting ASTM international standards. Sieve sizes N° 16, 30, 50, 100, and 200 were used. The granulometric curve was obtained from these data to estimate the tailings–gravel–sand ratio to prepare the mixture.

b. Chemical analysis

To characterize tailings chemically, pH and Eh were measured potentiometrically in triplicate in a 1:9 tailings–distilled water ratio.

c. Mineralogical Analysis

Quantitative evaluation of minerals by scanning electron microscopy (QEMSCAN) was used to analyze mineralogical mine tailings characterization. The results are expressed as a percentage of the total mass, as shown in Table 1. The tailings material was used for making concrete mixtures.

Table 1. Mineralogical tailings characterization.

Group	Mineral	Dam 1 Mass (%)	Dam 2 Mass (%)
Cu sulfides	Covellite	0.00	0.10
	Chalcocite/Digenite	0.002	0.09
Cu/Fe sulfides	Chalcopyrite	0.008	0.21
	Bornite	0.00	0.02
Cu oxidized minerals	Cuprite/Tenorite	0.01	0.03
	Other Cu Minerals	0.11	0.31
Gangue minerals containing Cu	Cu-bearing Phyllosilicates	2.80	2.09
	Cu-bearing Fe Oxide/Hydroxides	0.18	0.31
	Cu-bearing Wad	0.05	0.12
Sulfides	Pyrite	0.07	0.19
Oxides	Magnetite	3.09	8.23
	Hematite	4.50	8.32
	Goethite	0.56	0.65
	Other Fe Oxides/Hydroxides	1.00	0.98
	Ilmenite	0.58	0.46
	Rutile	0.24	0.29
	Corundum	0.00	0.01
Sulfates	Gypsum/Anhydrite/Bassanite	0.41	0.22
	Jarosite	0.01	0.02
	Alunite	0.01	0.02
Carbonates	Calcite/Dolomite	3.05	2.59
Phosphates	Apatite	1.01	0.90
Tectosilicates	Quartz	10.56	15.33
	K-Feldspars (Orthoclase, Anorthoclase)	4.83	4.88
	Ca, Na-Feldspars (Plagioclase Series)	33.67	26.25
Phyllosilicates	Kaolinite Group	0.30	0.46
	Biotite/Phlogopite	2.86	2.04
	Chlorite Group	11.53	7.08
	Muscovite/Sericite	3.26	4.25
	Illite	0.16	0.09
	Smectite Group (Montmorillonite, Nontronite)	3.15	1.60
	Pyrophyllite/Andalusite	0.06	0.10
Others ilicates	Hornblende	6.63	7.53
	Titanite	3.73	2.42
	Epidote	0.48	0.99
Others	Others	1.02	0.83
Total		100	100

2.1.2. Cement

CBB (Bío-Bío cement), a pozzolanic-type ordinary Portland cement was used. This type of cement is used due to its greater resistance to sulfates contained in the concrete.

2.1.3. Conventional Aggregates

Sand and gravel from "La Chimba" sector in Antofagasta were used. The sand is characterized as aeolian and the gravel as alluvial, with a 10 mm maximum size. Aggregates with a low content of fines smaller than mesh N° 200 (0.25%) were selected and mixed with tailings with a high fines content to provide a mixture with bigger particles. Sand and gravel granulometry was determined according to NCh 165 [30].

2.1.4. Water

Tap water was used. It was not tested in the lab because it was considered suitable for concrete production, according to standard NCh 1498 [32].

2.1.5. Superplasticizer Additive

A Sika Viscocrete 5100 polycarboxylate-based superplasticizer additive was used. The superplasticizer is an additive used as a highly effective reducer of the water necessary for making concrete mixtures manageable, thus increasing the initial and final concrete strength.

2.2. Cement Aggregate Mixture Assays

2.2.1. Materials Dosage

A dosage used for ordinary mortars, NCh 158 [33], was used for the mixtures: two parts of cement (weight), six parts of aggregates, and one part of water. The water–cement ratio of the mixture was 0.5. In this study, the aggregates of the reference mixture were substituted by tailings. For this mortar, the water dosage was increased because the previous dosage made the sample dry and unworkable. To prepare a mixture with the ratios previously established and a 0.5 water–cement ratio, superplasticizers were added only to this mixture, with the dosage being 1.5% with respect to the cement weight. The same ratios were used for the dosages of gravel–sand–tailings mixtures, by using a certain quantity of tailings to obtain fines percentages of 7% and 15% from the aggregate mixture with mesh N° 200. Aggregate adjustment due to humidity was considered when preparing the mixture.

2.2.2. Reference Specimens

After characterizing the samples, three reference specimens were made for each dam, using tailings as aggregates. They were made according to Chilean standard NCh158 [32]—that is, 4 × 4 × 16 cm specimens [32]. The same procedure was followed for tailings from both dams.

Table 2 shows the cement, water, and tailings dosages from Dams 1 and 2, used for making three reference specimens. The use of tailings with a big quantity of fines increased the amount of water required to maintain the sample's workability, thus changing the water–cement ratio. To prove that the superplasticizer additive normally employed in concrete could be used in tailings mixtures, an assay was made with 1% additive. The dosage is shown in Table 3.

Table 2. Cement, tailings, and water dosage for making reference specimens using materials from Dams 1 and 2.

Mortar Specimen	Dam 1 Mass (g)	Dam 2 Mass (g)
Cement	500	500
Water	400	365
Dry tailings	1500	1500

Table 3. Cement, tailings, water, and superplasticizer dosage for making reference specimens using materials from Dams 1 and 2.

Mortar Specimen	Dam 1 Mass (g)	Dam 2 Mass (g)
Cement	500	500
Water	250	264
Dry tailings	1500	1500
Superplasticizer	7.5	7.5

The method for specimen conservation required covering them with a non-absorbent plastic material to avoid water evaporation. Then, the mold was put into a wet chamber. The specimens were removed from the molds after 48 hours. The unmolded specimens were smoothly cleaned and classified. Once ready, the specimens were taken to the setting area and, later, to curing chambers for 28 days to ensure cement particle hydration.

Each specimen was subjected to bending and compression strength assays, following Chilean standards [32]. Strength values are expressed in kg cm^{-2}. The bending strength was determined for three specimens and compression strength for the corresponding six specimens. Both strengths, mortar and specimens, correspond to the mean value resulting from all assays.

2.2.3. Gravel-Sand-Tailings Specimens

Specimens were made by using tailings as aggregates. The tailings were mixed with gravel and/or sand, so that the fines (mesh N° 200) in the aggregates incorporated could be 7% and 15%. Only tailings from Dam 2 were used because, according to size distribution, they contain a lower percentage of fine particles smaller than sieve size N° 200 (29.6%), thus making their use as inputs more attractive because the mortars obtained show better mechanical characteristics. The mortars were made in triplicate. Table 4 shows the compositions used in each case.

Table 4. Mortar mix with 7% and 15% fine particles from Dam 2 mine tailings.

Material	Mass (g) 7% Fines	Mass (g) 15% Fines
Water	250	250
Sand	300	0
Cement	500	500
Gravel	825	720
Dam 2	394	820

To prepare 7% fines specimens, the following aggregate ratios were used: 25% tailings, 20% sand, and 55% gravel for a 1500 g total weight. To prepare 15% fines specimens, the following aggregate ratios were used: 52% tailings and 48% gravel for a total weight of 1500 g of aggregates. The ratios were obtained on the basis of tailings fines content below mesh N° 200 and the granulometry of the aggregates combined with tailings, gravel, and sand.

3. Results and Discussions

3.1. Characterization

3.1.1. Physical Characterization

Figure 3 shows the granulometric distribution of tailings. Dam 1 shows a higher ratio of fines particles compared with the samples from Dam 2—that is, 76.22% and 29.63%, respectively. For this reason, mine tailings from Dam 2 represent a better alternative as aggregates in the production of mortar, concrete, and paving stones for construction. As contents exceed standards, ways to reduce them must be found. This fines excess is responsible for significantly increasing the amount of water in the mixtures. The use of superplasticizer additives is one of the options, along with the use of other aggregates with less fines content.

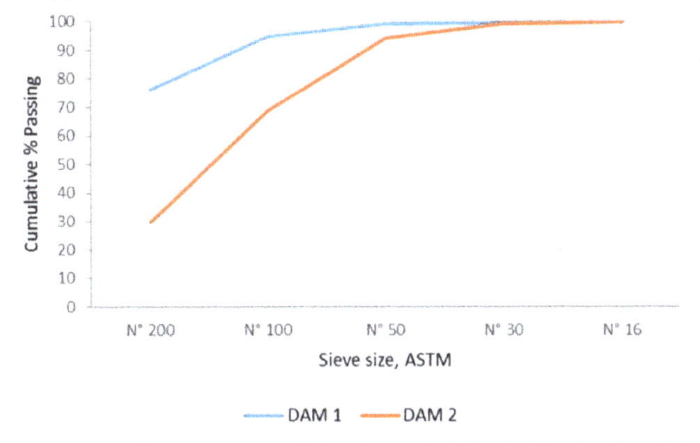

Figure 3. Particle size distribution of copper mine tailings samples.

For ground aggregates, Chilean standards indicate that the fines content must not be greater than 7% for fine aggregates and 1% for coarse aggregates. SERNAGEOMIN indicates that the percentage with sieve size N° 200 must be lower than 20% for the construction of deposit walls.

The real dry density of D1 and D2 tailings is 2420 and 2420 kg m^{-3}, respectively. The water absorption of D1 and D2 is 2.70% and 2.57%, respectively.

Figure 4 shows the granulometry of the sand and gravel used for gravel–sand–tailings mixtures. The actual density (superficially saturated and dried) of the gravel and sand is 2730 and 2720 kg m^{-3}, respectively. The gravel and sand water absorption is 1.43% and 0.53%, respectively. The amount of fines under mesh N° 200 for gravel and sand is 0.23 and 0.25 respectively.

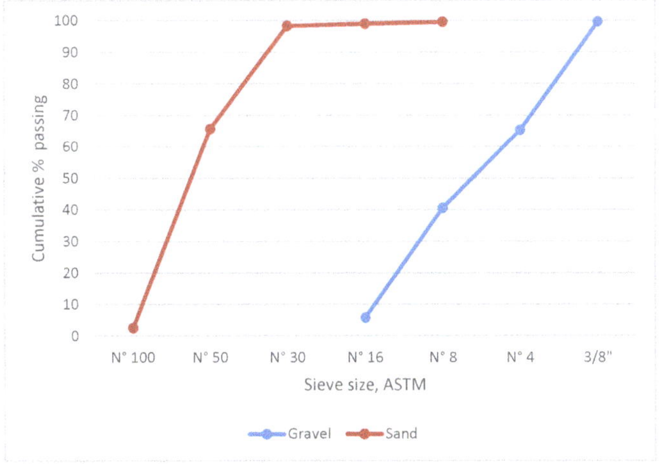

Figure 4. Granulometry of the sand and gravel used for gravel–sand–tailings mixtures.

3.1.2. Chemical Characterization

The results of mine tailings sample measurements are shown in Table 5. The mine tailings pH is almost neutral. This reveals that it should not produce any effects on the mortar or concrete. Under conditions of low pH and high Eh, primary minerals actively oxidize and secondary soluble mineral phases dissolve [33]. D1's average humidity is 9.90% and D2's is 5.02% (kg of water and kg of tailings^{-1}), considering five samples for each dam.

Table 5. Mine tailings sample measurement

Sample	pH	Eh Mvolt	Humidity (% kg_{water} $kg_{tailings}$$^{-1}$)
D1-Sample 1	7.28	−1200	10.69
D1-Sample 2	7.27	−1550	9.12
D2-Sample 1	6.99	−130	2.59
D2-Sample 2	7.05	−70	7.46

3.1.3. Mineralogical Characterization

The mineralogical characterization of tailings was made with QEMSCAN. The most significant mineralogical contents for both dams were tectosilicates, with almost 50% mass, with the highest ratio being provided by Ca, Na feldspars, and quartz. The second most significant groups in abundance were phyllosilicates, the chlorite group contributing the most, with 21.32% and 15.6% for Dams 1 and 2, respectively. These values are quite important because quartz has a high mechanical resistance to impact and is mainly responsible for the mechanical strength of tailings products [12,34].

Sodium feldspar improves resistance to bending and impact; it increases resistance to stress; and provides hardness and durability, two characteristics that make the use of these tailings quite attractive as inputs for making paving stone mixtures.

The amount of sulfates may exceed standards. So, the use of fewer amounts of this aggregate decreases the number of fines and sulfates in the mixture.

3.1.4. Mechanical Characterization

Table 6 shows the mean values of the results obtained from the bending strength assays made on the mortar by using tailings from D1 and D2. Since the sand was thoroughly replaced by tailings, the latter contain a high percentage of fines with sieve size N° 200—that is, 76.2% and 29.6% in tailings Dams 1 and 2, respectively. The aggregate and cement contents remained; however, the amount of water was increased until the proper consistency of the mixture was obtained. The mortar made from D1 tailings required 400 g of water—that is, 150 g greater than suggested by the ideal mixture, namely 60% greater than estimated. Meanwhile, the mortar made from D2 tailings as an aggregate, which contains a smaller amount of extremely fine particles, required 365 g of water—that is, 46% more than conventionally required. The mean strengths were obtained from two different sites in D1 and D2 tailings. The D1 tailings strength in Dams 1 and 2 was 44.37 and 42.46 kg cm^{-2}, respectively, resulting in an average of 43.42 kg cm^{-2} for D1. For the mortar using raw material from D2, the average was 68.70 kg cm^{-2}. These results reveal the influence of the percentage of fines on the bending strength of the mixtures since, in comparing mineralogy, it does not change as much between tailings.

Table 6. Bending strength assay results after 28 days.

Bending	Strength (kN)	Pressure (kg cm^{-2})
D1-1	1.86	44.37
D1-2	1.78	42.46
D2-1	2.79	66.55
D2-2	2.97	70.84

The compression results of the mortars made with both tailings are shown in Table 7. Compression assays showed that D2 tailings are more resistant to bending stress than D1 tailings. The results of the compression assays indicate that the mixture using D2 tailings is more resistant to compression than the mixture made with D1 tailings, a fact that is also attributed to the greater presence of fines in D1 tailings.

Table 7. Compression strength assay results after 28 days.

Compression	Strength (kN)	Pressure (kg cm^{-2})
D1-1	14.81	94.33
D1-2	14.40	91.72
D2-1	27.00	172.24
D2-2	27.00	175.61

The bending and compression results for tailings mortars with a superplasticizer are shown in Tables 8 and 9. By reducing the water content, the strength increases. Nevertheless, the strengths obtained are small for making paving stones [31]. Table 9 shows the deviations for the value accepted in Chile, 500 kg cm^{-2} [35], determined as the ratio between the experimental and theoretical values, divided by the theoretical value multiplied by 100 ($\Delta\varepsilon$).

Table 8. Bending assay results for 7% and 15% fine particles.

Bending	7% Fine Particles		15% Fine Particles	
	Strength (kN)	Pressure (kg cm^{-2})	Strength (kN)	Pressure (kg cm^{-2})
D2-A	4.21	100.42	3.40	81.10
D2-B	3.75	89.45	3.42	81.34
D2-C	2.93	69.89	3.40	81.10

Table 9. Compression assay results for 7% and 15% fine particles.

Compression	7% Fine Particles			15% Fine Particles		
	Strength (kN)	Pressure (kg cm^{-2})	$\Delta\varepsilon$ %	Strength (kN)	Pressure (kg cm^{-2})	$\Delta\varepsilon$ %
D2-A1	73.57	468.72	−6.3	66.58	424.19	−15.2
D2-A2	78.97	503.12	0.6	66.96	426.61	−14.7
D2-B1	77.58	494.27	−1.1	70.19	447.18	−10.6
D2-B2	77.61	494.46	−1.1	70.37	448.33	−10.3
D2-C1	74.67	475.73	−4.9	72.25	460.31	−7.9
D2-C2	74.62	475.41	−4.9	70.64	450.05	−10.0

The mechanical assays conducted on specimens with cement–gravel–sand–tailings mixtures with a fines content smaller than sieve size N° 200, amounting to 7% and 15% in the aggregates, are shown in Table 8 for bending and Table 9 for compression. Results show that bricks made with 7% fines particles are more resistant to bending stress than those made with 15% fines particles, with sieve size N° 200.

The results of the compression assays reveal that bricks made with aggregates containing 7% of fines particles smaller than 0.075 mm show a slightly greater compression strength than mixtures prepared with 15% of fines particles smaller than sieve size N° 200. The results obtained are slightly lower than indicated by [35], where individual strengths higher than 500 kg cm^{-2} are given. In the case of 7% fines the resulting average strength was a relative deviation of only −2.9%, while in the case of 15% fines, this deviation was higher, with an average value of −11.4%. The results in Table 9 correspond to mixtures with a higher docility than those used for making paving stones on an industrial basis. By using mixtures with either a smaller amount of water, a greater cement content or a superplasticizer, the most highly resistant paving stones could be made on an industrial basis.

4. Conclusions

In this study, copper mine tailings were used as aggregates for making cement–tailings and cement–gravel–sand–tailings mixtures to be used in the manufacture of paving stones for construction.

The cement–tailings mixtures did not reach the strengths required for making paving stones. On the other hand, tailings aggregates show fines contents higher than the standards with sieve size N° 200, thus greatly increasing water consumption. It is possible to add superplasticizer additives to the mixtures to reduce the amount of water required; however, although strengths increase when reducing the water/cement ratio, they are not enough to reach recommended values.

In using cement–gravel–sand–tailings mixtures, only material from D2 tailings could be utilized since D1 contained excess fines, thus making D1 ineffective.

The use of fines from 7% aggregates results in strengths 9% higher than using 15% aggregates; however, from an environmental and economic viewpoint, the second option is more attractive since it uses 53% of tailings—that is, more than twice the 7% alternative.

The potential use of tailings as part of building materials such as paving stones and blocks could greatly support physical and chemical tailings stabilization since they may be used as part of tailings walls and horizontal surfaces, thus reusing their waste, according to the circular economy.

The mine tailings from the two dams studied show a slightly neutral pH and contain a small quantity of sulfides. Thus, they do not represent a threat for developing acid mine drainage. The humidity contained in mine tailings is used to reduce water consumption during the production of paving stones from mine tailings.

Author Contributions: Conceptualization, E.J.L., V.Z. and Í.L.M; methodology, E.J.L.; validation, E.J.L., V.Z. and Í.L.M; formal analysis, Y.R.; investigation, E.J.L., V.Z., Y.R., Í.L.M. and F.P.; resources, E.J.L.; data curation, V.Z; writing—original draft preparation, E.J.L.; writing—review and editing, E.J.L.; visualization, E.J.L.; supervision, E.J.L.; project administration, E.J.L.; funding acquisition, V.Z. All authors have read and agreed to the published version of the manuscript.

Funding: This research was funded by "Rapid Implementation Projects-Pilot Scaling", Universidad Católica del Norte, Engineering Project 2030, 16ENI2-71940.

Acknowledgments: The authors are thankful to "Rapid Implementation Projects-Pilot Scaling", Universidad Católica del Norte, Engineering Project 2030, 16ENI2-71940, for the financial support of this project.

Conflicts of Interest: The authors declare no conflict of interest.

References

1. Adiansyah, J.S.; Rosano, M.; Vink, S.; Keir, G. A framework for a sustainable approach to mine tailings management: Disposal strategies. *J. Clean. Prod.* **2015**, *108*, 1050–1062. [CrossRef]
2. Lam, E.J.; Gálvez, M.E.; Cánovas, M.; Montofré, I.L.; Rivero, D.; Faz, A. Evaluation of metal mobility from copper mine tailings in northern Chile. *Environ. Sci. Pollut. Res.* **2016**, *23*, 11901–11915. [CrossRef] [PubMed]
3. Esquenazi, E.J.L.; Norambuena, B.F.K.; Bacigalupo, I.L.M.; Estay, M.E.G. Necessity of intervention policies for tailings identified in the Antofagasta Region, Chile. *Rev. Int. Contam. Ambient.* **2019**, *35*, 515–539. [CrossRef]
4. Lee, M.R.; Correa, J.A. Effects of copper mine tailings disposal on littoral meiofaunal assemblages in the Atacama region of northern Chile. *Mar. Env. Res.* **2005**, *59*, 1–18. [CrossRef]
5. Lam, J.; Cánovas, M.; Gálvez, M.E.; Montofré, I.L.; Keith, B.F.; Faz, A. Evaluation of the phytoremediation potential of native plants growing on a copper mine tailing in northern Chile. *J. Geoch. Expl.* **2017**, *182*, 210–217. [CrossRef]
6. Nehdi, M.; Tariq, A. Evaluation of sulfidic mine tailings solidified/stabilized with cement kiln dust and fly ash to control acid mine drainage, Mining. *Metall. Explor.* **2008**, *25*, 185–198. [CrossRef]
7. Favas, P.J.C.; Pratas, J.; Gomes, M.E.P.; Cala, V. Selective chemical extraction of heavy metals in tailings and soils contaminated by mining activity: Environmental implications. *J. Geochem. Explor.* **2011**. [CrossRef]

8. Neel, C.; Bril, H.; Courtin-Nomade, A.; Dutreuil, J.P. Factors affecting natural development of soil on 35-year-old sulphide-rich mine tailings. *Geoderma* **2003**, *111*, 1–20. [CrossRef]
9. Ahmari, S.; Zhang, L. Production of eco-friendly bricks from copper mine tailings through geopolymerization. *Constr. Build. Mater.* **2012**, *29*, 323–331. [CrossRef]
10. Davidovits, J. Geopolymers: Man-made rock geosynthesis and the resulting development of very early high strength cement. *J. Mater. Educ.* **1994**, *16*, 91.
11. Hardjito, D.; Wallah, S.E.; Sumajouw, D.M.; Rangan, B.V. On the development of fly ash-based geopolymer concrete. *Mater. J.* **2004**, *101*, 467–472.
12. Chen, Y.; Zhang, Y.; Chen, T.; Zhao, Y.; Bao, S. Preparation of eco-friendly construction bricks from hematite tailings. *Constr. Build. Mater.* **2011**, *25*, 2107–2111. [CrossRef]
13. Chou, M.-I.; Chou, S.-F.; Patel, V.; Pickering, M.; Stucki, J. Manufacturing Fired Bricks with Class F Fly Ash from Illinois Basin Coals. 2006. Available online: https://pdfs.semanticscholar.org/9ec8/5a6a34b51eb7d43db05df544e8528e508b71.pdf (accessed on 17 February 2020).
14. Morchhale, R.K.; Ramakrishnan, N.; Dindorkar, N. Utilisation of copper mine tailings in production of bricks. *J. Inst. Eng. Civ. Eng. Div.* **2006**, *87*, 13–16.
15. Roy, S.; Adhikari, G.R.; Gupta, R.N. Use of gold mill tailings in making bricks: A feasibility study. *Waste Manag. Res.* **2007**, *25*, 475–482. [CrossRef] [PubMed]
16. Shon, C.S.; Saylak, D.; Zollinger, D.G. Potential use of stockpiled circulating fluidized bed combustion ashes in manufacturing compressed earth bricks. *Constr. Build. Mater.* **2009**, *23*, 2062–2071. [CrossRef]
17. Majidi, B. Geopolymer technology, from fundamentals to advanced applications: A review. *Mater. Technol.* **2012**, *24*, 79–87. [CrossRef]
18. Duxson, P.; Fernández-Jiménez, A.; Provis, J.L.; Lukey, G.C.; Palomo, A.; van Deventer, J. Geopolymer technology: The current state of the art. *J. Mater. Sci.* **2007**, *42*, 2917–2933. [CrossRef]
19. Duxson, P.; Mallicoat, S.W.; Lukey, G.C.; Kriven, W.M.; van Deventer, J.S.J. The effect of alkali and Si/Al ratio on the development of mechanical properties of metakaolin-based geopolymers. *Colloids Surf. A Physicochem. Eng. Asp.* **2007**, *292*, 8–20. [CrossRef]
20. Drechsler, M.; Graham, A. Geopolymers—An Innovative Materials Technology Bringing Resource Sustainability to Construction and Mining Industries An Innovative Materials Technology Bringing Resource Sustainability to Construction and Mining Industries. In Proceedings of the 48th Institute of Quarrying Conference, Adelaide, Australia, 12–15 October 2005. [CrossRef]
21. Shi, C.; Fernández-Jiménez, A. Stabilization/solidification of hazardous and radioactive wastes with alkali-activated cements. *J. Hazard. Mater.* **2006**, *173*, 1656–1663. [CrossRef]
22. Choudhary, J.; Kumar, B.; Gupta, A. Application of waste materials as fillers in bituminous mixes. *Waste Manag.* **2018**, *78*, 417–425. [CrossRef]
23. Huang, I.B.; Keisler, J.; Linkov, I. Multi-criteria decision analysis in environmental sciences: Ten years of applications and trends. *Sci. Total Environ.* **2011**, *409*, 3578–3594. [CrossRef] [PubMed]
24. Kuity, A.; Jayaprakasan, S.; Das, A. Laboratory investigation on volume proportioning scheme of mineral fillers in asphalt mixture. *Constr. Build. Mater.* **2014**, *68*, 637–643. [CrossRef]
25. Pasandín, A.R.; Pérez, I.; Ramírez, A.; Cano, M.M. Moisture damage resistance of hot-mix asphalt made with paper industry wastes as filler. *J. Clean. Prod.* **2016**, *112*, 853–862. [CrossRef]
26. Wang, H.; Al-Qadi, I.L.; Faheem, A.F.; Bahia, H.U.; Yang, S.-H.; Reinke, G.H. Effect of Mineral Filler Characteristics on Asphalt Mastic and Mixture Rutting Potential. *Transp. Res. Rec. J. Transp. Res. Board* **2011**, *2208*, 33–39. [CrossRef]
27. Xu, W.; Cao, P.; Tian, M. Strength development and microstructure evolution of cemented tailings backfill containing different binder types and contents. *Minerals* **2018**, *8*, 167. [CrossRef]
28. Esquenazi, E.L.; Norambuena, B.K.; Bacigalupo, I.M.; Estay, M.G. Evaluation of soil intervention values in mine tailings in northern Chile. *PeerJ* **2018**, *6*, e5879. [CrossRef]
29. INN. *NCh 1223 of.77: Concrete and Mortar Aggregates Determination of Material Finer Than 0.0780 mm*; Instituto Nacional de Normalización: Santiago, Chile, 1977; pp. 1–7.
30. INN. *NCh 165 of.2009—Concrete and Mortar Aggregates—Sieving and Determination of Grading*; Instituto Instituto Nacional de Normalización: Santiago, Chile, 2009; pp. 1–17.
31. INN. *NCh 1239 Of.2009—Aggregates for Mortar and Concrete—Determination of Real and Net Density and Absorption of Water of Fine Aggregate*; Instituto Nacional de Normalización: Santiago, Chile, 2009; pp. 2–11.

32. INN. *NCh 1498 of.2012—Concrete and Mortar-Mixing Water—Classification and Requirements*; Instituto Nacional de Normalización: Santiago, Chile, 2012; pp. 1–12.
33. INN. *NCh 158 of.67: Compressive and Flexural Strength If Cement Mortar*; Instituto Nacional de Normalización: Santiago, Chile, 1999; pp. 1–13.
34. Li, R.; Zhou, Y.; Li, C.; Li, S.; Huang, Z. Recycling of industrial waste iron tailings in porous bricks with low thermal conductivity. *Constr. Build. Mater* **2019**, *213*, 43–50. [CrossRef]
35. Ministerio de Vivienda, M. *Código de Normas y Especificaciones Técnicas de Obras de Pavimentación*; División Técnica de Estudio y Fomento Habitacional-DITEC: Santiago, Chile, 2018; pp. 155–163.

International Journal of
Environmental Research and Public Health

Article

The Analysis of Green Areas' Accessibility in Comparison with Statistical Data in Poland

Joanna Wysmułek [1,*], Maria Hełdak [1] and Anatolii Kucher [2]

1 Institute of Spatial Management, Wrocław University of Environmental and Life Sciences, 55 Grunwaldzka Street, 50357 Wrocław, Poland; maria.heldak@upwr.edu.pl
2 V. N. Karazin Kharkiv National University, Svobody Square, 6, 61022 Kharkiv, Ukraine; anatoliy_kucher@ukr.net
* Correspondence: joanna.wysmulek@upwr.edu.pl

Received: 20 May 2020; Accepted: 16 June 2020; Published: 22 June 2020

Abstract: The study discusses the problem of public green areas' accessibility for the residents of large cities in Poland. The purpose of the research is to assess the possibility of applying the British Accessible Natural Greenspace Standard (ANGSt) method in determining the amount of natural green space available to residents in Polish conditions including, in particular, the assessment of accessibility using data collected by the Central Statistical Office and the verification of results based on detailed research. The identification of green areas for 18 voivodeship cities in Poland was prepared using the GIS programme, taking into account public green space, provided for general access and free of change. The verification of the ANGSt method consisted of mapping spatial barriers extending the route of access either on foot or by roads as well as closed private areas. The conducted research revealed that, after taking into account the access routes to selected areas, the distance to public green areas increased, on average, from 50 m in the smallest cities (Gorzów Wielkopolski and Olsztyn) to as much as 450 m in Warszawa. A detailed analysis showed that the discussed accessibility was reduced, on average, by almost 10% for the residents of the analysed cities. It was also found that the introduced barriers did not affect the accessibility of more distant, larger green space areas.

Keywords: natural green space accessibility; public green space; ANGSt method; spatial policy; environmental planning

1. Introduction

The quality of life in a city is affected, among other things, by physical, social, environmental and economic characteristics of a given space [1,2]. The places where residents can actively and passively rest include, e.g., sports facilities (swimming pools, fitness clubs, sports halls), cultural institutions (cinemas, theatres, operas, museums) and urban greenery (parks, green squares, gardens, street and residential area greenery, forests, green protected areas). The choice of a place for a short rest during the day and the type of relaxation often depend on the distance to cover and on certain natural factors (e.g., the presence of lawns, trees, shrubs, surface waters, varied terrain) and also non-natural ones (parking lots, paths, places for active recreation, playgrounds, picnic areas, elements of small architecture, historical buildings and educational facilities) [3].

Scientific research has shown that green areas provide people with a wide range of advantages that contribute to their health and well-being, as well as improve their quality of life [4–7]. They function as the green lungs in cities by absorbing pollution [8,9], thus providing people with space to enjoy nature and take advantage of it in the stressful moments of modern life [10–12].

Urban green areas also perform aesthetic, environmental, health and recreational functions. All these qualities show that greenery is an important element of sustainable urban development.

The positive contribution made by green areas in people's lives increases the demand for more reserves and access to green areas both in cities and around them [13]. It was also found that residing close to natural green space results in health benefits [2,14–16], in advantages for the environment [17,18] and also in economic profits [19]. The aforementioned compilation of research highlights numerous advantages of greenery and the way in which accessing it brings benefits to both public health and social well-being. In order to emphasize its value and advantages, urban standards regarding the arrangements for open areas were defined in Poland in the 1970s (e.g., within 800 m from the place of residence a recreational park, covering the area of 2 ha, should be provided).

Unfortunately, the set standards were not met, and in the 1990s of the 20th century and in the years to follow, the undeveloped areas were designated for commercial purposes [20].

Currently, no central regulations oblige designers of local plans to provide public green areas and recreation within housing structures. The applicable law applies only to the area covered by the investment, stating that the minimum 25% of its area should be a biologically active area, if a different percentage does not result from the findings of the local zoning regulations [21]. This creates a large field for over-interpretation of regulations.

Organizational units in most countries use indicators related to the surface share of greenery when planning spatial development strategies for cities. In practice, they most often use data published by the Central Statistical Office. However, it is limited to taking into account only a few categories of greenery. It does not take into account, among other things, the distance of inhabitants' residences from green areas. The actual share of areas of natural and social significance may be significantly different from the values published by the Central Statistical Office, depending on the land use structure. Reliance only on these data can have a disastrous effect on local government administration units when undertaking program and strategic activities related to urban development.

Many researchers indicate the great importance of urban green planting for the development of residential housing [22–24]. The current construction law is limited to providing guidelines regarding the presence of biologically active areas within a given investment in accordance with the local land development plan. There are no guidelines requiring the establishment of large-area greenery facilities. It results in limiting the green areas in the newly developed residential housing areas to the minimum and not using greenery with a significant abundance of leaves [3,20]. It also reduces their recreational potential, approached as the capacity of the environment to create conditions providing rest and physical and mental recovery to people [4,25,26]. An additional reason for the recreational potential decline is the chaotic way in which spatial development takes place [27]. Charles Landry [28] (p. 252) believes that "urban planning is suffocated by the absence of allegedly unprofitable investments in public infrastructure". Current inclusion of greenery, as areas in the spatial design of a modern city centre, becomes increasingly difficult to implement because there are either no free areas or they are too valuable for this type of investment. The green areas available in Polish cities are continuously shrinking, which concerns both the city centres and residential districts, public and semi-private spaces (e.g., as a result of the liquidation of front gardens, functioning in residential districts in the previous century, in favour of parking spaces). Green areas disappear, similarly to their usage methods [29].

The research covered the selected, publicly accessible, recreational and free of charge green spaces. The following green areas were distinguished: parks, squares, lawns, residential green spaces, street greenery and municipal woodland. The definition of urban greenery in accordance with the dictionary of basic land use and spatial planning terms has been adopted in the article, i.e., "green spaces in cities, covered with vegetation presenting ecological, protective, recreational and aesthetic features". Urban greenery occurs in the form of parks, squares, gardens, isolation and protective greenery; it occurs in inner-city areas, housing, leisure, industrial and suburban areas; it creates a band, wedge, radial, annular, irregular, compact or distracted system in the city; and it can be part of the city's natural system and fragments of the ecological system of protected areas [30] (p. 145).

Moreover, according to Piątkowska K. [31] (pp. 37–38), green areas are open and are consciously composed and multifunctional, including, inter alia, in the protection and shaping of the climate and the environment, as well as in the scope of social and service functions provided for the residents. They can occur both in the city, as urban green areas and recreation, as well as in areas associated with weekend and periodic rest, with the rural settlement network or in industrial areas.

According to the explanation from Bartosiewicz and Brzywczy-Kunińska [32] (pp. 11–12), green areas are shaped according to the plans for the development of cities and housing estates. In addition, there are public green territories available to everyone, such as parks and green areas, but also greenery closed or intended for a limited group of people, for example, gardens in schools, hospitals, factories or plots. Based on the diverse beneficial influence of green areas on urban units, many studies have focused on analysing access to natural green spaces. The importance of accessibility and access to green areas was analysed taking into account two perspectives: accessibility in terms of travel and distance and also the perception of such access. From a distance perspective, accessibility is defined as "the ease with which activities at one place may be reached from another via a particular travel mode" [33] (p. 105). Therefore, a number of studies have adopted geographic information systems (GIS) to analyse the accessibility of green areas based on distance, travel time and costs. For example, Neuvonen et al. [34] analysed the relationship between access to natural green areas and the frequency of visits. Oh and Jeong [35] studied the accessibility of pedestrians to city parks. There is also a large number of studies researching the accessibility of green space using both distance and travel time and focusing on equal access to green areas in different communities, supporting the concept of environmental justice [36–42].

There are many methods for analysing access to green areas. The methods used in the discussion presented in this study on the research results include, e.g., the method used by Feltynowski et al. in Poland [43]. It provided the comparison of data from five publicly available sources: (1) public statistics, (2) the national land surveying agency, (3) satellite imagery (Landsat data), (4) the Urban Atlas, and (5) the Open Street Map. The results reveal large differences in the total amount of urban green spaces in the cities as depicted in different datasets. Other studies cited in the discussion are the ones carried out in Romania by Badiu D.L et al. [44]. She emphasizes that the most used quantitative indicator to assess urban green infrastructure is urban green space (UGS) per capita. In order to address the problem of urban sustainable development, 26 m^2 were adopted per resident in all cities. Aerial photos were taken for 38 cities; next, the data were compared with three other databases (National Institute of Statistics, Environmental Protection Agencies and Urban Atlas) to check for differences. In addition, the significant differences between the surface of UGS reported by the administrative offices and those resulting from the spatial analysis were found. In China, Feng et al. examined the relationship between urban park accessibility and population distribution at different administrative levels [45].

The presented article focuses primarily on the assessment of the availability of natural green areas in Polish provincial cities. It takes into account the application of the British ANGSt (Accessible Natural Greenspace Standard) method used to determine the amount of natural green available to residents, adopting a modification to Polish conditions. It also verifies the results obtained with indicators regarding green areas published by the Central Statistical Office. The process of performing ANGSt includes determining the research area and relevant data sources, filtering data to meet the adopted criteria for available green areas. Mapping works are performed using the geographic information system (GIS) software. As the work progressed, it was also decided to carry out modified analyses imposing spatial barriers to the model.

The following research hypotheses were adopted in the study:

1. Residents of large cities have difficulty accessing the generally accessible public green spaces of 2 ha in area and more, located within 300 m distance from a housing estate.
2. The data collected by the Central Statistical Office are insufficient for planning and designating green areas in the urban space and should be supported by scientific research.

Nowadays, both in legal regulations and in the literature of the subject, there is no unambiguous interpretation for the areas covered with vegetation located in the city.

2. Materials and Methods

2.1. Study Area

The presented research is focused on the assessment of the amount of greenery available in terms of recreation for the residents of Polish voivodeship cities. Since 1999, the country has been divided into 16 voivodeships. In two regions, Lubusz and Kuyavian-Pomeranian, the seat of the marshal and governor was divided into two cities. Thus, the research covers 18 voivodeship capitals (Figure 1). The boundary of the studied areas was the 10 km buffer from the administrative borders of the cities.

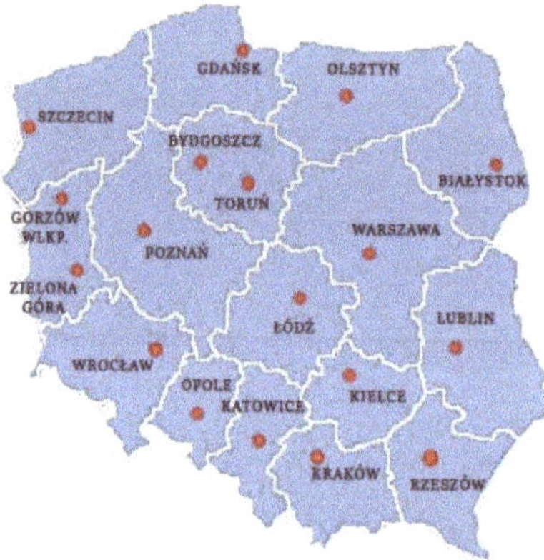

Figure 1. Location of study area with boundaries of voivodships.

Table 1 also presents the indicator of public green area per inhabitant (1 January 2019).

Table 1. The percentage of cemeteries' area against the area of public green space and indicator of green area in individual voivodeship cities in Poland.

City	Public Green Space [ha]		Cemeteries' Area [%]		Indicator of Green Area per Inhabitant [m²/os]
	2010	2018	2010	2018	2018
Białystok	896.9	1041.8	9.9	9.0	35.51
Bydgoszcz	1911.3	1927.1	5.2	4.9	54.44
Gdańsk	2092.0	2249.2	4.4	4.2	38.63
Gorzów Wlkp.	484.3	533.4	7.4	7.1	43.67
Katowice	1228.1	919.9	6.3	8.4	30.42
Kielce	584.3	593.3	7.5	8.0	30.28
Kraków	3230.2	3297.2	4.3	4.1	42.97
Lublin	1183.0	1380.6	6.4	5.5	40.36
Łódź	3644.3	4044.6	6.2	5.6	58.20
Olsztyn	2127.9	1851.3	3.8	4.4	29.29
Opole	643.0	585.1	7.0	9.3	49.46
Poznań	3739.6	4475.9	6.7	5.6	81.14
Rzeszów	658.2	636.1	7.1	8.6	34.59
Szczecin	3316.6	3497.0	5.6	6.1	86.46
Toruń	1008.9	1135.3	8.4	7.5	56.92
Warszawa	4634.6	4988.3	8.0	7.7	28.26
Wrocław	2930.6	3580.5	4.9	3.9	56.65
Zielona Góra	898.1	946.4	3.2	7.4	68.02

Based on: Local Data Bank from the Central Statistical Office [46].

2.2. Research Methods

The applied research method followed the steps listed below:

- the identification of the research problem and the selection of the research area;
- the subject literature analysis;

- the selection of public green areas' categories constituting the subject of conducted analyses;
- preparing statistical data on natural green areas in 18 capitals of voivodeships in Poland;
- spatial identification of natural green areas in individual voivodeship cities using the GIS programme;
- mapping accessibility buffers using the ANGSt method, following prior identification of housing estates;
- the obtained results verification against the data retrieved from the Central Statistical Office;
- the verification of assessment results regarding the accessibility to public green areas using the ANGSt method based on the identification of spatial barriers;
- re-assessment of the accessibility to public green areas;
- and conclusions.

Due to the low popularity of recreational activities in cemeteries, as well as their variety in terms of vegetation, and also a fairly low percentage and qualitative diversity in relation to total green zone, it was decided to skip this component.

Family allotment gardens were also left out due to their limited, in most cases, accessibility for the general public (the majority of gardens are closed and only the owners of individual gardens have access to them). Pastures, arable land, meadows or undeveloped riverside and post-industrial areas were not taken into account either.

The subject of the analyses was verified in terms of the green area defined in the Nature Conservation Act [47]: "green space stands for the arranged areas along with technical infrastructure and buildings functionally connected with them, covered with vegetation, performing public functions and in particular: parks, lawns, promenades, boulevards, botanical and zoological gardens, children game parks and historic gardens, cemeteries, greenery along the roads in built-up areas, green squares, historic fortifications, constructions, storage sites, airports, railway stations and industrial facilities". The verification also covered the accessibility of selected areas for residents.

After the units were separated for further research, a compilation of statistical data made available by the Central Statistical Office for green areas in 18 voivodeship cities in Poland was carried out. The availability of data and the method of interpretation of green areas were the basis for selecting the data source: "Green urban areas consist of parks, urban forests, green squares, green residential spaces, street greenery, zoological and botanical gardens, as well as cemeteries. The key to choosing this type of green areas is that they are managed by the municipal authorities" [43] (p. 61). In order to determine the tendency for the transformation of green areas, the values from three different years, i.e., 2004, 2010 and 2018, were compared. The areas of cities from the same periods were also compared.

The data provided by the Central Statistical Office are available on the website www.stat.gov.pl. It collects and provides information on, among others: residents, the environment, living conditions, education, culture, work, finance, economy and natural goods. City authorities and business intelligence services often use this information. Their advantage is high data availability. Thanks to this, analyses can be conducted on full material collections. Another advantage is the lack of influence of third parties on the data creation process. On the other hand, the main disadvantages include the reliability and method of obtaining data.

The next stage of the research was the spatial identification of green areas in individual voivodeship cities in Poland. Taking into account a rather limited concept of public green space represented by public statistics and the discrepancies in the occurrence of green areas in cities [43], municipal spatial information systems and the topographic database were used for the purpose of mapping green areas. In order to verify green areas more accurately, the cities were divided into smaller units (specified individually for each centre). The minimum mapping unit was 50 m. In case of doubts regarding free accessibility, the areas were verified individually by searching for information on the local authorities' websites. Next, the particular layers were mapped in the GIS and green areas were selected using the aforementioned guidelines. The underlay for each city was then buffered using the recommendations provided in the ANGSt method. This technique was developed in 1998 to study the extent to which the population living in the southeastern part of Great Britain has access to public green areas.

The adapted method determined the minimum area of greenery to be 2 ha. According to many researchers [48,49], expanses with smaller areas are insufficient to provide adults with adequate support for physical activity. The size of green areas can also slow down the phenomenon of appropriation of land by one group of people [50–52].

The following standards specify the distance between the place of residence and green areas presenting particular surfaces, in accordance with the ANGSt method [53]:

- No person should live more than 300 m from their nearest area of accessible natural green space of at least 2 ha in size.
- There should be at least one 20 ha accessible natural green area within 2 km from home.
- There should be one accessible natural green zone 100 ha site within 5 km.
- There should be one accessible natural green space 500 ha within 10 km.

Following the above recommendations, green areas of less than 2 ha in size were excluded from the analysis. In addition to the areas within the administrative borders of individual cities, a 10 km buffer around them was also taken into account (maximum value for the ANGSt model). Buffers were deleted using ArcGIS. Due to the concentration of the objects, Euclidean buffers were used. Considering the fact that a city constitutes the subject of research, the earlier defined green areas were adopted as natural green areas.

Then, using the data contained in the spatial information system of provincial cities, a layer presenting the estate was applied. Prior distinguishing smaller units, for which it was easier to determine the number of households, turned out to be helpful. The population was counted using an average statistical size of a household—3 people in a dwelling [46]. On this basis, the percentage of the population having access to individual green areas was calculated. The next step was the verification of the obtained results against the data retrieved from the Central Statistical Office [46]. The final research stage consisted of mapping spatial barriers. Major line infrastructures (railways, highways, roads with more than 3 lanes, bridges, and rivers) as well as private premises were mapped using the topographic database. It was decided that in the city centre, where pedestrian crossings are quite densely designated, the occurring barriers would not be taken into account, except for railway lines and waterways [54]. The research was finalised with reassessing residents' accessibility to the generally accessible green areas.

The presented results were divided into three sections. The first of them verifies the data retrieved from the Central Statistical Office based on the areas selected for further analyses. The second focuses on the results of the ANGSt method application for the selected cities and analyses how population levels in individual cities relate to them. The third section contains maps summarizing the generally accessible green areas with the additionally designated spatial barriers.

3. Results

3.1. Descriptive Statistics

Prior to the spatial identification of green areas in individual cities, the general characteristics of cities were determined based on the data from the Central Statistical Office. The ratio of the area of generally accessible green spaces (the area of municipal forests, parks, lawns, residential green spaces and street greenery was summed up) against the area of particular cities was calculated (Table 2).

Table 2. The area of generally accessible green space against the area of individual voivodeship cities in Poland.

City	Greenery [ha]			The Ratio of the Area of Generally Accessible Green Spaces to the City's Surface [%]		
	2004	2010	2018	2004	2010	2018
Białystok	727.3	896.9	1041.8	8	8	9
Bydgoszcz	1662.7	1911.3	1927.1	10	10	10
Gdańsk	1789.0	2092.0	2249.2	7	8	8
Gorzów Wlkp.	525.5	484.3	533.4	6	5	6
Katowice	1266.1	1228.1	919.9	8	7	5
Kielce	483.0	584.3	593.3	4	5	5
Kraków	2548.1	3230.2	3297.2	8	9	10
Lublin	1302.3	1183.0	1380.6	9	8	9
Łódź	3318.2	3644.3	4044.6	11	12	13
Olsztyn	1893.0	2127.9	1851.3	22	23	20
Opole	362.8	643.0	585.1	4	6	4
Poznań	3516.9	3739.6	4475.9	13	13	16
Rzeszów	426.2	658.2	636.1	8	5	5
Szczecin	3030.8	3316.6	3497.0	10	10	11
Toruń	879.4	1008.9	1135.3	8	8	9
Warszawa	5129.7	4634.6	4988.3	10	8	9
Wrocław	2662.7	2930.6	3580.5	9	10	12
Zielona Góra	758.9	898.1	946.4	13	15	3 *

* The administrative area of the city was extended on 2 January 2015.

The above table shows changes taking place in the share of the generally accessible green areas over the period of 13 years. They can result from alterations in the interpretation of urban green space as well as functional transformations of the areas covering both the creation of new green areas and changing the existing ones into either residential housing or service and industrial buildings. No correlation was identified between the size of individual cities, the number of residents and the occupied green area (Figure 2).

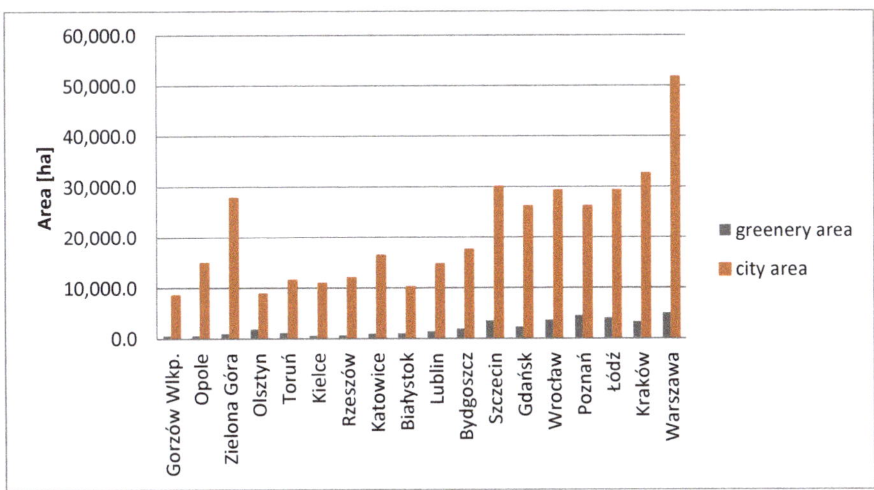

Figure 2. The area of cities and green spaces by the increasing number of residents.

Furthermore, the areas selected for further analysis on the basis of urban geoportals did not show any connection between the areas of cities and those of the generally accessible green zones located in their boundaries.

3.2. Accessible Natural Greenspace Standard

Following the standards developed based on the British ANGSt method, the residents' distances to green areas of particular surfaces in 18 cities were presented (Table 3). After compiling four layers for each city, the percentage of residents for which all criteria were met (column "All of the ANGSt Requirements Met") was obtained.

Table 3. The accessibility of green areas according to ANGSt standards in voivodeship cities in Poland.

City	All Households	Access to at Least 2 ha within 300 m [%]	Access to at Least 20 ha within 2 km [%]	Access to at Least 100 ha within 5 km [%]	Access to at Least 500 ha within 10 km [%]	All of the ANGSt Requirements Met [%]	None of the ANGSt Requirements Met [%]
Białystok	297,288	9	58	92	100	7	0
Bydgoszcz	352,313	20	48	89	100	17	0
Gdańsk	464,254	19	56	72	100	12	0
Gorzów Wlkp.	124,295	21	47	72	100	16	0
Katowice	281,950	12	45	68	100	5	0
Kielce	188,495	38	85	76	100	14	0
Kraków	76,748	26	60	79	100	12	0
Lublin	322,930	25	64	80	100	11	0
Łódź	687,702	18	56	78	100	10	0
Olsztyn	173,070	29	64	82	100	14	0
Opole	128,140	31	50	79	100	18	0
Poznań	671,233	20	62	76	100	12	0
Rzeszów	191,009	13	42	54	100	7	0
Szczecin	403,883	32	65	78	100	12	0
Toruń	182,725	21	54	62	100	10	0
Warszawa	1,764,615	20	67	72	100	6	0
Wrocław	638,600	22	66	87	100	8	0
Zielona Góra	139,819	12	62	85	100	10	0

The above table shows that the residents of Kielce, Szczecin and Opole have the relatively largest access to green spaces of 2 ha and more within a 300 m radius. In turn, the smallest access refers to the residents of Białystok, Katowice and Rzeszów. The obtained results indicate that the situation of the residents in the analysed cities does not seem to be good. In the largest Polish cities, such as Warszawa, Gdańsk, Wrocław and Poznań, the comfort of using the generally accessible green spaces within a 300 m radius and offering 2 ha of area is available for only about 20% of the residents.

These studies, however, require more detailed analyses addressing the access to smaller squares and playgrounds, which has been indicated as the subject matter of the planned research.

The most noteworthy research result is the absence of accessibility to green areas closest to residential estates. In the analysed cities, the accessibility of areas covered by the generally accessible vegetation presenting the surface of 2 ha, located in the immediate vicinity (300 m), ranges from 9% to 38% of the residents' number. Figure 3 shows the availability of greenery for residents based on the examples of Białystok and Kielce, respectively the largest and smallest percentages of access to green areas within 300 m.

The conducted studies have also shown that the accessibility of greenery does not depend on the city size. The surface of the areas featuring natural values does not show any correlation with the number of residents. Furthermore, the statistical data, showing the ratio of green areas against the size of individual cities, do not affect residents' accessibility to green areas, primarily the ones in the immediate vicinity (Table 4). After comparing the data in the table, it can be seen that cities with a higher ratio of green areas to their entire area do not have greater availability of green areas in the closest distance for residents. After analysing the types of areas found in individual cities, it was noticed that the higher ratio of green areas to the city surface is often due to the distribution of greenery in the peripheral areas of the city—most often, they are large forest areas. However, they do not translate into an increase in available green areas in the immediate vicinity of residential areas. As shown below (Table 4), e.g., Kielce, despite one of the smaller relations of green areas to the surface of the whole city, has the highest access of green areas in the immediate vicinity of residential areas.

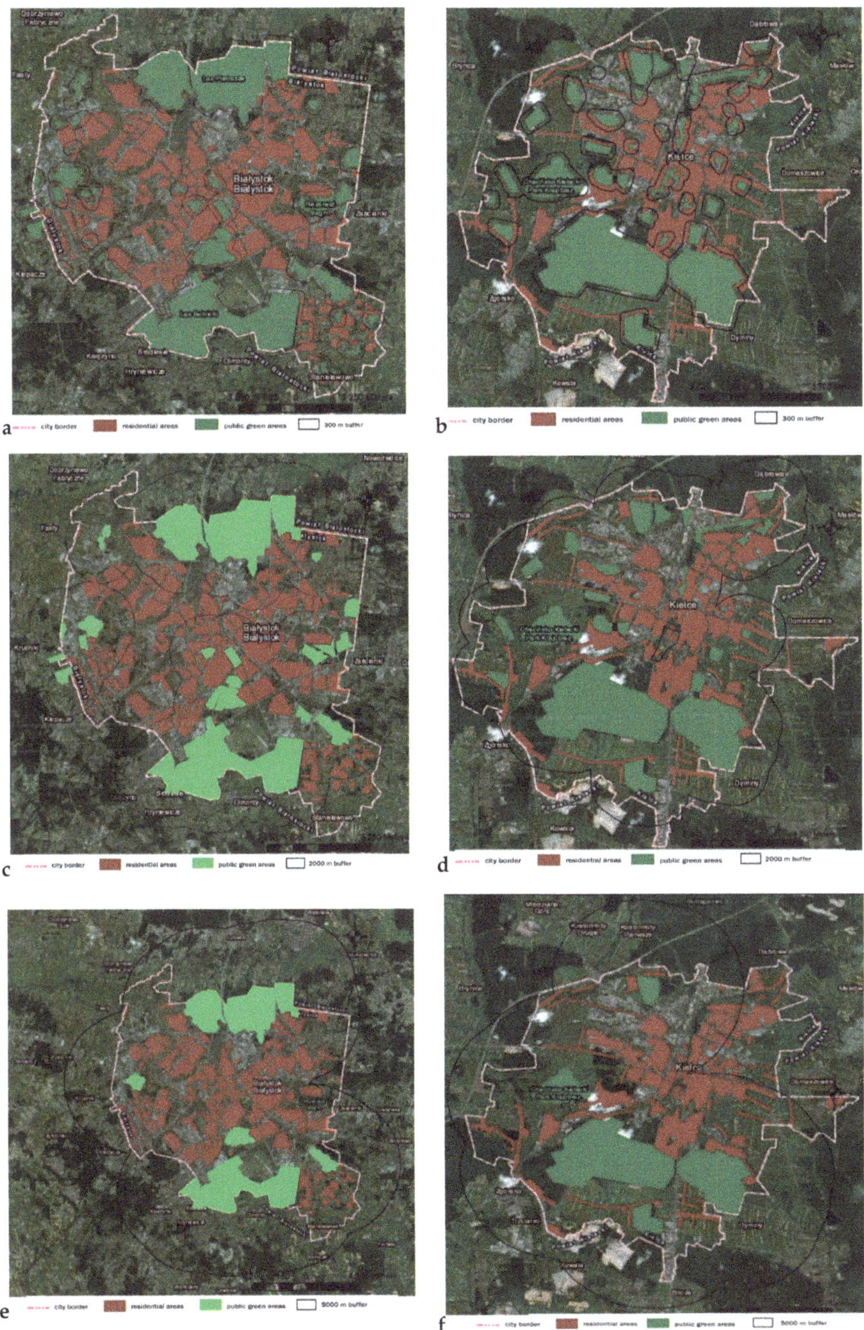

Figure 3. Access of residents to green spaces: (**a**) 300 m buffer in Białystok, (**b**) 300 m buffer in Kielce, (c) 2 km buffer in Białystok, (**d**) 2 km buffer in Kielce, (**e**) 5 km buffer in Białystok, and (**f**) 5 km buffer in Kielce.

Table 4. Comparison of the value of the green area indicator with the ratio of green areas to the city area according to the availability of at least 2 ha of green area at a distance of 300 m in provincial cities in Poland.

d	Access to at Least 2 ha within 300 m [%]	Indicator of Green Area per Inhabitant [m^2/os]	The Ratio of Green Areas to the City's Surface in 2018 [%]
Białystok	9	35.53	9
Katowice	12	30.42	5
Zielona Góra	12	68.02	3 *
Rzeszów	13	34.59	5
Łódź	18	58.2	13
Gdańsk	19	38.63	8
Warszawa	20	28.26	9
Bydgoszcz	20	54.44	10
Poznań	20	81.14	16
Gorzów Wlkp.	21	43.67	6
Toruń	21	56.92	9
Wrocław	22	56.65	12
Lublin	25	40.36	9
Kraków	26	42.97	10
Olsztyn	29	29.29	20
Opole	31	49.46	4
Szczecin	32	86.46	11
Kielce	38	30.28	5

* The administrative area of the city was extended on 2 January 2015.

3.3. Spatial Barriers

The last step consisted of mapping spatial barriers. After taking into account the access routes to the selected areas, it occurred that the average distance was extended from 50 m in the smallest cities (Gorzów Wielkopolski and Olsztyn) to as much as 450 m in Warszawa, resulting in reduction of accessibility for residents by 9% (see Table 5).

Table 5. Comparison of access to green areas including spatial barriers.

Cities	Access to at Least 2 ha within 300 m [%]	Access to at Least 2 ha within 300 m with Spatial Barriers [%]	Differences Due to the Consideration of Barriers [%]
Białystok	6	5	−1
Katowice	12	8	−4
Zielona Góra	12	10	−2
Rzeszów	13	11	−2
Łódź	18	13	−5
Gdańsk	19	16	−3
Warszawa	20	11	−9
Bydgoszcz	20	17	−3
Poznań	20	16	−4
Gorzów Wlkp.	21	19	−2
Toruń	21	18	−3
Wrocław	22	17	−5
Lublin	25	22	−3
Kraków	26	20	−6
Olsztyn	29	26	−3
Opole	31	28	−3
Szczecin	32	29	−3
Kielce	38	35	−3

To compare the data collected before and after the barriers were applied, tests were performed to see if there was a statistically significant change. Nonparametric tests were used. To compare two paired groups for variables that do not have a normal distribution, the Wilcoxon signed-rank test and the sign test were used. The calculations were made in the STATISTICA program.

The Wilcoxon test consists in ranking differences in measurements for subsequent observations; the differences between measurement 1 and 2 are calculated. Then, these differences are ranked, i.e., the results obtained are ranged from smallest to highest and given subsequent ranks. Then, the sum of ranks is counted for differences that were negative and for differences that were positive (results with no differences are not significant). Then selected the sum (negative or positive differences) that was greater, and this result is the result of the Wilcoxon test (up to 25 observations). Then, the units are converted to Z values: Wilcoxon test result (i.e., the highest sum of ranks), the average divided by the standard deviation. This result is checked in the normal distribution tables [55].

H_0—The results of trial 1 (access at a distance of 300 m) and trial 2 (access at a distance of 300 m with barriers) are the same/statistically insignificant.

H_1—The results of trial 1 (access at a distance of 300 m) and trial 2 (access at a distance of 300 m with barriers) are statistically significant.

Values read in the program: Z value: 3.724; $p = 0.00020 < 0.05 \propto$

Results in the sign test:

$$Z = 4.01\ p = 0.00006 < 0.05 \propto \tag{1}$$

$\propto -level\ of\ significance.$

The tests before and after applying the barriers differ from each other with statistical significance. Introducing barriers did not affect the accessibility of further located larger green areas.

4. Discussion

The selected type of analysis can never provide all the answers to the complex range of problems with access to green areas. One of the important reasons for undertaking the work was to draw attention to the problem of urban planning.

The need of green areas' accessibility is discussed with increasing frequency, e.g., [34,36,40,41,56–59]. The problems are the set of data regarding green areas, the coherence of such data and the absence of general classification in terms of green space. The data collected by the Central Statistical Office depend on, e.g., changes in land ownership or quite general and diverse interpretation of the problem over the years. It poses a major challenge for urban spatial planning and management of green areas [44]. Raw data not subjected to an in-depth analysis can present a false image of the existing situation. In addition, the interpretation of the term "green areas" in different cities differs, even if they are to comply with the general definitions used by the Central Statistical Office.

The conducted research placed emphasis only on the generally accessible areas with surface exceeding 2 ha, in accordance with the ANGSt standards. It is, however, necessary to pay attention to the fact that, while planning and managing urban space, such areas as arable land or private and informal green zones should be taken into account. They can constitute an additional place for meetings and recreation for a wider population, and can also play the role of an ecological passage. Moreover, smaller areas, e.g., the increasingly popular pocket parks in built-up areas, can represent a good tool in creating new public spaces [23,60].

The additional problem in conducting the discussed research was the absence of a unified classification covering green fields for all voivodeship cities. In some of them, the inventory data concerning urban green areas were missing. This means that there is a need for introducing land use classes or land cover as the indicators of what green space stands for and, thus, how to combine it with the rest of ecosystem services. However, as shown by the latest comparative studies carried out by Kremer et al. [61], due to the differences in urban morphology and its heterogeneity, even such an approach has its own limitations. As a result, the research that was conducted provides vital information on urban ecosystem services only in these areas where local or regional measurements were performed [62].

In addition, spatial data from public statistics are collected only for a municipality or an equivalent unit and cannot be divided into smaller administrative units (e.g., districts or housing estates), whereas green space data are primarily important in a local scale. Other researchers also compared various information sources about green areas in urban space and came to similar conclusions. For example, Feltynowski et al. [43] compared satellite imagery with the Lodz public statistics, the national land surveying agency, Open Street Map and the Urban Atlas, and identified significant differences between them. They noticed that the most comprehensive set of data from the country is the set from a geodetic agency, which shows that greenery constitutes 61.2% of the city's area. In their research, however, they did not take into account the accessibility to green areas by residents. Similar studies were conducted in Romania, although, in the case of those cities, official statistics overestimated green areas in comparison to aerial photographs [44].

Research studies carried out, among others, in China also show that the availability of public parks for city dwellers is often unevenly distributed. Analyses conducted by Feng et al. [45] were carried out in parks in Beijing, and the relationship between the availability of city parks and population distribution at various administrative levels was analysed. The results of their research showed that the availability of city parks is different at the administrative level and that places with more city parks usually have more accessibility. Zhang et al. [63] also conducted research related to equal access of residents to green areas. The results obtained show, similarly to those presented in this article, that the current planning of urban public greenery using only the ratio of public greenery to urban land and public greenery per inhabitant is insufficient. Currently, planning of open space areas in Polish cities is based on a narrow classification used for public statistics purposes, covering formal green areas managed by the authorities. As in the case of Chinese cities, in Poland, there is also a lack of information about recreational areas covering smaller ranges (districts or housing estates). It poses a barrier for green areas' planning in urban infrastructure and making them accessible to residents. It also introduces false information about the anticipated needs, e.g., large forest areas located on the outskirts of cities increase the green areas' ratio, although they do not fully affect their share in the accessibility among residents.

The authors concluded that a more consistent approach to green space qualification could turn out to be helpful for reporting purposes throughout the entire European Union—particularly, unification of the green area terminology and the approach to private, semi-private and open spaces. In order to facilitate management and research in the field of green zones, as well as further development of urban areas, data sets should provide information who exactly is responsible for each space management. Predominantly, various municipal institutions are interested in those green regions for which they are formally responsible and, in the situation of poor cooperation between various stakeholders, they neglect other green areas and connections between them [43,64].

Another question asked by the authors referred to the selection of size and distance from green areas. Within the framework of the research continuation, it is worth reversing the question—what is the distance from residential areas to green areas and what is their surface size? The research should also be extended by the verification of human behaviour—are the residents willing to spend time in the selected green units?

The ANGSt model does not require assessing the quality components which the conducted research did not provide information about, e.g., such facilities as car parks or access for persons with disabilities, as well as proper markings and promotion of the areas. It mainly applies to the areas distant from the place of residence by 5 km and more. The quality of landscape or the sense of peace and quiet were not taken into account either. Yet another limitation of this study, at least in terms of the population pressure model, is the fact that it does not consider the population visiting a given site but not residing in the analysed region. In terms of the ANGSt model, this fact is irrelevant, as its requirements were established for the green space at the local level (i.e., up to 10 km from the place of residence). However, the protected landscapes, and national parks in particular, will always have visitors, often in large numbers, coming from outside the distance provided for in the ANGSt model, which can negatively affect the quality of the user experience and also increase the likelihood of harmful consequences resulting from a high degree of visitor pressure.

5. Conclusions

The conducted research allowed for formulating the following conclusions:

- The accessibility assessment of the selected large cities' residents in Poland to green areas of 2 ha and more within a 300 m radius does not seem to be good. Based on the applied Accessible Natural Greenspace Standard (ANGSt) method, it was found to range between 6% and 38%, with a median of 20.5%. This confirms the first hypothesis—that the residents of large cities have difficulty accessing public green areas of general public accessibility with an area of 2 ha and more, located within 300 m distance from a housing estate.
- Detailed analysis based on the Accessible Natural Greenspace Standard (ANGSt) method in Poland have revealed some of its shortcomings. Namely, determining the buffers of the required distance from the areas of public green zone, without including spatial barriers, distorts the actual route of reaching them; the method should be modified to assess the distance from green areas and take into account existing spatial barriers. In the course of the conducted analyses, it has been found that the route covered between "the place of residence and the green space" can, in fact, be much longer. This reduces the quality of life measured by the distance from green spaces of varying area and, additionally, reduces the share of residents' access to the publicly accessible green areas.
- The research revealed changes in the assessment of green areas' accessibility in the case of the ANGSt method modification, in particular, regarding the green space located in the close proximity of residential areas (2 ha surface size, at a distance of 300 m). The detailed analysis of voivodeship capitals, in terms of the discussed problem, did not present favourable results (the route extension from 50 m in Gorzów Wielkopolski and Olsztyn, to 450 m in Warszawa, taking into account spatial barriers, was recorded). At the same time, no changes in the assessment of accessibility to large-scale urban green areas (500 ha surface size at a distance of 10 km) were observed. The method modification revealed some of its shortcomings; spatial barriers (rivers, railways, and communication arteries) are frequently impossible to get through for a mother with a child or a senior citizen. In turn, extending the route can result in making such distance impossible to cover on foot.
- The analysis of existing legal acts as well as standards and design guidelines for housing estates confirms the first thesis, i.e., in Poland, there are no standards defining the distance to the generally accessible green areas to be met per one resident. Taking into account the accessibility of green areas, especially the ones in the immediate vicinity, is important for further discussions on the delimitation of green areas in cities, primarily from the perspective of urban planning and management of urban green areas in order to provide ecosystem services and ensure adequate access to these places. Certain works have been undertaken many times; however, due to the absence of possibility for their implementation, they usually remained within the sphere of projects.
- The problems encountered during the research that was conducted in finding the reliable statistical data also confirmed the second thesis. The data collected by the Central Statistical Office are insufficient for urban green space planning and management. They should be supplemented by scientific research for the proper development of urban areas. The problem refers to, e.g., the area land classified in the Land and Property Register as forest and woodland in cities is not always used for recreational greenery purposes. The difference between the registered status and the actual development is often significant. An ongoing update of the land and property records by means of site visits should be performed. Scientific research introduces some modernization in space management; therefore, simulations and solutions preceded by the respective analyses of variants aimed at improving access to the organized forms of recreation would certainly turn out crucial in this case.

The research results can be used to determine priorities in the analysed regions as well as identify the key areas lacking accessibility to green spaces and take them into account in urban spatial policy.

Planned research: The research should be continued and focus on the behaviour of residents and their willingness to take advantage of green space, and also on determining whether and at what distance from the place of residence the areas smaller than 2 ha in size (not covered by the presented research) are located. In addition, the access to green spaces targeted at different age groups remains an important problem for future research. The authors will conduct relevant analyses taking into account the needs of younger children, older children, young people, adults and older people within smaller than 2 ha of generally accessible green areas. Following the above, they intend to analyse access to playgrounds with various equipment, sports fields, walking areas, open air gyms, walking paths, etc.

Author Contributions: Conceptualization, A.K., M.H. and J.W.; methodology, J.W.; software, J.W.; validation, M.H. and J.W.; formal analysis, J.W.; investigation, A.K. and J.W.; resources, J.W.; data curation, J.W.; writing—original draft preparation, J.W.; writing—review and editing, M.H. and J.W.; visualization, J.W.; supervision, M.H.; project administration, J.W.; and funding acquisition, M.H. All authors have read and agreed to the published version of the manuscript.

Funding: The research is financed/co-financed under the Leading Research Groups support project from the subsidy increased for the period 2020–2025 in the amount of 2% of the subsidy referred to Art. 387 (3) of the Law of 20 July 2018 on Higher Education and Science, obtained in 2019.

Conflicts of Interest: The authors declare no conflict of interest.

References

1. Sirgy, M.J.; Cornwell, T. How neighborhood features affect quality of life. *Soc. Indic. Res.* **2002**, *59*, 79–114. [CrossRef]
2. Weziak-Bialowolska, D. Quality of life in cities—Empirical evidence in comparative European perspective. *Cities* **2016**, *58*, 87–96. [CrossRef]
3. Szumacher, I.; Malinowska, E. Recreation potential of chosen types of urban green spaces. *Probl. Landsc. Ecol.* **2013**, *34*, 229–236. (In Polish)
4. Zhang, L.; Tan, P.Y.; Diehl, J.A. A conceptual framework for studying urban green spaces effects on health. *J. Urban Ecol.* **2017**, *3*, 1–13. [CrossRef]
5. Hełdak, M.; Raszka, B. Evaluation of the local spatial policy in Poland with regard to sustainable development. *Pol. J. Environ. Stud.* **2013**, *22*, 395–402.
6. Hansmann, R.; Hug, S.-M.; Seeland, K.; Hansmann, R. Restoration and stress relief through physical activities in forests and parks. *Urban For. Urban Green.* **2007**, *6*, 213–225. [CrossRef]
7. Nielsen, T.A.S.; Hansen, K.B. Do green areas affect health? Results from a Danish survey on the use of green areas and health indicators. *Health Place* **2007**, *13*, 839–850. [CrossRef] [PubMed]
8. Lennon, M.; Douglas, O.; Scott, M. Urban green space for health and well-being: Developing an 'affordances' framework for planning and design. *J. Urban Des.* **2017**, *22*, 778–795. [CrossRef]
9. Baycan-Levent, T.; Nijkamp, P. Urban green space policies: A comparative study on performance and success conditions in European cities. In Proceedings of the 44th European Congress of the European Regional Science Association, Porto, Portugal, 25–29 August 2004. Available online: https://www.researchgate.net/profile/Tuzin_Baycan/publication/4874463_Urban_Green_Space_Policies_A_Comparative_Study_on_Performance_and_Success_Conditions_in_European_Cities/links/02bfe5140ef925eb0f000000/Urban-Green-Space-Policies-A-Comparative-Study-on-Performance-and-Success-Conditions-in-European-Cities.pdf?origin=publication_detail (accessed on 20 February 2020).
10. Sugimoto, K. Quantitative measurement of visitors' reactions to the settings in urban parks: Spatial and temporal analysis of photographs. *Landsc. Urban Plan.* **2013**, *110*, 59–63. [CrossRef]
11. Przybyła, K.; Kulczyk-Dynowska, A.; Kachniarz, M. Quality of life in the regional capitals of Poland. *J. Econ. Issues* **2014**, *48*, 181–196. [CrossRef]
12. Kazak, J.K.; Chruściński, J.; Szewrański, S. The development of a novel decision support system for the location of green infrastructure for stormwater management. *Sustainability* **2018**, *10*, 4388. [CrossRef]

13. De Ridder, K.; Adamec, V.; Bañuelos, A.; Bruse, M.; Burger, M.; Damsgaard, O.; Dufek, J.; Hirsch, J.; Lefebre, F.; Pérez-Lacorzana, J.; et al. An integrated methodology to assess the benefits of urban green space. *Sci. Total Environ.* **2004**, *334*, 489–497. [CrossRef]
14. Grahn, P.; Stigsdotter, U.K. The relation between perceived sensory dimensions of urban green space and stress restoration. *Landsc. Urban Plan.* **2010**, *94*, 264–275. [CrossRef]
15. Lee, A.C.; Maheswaran, R. The health benefits of urban green spaces: A review of the evidence. *J. Public Health* **2010**, *33*, 212–222. [CrossRef] [PubMed]
16. Richardson, E.A.; Mitchell, R. Gender differences in relationships between urban green space and health in the United Kingdom. *Soc. Sci. Med.* **2010**, *71*, 568–575. [CrossRef] [PubMed]
17. Nowak, D.J.; Crane, D.E.; Stevens, J.C. Air pollution removal by urban trees and shrubs in the United States. *Urban For. Urban Green.* **2006**, *4*, 115–123. [CrossRef]
18. Chiesura, A. The role of urban parks for the sustainable city. *Landsc. Urban Plan.* **2004**, *68*, 129–138. [CrossRef]
19. Jim, C.; Chen, W.Y. External effects of neighbourhood parks and landscape elements on high-rise residential value. *Land Use Policy* **2010**, *27*, 662–670. [CrossRef]
20. Dąbrowska-Milewska, G. Urban planning standards for residential areas-chosen issues. *Archit. Artibus* **2010**, *2*, 17–31. (In Polish)
21. Regulation of the minister for infrastructure of 12 April 2002 on the technical conditions that shall be met by buildings and their location. *J. Laws* **2019**. Item 1065. (In Polish)
22. Khoshtaria, T.; Chachava, N. The planning of urban green areas and its protective importance in resort cities (case of Georgian resorts). *Ann. Agrar. Sci.* **2017**, *15*, 217–223. [CrossRef]
23. Shafray, E.; Kim, S. A study of walkable spaces with natural elements for urban regeneration: A focus on cases in Seoul, South Korea. *Sustainability* **2017**, *9*, 587. [CrossRef]
24. Naeem, S.; Cao, C.; Fatima, I.S.A.K.; Najmuddin, O.; Acharya, B.K. Landscape greening policies-based land use/land cover simulation for beijing and islamabad—An implication of sustainable urban ecosystems. *Sustainability* **2018**, *10*, 1049. [CrossRef]
25. Xing, Y.; Jones, P.; Donnison, I. Characterisation of nature-based solutions for the built environment. *Sustainability* **2017**, *9*, 149. [CrossRef]
26. Liu, H.-L.; Shen, Y.-S. The impact of green space changes on air pollution and microclimates: A case study of the Taipei Metropolitan Area. *Sustainability* **2014**, *6*, 8827–8855. [CrossRef]
27. Ward, C.D.; Parker, C.M.; Shackleton, C. The use and appreciation of botanical gardens as urban green spaces in South Africa. *Urban For. Urban Green.* **2010**, *9*, 49–55. [CrossRef]
28. Landry, C. *The Creative City: A Toolkit for Urban Innovators*, 1st ed.; Earthscan: London, UK, 2000; pp. 163–257.
29. Bardzińska-Bonenberg, T. Changes in the concept of modernity in housing construction in the 20th century based on the example of Max John in Poznań. In *Modernity in Architecture. Transformation-Technology-Identity*; Pallado, J., Ed.; Monograph 6; Faculty of Architecture of the Silesian University of Technology: Gliwice, Poland, 2012; pp. 19–27. (In Polish)
30. Kachniarz, T.; Niewiadomski, Z.; Suliga, J.; Sumień, T. Słownik podstawowych pojęć z dziedziny planowania przestrzennego. In *Nowe Podstawy Prawne Zagospodarowania Przestrzennego*; Kachniarz, T., Niewiadomski, Z., Eds.; Agencja Wydawnicza Instytutu Gospodarki Przestrzennej i Komunalnej: Warsaw, Poland, 1994; pp. 109–146.
31. Piątkowska, K. *Kształtowanie Obiektów i Zespołów Usługowych*; Zieleń i Wypoczynek, Przemysłowy Instytut Elektroniki Branżowy Ośrodek Informacji Naukowej, Technicznej i Ekonomicznej: Warsaw, Poland, 1983.
32. Bartosiewicz, A.; Brzywczy-Kunińska, Z. *Urządzanie i pielęgnacja terenów zieleni*; Państwowe Wydawnictwo Szkolnictwa Zawodowego: Warsaw, Poland, 1973.
33. Liu, S.; Zhu, X. Accessibility analyst: An integrated GIS tool for accessibility analysis in urban transportation planning. *Environ. Plan. B Plan. Des.* **2004**, *31*, 105–124. [CrossRef]
34. Neuvonen, M.; Sievänen, T.; Tönnes, S.; Koskela, T. Access to green areas and the frequency of visits—A case study in Helsinki. *Urban For. Urban Green.* **2007**, *6*, 235–247. [CrossRef]
35. Oh, K.; Jeong, S. Assessing the spatial distribution of urban parks using GIS. *Landsc. Urban Plan.* **2007**, *82*, 25–32. [CrossRef]
36. Comber, A.; Brunsdon, C.; Green, E. Using a GIS-based network analysis to determine urban greenspace accessibility for different ethnic and religious groups. *Landsc. Urban Plan.* **2008**, *86*, 103–114. [CrossRef]

37. Omer, I. Evaluating accessibility using house-level data: A spatial equity perspective. *Comput. Environ. Urban Syst.* **2006**, *30*, 254–274. [CrossRef]
38. Wolch, J.; Wilson, J.P.; Fehrenbach, J. Parks and park funding in Los Angeles: An equity-mapping analysis. *Urban Geogr.* **2005**, *26*, 4–35. [CrossRef]
39. Omer, I.; Or, U. Distributive environmental justice in the city: Differential access in two mixed Israeli cities. *J. Econ. Soc. Geogr.* **2005**, *96*, 433–443. [CrossRef]
40. Smoyer-Tomic, K.E.; Hewko, J.N.; Hodgson, J. Spatial accessibility and equity of playgrounds in Edmonton, Canada. *Can. Geogr.* **2004**, *48*, 287–302. [CrossRef]
41. Hewko, J.; Smoyer-Tomic, K.E.; Hodgson, M.J. Measuring neighbourhood spatial accessibility to urban amenities. Does aggregation error matter? *Environ. Plan. A Econ. Space* **2002**, *34*, 1185–1206. [CrossRef]
42. Hoffimann, E.; Barros, H.; Ribeiro, A.I. Socioeconomic inequalities in green space quality and accessibility—Evidence from a Southern European city. *Int. J. Environ. Res. Public Health* **2017**, *14*, 916. [CrossRef]
43. Feltynowski, M.; Kronenberg, J.; Bergier, T.; Kabisch, N.; Łaszkiewicz, E.; Strohbach, M. Challenges of urban green space management in the face of using inadequate data. *Urban For. Urban Green.* **2018**, *31*, 56–66. [CrossRef]
44. Badiu, D.L.; Iojă, C.I.; Pătroescu, M.; Breuste, J.; Artmann, M.; Niță, M.R.; Grădinaru, S.R.; Hossu, C.A.; Onose, D.A. Is urban green space per capita a valuable target to achieve cities' sustainability goals? Romania as a case study. *Ecol. Indic.* **2016**, *70*, 53–66. [CrossRef]
45. Feng, S.; Li, J.; Sun, R.; Feng, Z.; Li, J.; Khan, M.S.; Jing, Y. The distribution and accessibility of Urban Parks in Beijing, China: Implications of social equity. *Int. J. Environ. Res. Public Health* **2019**, *16*, 4894. [CrossRef]
46. Central Statistical Office of Poland (GUS) (2020). Available online: http://www.gus.pl/ (accessed on 6 May 2019).
47. The Act on Nature Conservation. *J. Laws* **2004**. No. 92 Item 880. (In Polish)
48. Nieuwenhuijsen, M.J.; Khreis, H.; Triguero-Mas, M.; Gascon, M.; Dadvand, P. Fifty Shades of Green: Pathway to healthy urban living. *Epidemiology* **2017**, *28*, 63–71. [CrossRef] [PubMed]
49. Bosch, M.V.D.; Nieuwenhuijsen, M.J. No time to lose—Green the cities now. *Environ. Int.* **2017**, *99*, 343–350. [CrossRef]
50. Marmot, M. *Sustainable Development: The Key to Tackling Health Inequalities*; Sustainable Development Commission: London, UK, 2010; Available online: http://www.sd-commission.org.uk/publications.php@id=1053.html (accessed on 22 March 2020).
51. Mitchell, R.; Popham, F. Effect of exposure to natural environment on health inequalities: An observational population study. *Lancet* **2008**, *372*, 1655–1660. [CrossRef]
52. Nieuwenhuijsen, M.; Khreis, H. Car free cities: Pathway to healthy urban living. *Environ. Int.* **2016**, *94*, 251–262. [CrossRef] [PubMed]
53. Report 2007: An Analysis of Accessible Natural Greenspace Provision in the South East. Available online: http://www.forestry.gov.uk/southeastengland (accessed on 18 March 2020).
54. La Rosa, D. Accessibility to greenspaces: GIS based indicators for sustainable planning in a dense urban context. *Ecol. Indic.* **2014**, *42*, 122–134. [CrossRef]
55. Kerby, D.S. The simple difference formula: An approach to teaching nonparametric correlation. *Compr. Psychol.* **2014**, *3*, 11. [CrossRef]
56. Chen, Z.; Xu, B.; Gao, B. Assessing visual green effects of individual urban trees using airborne Lidar data. *Sci. Total Environ.* **2015**, *536*, 232–244. [CrossRef]
57. Capotorti, G.; Del Vico, E.; Anzellotti, I.; Celesti-Grapow, L. Combining the conservation of biodiversity with the provision of ecosystem services in urban green infrastructure planning: Critical features arising from a case study in the Metropolitan Area of Rome. *Sustainability* **2016**, *9*, 10. [CrossRef]
58. Korwel-Lejkowska, B.; Topa, E. Accessibility of urban parks, as elements of green infrastructure in Gdańsk. *Rozwój Regionalny I Polityka Regionalna* **2017**, *37*, 63–75. (In Polish) [CrossRef]
59. Landor-Yamagata, J.L.; Kowarik, I.; Fischer, L.K. Urban foraging in Berlin: People, plants and practices within the metropolitan green infrastructure. *Sustainability* **2018**, *10*, 1873. [CrossRef]
60. Noszczyk, T.; Hernik, J. Understanding the cadastre in rural areas in Poland after the socio-political transformation. *J. Spat. Sci.* **2019**, *64*. [CrossRef]

61. Kremer, P.; Hamstead, Z.; Haase, D.; McPhearson, T.; Frantzeskaki, N.; Andersson, E.; Kabisch, N.; Larondelle, N.; Rall, E.L.; Voigt, A.; et al. Key insights for the future of urban ecosystem services research. *Ecol. Soc.* **2016**, *21*, 21. [CrossRef]
62. Haase, D.; Larondelle, N.; Andersson, E.; Artmann, M.; Borgström, S.; Breuste, J.; Gómez-Baggethun, E.; Gren, Å.; Hamstead, Z.; Hansen, R.; et al. A quantitative review of urban ecosystem service assessments: Concepts, models, and implementation. *AMBIO* **2014**, *43*, 413–433. [CrossRef] [PubMed]
63. Zhang, J.; Cheng, Y.; Wei, W.; Zhao, B. Evaluating spatial disparity of access to public parks in gated and open communities with an improved G2SFCA model. *Sustainability* **2019**, *11*, 5910. [CrossRef]
64. Kronenberg, J.; Pietrzyk-Kaszyńska, A.; Zbieg, A.; Żak, B. Wasting collaboration potential: A study in urban green space governance in a post-transition country. *Environ. Sci. Policy* **2016**, *62*, 69–78. [CrossRef]

International Journal of
Environmental Research and Public Health

Article

Overcoming Barriers to Agriculture Green Technology Diffusion through Stakeholders in China: A Social Network Analysis

Wenke Wang [1,2], Jue Wang [1], Kebei Liu [1] and Yenchun Jim Wu [3,4,*]

1 Business School, Sichuan Normal University, Chengdu 610101, China; wangwk@sicnu.edu.cn (W.W.); 2017190342@stu.sicnu.edu.cn (J.W.); 2018190920@stu.sicnu.edu.cn (K.L.)
2 Sichuan Provincial Key Laboratory of Sci-tech Finance and Mathematical Finance, Sichuan University, Chengdu 610064, China
3 Graduate Institute of Global Business and Strategy, National Taiwan Normal University, Taipei 106, Taiwan
4 College of Management, National Taipei University of Education, Taipei 106, Taiwan
* Correspondence: ycwu@ntnu.edu.tw or wuyenchun@gmail.com

Received: 22 August 2020; Accepted: 23 September 2020; Published: 24 September 2020

Abstract: It is crucial to actively encourage the development of agriculture green technology, which has been regarded as one of the most effective solutions to the environmental degradation caused by agricultural activities. However, agriculture green technology diffusion is indeed a challenging task and still faces numerous barriers. The stakeholders who can potentially deal with these barriers, however, have been overlooked by previous studies. To address these issues, social network analysis was performed to identify critical stakeholders and barriers. Their interactions in agriculture green technology diffusion were analyzed based on the literature, a questionnaire survey and expert judgments. A two-mode network and two one-mode networks were used to analyze the relationships among the identified 12 barriers and 14 stakeholders who can influence these 12 barriers identified. The results show that agricultural research institutes, universities, agribusiness, agencies of township promotion, the government and farmers' relatives are key stakeholders and that the limited market demand for green technology and the high cost of its diffusion are two main barriers. However, poor green technology operability and farmer families in distress are factors that are not as important as previously perceived. Finally, some recommendations and suggestions are provided to promote agriculture green technology diffusion in China.

Keywords: agriculture green technology diffusion; stakeholders; social network analysis; barriers

---

1. Introduction

With the implementation of the rural revitalization strategy, China has experienced rapid development of agriculture and realized socialized etiquette and civility, effective governance and prosperous life. However, agriculture is the second-largest contributor to greenhouse gas emissions. To a large extent, its high-energy-consuming production technology has caused various environmental problems, which have greatly affected the development of agriculture [1]. Therefore, many scholars have conducted research related to this subject, and their results have shown that green technology can help reduce resource consumption and environmental pollution, while optimizing output [2]. Agricultural green technology aims to produce safe, pollution-free, high-quality, nutritious and green agricultural products. It is a group of technologies composed of multiple sub-technologies, including those related to pre-production (new variety of technology), mid-production (formula fertilization by soil testing, biological control technology of plant diseases and pests) and post-production (straw recycling technology). The diffusion of these technologies is a necessary condition for realizing food safety, and it provides the foundation and guarantee for the development of green agriculture [3]. Agriculture green technology can be adopted throughout the entire process of production, development and use of agricultural products. In recent years, developing countries have attached great importance to the development of agriculture green technology, which has promoted the creation of a green environment [4].

As a result of the numerous sustainable benefits of agriculture green technology, many governments and organizations have taken measures to promote its diffusion and adoption. However, it is still a challenging task. Based on the experience of diffusing green agricultural technology—such as green pest control, soil nutrient management (such as soil testing formula technology), conservation tillage and precision agriculture—it has been concluded that, even if a technology is useful, there are different factors that limit its effectiveness [5]. In research on the diffusion of agricultural technology, factors that inhibit its diffusion include uncertainty, information asymmetry, the number of adopters, land transfer and the credibility and authority of policymakers [6]. Lambrecht et al. analyzed the process of agricultural technology diffusion in eastern Congo, which showed that only 13% of farmers with an awareness of new technology adoption were willing to try it, and as many as 70% of farmers who tried a new technology were willing to continue using it [7]. To promote the diffusion and adoption of green technology, scholars have conducted many studies from different perspectives. These studies include (1) the impact of individual characteristics of farmers on the adoption of agriculture green technology [8]; (2) the impact of household resource endowment on the adoption of agriculture green technology [9]; (3) the impact of green agriculture promotion systems on the adoption of agriculture green technology [10]; and (4) the impact of economic risks on the adoption of agriculture green technology [11]. These studies have provided valuable information and ideas to guide developing countries in formulating policies that promote the diffusion and adoption of agriculture green technology.

In the process of promoting the diffusion of agriculture green technology, stakeholders play a vital role. Collaboration among stakeholders is an important step and integral in promoting the diffusion of agriculture green technology [4]. However, previous studies have overlooked stakeholder collaboration as a key to promoting agriculture green technology diffusion. The diffusion of agriculture green technology is increasingly influenced by diverse stakeholders due to factors such as farmers' demand, public pressure, market and supply chain pressure and government policies. Only the effective interaction of multiple stakeholders can promote the diffusion of agriculture green technology and improve the overall environmental performance. The organization, development and implementation of agriculture green technology diffusion require stakeholders to have inner motivation to stimulate the development of innovative diffusion activities. Inner motivation is mainly derived from objectives, diffusion capabilities and the diffusion power of stakeholders [12]. However, driven by self-interest, stakeholders may invest resources in individual goals, rather than collective needs. If serious consequences occur, stakeholders tend to shirk responsibility and pass it on to others [13]. Therefore, problems related to the process of green technology diffusion are mainly caused by insufficient collaboration among stakeholders [14]. To build effective collaboration among stakeholders in the diffusion of agriculture green technology, it is necessary to understand the power of each stakeholder to remove barriers. Power reflects the abilities of stakeholders to overcome specific barriers. Stakeholders with a high degree of power should be primarily responsible for related barriers [12]. Otherwise, conflicts may arise between stakeholders due to a mismatch of power and responsibility. Stakeholders should actively assume relevant responsibilities according to their power and abilities, which is a prerequisite for effective stakeholder collaboration in agriculture green technology diffusion.

At present, a limited amount of research has been conducted on stakeholder collaboration in the diffusion of agriculture green technology and the removal of barriers to diffusion by stakeholders. Researchers and practitioners agree that the diffusion of agriculture green technology is more complex and challenging because agriculture is "traditionally very conservative" and changes slowly. Therefore, it is of vital importance to clearly identify the relationship between stakeholders and diffusion barriers and identify crucial factors that affect the diffusion. In order to fill the gap in this aspect, this paper first analyzes barriers to the diffusion of agriculture green technology from the perspective of stakeholders. Then, a comprehensive analysis of stakeholders and barriers is provided, and their impact on the diffusion of agriculture green technology is quantified through the social network analysis (SNA) method. Finally, the relationship between different stakeholders and barriers is explored, and key stakeholders and barriers, as well as their impact, are also identified. To the best of our knowledge, this paper fills gaps in previous studies on green agriculture, which have rarely involved the relationship between different stakeholders and barriers. In our research, in contrast to traditional social network analysis, the one-mode network analysis examines the internal relationship in the same group of entities, while the two-mode network analysis explores the relationship between two different groups. The results of these analyses reveal interlinked stakeholders through the barriers they can jointly address and interlinked barriers through the stakeholders who have the power to influence them. Furthermore, the analysis of barrier factors from the perspective of stakeholders provides a useful reference for future research. In the process of agriculture green technology diffusion, the relationship between barriers and stakeholders, as well as the network characteristics of stakeholder collaboration, can be understood through this network model, and the cooperative relationship between stakeholders who overcome these barriers is also determined. Innovative viewpoints in this research can guide stakeholders to make correct decisions. This research not only enriches the theory of agriculture green technology diffusion and adoption but also provides insights into how China can effectively promote it and thereby facilitate agriculture development.

The rest of the paper is structured as follows: Section 2 reviews barriers to the diffusion of agriculture green technology and the related stakeholders, and Section 3 introduces social network analysis and the research methods adopted. Section 4 reports the research results and related indicator analysis, and Section 5 presents a discussion of related research. The conclusions and limitations of this study and future research directions are described in Section 6.

2. Literature Review

2.1. Barriers to the Diffusion of Agriculture Green Technology

The diffusion of green technologies refers to the process in which the source of innovation disseminates innovative green technologies among individuals and diffuses them throughout all of society [15]. It is a process in which new agricultural technologies, inventions and results are widely adopted by farmers over a certain period of time and by several channels, after going through the stages of cognition, decision-making, participation and implementation and acknowledgment [4]. The diffusion of agriculture green technology not only drives the spreading efficiency of general technological innovation but also purposefully integrates human technological innovation activities into the natural ecological cycle. As a result, it will not only increase the income of farmers but will also reduce the damage to the environment caused by current agricultural technologies and protect natural resources to a certain extent [16]. However, agriculture green technologies are developing and diffusing slowly [17].

At present, excessive consumption of the Earth's resources, a deteriorating environment and severe ecological damage have led to an increasing demand for agriculture green technology [2]. Although increasing importance has been attached to the development of green agriculture, and great efforts have been made to promote green technologies, problems still arise during the diffusion process and need to be resolved. Drawing on previous research, this paper systematically analyzes the content of the literature and determines corresponding barriers [18]. First, related journals and conference articles were searched in the databases of "Web of Science", "Compendex Engineering Village", ASCE library and CNKI, using keywords such as "barrier", "green", "agriculture green technology", "agriculture technology", "eco-friendly technology" and "ecological technology". By reviewing the abstracts and conclusions, the articles that met relevant requirements were assessed to determine their relevance to this study. When the objectives were identified, a content analysis was performed based on the TOE (technology–organization–environment) framework, which has been widely applied in the technology acceptance field [19]. Taking the various barriers that affect agriculture green technology as the research object, we identified and analyzed barriers to the diffusion of green technologies in terms of technology, organization and the environment.

2.1.1. Technological Barriers

The speed and effect of the diffusion of agriculture technologies are closely related to their nature. Currently, developing countries are putting more effort into innovating single agriculture technologies, while having relatively few supporting technical achievements and lacking effective technological support [10]. Therefore, it is difficult for them to adapt to the requirements of the development of green agriculture. This is also one of the important reasons that the diffusion of green technologies is difficult. As a result of the uncertainty and market risk of green technologies, many small and medium-sized enterprises are reluctant to make the investment in this field, which leads to insufficient investment in scientific research [20]. Insufficient technological innovation has led to a shortage in the supply of agriculture green technology.

During the diffusion of agriculture green technology, quite a few technology research institutes, limited by their own problems [21], do not have a comprehensive and in-depth understanding of the application of agriculture green technology, such as scarce scientific and technological resources, serious brain drain on green technologies and less tightly-bounded integration of industry, education and research. As a result, current technologies cannot meet market demand, and the diffusion of agriculture green technology faces difficulties. Furthermore, the level of informatization in green technologies also affects the spread of agriculture green technology [22]. In rural regions, geographic and information barriers make the channels through which farmers can obtain agriculture green technology relatively unidimensional. In addition, the poor applicability of present agriculture green technology and high diffusion costs have further impeded the spread of agriculture green technology.

2.1.2. Organization Barriers

The diffusion of agriculture green technology involves the active organization and participation of different stakeholders. Decisions made by enterprises and governments are of great significance to the spread of agriculture green technology. Relevant policies under implementation lack substantive supporting policies that could effectively promote the diffusion of green technologies [23], which has led to incomplete systems of research and science and promotion. Moreover, they face problems such as the disconnection between technology and production, rigid mechanisms and insufficient guarantees, so they cannot realize the development of modernized green agriculture. The diffusion of agriculture green technology will be seriously hindered by inadequate government policies and measures and poor organization [24].

Demonstration, application and promotion by agriculture technology workers are needed to transform agriculture green technology into productivity. As the work to promote agriculture technologies is not fully implemented by the associated government departments, promotion personnel have not received complete training in promoting agriculture green technology or in providing the necessary education. As a result of the aging of knowledge and a low level of personnel quality, farmers could not receive adequate technical guidance and technological upgrades, which will have an influence on the diffusion of agriculture green technology [25]. Moreover, as the farmers themselves have a low level of literacy, they are not capable of recognizing and applying agriculture green technology. This will further restrict farmers from using emerging agriculture green technology and restrain the spread of agriculture green technology [26].

2.1.3. Social and Market Barriers

Developing countries are still lagging behind in the market of agriculture green technology. On the one hand, farmers in these regions have a relatively low level of literacy [27]. They are conservative, show little interest in new technologies and new things, have a poor capability of acceptance and want to avoid all the associated risks. Therefore, they are skeptical of agriculture green technology, thus reducing their need in this regard [28].

The diffusion of agriculture green technology is also subject to the market environment. Great market risks and uncertainties in the innovation of agriculture green technology [29] make it hard to achieve a return on investment, thus making it difficult to obtain investment from financial institutions. Meanwhile, changes in farmers' preferences and reference groups, such as farmer households and family members, as well as market factors, have also influenced the adoption and diffusion of agriculture green technology. Familiarity with the product and the brand effect are key factors that influence the diffusion of agriculture green technology [30].

2.2. *Necessity of Research on Stakeholders*

With the transition of stakeholder theory from the dyadic perspective to the network perspective, cooperation among stakeholders has become a new goal in the field of stakeholder management [31]. Stakeholder cooperation means that a group of self-governing stakeholders take actions and make decisions on related issues by using shared rules and standards [32]. Factors such as conflict of interest, different opinions and complicated relations will impede cooperation [33]. A good collaboration of interest and relationships among stakeholders is helpful for achieving full integration and optimization of resources so as to realize successful coordination among stakeholders.

Power is one of the most salient attributes of stakeholders [34]. Power means the ability of social actors to change others' behaviors, so as to achieve their own objectives, regardless of opposition [35]. The power of stakeholders depends on a number of factors, such as the resources that they own [13] and their personal attributes and social status [36]. During their collaboration, stakeholders who have the most important resources will have more power [34]. Furthermore, the social network also highlights the strength of the structural positions of stakeholders [37], and the resources of stakeholders interact with their network positions. Therefore, it is necessary to understand the power of stakeholders and to determine their positions and relationships in the collaboration network, their attributes and the resources that they own.

In terms of research methods, the past decade has witnessed a number of studies that examined the factors of agriculture technology adoption using social network analysis. Wang et al. analyzed the efficiency of farmers' agricultural technology adoption and the interaction effect between the social network and extension service [38]. In their research, interaction and trust had positive effects on farmers' technology adoption efficiency, which was also reflected in our research. Ramirez discussed the influence of social networks on agricultural technology adoption; the study addressed the professional collaboration found in tenure relations and social and professional organizations, and the analysis showed that participation in organizations and kinship relationships are factors that influence agricultural technology adoption [39]. An interesting finding is that kinship relationships and township promotion agencies are also very important factors of stakeholders in our research. Kuang analyzed the reliability of interpersonal communication based on social networks and their information costs and found that the interpersonal communication of information established in social networks is most effective in the diffusion of agricultural technology [22], which is also a key stakeholder in our research.

Previous studies have shown that various barriers related to technologies, organizations, markets and social environments in the spread of agriculture green technology have inhibited the adoption and diffusion of agriculture green technology. At the same time, the diffusion of green technologies involves many stakeholders. Prior research and studies on the power and collaboration of stakeholders have suggested that the more powerful these stakeholders are, the more influence they have on the spread of green technologies. Effective collaboration among stakeholders will definitely have positive effects on the efficiency and performance of the diffusion of agriculture green technology. The existing research has barely explored how to effectively eliminate barriers to the diffusion of agriculture green technology from the perspective of stakeholders. Different resources provided by different stakeholders are needed to overcome these barriers. As a matter of fact, it is essential to study the influence of stakeholders on barriers to the diffusion of agriculture green technology, so as to improve the performance of this process.

3. Research Methods

This paper analyzes the complicated relationship between stakeholders and adoption barriers during the diffusion of agriculture green technology through social network analysis. Social network analysis is applied to examine the existence and strength of the relationship between a pair of stakeholders and adoption barriers. This theory emphasizes the "structure and mode" of the relationship and tries to identify the reasons behind adoption constraints and their corresponding influence [40].

3.1. Identification of the Stakeholders and Barriers

The establishment of social networks is based on nodes and links. This step mainly focuses on recognizing the network nodes, that is, finding the stakeholders of and barriers to the spread of agriculture green technology. First, the literature research method was used to define stakeholders and barriers [12]. Then, the Delphi method was used to identify more comprehensive and reasonable stakeholders and barriers through expert interviews using the data obtained in the first step [14]. The experts had to answer the following questions: (1) Does the literature research cover barriers or stakeholders that are not related to the diffusion of green technologies? (2) Can people understand the chosen barriers and stakeholders? (3) Are there any other barriers and stakeholders involved in the diffusion of green technologies?

In order to verify the validity and reliability of the chosen barriers and stakeholders, the experts interviewed all have rich research experience in this field and have unique insights into green technologies and the stakeholders involved. S* represents stakeholders and F# represents barriers, where * denotes the number of stakeholders and # is the number of barriers.

3.2. Determination of the Relationship between Stakeholders and Barriers

In order to analyze the influence of stakeholders on barriers to the spread of agriculture green technology, the snowball sampling technique, as well as questionnaires and expert seminars, were used to determine the relationship between stakeholders and barriers. Snowball sampling is particularly designed for cases that require professionals to give relevant opinions on issues in their field [41]. The questionnaires are mainly for researchers, managers and practitioners who have rich knowledge and experience in agriculture green technology [12]. The questionnaire was first distributed in November 2019. After modifying the questionnaire according to the opinions of respondents in a preliminary test, 297 respondents participated in the updated survey through a program named Questionnaire Star. The respondents include government administrators, company executives and workers at intermediary agencies, whose work is related to agriculture green technology, as well as research scholars and adopters of agriculture green technology. To ensure the reliability of the survey results, it was ensured that all respondents had more than three years of experience in studying, manufacturing and using agriculture green technology. In order to simplify the process, we used unified standards and determined the power of stakeholders on the barrier by applying an agreed-upon threshold value [42]. After the survey, a total of 11 expert representatives of agriculture green technology diffusion were invited to conduct workshops [14], including two developers, two market promotion representatives, two government administrators, two adopters and three academic researchers. All of them have more than 10 years of working experience and an in-depth understanding of agriculture green technology. A variety of methods were combined to expand the sample data. The process, including a survey with critical stakeholders and barriers and workshops with relevant experts, aims to minimize bias effects of dominant individuals and group pressure to conform. Therefore, the survey results and findings are of high accuracy from a scientific perspective. The result of this step is the definition of links in the stakeholder–barrier network.

An element in the stakeholder–barrier matrix refers to the condition in which a stakeholder handles barriers. It is composed of a group of stakeholders (X) and a group of barriers (Y). X_i represents one of the 14 stakeholders, Y_j denotes one of the 12 barriers, a_{ij} indicates stakeholders and X_i refers to whether one has the ability to solve barrier Y_j. The definitions are as below:

If $a_{ij} = 1$, then stakeholder X_i can resolve barrier Y_j;

If $a_{ij} = 0$, then stakeholder X_i cannot resolve barrier Y_j.

The stakeholder–barrier matrix was transformed into a stakeholder–stakeholder matrix and a barrier–barrier matrix. This can highlight the key stakeholders and barriers and evaluate the strength of the correlation on a quantitative basis.

3.3. Social Network Analysis

In contrast to traditional social network analysis, both the two-mode and one-mode network analysis were used to analyze the relationships between stakeholders and barriers, as well as the interrelationships between stakeholders or barriers, so as to build a good collaboration among stakeholders to address those barriers that are hard to solve.

3.3.1. Two-Mode Social Network Analysis

In this study, the stakeholders can be regarded as the subject of the action of the network, and different kinds of barriers can be seen as the "organization" or the "case" affiliated to it. The two-mode social network analysis method analyzes the power of stakeholders in the social network to determine its impact and the means of overcoming obstacles. Using a two-mode social network analysis, Gan et al. provided a creative perspective on overcoming barriers to off-site construction by engaging stakeholders [12]. However, in our research, the closeness centrality of the two-mode network was further analyzed to explore stakeholders' influence on the identified barriers.

1. Centrality analysis

(1) Degree centrality Degree centrality measures the number of nodes directly linked to one node. It focuses on the degree of the power of nodes only. In the stakeholder–barrier network, this implies that stakeholders have the power to handle the number of barriers. As for barriers, the degree centrality is equal to the number of involved stakeholders [43]. The degree centrality can be represented by Equation (1). Therefore, the greater the degree centrality, the more associated the points, and the more important the position in the affiliated network.

$$C_D(K) = deg_k = \sum_{j}^{N} A_{kj} \quad (1)$$

here, K represents the key node, j represents other nodes and N represents the total number of nodes. A_{kj} represents an element in the matrix of stakeholders and barriers.

(2) Closeness centrality The closeness centrality indicates the proximity of a network node to other nodes. The closeness centrality of a stakeholder refers to a function that represents the shortest network distance from the affected barrier to other stakeholders and barriers [44], which can be calculated by Equation (2). The greater the closeness centrality, the shorter

the average distance to other nodes, and the more advantageous its position in the process of information transmission.

$$C_C^{NM}(n_i) = \left[1 + \frac{\sum_{j=1}^{g+h} \min_k d(k,j)}{g+h+1}\right]^{-1} \quad (2)$$

here, $C_C^{NM}(n_i)$ represents the closeness centrality of stakeholders, g represents the number of stakeholders, h represents the number of barriers, k represents the barrier affected by a stakeholder, j represents some other stakeholder or barrier and $\min_k d(k,j)$ represents the shortest network distance between k and j. The closeness centrality of a barrier refers to a function that represents the shortest network distance from the affected stakeholder to other stakeholders and barriers [45], as shown in Equation (3).

$$C_C^{NM}(m_k) = \left[1 + \frac{\sum_{i=1}^{g+h} \min_j d(i,j)}{g+h+1}\right]^{-1} \quad (3)$$

Generally speaking, the closeness centrality reflects the proximity of members of the network. The shorter the distance, the more likely it is to be at the center of the whole network. In Equation (3), $C_C^{NM}(m_k)$ represents the closeness centrality of a barrier, g represents the number of stakeholders, h represents the number of barriers, i represents the stakeholder affected by a barrier, j represents some other stakeholder or barrier and $\min_j d(i,j)$ represents the shortest network distance between i and j.

(3) Betweenness centrality Betweenness centrality demonstrates the ability of a participant to play an intermediary role in the network [46], as is shown in Equation (4). It establishes the situation in which a node is located between two nodes and concentrates on the ability of a node to control the nodes at both ends. Only when each barrier affected by the stakeholder reaches a certain point will the stakeholder obtain betweenness centrality.

$$\frac{1}{2}\sum n_i, n_j \in m_k \frac{1}{X_{ij}^N} \quad (4)$$

here, n_j represents the other stakeholder of n_i and the co-shared barrier m_k, and X_{ij}^N represents the number of barriers shared by n_i and n_j. In the same way, if a barrier only relates to one stakeholder, then the barrier will reach the middle "point" of $\boldsymbol{g + h + 2}$.

2. Core–periphery analysis

The core–periphery network has an ideal structure in which the rows and columns are divided into two categories. The block in the main diagonal represents the core, and the other one is the periphery [44]. In the stakeholder–barrier network, the "core" is composed of subdivisions of a series of stakeholders. These stakeholders are closely linked to each barrier in the subdivisions. The stakeholder at the core can be regarded as the key stakeholder of the network. The "periphery", on the one hand, is composed of subdivisions that constitute a series of stakeholders that will not work together on the same barrier; on the other hand, it is composed of subdivisions of a series of barriers that are not correlated with each other [47]. This is because they are not connected by the same stakeholders. The core–periphery network structure can be expressed as Equations (5) and (6):

$$\rho = \sum_{i,j} a_{ij}\sigma_{ij} \quad (5)$$

$$\sigma_{ij} = \begin{cases} 1, if\ c_i = core\ or\ c_j = core \\ 0, other \end{cases} \quad (6)$$

where a_{ij} indicates whether there is a connection in the observed data, c_i represents the category (core or periphery) of node i, c_j denotes the category (core or periphery) of node j and σ_{ij} indicates whether there is a connection in the real structure.

3.3.2. One-Mode Social Network Analysis

During the process of agriculture green technology diffusion, the analysis of the stakeholder-barrier network can help identify the power of stakeholders over the barriers. Furthermore, it is a challenge for stakeholders to address multiple barriers as their resources are limited. By analyzing the stakeholder-stakeholder network and barrier-barrier network, the main stakeholders and barriers can be observed more clearly. The powerful stakeholders can be identified through the stakeholder-stakeholder network, thus the effective collaboration among them can be established to maximize their resources, so as to better address the barriers of agriculture green technology diffusion. While the barrier-barrier network can help stakeholders have a full understanding of the relationships between the barriers, and make the contributes to the decision making of agriculture green technology diffusion.

1. Network measures

   (1) Density analysis of the stakeholder/barrier network Density indicates the closeness of network members [48]. When the entire network has a greater density, it is more likely to influence the participants in terms of attitude and behavior. In return, the network is more complicated. The calculation of density is shown in Equation (7), with 0 ≤ density ≤ 1.

   $$D(G) = \frac{K}{N(N-1)} \quad (7)$$

   here, $D(G)$ represents the density of network G, K indicates the number of existing relationships and N represents the total node number of network G.

   (2) Cohesion analysis of the stakeholder/barrier network Cohesion indicates the relationship between two contacts of an action provider [48]. When cohesion is 0, there is no connection among the nodes; when cohesion is 1, there is a strong connection among the nodes, and any of them is capable of obtaining the same information and network benefits. Stronger cohesion means a closer relationship and a more complicated network. The calculation of cohesion is shown in Equation (8), with 0 ≤ cohesion ≤ 1.

   $$C(G) = \frac{\sum Adjm^2}{N(N-1)} \quad (8)$$

   here, $C(G)$ represents the cohesion index, $Adjm^2$ represents the shortest distance between two nodes and N represents the total node number of the network.

2. In/out-degree analysis In/out-degree analysis reflects the strength of the influence of nodes. It divides the correlation strength between two nodes into outdegree and indegree. The outdegree of a node indicates the total number of nodes that it directly points to, while indegree indicates the total number of dots that other nodes point to [48]. The calculations are shown in Equations (9) and (10):

   $$D_{D_o}(n_i) = \sum_{j}^{n} X_{ij} \quad (9)$$

$$D_{D_i}(n_i) = \sum_{j}^{n} X_{ji} \tag{10}$$

where $D_{Do}(n_i)$ represents the outdegree, X_{ij} represents the correlation between node i and node j, $D_{Di}(n_i)$ represents the indegree, and X_{ji} represents the correlation between node j and node i. If there is a correlation, then the value is 1; if there is no correlation, then the value is 0.

3. Structural Hole analysis of the stakeholder/barrier network The structural hole indicates a non-redundant connection among actors. It offers opportunities for holders to obtain an "information benefit" and "control benefit", so as to make them more competitive than members in another position of the network [48]. The parameters, range and formula are described below.

(1) Effective size The effective size of an actor equals the non-redundant factors of the network in which the actor's individual network deducts the redundancy of the network. It can measure the overall strength of the influence and the importance of nodes and indicates the control that a node has over the relationship between other nodes. The more effective the size of a node, the freer its action. The calculation of effective size is shown in Equation (11), with effective size > 0.

$$ES_i = s_i - \frac{2t}{s_i} \tag{11}$$

here, t represents the number of correlations in the network, and s_i represents the size of node i.

(2) Efficiency Efficiency can be used to describe the influence of a node on other related nodes in the network. The efficiency of a node is equal to the ratio of its effective size to its actual size. The higher the efficiency of the node, the greater its influence in the control process. The efficiency of node i is calculated by Equation (12), with $0 \leq$ efficiency ≤ 1.

$$EF_i = \frac{ES_i}{n} \tag{12}$$

here, n represents the number of nodes and ES_i represents the effective size.

(3) Aggregate constraint Aggregate constraint indicates the ability of a node to use a structural hole. When a node of a single network is directly or indirectly connected to another one, it causes an aggregate constraint. The more aggregate constraint that a node has, the greater constraint that it has over the node, with $0 \leq$ aggregate constraint ≤ 1. The aggregate constraint can be calculated by Equation (13):

$$c_{ij} = \left(p_{ij} + \sum p_{iq}p_{qj}\right), \; i \neq q \neq j. \tag{13}$$

In this formula, node i and node j both adjoin node q, p_{ij} represents the weight ratio of node j among all the neighboring points of node i, and p_{iq} and p_{qj} represent the weight ratio of node q among all neighboring points of node i and node j, respectively.

4. Results

4.1. Establishment of the Relation Network

In this study, we performed a literature review and expert interviews to screen out stakeholders and barrier factors for agriculture green technology diffusion after two rounds of the Delphi survey. We invited 20 experts to attend the discussion group, including eight scholars who excel in green agriculture and 12 experts in green production and management of agriculture and agriculture technology. In round I, the background and purpose of this research was sent to respondents by e-mail. The respondents were instructed to screen out the stakeholders and barrier factors of agriculture green technology diffusion that had been collected and to list their reasons. In round II, all the data and information collected in round I were analyzed, and the results were provided to these experts to allow them to re-evaluate and judge the influencing factors. A total of 15 barriers and 20 stakeholders were determined preliminarily, but after documentary review and expert selection, a total of 12 barriers and 14 stakeholders remained, which are listed in Table 1.

Table 1. Barriers and stakeholders affecting agriculture green technology diffusion.

Code	Barriers	Code	Stakeholders
F1	High cost of green technology innovation	S1	Agricultural Research Institute
F2	Poor green technology applicability	S2	University
F3	Poor green technology operability	S3	Agribusiness
F4	Non-significant benefits of green technology	S4	Supplier
F5	Farmer families in distress	S5	Researcher
F6	Limited knowledge of farmer	S6	Agency of township promotion
F7	Unscientific green technology promotion	S7	Media
F8	Inadequate implementation of policies	S8	Seller
F9	Lack of codes and standards of green technology	S9	Technical promoter
F10	High cost of green technology diffusion	S10	Government
F11	Limited market demand for green technology	S11	Intermediary institutions
F12	Low level of rural economy development	S12	Farmer
		S13	Competitor
		S14	Farmers' relatives

From the results of questionnaires and symposiums, the stakeholder–barrier network of agriculture green technology diffusion can be established by analyzing a relationships matrix between stakeholders and barriers and can be visualized through software [44], which is shown in Figure 1. The red circular nodes and blue rectangular nodes represent stakeholders and barriers, respectively, while the size of the nodes reflects the centrality. The key stakeholders and barriers are in the center of the network, so they have more links than other nodes. In a complicated system, a high cost of green technology diffusion and limited market demand for green technology are two major barriers that must be addressed by more stakeholders compared with other barriers. The main stakeholders that can influence barriers include agricultural research institutes, universities, agribusiness, agencies of township promotion, the government and farmers' relatives.

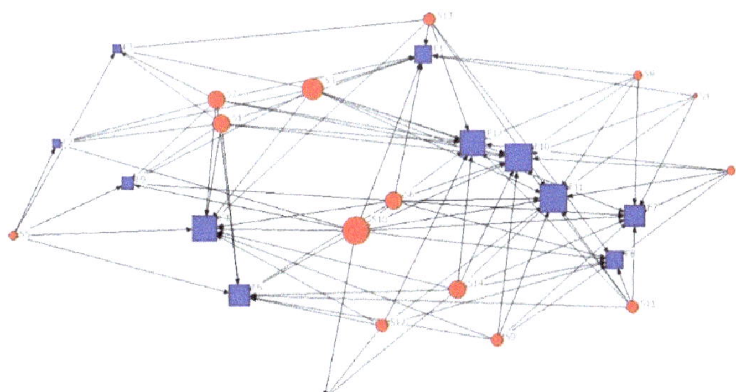

Figure 1. Visualization of the stakeholder–barrier network.

4.2. Stakeholder–Barrier Network Analysis

4.2.1. Centrality Analysis of Stakeholder–Barrier Network

Table 2 lists the centralities of the 14 stakeholders and 12 barriers that remained after analyzing the centrality of the stakeholder–barrier two-mode network. The results include degree centrality, betweenness centrality and closeness centrality

Table 2. The centrality of nodes in the stakeholder–barrier network.

Stakeholders	DC	Rank	CC	Rank	BC	Rank	Barriers	DC	Rank	CC	Rank	BC	Rank
S1	0.667	3	0.826	3	0.047	4	F1	0.571	7	0.750	7	0.046	7
S2	0.667	3	0.826	3	0.047	4	F2	0.357	10	0.667	10	0.013	11
S3	0.750	2	0.864	2	0.069	2	F3	0.357	10	0.643	11	0.014	10
S4	0.333	14	0.679	13	0.006	14	F4	0.714	3	0.818	3	0.077	3
S5	0.417	11	0.655	14	0.014	11	F5	0.214	12	0.600	12	0.003	12
S6	0.667	3	0.826	3	0.042	6	F6	0.643	5	0.783	5	0.064	5
S7	0.417	11	0.704	11	0.009	13	F7	0.643	5	0.783	5	0.052	6
S8	0.417	11	0.704	11	0.010	12	F8	0.571	7	0.750	7	0.037	8
S9	0.500	7	0.760	7	0.018	9	F9	0.429	9	0.692	9	0.020	9
S10	0.917	1	0.950	1	0.116	1	F10	0.786	1	0.857	1	0.090	1
S11	0.500	7	0.760	7	0.017	10	F11	0.786	1	0.857	1	0.089	2
S12	0.500	7	0.760	7	0.027	8	F12	0.714	3	0.818	3	0.076	4
S13	0.500	7	0.760	7	0.030	7							
S14	0.667	3	0.826	3	0.048	3							

DC = degree centrality, CC = closeness centrality, BC = betweenness centrality.

As shown in Table 2, the top three stakeholders in terms of degree centrality, betweenness centrality and closeness centrality are S10 (government), S3 (agribusiness) and S14 (farmers' relatives), indicating that these stakeholders secure an important position in the network. The top one is S10 (government), indicating that this kind of stakeholder can solve many barriers with a variety of resources, and is closely related to other objects. It plays the most important role in the network. In addition, these three kinds of stakeholders come from three levels: the diffusion source level, the promotion channel level and the adopter level, which means that agriculture green technology diffusion involves various stakeholders and that stakeholders at each level play a part in this process.

Furthermore, the degree centrality, betweenness centrality and closeness centrality of F10 (high cost of green technology diffusion) and F11 (limited market demand for green technology) are the top two factors in the centrality analysis of barriers. This signifies that these two barriers are key difficulties in agriculture green technology diffusion and should be paid more attention to in future research and development. In addition, other barriers, such as F4 (non-significant benefits of green technology), F6 (limited knowledge of farmer), F7 (unscientific green technology promotion) and F12 (low level of rural economy development), are among the top barriers, which indicates that more stakeholders are needed to address these barriers together.

4.2.2. Core–Periphery Network Structure of the Stakeholder–Barrier Network

The core–periphery analysis result of the stakeholder–barrier network is displayed in a density matrix (Table 3), and the final fitness is 0.624. The interactive density between the stakeholders and barriers is 0.778, which shows that stakeholders and barriers at the core are closely related. The density between core stakeholders and peripheral barriers and that between the core barriers and peripheral stakeholders are 0.545 and 0.593, respectively, which are significantly lower than the density index of core sub-divisions. This indicates a loose relationship between core stakeholders and peripheral barriers and a distant relationship between core obstacles and peripheral stakeholders. Therefore, the stakeholder–barrier network exhibits a core–periphery structure.

Table 3. Density matrix.

		Barrier	
		Core	**Periphery**
Stakeholder	Core	0.778	0.545
	Periphery	0.593	0.167
	Overall network density: 0.526		
	Final fitness: 0.624		

The core stakeholders and barriers (Table 4) are determined: there are six stakeholders and eight barriers at the core. The six stakeholders are S1 (Agricultural Research Institute), S2 (university), S3 (agribusiness), S10 (government), S6 (agency of township promotion) and S14 (farmers' relatives). The eight barriers at the core account for two-thirds of the total, with the exclusion of F2 (poor green technology applicability), F3 (poor green technology operability), F5 (non-significant benefits of green technology) and F9 (lack of codes and standards of green technology). Since core stakeholders are more capable of tackling these barriers, intensive interactions between core stakeholders are likely to occur. This boosts information exchange between core stakeholders, who may contribute to the formation of common values, attitudes and benefits in agriculture green technology.

Table 4. Core–periphery structure model of the stakeholder–barrier network.

	F1	**F8**	**F6**	**F4**	**F11**	**F12**	**F7**	**F10**	**F2**	**F3**	**F5**	**F9**
S1	1		1	1		1		1	1	1		1
S2	1		1	1		1		1	1	1		1
S3	1			1	1	1	1	1	1	1		1
S14		1	1	1	1	1	1	1			1	
S6	1	1	1	1	1		1	1				1
S10	1	1	1	1	1	1	1	1	1		1	1
S5			1	1					1	1		1
S8	1				1	1	1	1				
S9		1	1	1	1		1	1				
S4	1				1		1	1				
S11		1			1	1	1	1				
S12		1	1	1	1	1					1	
S13	1	1	1	1	1	1				1		
S7		1			1	1	1	1				

The core–periphery structure in this study can provide guidance for establishing an active collaboration network for stakeholders. For instance, S10 (government), located in the core, is able to solve 11 barriers, among which eight are in the core and four are in the periphery. As shown in Table 4, the eight core barriers may also be addressed by other core stakeholders. As a result, a collaboration between the core stakeholders and S10 (government) should be established. Given its high centrality degree, S10 (government) should play a key mediating role in the partnership between these core stakeholders. According to Table 4, the eight peripheral stakeholders listed may be able to deal with these core barriers even if they have a weaker influence than core stakeholders. Therefore, core stakeholders can team up with peripheral counterparts to solve core barriers. The four peripheral barriers can be solved by both core and peripheral stakeholders. For instance, F11 (limited market demand of green technology) can be affected and addressed by the peripheral stakeholder S8 (Seller), as well as by the four core stakeholders, namely, S10 (government), S3 (agribusiness), S6 (agency of township promotion) and S14 (farmers' relatives).

4.3. Stakeholder/Barrier Network Analysis

4.3.1. Stakeholder/Barrier Network Measures

It is indispensable to transform the two-mode stakeholder–barrier network into two one-mode networks of barrier–barrier and stakeholder–stakeholder. The density and cohesion of the one-mode networks of barrier–barrier and stakeholder–stakeholder are analyzed to understand the characteristics of the whole networks.

The density of the barrier network and stakeholder network is 0.999 and 0.967, respectively, indicating that they have relatively high density and that the relationships between barriers and stakeholders are complicated. The cohesion index of the barrier network is 0.985, signifying a close and complicated relationship among barriers. The cohesion index of the stakeholder network is 0.973, showing that stakeholders can contact each other. When a barrier emerges, stakeholders participating in the program of agriculture green technology may all be involved. Statistics of the cohesion index show that the relationship among stakeholders is close and intricate.

4.3.2. Analysis of the in/out-Degree of the Stakeholder/Barrier Network

As a result of the symmetrical nature of the barrier network, the indegree of each node equals the outdegree. In addition, the standard deviation is 0.373, indicating that the influence of each barrier is relatively concentrated. Figure 2 shows the degree of influence of each barrier. As shown in the figure, the highest degree of the barriers below is 11, including F1 (high cost of green technology innovation), F2 (poor green technology applicability), F4 (non-significant benefits of green technology), F6 (limited knowledge of farmer), F7 (unscientific green technology promotion), F8 (inadequate implementation of policies), F9 (lack of codes and standards of green technology), F10 (high cost of green technology diffusion), F11 (limited market demand of green technology) and F12 (low level of rural economy development). This reveals that they are more likely to lead to the occurrence of other barriers. By contrast, the degree of influence of F3 (poor green technology operability) and F5 (farmer families in distress) is 10, which is lower than the average level of 11, showing that they will cause a less direct impact on other barriers.

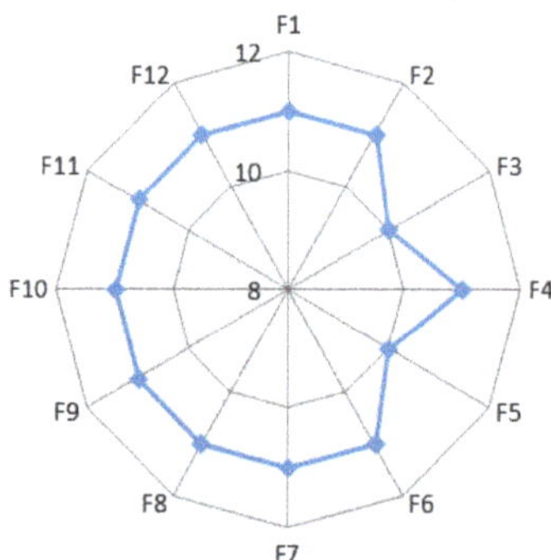

Figure 2. In/out-degree for the one-mode barrier social network.

Similar to the barrier network, the in/out-degree of each node of the stakeholder network remains the same. The overall degree is 176, while the average degree of the stakeholders is 12.57. Thus, a given stakeholder will influence an average of 12.57 stakeholders, including the original stakeholder, and, in turn, these stakeholders will also influence the original stakeholder. The influence of a particular stakeholder on others is relatively concentrated and vice versa. The standard deviation is 0.821.

Figure 3 shows the in/out-degree of stakeholders. S1 (Agricultural Research Institute), S2 (university), S3 (agribusiness), S6 (agency of township promotion), S9 (technical promoter), S10 (government) S11 (intermediary institutions), S12 (farmer), S13 (competitor) and S14 (farmers' relative/relatives of farmer) all reach the highest level of 13 (higher than the average of 12.57). Therefore, these stakeholders will have the greatest impact on barriers and play an important part in the stakeholder network. In contrast, S8 (seller), S4 (supplier), S7 (media) and S5 (researcher) have less impact on other stakeholders; that is, they have little impact on addressing barriers. Among them, sellers, suppliers and media, of which the in/out-degree is 12, play a more important role than scientific researchers, of which the in/out-degree is 10, in addressing barriers.

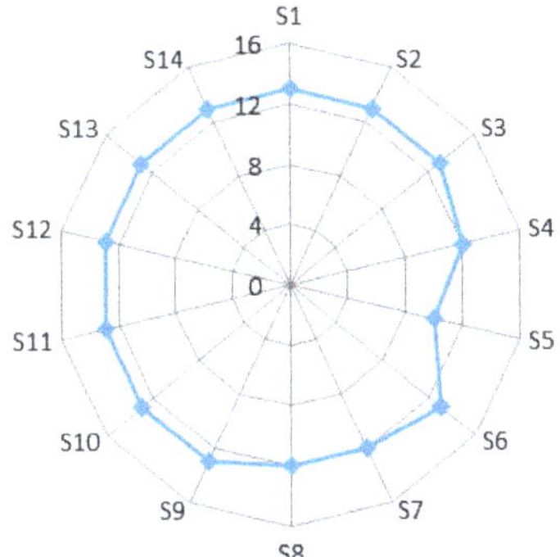

Figure 3. In/out-degree for the one-mode stakeholder social network.

4.3.3. Structural Hole Analysis of the Stakeholder/Barrier Network

The measurement of a structural hole will help identify barriers that have a great impact on other non-associated barriers in the barrier–barrier network, as well as stakeholders that have a stronger influence on other non-associated stakeholders in the stakeholder–stakeholder network.

Figure 4 shows the measurement results of the structural holes in the barrier–barrier network. The values of Y-axis are the results for measuring the structural hole by effective size, efficiency and aggregate constraint. It is clear that the efficiency of the structural holes in the barrier network is consistent with the distribution of effective size, while it has a negative correlation with the aggregate constraint. F3 (poor green technology operability) and F5 (farmer families in distress) are among the weaker barriers in the barrier network. Their small effective size and efficiency indicate a relatively low level of controllability. However, the aggregate constraint of these two barriers is relatively high, which means that they play a weaker role in the barrier network, as they are constrained by the other ten barriers.

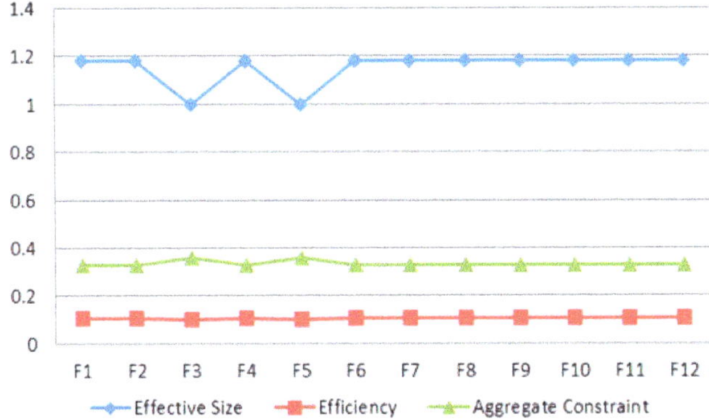

Figure 4. Structural holes for the barrier network.

The measurement results of the structural holes in the stakeholder network are shown in Figure 5. To be specific, the effective size of S4 (supplier), S5 (researcher), S7 (media) and S8 (seller) is relatively small, while the efficiency is relatively low. This demonstrates weak controllability. In addition, they have a strong aggregate constraint, showing that they do not play a key role in the control of barriers.

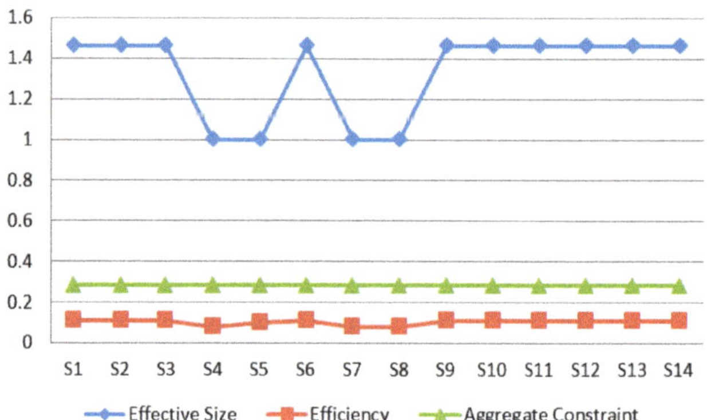

Figure 5. Structural holes for the stakeholder network.

5. Discussion

Agriculture is being upgraded with the advancement of the era. Green agriculture has gained unprecedented attention in China. Quite a few scholars have conducted research into agriculture green technology diffusion from different perspectives. However, most of them only analyzed barriers or stakeholders and put forward advice in this regard. Only a few of them have focused on the key subject of stakeholders overcoming barriers. This study offers new perspectives for exploring collaboration among stakeholders in agriculture green technology diffusion. The results of this analysis are discussed in this section.

The core and most important stakeholders identified in the two-mode network analysis are agricultural research institutes, universities, agribusiness, the government, agencies of township promotion and farmers' relatives, which can influence other stakeholders with their power and reputation, and they are more likely to address core barriers. The key barriers are the limited market demand for green technology and the high cost of its diffusion, which are also found in the one-mode network analysis. Furthermore, the results of the one-mode network analysis also show that poor green technology operability and farmer families in distress are factors that are not as important as previously perceived, and the relatives of farmers and township promotion agencies are becoming very important in social networks. From the analysis of both the two-mode and one-mode networks, it is apparent that the government has the strongest power over the barriers and can affect other stakeholders. Therefore, it is important for the government to use its power to establish collaboration among the stakeholders, especially those who are in the core position and can best address the barriers.

(1) Survey results and findings show that "high cost of green technology diffusion" and "limited market demand for green technology" are deemed important barriers in the social network. Although agriculture green technology has gained increasing attention from all walks of society, it is still developing slowly. The majority of stakeholders are accustomed to traditional agriculture technologies. Many of them, such as agribusinesses and farmers, hold a negative attitude toward adopting agriculture green technology. They pay considerable attention to the cost of agriculture green technology diffusion during the research, innovation, production, promotion and application process [45], which causes the cost of agriculture green technology adoption to become a key factor of whether they will adopt agriculture green technology

or not. In remote rural areas and ethnic minority-inhabited regions in West China in particular, the increase in diffusion costs has directly affected the adoption of agriculture green technology. Therefore, high diffusion costs will not only lead to negative attitudes of most stakeholders toward agriculture green technology, but also increase the proportion of farmers who adopt traditional agriculture technology. These two factors may significantly reduce the market demand for agriculture green technology, thus reducing its supply [1]. This could also explain why the "high cost of green technology diffusion" and "limited market demand for green technology" are regarded as major barriers to the diffusion of green technologies during the promotion and diffusion process.

(2) As a key stakeholder in the agriculture green technology diffusion in China, the "government" has a clear central position in the network and has a strong influence. It can influence more barriers than other stakeholders and plays an important role in promoting stakeholder collaboration. Although the Chinese government has attached great importance to the development and diffusion of agriculture green technology in recent years, it has not diffused as well as expected. As a result of externalities and the path dependence effect [49], the innovation and diffusion of agriculture green technology require policy support from the government to reach an optimal level. Effective government environmental policy is a necessary intervention that will exert pressure on enterprises to reduce emissions [50], form shadow prices of emissions and help induce the innovation and diffusion of green technology by many stakeholders. As the main advocate of agriculture green technology, the government's incentives, supervision and management have a direct impact on its diffusion and promotion. However, as a result of imperfect laws, regulations and supervision systems, government power is limited, and cannot reach all areas in many sparsely populated rural areas, resulting in the failure to implement agriculture green technology [24].

(3) Two barriers—poor green technology operability and farmer families in distress—are not as important as previously perceived, which can be seen from the results of social network analysis. In recent years, China has attached great importance to the development of green agriculture [4], and the trading market for agriculture green technology has gradually matured. The strength of research and development, maturity, usefulness and ease of use of agriculture green technology have also been reinforced by rigorous market testing and screening [51]. For farmers with a low level of cultural knowledge, its operability can still be effective, so it is less important than public awareness. Second, in order to achieve the "two centenary goals", China has made two strategic decisions—rural revitalization and poverty alleviation—at the national level, greatly improving the living standards of farmers and gradually realizing common prosperity by eliminating poverty and improving people's livelihood. Therefore, the barrier "farmer families in distress" is becoming less important, and is no longer the main barrier to the diffusion of agriculture green technology.

(4) As expected, six core stakeholders, namely, agricultural research institutes, universities, agribusiness, government, agencies of township promotion and farmers' relatives, can influence other stakeholders with their power and reputation, and they are more likely to address core barriers. Whether the core stakeholders can achieve effective cooperation and collaboration has greatly affected the development of agriculture green technology. The entire process of the diffusion of agriculture green technology, from the establishment of goals to the realization of technology and the determination of the final operational flow, requires effective coordination among multiple stakeholders. The information exchange, intensive interaction and cooperation among the core stakeholders, including agricultural research institutes, universities, agribusiness, government, agencies of township promotion and farmers' relatives, will facilitate common values, attitudes and interests in agriculture green technology, thus improving its acceptance by other stakeholders and further promote its diffusion [52]. Furthermore, given its high centrality degree, the government should play a key intermediary role in promoting partnerships among these

core stakeholders. Rural areas need the government to design better rules and regulations to greatly enhance the level of agriculture green technology, because of its system, scale and industrial characteristics.

(5) An interesting finding is that relatives of farmers and township promotion agencies are becoming very important in social networks. At present, farmers' relatives and township promotion agencies need to further deepen their understanding of agriculture green technology. Although an increasing number of adopters and farmers realize that ecological and environmental damage is a serious problem, they often consider it to be an issue that needs to be addressed by the government, and few farmers are concerned about the adoption of agriculture green technology. As a result of the structure of rural society, a relatively fixed social network is established in relatively closed spaces, social norms are relatively easy to form, and the social relationship between individuals is more affected by factors such as blood kinship and geographical scope, which can significantly affect the adoption of green technology by farmers [53,54]. At the same time, because of dispersed settlement, decentralized management and a low degree of organization in rural areas, the diffusion of agriculture green technology is hindered. Therefore, farmers' relatives and township promotion agencies play an increasingly important role in the diffusion and promotion of agriculture green technology.

6. Conclusions

Agriculture green technology is considered to be a useful tool to address various development issues, but it is still in its infancy in many countries, especially in developing countries like China. In the process of studying agriculture green technology diffusion, stakeholder collaboration as a key factor has been neglected by scholars, and it is not clear who among the stakeholders can solve which barrier. Through the analysis of the stakeholder–barrier network, this paper links the identified barriers to stakeholders and provides an innovative perspective for exploring stakeholder collaboration in the process of agriculture green technology diffusion.

With the results of the social network analysis method, this paper discusses the influence of stakeholders on various barriers to agriculture green technology diffusion. Two-mode social networks, one-mode stakeholder networks and one-mode barrier networks are all used to study the relationship between stakeholders and barriers. This expands the traditional research model on stakeholders and barriers and provides theoretical and practical significance for promoting the diffusion of agriculture green technology. The following important findings are drawn from this study: (1) Compared with other barriers, "high cost of green technology diffusion" and "limited market demand of green technology" have more serious impacts on the diffusion of agriculture green technology. (2) The government, as the most influential stakeholder, plays an important role in overcoming barriers to the diffusion of agriculture green technology and promoting stakeholder collaboration. (3) Some barriers, such as poor green technology operability and farmer families in distress, are not as important as usually perceived. (4) Collaboration among core stakeholders, such as agricultural research institutes, universities, agribusiness, the government, agencies of township promotion and farmers' relatives, can influence other stakeholders and facilitate the resolution of barriers. Effective promotion of collaboration among core stakeholders plays an extremely important role in overcoming barriers to the diffusion of agriculture green technology. (5) Farmers' relatives and agencies of township promotion are becoming very important in social networks. Although these research results need to be tested with more in-depth practical investigations and academic research, the results of this study can provide stakeholders with new ideas to remove barriers to the diffusion of agriculture green technology, and can serve as a policy- and decision-making reference for government management departments and enterprises to facilitate this diffusion.

On the basis of the results of the analysis, this paper proposes some suggestions for promoting the diffusion of agriculture green technology. First, in order to successfully diffuse and widely implement agriculture green technologies in China, it is necessary to further actively respond to and implement the two strategic decisions of rural revitalization and poverty alleviation, continuously improve the living standards of farmers' families, promote the development of green agriculture as a whole, boost the performance of agriculture green technology diffusion, reduce the cost of diffusion and vigorously tap the green technology market, thereby increasing the demand for agriculture green technology.

Second, as the main advocate of agriculture green technology, the government's environmental policies can help encourage many stakeholders to cooperate in the diffusion of green technology innovation, thus addressing many diffusion barriers. If environmental policy tools, such as carbon tax and subsidies, can be integrated and synergistically combined, then the results will be better.

Third, the government should actively respond to the trend of agricultural development and promote collaboration among core stakeholders, such as agricultural research institutes, universities, agribusiness, government, agencies of township promotion and farmers' relatives through policy incentives and regulation design, and it should give full play to the basic role of market mechanisms in resource allocation, build an innovation and diffusion system of agriculture green technology based on stakeholder coordination, promote the supply of green technology innovation, reduce the cost, improve efficiency and further promote the diffusion of agriculture green technology.

Finally, it is necessary to attach importance to the role of farmers' relatives and agencies of township promotion in the diffusion of agriculture green technology, raise farmers' awareness of green development, promote awareness of agriculture green technology among farmers' relatives and township promotion agencies and strengthen their communication to promote barrier removal with policy guidance and social education. Farmers' relatives and township promotion agencies should be encouraged to actively participate in the diffusion of agriculture green technology with policy incentives and benefit-sharing mechanisms.

This study also has some limitations. First, one limitation of this study is that the data are from China, where agriculture green technology is in its initial stage of development, so the results of this study can only reflect the role of stakeholders in the diffusion of agricultural green technology in China, and not in countries in which the development of agriculture green technology is relatively mature.

For future research, international comparison research can be conducted in these countries in order to establish international benchmarks for the acceptance of agriculture green technology. As the diffusion of agricultural green technology is a complex process, involving basic research, applied research and on-farm research to adapt new techniques to local conditions, participatory extension strategies, the establishment of farmer-to-farmer networks and the acquisition of an in-depth understanding of household dynamics, decision-making and community socioeconomic conditions, a modeling approach can be considered the first step to assess the dynamics of technology transfer and adoption. Future research may place a focus on carrying out follow-up on-farm research and surveys to validate, strengthen or fine-tune the findings from a valuable modeling approach, so as to better understand interactions among stakeholders and barriers to diffusion. Additionally, agriculture green technologies may be better adopted among some groups of farmers compared with those with less education or more limited resources. Future research can be carried out on target farm groups with similar socioeconomic backgrounds, to clarify the relationship between stakeholders and diffusion barriers and identify crucial factors that affect adoption and diffusion among groups of farmers with different socioeconomic backgrounds.

Author Contributions: Conceptualization, Y.J.W. and W.W.; Methodology, W.W.; Formal Analysis, Y.J.W.; Investigation, W.W. and J.W.; Data Curation, W.W., J.W. and K.L.; Writing—Original Draft Preparation, W.W. and K.L.; Writing—Review & Editing, Y.J.W. and W.W. All authors have read and agreed to the published version of the manuscript.

Funding: This research was funded by the International Science and Technology Innovation Cooperation Project of Sichuan Province (20GJHZ0039), The Social Science Planning Project of Sichuan Province (SC19B111), 2020 Philosophy and Social Science Planning Project of Chengdu (YY0520200707), and the Ministry of Science and Technology, Taiwan (MOST 108-2511-H-003-034-MY2).

Acknowledgments: The acquisition of data was a complex process in this study. The authors are grateful to all the respondents in the questionnaire and the participation of different industry experts. In addition, we appreciate editors and anonymous reviewers for the invaluable comments and suggestions that helped in improving this paper.

Conflicts of Interest: The authors declare no conflict of interest.

References

1. Xu, X.C.; Huang, X.Q.; Huang, J.; Gao, X.; Chen, L.H. Spatial-Temporal Characteristics of Agriculture Green Total Factor Productivity in China, 1998–2016: Based on More Sophisticated Calculations of Carbon Emissions. *Int. J. Environ. Res. Public Health* **2019**, *16*, 3932. [CrossRef] [PubMed]
2. Krawczyk, A.; Domagala-Swiatkiewicz, I.; Lis-Krzyscin, A. The effect of substrate on growth and nutritional status of native xerothermic species grown in extensive green roof technology. *Ecol. Eng.* **2017**, *108*, 194–202. [CrossRef]
3. Bukchin, S.; Kerret, D. Food for Hope: The Role of Personal Resources in Farmers' Adoption of Green Technology. *Sustainability* **2018**, *10*, 1615. [CrossRef]
4. Rupani, P.F.; Delarestaghi, R.M.; Asadi, H.; Rezania, S.; Park, J.; Abbaspour, M.; Shao, W. Current Scenario of the Tehran Municipal Solid Waste Handling Rules towards Green Technology. *Int. J. Environ. Res. Public Health* **2019**, *16*, 979. [CrossRef] [PubMed]
5. Chen, M.P.; Zhang, L.; Teng, F.; Dai, J.J.; Li, Z.; Wang, Z.Q.; Li, Y.T. Climate technology transfer in BRI era: Needs, priorities, and barriers from receivers' perspective. *Ecosyst. Health Sustain.* **2020**, *6*, 1–12. [CrossRef]
6. Johnson, M. Barriers to innovation adoption: A study of e-markets. *Ind. Manag. Data Syst.* **2010**, *110*, 157–174. [CrossRef]
7. Lambrecht, I.; Vanlauwe, B.; Merckx, R.; Maertens, M. Understanding the Process of Agricultural Technology Adoption: Mineral Fertilizer in Eastern DR Congo. *World Dev.* **2014**, *59*, 132–146. [CrossRef]
8. Zhang, H.; Zhang, Y.M.; Wu, S.; Cai, R. The Effect of Labor Migration on Farmers' Cultivated Land Quality Protection. *Sustainability* **2020**, *12*, 2953. [CrossRef]
9. Xie, R.Z.; Li, S.K.; Li, X.J.; Jin, Y.Z.; Wang, K.R.; Chu, Z.D.; Gao, S.J. The analysis of conservation tillage in China-conservation tillage and crop production: Reviewing the evidence. *Sci. Agric. Sin.* **2007**, *40*, 1914–1924.
10. Abate, G.T.; Rashid, S.; Borzaga, C.; Getnet, K. Rural Finance and Agricultural Technology Adoption in Ethiopia: Does the Institutional Design of Lending Organizations Matter? *World Dev.* **2016**, *84*, 235–253. [CrossRef]
11. Carter, M.R.; Cheng, L.; Sarris, A. Where and how index insurance can boost the adoption of improved agricultural technologies. *J. Dev. Econ.* **2016**, *118*, 59–71. [CrossRef]
12. Gan, X.L.; Chang, R.D.; Wen, T. Overcoming barriers to off-site construction through engaging stakeholders: A two-mode social network analysis. *J. Clean. Prod.* **2018**, *201*, 735–747. [CrossRef]
13. Saito, H.; Ruhanen, L. Power in tourism stakeholder collaborations: Power types and power holders. *J. Hosp. Tour. Manag.* **2017**, *31*, 189–196. [CrossRef]
14. Huang, N.; Bai, L.B.; Wang, H.L.; Du, Q.; Shao, L.; Li, J.T. Social Network Analysis of Factors Influencing Green Building Development in China. *Int. J. Environ. Res. Public Health* **2018**, *15*, 2684. [CrossRef]
15. An, X.; Wen, Y.L.; Zhang, Y.Q.; Xu, S. Evaluation of the forestry and environmental conservation policies in Western China with multi-output regression method. *Comput. Electron. Agric.* **2019**, *157*, 239–246. [CrossRef]
16. Ghadiyali, T.R.; Kayasth, M.M. Contribution of green technology in sustainable development of agriculture sector. *J. Environ. Res. Dev.* **2012**, *7*, 590–596.
17. Elmustapha, H.; Hoppe, T.; Bressers, H. Consumer renewable energy technology adoption decision-making; comparing models on perceived attributes and attitudinal constructs in the case of solar water heaters in Lebanon. *J. Clean. Prod.* **2018**, *172*, 347–357. [CrossRef]
18. Chang, R.D.; Zuo, J.; Soebarto, V.; Zhao, Z.Y.; Zillante, G.; Gan, X.L. Sustainability Transition of the Chinese Construction Industry: Practices and Behaviors of the Leading Construction Firms. *J. Manag. Eng.* **2016**, *32*, 05016009. [CrossRef]
19. Xue, X.L.; Zhang, X.L.; Wang, L.; Skitmore, M.; Wang, Q. Analyzing collaborative relationships among industrialized construction technology innovation organizations: A combined SNA and SEM approach. *J. Clean. Prod.* **2018**, *173*, 265–277. [CrossRef]
20. Yao, Y.T.; Chen, W.M.; Li, X.N. Research of relationship between environmentally-friendly agricultural technology innovation and agricultural economic growth. *China Popul. Environ.* **2014**, *24*, 122–130.

21. Yan, G.Q.; He, Y.C.; Zhang, X.H. Green technology progress, agricultural economic growth and pollution space spillover effect: Evidence og agricultural water utilization progress in China. *Resour. Environ. Yangtze Basin* **2019**, *28*, 2921–2935.
22. Kuang, H.Y. Analysis of information dissemination in agricultural technology's diffusion based on social network. *J. Anhui Agric. Sci.* **2015**, *43*, 376–378.
23. Yu, D.K.; Peng, J.; Tu, G.P.; Liu, Y.H. The evaluation of agricultural technology diffusion environment of China regions. *Math. Pract. Theory* **2018**, *48*, 43–57.
24. Xu, L.Y.; Zhou, Z.Y.; Du, J.G. An Evolutionary Game Model for the Multi-Agent Co-Governance of Agricultural Non-Point Source Pollution Control under Intensive Management Pattern in China. *Int. J. Environ. Res. Public Health* **2020**, *17*, 2472. [CrossRef]
25. Sally, B. Enabling adaptation? Lessons from the new 'Green Revolution' in Malawi and Kenya. *Clim. Chang.* **2014**, *122*, 15–26.
26. Pratt, O.J.; Wingenbach, G. Factors affecting adoption of green manure and cover crop technologies among Paraguayan smallholder farmers. *Agroecol. Sustain. Food Syst.* **2016**, *40*, 1043–1057. [CrossRef]
27. Zhang, S.L. Analysis of stakeholders in the structural reform of agricultural supply side. *J. Shandong Agric. Univ. Sci. Ed.* **2019**, *21*, 40–44.
28. George, T. Why crop yields in developing countries have not kept pace with advances in agronomy. *Glob. Food Sec.* **2014**, *3*, 49–58. [CrossRef]
29. Wicki, S.; Hansen, E.G. Green technology innovation: Anatomy of exploration processes from a learning perspective. *Bus. Strateg. Environ.* **2019**, *28*, 970–988. [CrossRef]
30. Samantha, S.; Angela, P. Eating clean and green? Investigating consumer motivations towards the purchase of organic food. *Australas. Mark. J.* **2010**, *18*, 93.
31. Stephanie, M.; Sabrina, L.F. Stakeholder analysis and engagement in projects: From stakeholder relational perspective to stakeholder relational ontology. *Int. J. Proj. Manag.* **2014**, *32*, 1108–1122.
32. Wood, D.J.; Gray, B. Toward a comprehensive theory of collaboration. *J. Appl. Behav. Sci.* **1991**, *27*, 139–162. [CrossRef]
33. Liang, X.; Shen, G.Q.; Guo, L. Improving Management of Green Retrofits from a Stakeholder Perspective: A Case Study in China. *Int. J. Environ. Res. Public Health* **2015**, *12*, 13823–13842. [CrossRef] [PubMed]
34. Mitchell, R.K.; Agle, B.R.; Wood, D.J. Toward a theory of stakeholder identification and salience: Defining the principle of who and what really counts. *Acad. Manag. Rev.* **1997**, *22*, 853–886. [CrossRef]
35. Gaski, J.F. The theory of power and conflict in channels of distribution. *J. Mark.* **1984**, *48*, 9–29. [CrossRef]
36. Tiew, F.; Homles, K.; Bussy, N.D. Tourism events and the nature of stakeholder power. *Event Manag.* **2015**, *19*, 525–541. [CrossRef]
37. Rowley, T.J. Moving beyond dyadic ties: A network theory of stakeholder influences. *Acad. Manag. Rev.* **1997**, *22*, 887–910. [CrossRef]
38. Wang, G.L.; Lu, Q.; Capareda, S.C. Social network and extension service in farmers' agricultural technology adoption efficiency. *PLoS ONE* **2020**, *15*, e0235927. [CrossRef]
39. Ramirez, A. The Influence of Social Networks on Agricultural Technology Adoption. *Procedia Soc. Behav. Sci.* **2013**, *79*, 101–116. [CrossRef]
40. Yang, R.J.; Zou, P.X. Stakeholder-associated risks and their interactions in complex green building projects: Asocial network model. *Build. Environ.* **2014**, *73*, 208–222. [CrossRef]
41. Mao, C.; Shen, Q.P.; Pan, W.; Ye, K.H. Major Barriers to Off-Site Construction: The Developer's Perspective in China. *J. Manag. Eng.* **2015**, *31*, 04014043. [CrossRef]
42. Shen, L.Y.; Song, X.N.; Wu, Y.; Liao, S.J.; Zhang, X.L. Interpretive Structural Modeling based factor analysis on the implementation of Emission Trading System in the Chinese building sector. *J. Clean. Prod.* **2016**, *127*, 214–227. [CrossRef]
43. Malik, H.A.M.; Mahesar, A.W.; Abid, F.; Waqas, A.; Wahiddin, M.R. Two-mode network modeling and analysis of dengue epidemic behavior in Gombak, Malaysia. *Appl. Math. Model.* **2017**, *43*, 207–220. [CrossRef]
44. Liu, J. *Lectures on Whole Network Approach: A Practical Guide to UCINET.*; Truth&Wisdom Press: Shanghai, China; People's Publishing House: Shanghai, China, 2014.
45. Gan, S.S.; Pujawan, I.N.; Wahjudi, D.; Tanoto, Y.Y. Pricing decision model for new and remanufactured short life-cycle products with green consumers. *J. Revenue Pricing Manag.* **2019**, *18*, 376–392. [CrossRef]
46. Faust, K. Centrality in social networks. *Soc. Netw.* **1997**, *19*, 157–191. [CrossRef]

47. Borgatti, S.P.; Everett, M.G. Models of core/periphery structures. *Soc. Netw.* **1999**, *21*, 375–395. [CrossRef]
48. Yuan, J.F.; Chen, K.W.; Li, W.; Ji, C.; Wang, Z.R.; Skibniewski, M.J. Social network analysis for social risks of construction projects in high-density urban areas in China. *J. Clean. Prod.* **2018**, *198*, 940–961. [CrossRef]
49. Ley, M.; Stucki, T.; Woerter, M. The Impact of Energy Prices on Green Innovation. *Energy J.* **2016**, *37*, 41–75. [CrossRef]
50. Turken, N.; Carrillo, J.; Verter, V. Strategic supply chain decisions under environmental regulations: When to invest in end-of-pipe and green technology. *Eur. J. Oper. Res.* **2020**, *283*, 601–613. [CrossRef]
51. Chan, Y.H.; Cheah, K.W.; How, B.S.; Loy, A.C.M.; Shahbaz, M.; Singh, H.K.G.; Yusuf, N.R.; Shuhaili, A.F.A.; Yusup, S.; Ghani, W.A.W.A.K.; et al. An overview of biomass thermochemical conversion technologies in Malaysia. *Sci. Total Environ.* **2019**, *680*, 105–123. [CrossRef]
52. Lederer, J.; Ogwang, F.; Karungi, J. Knowledge identification and creation among local stakeholders in CDM waste composting projects: A case study from Uganda. *Resour. Conserv. Recycl.* **2017**, *122*, 339–352. [CrossRef]
53. Adnan, N.; Nordin, S.M.; Rasli, A.M. A possible resolution of Malaysian sunset industry by green fertilizer technology: Factors affecting the adoption among paddy farmers. *Environ. Sci. Pollut. Res.* **2019**, *26*, 27198–27224. [CrossRef] [PubMed]
54. Lu, H.; Hu, L.; Zheng, W.; Yao, S.; Qian, L. Impact of household land endowment and environmental cognition on the willingness to implement straw incorporation in China. *J. Clean. Prod.* **2020**, *262*, 121479. [CrossRef]

International Journal of
Environmental Research and Public Health

Article

Simulating Urban Growth Scenarios Based on Ecological Security Pattern: A Case Study in Quanzhou, China

Xiaoyang Liu [1], Ming Wei [2,3,*] and Jian Zeng [1,*]

1 School of Architecture, Tianjin University, Tianjin 300072, China; liuxiaoyang@tju.edu.cn
2 School of Transportation Science and Engineering, Harbin Institute of Technology, Harbin 150090, China
3 School of Earth and Environmental Sciences, The University of Queensland, Brisbane 4072, QLD, Australia
* Correspondence: m.wei@uq.edu.au (M.W.); 13602058416@vip.163.com (J.Z.)

Received: 6 August 2020; Accepted: 2 October 2020; Published: 5 October 2020

Abstract: In recent decades, the ecological security pattern (ESP) has drawn increasing scientific attention against the backdrop of rapid urbanization and worsening ecological environment. Despite numerous achievements in identifying and constructing the ecological security pattern, limited attention has been paid on applying ESP to predict urban growth. To bridge the research gap, this paper took Quanzhou, China as a study case and incorporated the identified ESP into an urban growth simulation with three distinct scenarios. Following the "ecological source–ecological corridor–ecological security pattern" paradigm, the ESP identification was carried out from four single aspects (i.e., water, geology, biodiversity, and recreation) into three levels (i.e., basic ESP, intermediate ESP, and optimal ESP). Grounded in an equally weighted superposition algorithm, the four single ESPs were combined as an integrated ESP (IESP) with three levels. Taking IESP as an exclusion element, urban growth simulation in 2030 was completed with thee SLEUTH model. Drawing on the three levels of IESP, our urban growth simulation contained three scenarios. In terms of urban sprawl distribution coupled with urban growth rate, an optimal urban growth scenario is recommended in this paper to balance both urban development and eco-environment protection. We argue that our ESP-based urban growth simulation results shed new light on predicting urban sprawl and have the potential to inform planners and policymakers to contribute to more environmentally-friendly urban development.

Keywords: ecological security pattern; SLEUTH model; urban growth simulation

---

1. Introduction

Human activity and ecological environment are intrinsically linked. In recent decades, the rapid development of urbanization has induced a series of environmental problems including the overconsumption of natural resources, environmental degradation, air pollution, and soil erosion [1–4]. These environmental problems, in return, intensify the risk of natural disasters and jeopardize human well-being in a wide variety of aspects from health and safety to living standards [5,6]. Sustainable development entails a stable eco-environment and vibrant ecosystem service [4,7]. As such, to achieve the trade-off between economic development and eco-environment protection, the concept of ecological security was proposed in the early 1980s [8,9]. Since then, from national governments to worldwide organizations, plenty of strategies to maintain ecological security have been enacted and implemented [10–12].

So far, in terms of different research perspectives, there are numerous definitions of ecological security. For example, Ezeonu and Ezeonu [13] deemed that ecological security refers to the states in which the health and integrity of the Earth's ecosystem are well conserved, protected, and restored.

Pirages [14] argued that ecological security is the balance between human need and environmental affordability for sustainable development. Synthesizing the definitions of ecological security in previous scholarship, the core element of this concept is in relation to coordinating the relationship between the natural ecosystem and economic society [8,13–17]. On account of the importance of ecological security, a growing body of literature on identifying ecological security patterns (ESPs) has emerged. To date, press-state-response (PSR) theory and circuit theory are the two conceptual frameworks most applied in extant literature [2–5,8,12,18,19]. Grounded in RSP theory, researchers first identify the pressure indicators allied with pressure-bearing indicators, then apply mathematical modeling and landscape ecology methods such as the artificial neural network model, entropy matter–element model, and grey model to calculate the weight of each indicator's influence and integrate the weighted indicators to identify the ESP [8,18,20–22]. With respect to the studies based on circuit theory, researchers mainly follow the "ecological sources–ecological corridors–ecological nodes–ecological security pattern" paradigm, in which a series of ecological and spatial analytical approaches such as resistance surface construction, overlay analysis, and network analysis could be applied to achieve the construction of ESP [1–5,10,19].

Despite a fruitful line of research in analyzing ESP, there seems to be a lack of ESP application, particularly in incorporating ESP into urban growth simulation [1–9,15–19,23–25]. On one hand, most ESP-related studies have focused on ESP identification and construction. The effects of those identified ESPs in influencing or guiding urban growth have been neglected in the existing scholarship. Therefore, our knowledge of ESPs is limited in the scope of the current situation rather than future planning. On the other hand, a series of state-of-the-art urban growth simulation models has been dedicated to conducting future urban planning, however, these simulation models fail to take the potential effects of ESPs into account. Consequently, ESP, as an important factor in balancing urban development and eco-environment protection, has been arguably yet to be fully considered in urban simulation and its potential effects on urban planning remain unclear.

On the basis of the mechanism of cellular automata, a diverse range of models have been developed and improved to simulate and predict future urban growth over the past few decades [23]. Among these, the SLEUTH model has been widely applied due to its comprehensive evaluation and high accuracy [23–25]. SLEUTH is an acronym made up of its input datasets from six dimensions, namely, slope, land use/land cover (LULC), exclusion, urban extent, transportation, and hill-shade. Utilizing the input datasets, the SLEUTH model sets out five growth parameters (i.e., diffusion, breed, spread, slope resistance, and road gravity) and four growth types (i.e., spontaneous growth, new spreading center growth, edge growth, and road-influenced growth) to refine the calibration and finalize the urban growth prediction [26,27]. In building the exclusion layer, water bodies, wetlands, and reserved forest have mostly been used in previous studies, while, factors related to ecological security are scarcely mentioned [23–26,28,29]. As a matter of fact, it is evident that ESP covers a much larger extent than the conservation area [10,11]. In light of the importance of ESP in achieving sustainable development [1–12,15–19], there is a compelling need to incorporate ESP in urban growth simulation. It follows that if we, as a society, call for more eco-friendly and sustainable urban development, then ESP is required in simulating urban growth as a key facet in better informing planners and policymakers in landscape design and urban planning.

To bridge the aforementioned research gap, this paper took Quanzhou, a famous coastal city in Southeast China, as the study context and employed a suite of analytical approaches to fulfil the following objectives: (1) to identify the ESP from a single aspect to integrated aspect; (2) to classify the ESP into three levels, namely, basic ESP, intermediate ESP, and optimal ESP; and (3) to simulate urban growth with the incorporation of ESP to provide recommendations for future urban planning in terms of the simulation results.

The rest of this paper is organized as follows. Section 2 outlines the study context allied with data sources. Section 3 presents the analytical approaches plus the process of the SLEUTH model calibration. Section 4 provides the three levels of ESP from the single aspect to integrated aspect and the ESP-based urban growth simulation results. Section 5 discusses the implications of the modeling results and Section 6 presents avenues for future research before drawing some concluding remarks.

2. Study Context and Data Sources

2.1. Study Context

Quanzhou (117°25′ E–119°05′ E, and 24°30′ N–25°56′ N) is situated in the eastern Fujian Province, China and is a prefecture-level port city known as the starting point of the Maritime Silk Road. In the humid subtropical climatic context, Quanzhou has rich water resources with an average annual rainfall up to 1400–2000 mm, and both the Jin and Luo Rivers flow into Quanzhou Bay in the Taiwan Strait (Figure 1). As one of the most developed cities in Fujian Province, Quanzhou has had the highest GDP (Gross Domestic Product) in the Fujian Province for 21 years [30]. Quanzhou is also the most populous city in Fujian Province, with over 8.5 million residents reported in the four districts, three county-level cities, four counties, and two special economic districts within Quanzhou [30].

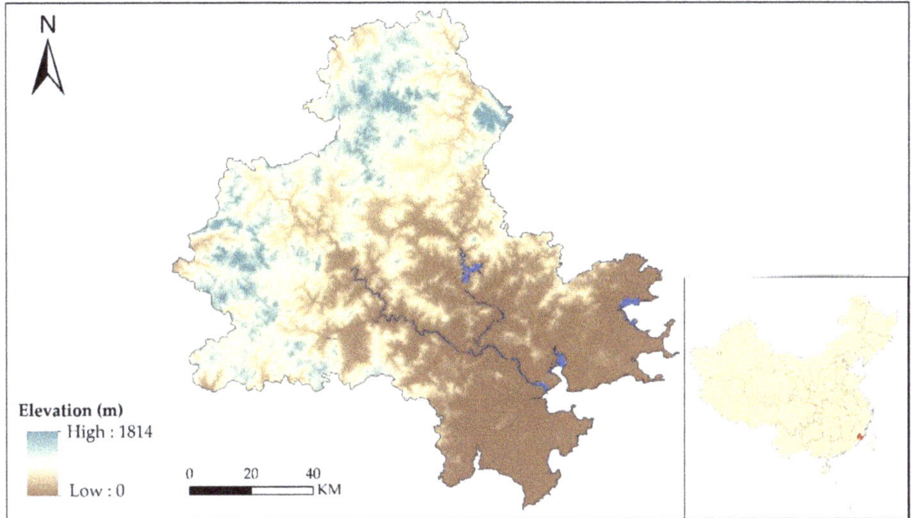

Figure 1. Location of Quanzhou.

2.2. Data

This study contained two major analytical components, namely, ESP identification and urban growth simulation. To accomplish these analyses, a wide range of datasets across transportation, land use, climatic context, and soil type were applied (Table 1). Our research data were mainly derived from official open sources and a series of satellite images. For example, land use data, the core dataset for both ESP identification and urban growth simulation, were obtained from the Resource and Environment Cloud Platform (http://www.resdc.cn/) across four periods (i.e., 2000, 2005, 2010, and 2015). Digital Elevation Model (DEM) data were achieved from the Geospatial Data Cloud (https://www.gscloud.cn/), and the corresponding information such as slope, curvature, and hill-shade was computed based on DEM data using an ArcGIS platform. Normalized difference vegetation index (NDVI), an indicator of vegetation greenness, was computed using Landsat 8 (https://earthexplorer.usgs.gov). It is noteworthy that in this study, nighttime light data were applied to revise the resistance surface in identifying the biodiversity ESP. More detailed data information is presented in the following table (Table 1).

Table 1. Data information.

Data	Utility	Data Source
Land use	ESP identification; Urban growth simulation	Resource and Environment Cloud Platform http://www.resdc.cn/
Annual rainfall	Water & Geology ESP identification	National Meteorological Information Center http://data.cma.cn/; Local weather station records
NDVI	Biodiversity & Geology ESP identification	U.S. Geological Survey Landsat image
Slope	Water & Geology & Biodiversity ESP identification; Urban growth simulation	Calculated from DEM data; Geospatial Data Cloud https://www.gscloud.cn/
Elevation	Water & Geology & Biodiversity ESP identification	Derived from DEM data; Geospatial Data Cloud https://www.gscloud.cn/
Curvature	Geology ESP identification	Calculated from DEM data; Geospatial Data Cloud https://www.gscloud.cn/
Hill-shade	Urban growth simulation	Calculated from DEM data; Geospatial Data Cloud https://www.gscloud.cn/
Road	Geology & Biodiversity ESP identification; Urban growth simulation	National Geoinformation Service http://www.webmap.cn
Geological hazards	Geology ESP identification	Fujian Seismological Bureau; Fujian Water Conservancy Bureau
Soil type	Geology ESP identification	Fujian Agriculture Department
Nighttime light	Biodiversity ESP revision	Luojia-1 Satellite image http://www.hbeos.org.cn
Recreation resources	Recreation ESP identification	Ministry of Ecology and Environment of China; Fujian Forestry Bureau

3. Methodology

As discussed in Section 2.2, this study mainly comprises two analytical components, one identifies three levels of ESP in accordance with the "ecological sources–ecological corridor–ecological pattern" paradigm; the other uses the SLEUTH model to predict urban growth with the consideration of the identified ESPs. Figure 2 illustrates the completed methodological framework in this study and the following subsections outline the detailed analysis process to achieve the specific goals in ESP identification and SLEUTH model calibration.

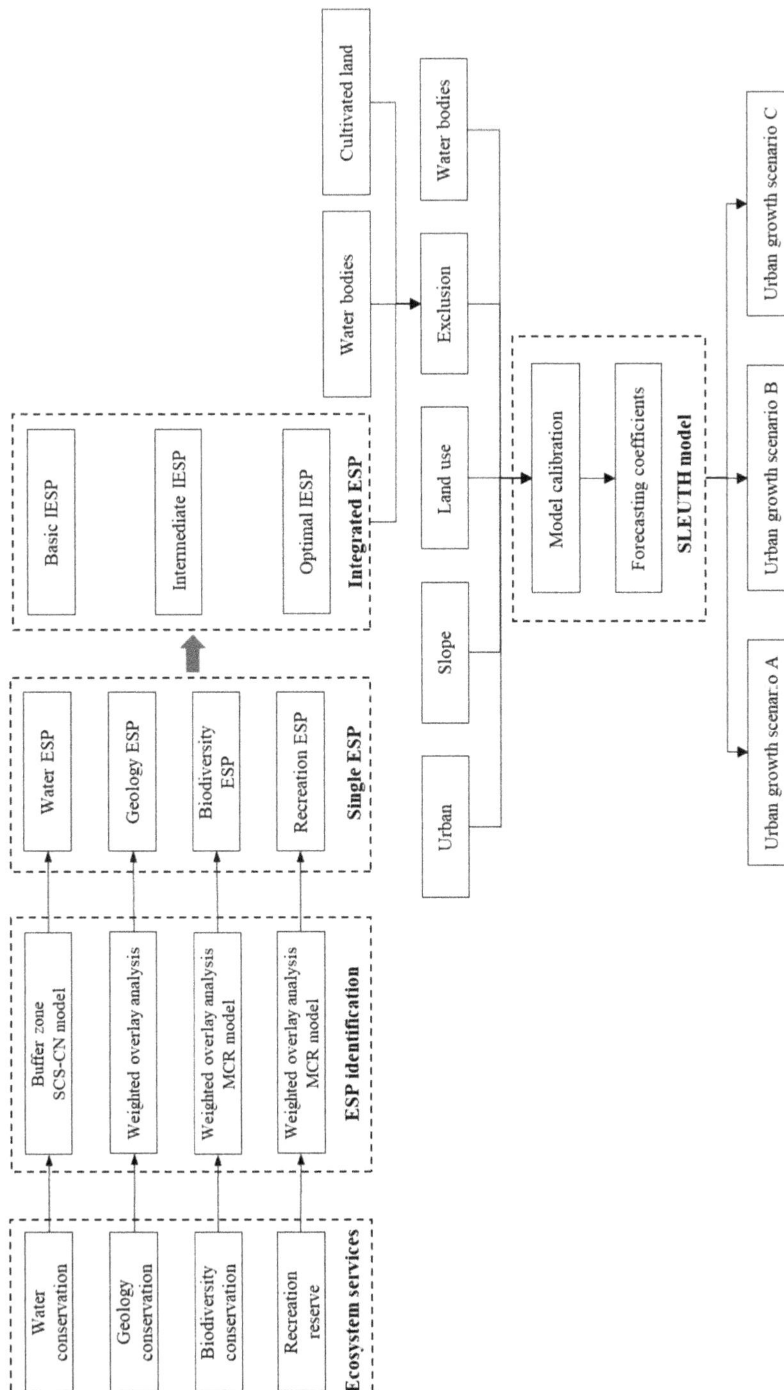

Figure 2. Methodological framework.

3.1. ESP Identification and Construction

3.1.1. Overlay Analysis

Overlay analysis is one of the most popular techniques in spatial analysis. Drawing on different spatial datasets equipped with diverse attribute information, overlay analysis creates an integrated layer by spatially combining information from different datasets. Spatial join is a core operation in overlay analysis, especially when layers are stored with different spatial resolutions or in different data types. By computing the information derived from different datasets, new attribute information can be added in the integrated layer. It is noteworthy that in adding new attribute information, certain selected factors can be weighted based on their importance to the overall goal.

In this study, overlay analysis was applied in identifying ESP for all four aspects and concerning identifying geology ESP, biodiversity ESP and recreation ESP, different weights were assigned in terms of the corresponding criteria (Figure 2).

3.1.2. Soil Conservation Service Curve Number (SCS-CN) Model

Flood inundation area is a key aspect related to water ESP [19,31]. To incorporate this important information in water ESP identification and construction, the Soil Conservation Service Curve Number (SCS-CN) model was applied to estimate the surface runoff (Figure 2). Based on water balance and soil–cover complexes, the SCS-CN model assumes that the ratio of real soil retention after runoff to potential maximum retention is equal to the ratio of direct runoff to the maximum possible runoff. The detailed expression is as follows:

$$\frac{Q}{P - I_a} = \frac{F}{S} \quad (1)$$

where Q is the gathered runoff (mm); P denotes rainfall (mm); I_a is initial abstraction (mm); F is cumulative infiltration (mm); and S is the potential maximum retention (mm).

On account of the empirical findings, I_a is equal to 0.2S. As such, the SCS-CN model can be rewritten as follows:

$$Q = \begin{cases} \frac{(P-0.2S)^2}{P+0.8S}, & P \geq 0.2S \\ 0, & P < 0.2S \end{cases} \quad (2)$$

In addition, according to the USDA (U.S. Department of Agriculture) handbook on hydrology section, the potential maximum retention can be calculated with the runoff curve number (CN), which varies by soil type [31]. The formula is expressed as follows:

$$S = \frac{25400}{CN} - 254 \quad (3)$$

With the potential maximum retention (S) and gathered runoff (Q), the overall water volume (V) can be achieved as V = S × Q. Then, drawing on the possible rainfall occurrence, CN and DEM features, and the inundated area, a critical indicator of identifying waster ESP can thereby be computed.

3.1.3. Minimum Cumulative Resistance (MCR) model

The minimum cumulative resistance (MCR) model was applied to extract the ecological corridor. The MCR model can build a resistance surface to estimate the difficulty of species diffusion from origin to destination. The mathematical expression is as follows:

$$MCR = f_{min} \sum_{j=n}^{i=m} D_{ij} \times R'_{ij} \quad (4)$$

where MCR represents the cumulative resistance value of a complete diffusion process; D_{ij} denotes the distance from location i to location j; R'_{ij} is the revised resistance value from location i to location j; and

f_{min} as a positive function examines the minimum value of all diffusion process. In this study, the MCR model is completed by the function of cost distance on the ArcGIS platform.

Drawing on the resistance surface, the ecological corridor can be extracted by the least cost path method on the ArcGIS platform. Based on the identified ecological sources, plus the extracted ecological corridors, the ecological security pattern can thereby be identified and established. In this study, the MCR model was applied for the identification of the biodiversity ESP and recreation ESP (Figure 2), and the original resistance surface was generated based on land use type.

3.1.4. Resistance Surface Revision

Traditionally, the resistance surface construction only takes land use into account, without considering the potential influence of human activity on the process of species diffusion. To incorporate the effects of both human activity and natural environment, nighttime light data were employed in this study as a way to revise the resistance surface derived from land use. Given the different spatial resolution between land use (30 m) and nighttime data (100 m), the average nighttime light is resampled and recomputed for each land parcel of land use map. The detailed mathematical expression is as follows:

$$NLI_i = \sum_{j=1}^{n}\left(\frac{Area_{i(j)}}{Area_i} \times NLI_j\right) \quad (5)$$

where NLI_i is the average nighttime light for land parcel i, as every land parcel is divided into n pieces; $Area_i$ is the area of land parcel i; and $Area_j$ is the area of piece j within land parcel i, and NLI_j is the nighttime light for piece j. With the nighttime light for each land parcel computed, the resistance surface revision can be expressed as follows:

$$R_i' = \frac{NLI_i}{\left(\sum_{i=1}^{n} NLI_{a(i)}\right)/n} \times R_i \quad (6)$$

where R_i' is the revised resistance value for land parcel I; R_i is the original resistance value of land parcel i; NLI_i is the average nighttime light for land parcel i; and $(\sum n\ i = 1\ NLIa(i))/n$ represents the average nighttime light for all land parcels in land use type a. In this study, this revised resistance surface was applied in the MCR model to identify biodiversity ESP.

3.1.5. Integrated ESP Construction

Following the "ecological sources–ecological corridor–ecological pattern" paradigm, the ESP was identified and constructed into three levels (i.e., basic ESP level, intermediate ESP level, and optimal ESP level) concerning four single aspects (i.e., water ESP, geology ESP, biodiversity ESP, and recreation ESP). To combine the four single ESPs into one integrated ESP (IESP), first, the three distinct ESP scenarios were assigned values of 1 for basic ESP level, 2 for the intermediate ESP level, and 3 for the optimal ESP level; second, drawing on the equally weighted superposition algorithm, the four identified ESPs were integrated with the maximum value among the four single ESPs. For example, a land parcel in basic water ESP (value 3), intermediate geology ESP (value 2), intermediate biodiversity ESP (value 2), and optimal recreation ESP (value 1) was identified as a basic IESP (value 3). The formula is expressed as follows:

$$\text{IESP} = \text{Max}(ESP_i),\ i = 3, 2, 1 \quad (7)$$

where basic ESP level, intermediate ESP level and optimal ESP level were assigned values of 3, 2, 1, respectively. In this study, the integration model was completed using the mosaic function in ArcGIS raster analysis.

3.2. SLEUTH Model Calibration

The SLEUTH model requires input data from five aspects, namely, slope, land use, exclusion, urban extent, transportation, and hill-shade [32]. In this study, four land-use layers (i.e., land use map in 2000, 2005, 2010, and 2015), two road layers (i.e., road map in 2010 and 2015), one slope layer and one exclusion layer with three elements (i.e., IESP, water bodies, and cultivated land) were equipped to run the SLEUTH model. The SLEUTH model mainly contains two parts: calibration and simulation. Based on the historical data coupled with customized settings, the model was calibrated from coarse level, fine level to final level. In the process of the calibration, five controlling parameters were generated to determine the mechanism of urban growth (i.e., dispersion coefficient, breed coefficient, road-gravity coefficient, slope resistance coefficient, and spread coefficient). In terms of the five controlling parameters, four types of urban growth are predicted, namely, spontaneous growth, new spreading urban centers, edge growth around existing urban areas, and road-influenced growth [33]. To examine the accuracy of model calibration and attain a more reliable prediction result, the optimal SLEUTH metric (OSM) was employed to present the best goodness-of-fit:

$$OSM = compare \times pop \times edges \times clusters \times slope \times xmean \times ymean \quad (8)$$

where compare is the ratio of simulated urban population over actual urban population; pop refers to the least square regression score for modeled urbanization; edge refers to the least square regression score for modeled urban edges count; cluster refers to the regression score for modeled urban clusters; slope indicates the terrain surface; xmean is average x values for modeled urbanized cells; and ymean is average y values for modeled urbanized cells.

Drawing on the OSM expression, the accuracy of urban growth simulation contains four aspects: the size of urban growth (compare and pop), the shape of urban growth (edges and clusters), the position of urban growth (xmean and ymean), and the terrain of the study area (slope). With all these parameters from controlling coefficients, growth types, and growth accuracy equipped, the Monte Carlo iteration was conducted to gradually narrow the range of growth coefficients and gain better simulation goodness.

In this study, based on historical data, Table 2 shows the results of the SLEUTH model calibration. Subsequently, in line with the growth coefficients in the final calibration, we examined the simulation accuracy for 2005, 2010, and 2015, respectively (Table 3). The high simulation accuracy proved that our model calibration was qualified to run the simulation process. It is noteworthy that with the consideration of the three levels of IESPs, our urban growth model also incorporated three scenarios (Figure 2). Table 4 shows the exclusion scheme of the three IESPs in the urban growth simulation.

Table 2. Calibration coefficients for the SLEUTH model.

Growth Coefficients	Coarse		Fine		Final		BFC
	MCI = 5		MCI = 7		MCI = 9		
	NI = 3163		NI = 7851		NI = 7796		
	OSM = 0.4679		OSM = 0.4723		OSM = 0.5034		
	Range	Step	Range	Step	Range	Step	
Dispersion	0–100	25	25–100	15	40–75	7	81
Breed	0–100	25	50–100	10	50–75	5	51
Road gravity	0–100	25	0–75	15	30–75	9	66
Slope	0–100	25	25–70	9	25–40	3	35
Spread	0–100	25	25–100	15	25–55	6	42

BFC: Best Fit Coefficient, MCI: Monte Carlo Iterations, NI: Number of Iterations.

Table 3. The SLEUTH model simulation accuracy.

Modelling Results	2005	2010	2015
Actual value (number of pixels)	64,455	106,264	127,652
Simulation value (number of pixels)	56,427	94,628	11,6457
Simulation accuracy (%)	87.54	89.05	91.23

Table 4. Exclusion probability of the selected layers in the three urban growth scenarios (%).

Exclusion Layers	Urban Growth Scenario A	Urban Growth Scenario B	Urban Growth Scenario C
Built-up area in 2015	0	0	0
Water body	100	100	100
Cultivated land	100	100	100
Basic IESP	100	100	100
Intermediate IESP	70	70	0
Optimal IESP	50	0	0

BFC: Best Fit Coefficient, MCI: Monte Carlo Iterations, NI: Number of Iterations.

4. Results

4.1. ESP Identification and Construction

4.1.1. Water ESP Identification

The ultimate purpose of water ESP identification and construction is to restore natural hydrological processes and avoid or mitigate flood risk. Thereby, drawing on previous scholarship [3–5], four evaluation factors were selected to comprehensively identify water ESP, namely, river and lake systems, surface water sources, flood storage area, and inundation area (Table 5). According to the criteria listed in Table 5, water ESP with three different levels was estimated by using the buffer zone and SCS–CN model (Figure 3a). As the water ESP in three different levels was spatially nested with similar shapes, the area became a major aspect of dissimilarity. Basic water ESP covers an area of 517 km^2, intermediate water ESP covers 759 km^2, and optimal water ESP covers 1115 km^2.

Table 5. Criteria for water ESP identification.

Evaluation Factor	Basic Water ESP	Intermediate Water ESP	Optimal Water ESP
Distance to river and lake (m)	≤50	50–150	150–500
Distance to surface water (m)	≤500	500–1000	1000–1500
Flood storage area (m^3)	3rd level of water storage area	2nd level of water storage area	1st level of water storage area
Distance to inundation area (km^2)	10–Year rain event	50–Year rain event	1000–Year rain event

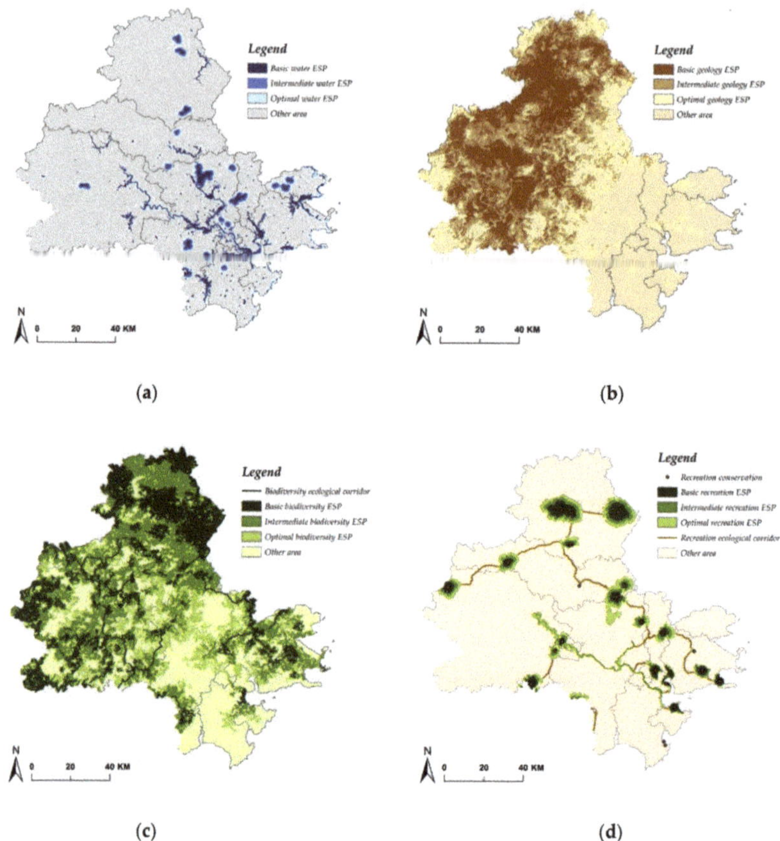

Figure 3. (**a**) Water ESP; (**b**) Geology ESP; (**c**) Biodiversity ESP; (**d**) Recreation ESP.

4.1.2. Geology ESP Identification

Considering the topographic characteristics and the influence of human activities, factors such as average annual rainfall, slope, elevation, surface curvature, soil type, NDVI, land use type, distance to major roads, and the number of geological hazards were selected to evaluate geology ESP in three levels, and their weights, sensitivity coupled with the corresponding criteria, were assigned in line with previous scholarship [1,4,5,18] and are presented in Table 6. The spatial distribution of geology ESP is illustrated in Figure 3b. It shows that the basic geology ESP is mainly located in the western part of the city, relatively distant from the built-up area, while optimal geology ESP covers a much larger area, except the western part, where some built-up adjacent areas are also highlighted.

Table 6. Criteria for geology ESP identification.

Evaluation Factor	Standardized Value					Weight
	No Impact Area	Optimal ESP	Intermediate ESP	Basic Security Pattern		
	Insensitive (1)	Mildly Sensitive (3)	Moderately Sensitive (5)	Sensitive (7)	Highly Sensitive (9)	
Average annual rainfall (mm)	<1300	1300–1400	1400–1500	1500–1600	>1600	0.15
Slope (°)	<5	5–15	15–25	25–35	>35	0.1
Elevation (m)	<200	200–500	500–800	800–1000	>1000	0.1
Curvature	−0.5–0.5	(−1.5, −0.5], [0.5–1.5)	(−2.5, −1.5], [1.5–2.5)	(−3.5, −2.5], [2.5–3.5)	(−∞,−3.5], [3.5, ∞)	0.1
Soil type	Paddy soil Calcareous soil	Saline soil Sandy soil Meadow soil Limestone soil	Lateritic soil	Yellow soil Yellow–red soil Rhogosol Lithosol	Purple soil	0.1
Normalized difference vegetation index (NDVI)	<0.55	0.4–0.55	0.25–0.4	0.1–0.25	<0.1	0.1
Land cover	Construction land Waterbody Wetland	Forest Natural grassland Improved grassland	Irrigable land Dryland Garden plot	Artificial grassland Wild grassland Saline–alkali land	Slash land Bare land Sandy land Gravel land	0.1
Distance to major road (m)	>5000	3000–5000	1500–3000	500–1500	<500	0.1
Geological hazards number (per 25 km^2)	<2	2–4	4–6	6–8	>8	0.15

4.1.3. Biodiversity ESP Identification

According to the basics of habitat and the features of biological community, the ecological sources were evaluated by the criteria for biological habitat suitability (Table 7), in which both natural factors and human factors were incorporated and weighted. Drawing on land cover resistance (Table 8) and the MCR model, the resistance surface for biological migration was first built and subsequently revised by the nighttime light data (Figure 4). With the revised resistance surface (Figure 4b), the Jenks Natural Breaks Classification was applied to stratify the resistance value into four levels, namely, level 1 (0–271), level 2 (271–881), level 3 (881–1954), and level 4 (1954–18,077). In addition, in terms of the revised resistance surface and the identified ecological sources, the ecological corridors were extracted using the least–cost–path on the ArcGIS platform. As ecological sources, ecological corridors, and resistance surface were all estimated, the biodiversity ESP in three levels were identified in accordance with the criteria listed in Table 9. From Figure 3c, it is obvious that compared to the geology ESP, the spatial distribution of biodiversity ESP was less concentrated and much closer to the built–up area.

Table 7. Criteria for biological habitat suitability.

Evaluation Factor	Classification	Value	Weight
Land cover	Urban and other construction lands	0	0.35
	Rural residential land	1	
	Bare land	2	
	Lowly covered grassland	3	
	Dryland and medium covered grassland	5	
	Sparse forest and waterway	6	
	Shrub and highly covered grassland	7	
	Closed forest, lake, reservoir, wetland	8	
	Paddy filed and mudflats	10	
Elevation (m)	0–100	5	0.10
	100–800	10	
	800–1500	5	
	>1500	1	
Distance to water sources (m)	0–2000	6	0.25
	2000–7000	8	
	7000–15000	10	
	15000–30000	5	
	>30000	2	
Distance to Built–up area (m)	>6000	10	0.15
	4000–6000	5	
	2000–4000	3	
	0–2000	1	
	0	0	
Distance to road (m)	0–500	0	0.15
	500–1000	1	
	1000–2000	3	
	2000–4000	5	
	>4000	10	

Table 8. Land cover resistance for biodiversity and recreation ESP identification.

Land Cover	Resistance Coefficient	Land Cover	Resistance Coefficient
Closed forest	1	Mudflats	100
Shrub forest, highly covered grassland	10	Dryland	200
Medium covered grassland	20	Bare land and saline–alkali land	300
Sparse forest	30	Rural residential land	400
Paddy field	50	Urban land	500
Waterbody	50	Other construction lands	500

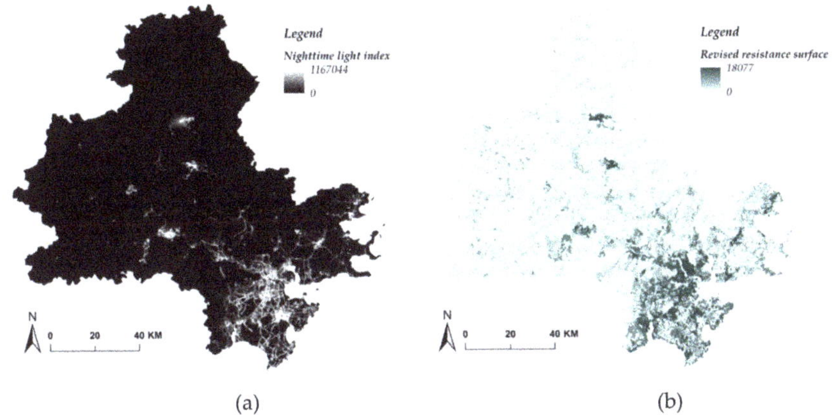

Figure 4. (**a**) Nighttime light in Quanzhou city; (**b**) Revised resistance surface.

Table 9. Criteria for biodiversity ESP identification.

Evaluation Factor	Basic Biodiversity ESP	Intermediate Biodiversity ESP	Optimal Biodiversity ESP
Distance to biodiversity source (m)	0	0–200	200–300
MCR value	Level 1	Level 2	Level 3
Distance to biodiversity corridor (m)	<100	100–200	200–300

4.1.4. Recreation ESP Identification

Similar to the process of biodiversity ESP identification, recreation ESP identification contains recreation sources identification, surface resistance establishment, and recreation corridor extraction. In the study area, recreation sources include two national parks, three provincial nature reserves, twelve provincial forest parks, one national scenic spot, and one provincial scenic spot. With respect to recreation corridors, 19 important corridors were extracted by applying the least–cost–path and gravity model [3,34]. Under the criteria in Table 10, the recreation ESP in three scenarios are identified in Figure 3d.

Table 10. Criteria for recreation ESP identification.

Evaluation Factor	Basic Recreation ESP	Intermediate Recreation ESP	Optimal Recreation ESP
Distance to recreation source (m)	0	0–200	200–300
MCR value	Level 1	Level 2	Level 3
Distance to recreation corridor (m)	<100	100–200	200–300

4.1.5. Integrated ESP Identification

Combining the four single ESP with the principle of IESP (see Section 3.1.3), IESP in three basic, intermediate, and optimal scenarios were identified (Figure 5). Basic IESP represents the core areas of ecological restoration and ecological protection and plays an important role in maintaining eco–system service. The boundary of this area should form an ecological conservation redline and urban development should not invade into this area. Basic IESP covered an area of 4708 km^2, accounting for 42.6% of the whole study area. The spatial distribution of basic IESP was mainly concentrated in the western mountainous area and water resource conservation, with the land cover of forest, wetlands, and grasslands. With regard to intermediate IESP, it is distributed as a surrounding belt of basic IESP with an area of 7817 km^2 (70.8% of all study area). Given that the intermediate IESP encompasses peri–urban areas with high ecological significance, urban sprawl should be appropriately restricted in this intermediate IESP. Optimal IESP, with an area of 8964 km^2 (81.2% of all study area), distributes the outer edge of built–up area. This area usually has less sensitivity to human activities and acts as a buffer zone between urban construction and ecological space. Under the primary goal of ecological protection, optimal IESP represents the ideal area for future development. In practical urban planning, optimal IESP can be moderately developed with conditional construction activities.

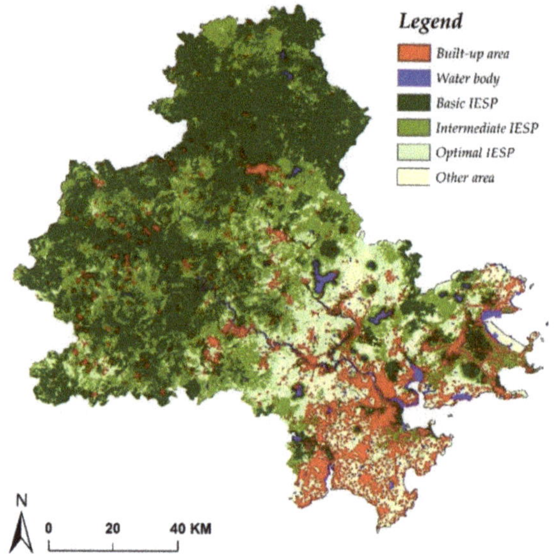

Figure 5. Integrated ESP with three different scenarios in Quanzhou city.

4.2. Urban Growth Modeling Results

Incorporating IESP into exclusion layers of the SLEUTH model, the urban growth simulation from 2015–2030 also contained three distinct scenarios (Figure 6). Urban growth scenario A (Figure 6a) only takes basic IESP as the exclusion element, indicating apart from the basic IESP and the other two fixed exclusion layers (i.e., water body and cultivated land), all areas were available for urban sprawl. In comparison with urban growth scenario A, urban growth scenario B had relatively more restrictions, in addition to totally prohibited basic IESP, intermediate IESP could only be urbanized with a 30% area (Figure 6b). With respect to urban growth scenario C, it had the most restrictions on urbanization with the primary goal to restore the eco–environment (Figure 6c). Apart from 100% basic IESP and 70% intermediate IESP, 50% of optimal IESP was also excluded in urban development. From Table 11, it was observed that scenario A had the largest urban growth area and scenario C took the smallest urban sprawl. In terms of urban growth rate, it showed that all three urban growth scenarios had a lower rate than the historical record from 2000–2015. As for the spatial distribution, scenario C, in line with the principle of ecological priority, had the least invasion into the forest, grassland, and wetland. However, those urban growth areas were small and fragmented without spatial continuity, which might result in urban development being less efficient. Scenario A, following the principle of urban development, has shown a trend of integration of urban–rural development, while, at the price the large peri–urban eco–environment is occupied. In contrast with scenario A and scenario C, scenario B was shown to spatially satisfy the two–way demand of both eco–environment protection and urban development. In general, given the urban growth rate derived from the SLEUTH model simulation, taking ESP as a factor, it is shown to be an effective way to avoid disorderly urban sprawl and mitigate the ecological system pressure.

Table 11. Statistics of urban growth simulation.

Period	Urban Growth Scenarios	Urban Growth Area (km^2)	Annual Urban Growth Rate (%)
2000–2015	Historical record	538.8	4.4%
2015–2030	Urban growth scenario A	750.5	3.6%
	Urban growth scenario B	677.7	3.3%
	Urban growth scenario C	498.4	2.5%

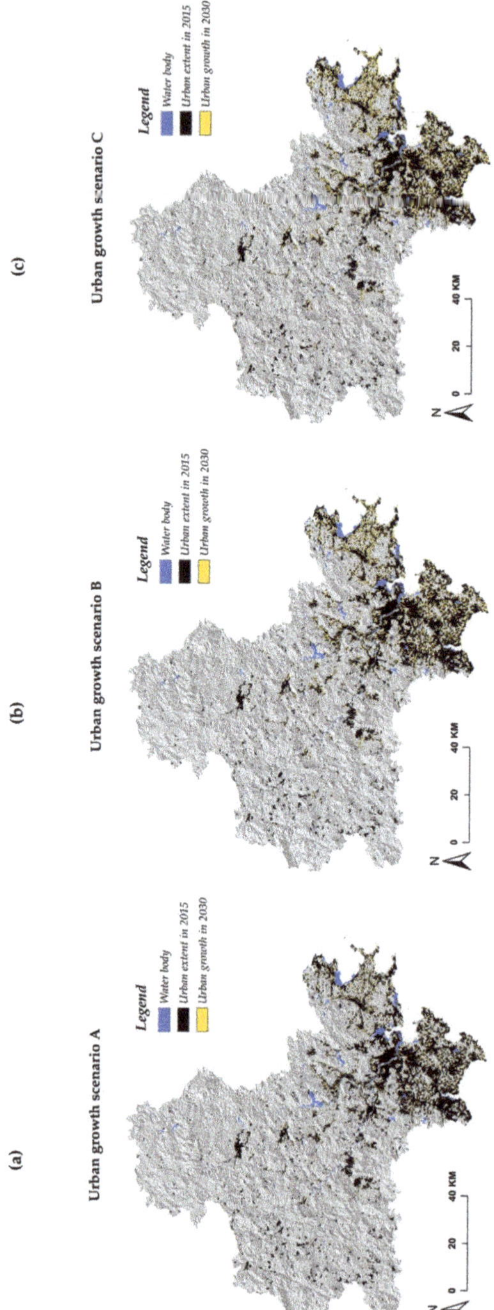

Figure 6. (**a**) Urban growth scenario A; (**b**) Urban growth scenario B; (**c**) Urban growth scenario C.

5. Discussion

The main contribution of this study was to incorporate multiple levels of ESPs to simulate urban growth into different scenarios. With this, one can quantitatively predict and evaluate the interaction between urban development and eco–environment protection, thereby shedding new light on a macroscopic planning scheme with the ultimate goal of safeguarding regional ecological security and satisfying human demand for ecosystem services. As we know, following the initiation of the reform and opening policy, China has experienced tremendous urbanization in recent decades, urban population percentage was up to over 60% in 2020 against 17.92% in 1978 [35]. On one hand, the booming population has brought dramatic ecological land transformed into construction land. On the other hand, along with the increasing human demand for ecological services, diverse pollution generated from urban population aggravates the pressure on the ecological system. Consequently, with declining ecological service, increasing urban environment issues such as urban heat islands have popped up and poses a danger to human health [36]. In this urgent situation, there has been growing scholarly attention to applying ESP in coordinating urban development and ecological protection [36]. On account of its strength in outlining spatial division, ESP is regarded as an effective tool to identify the important patches for ecological service where urban development should be restricted [1,2]. However, most of the ESP studies so far have remained at the stage of 'how to identify the ESP and where the ESP is'; the application of ESP, particularly in simulating urban growth and guiding future urban planning, has rarely been investigated. Given the importance of ESP, this study proposes ESP–based urban growth simulation to address this research gap.

In an attempt to achieve sustainable urban development with ecological protection being the priority, in 2019, China established the national territory spatial planning system and stressed three red lines in urban construction including the red line of the permanent basic farmland, the red line of ecological protection, and the red line of urban growth boundary. In this ESP–based urban growth simulation, those red lines related patches and their potential evolution are shown in different scenarios. The red lines are rigid, which does not mean that the outer area is fixed and invariant. As a matter of fact, under the requirements of redlines, there are plenty of possibilities for the interaction between urban development and eco–environment protection. Drawing on the scenarios presented in this study, we can foresee the consequences in terms of different urban development strategies. These potential consequences would be valuable to suggest to urban planners 'what can be achieved' and 'what should be avoided'. As such, the ESP-based urban growth simulation proposed in this study can provide a new understanding of guiding urban development in an orderly fashion and prevent uncontrolled urban sprawl. In addition, the paradigm proposed by this study for identifying ESP and simulating urban sprawl can easily be adapted to many other regions across the globe, especially for regions undergoing rapid urbanization in developing countries.

6. Conclusions

This paper applied the SLEUTH model to predict urban growth in three different scenarios with the consideration of ESP. Following the "ecological sources–ecological corridors–ecological security pattern" paradigm, ESP was respectively identified from four aspects, namely, water, biodiversity, geology, and recreation. Combining the four single ESPs, one integrated ESP was established and classified into basic IESP, intermediate IESP, and optimal IESP. Separately incorporating the three IESP levels as an exclusion element, urban growth simulation was carried out into three scenarios. The simulation results illustrate the possible urban growth in 2030 under China's Ecological Redline Policy.

In terms of the different restrictions of basic IESP, intermediate IESP, and optimal IESP, three urban growth scenarios have different emphasis between eco-system protection and urban development. Urban growth scenario C was shown to cover large peri–urban eco–service areas with relatively intense urban sprawl. Urban growth scenario A was found to markedly maintain the ecological system, however, its urban sprawl was scattered and fragmented. Urban growth scenario B was shown to be an ideal scheme with reasonable urban development and ecological system safeguarded. Considering the significance of ESP and necessity of eco–environment protection, our results in the urban growth simulation were arguably important with the capacity of shedding new light on smart growth, urban growth boundary control, urban design, and master planning. This paper also presented a paradigm for future studies to simulate environmentally-friendly urban growth.

Despite the innovative design of this study, two aspects form the avenue for future studies. First, at the different stages of urban development, the influence of human activities on ecological services may not stay the same, and more sociodemographic factors such as income, GDP, and population can be considered in both ESP construction and urban growth. Second, in the context of rapid urbanization, urban growth has apparent features of complexity and uncertainty, so both natural factors and human factors should be spatiotemporally evaluated to capture the driving factors and achieve a better urban growth simulation. To conclude, from a particular view of eco–environment protection, we took Quanzhou, China, as the study case and simulated its urban growth in 2030. Our simulation results add to the growing body of both ESP construction and urban growth simulation, supplementing the evidence base for planners and policymakers to make better landscape designs and master planning.

Author Contributions: Conceptualization, X.L. and J.Z.; Methodology, X.L., J.Z., and M.W.; Formal analysis, X.L. and J.Z.; Investigation, X.L. and M.W.; Data curation, X.L. and M.W.; Writing—Original draft preparation, X.L. and M.W.; Writing—Review and editing, M.W. and X.L.; Visualization, X.L. and M.W.; Supervision, J.Z.; All authors have read and agreed to the published version of the manuscript.

Funding: This research was funded by the National Key Research and Development Program of China (2016YFC0502903).

Conflicts of Interest: The authors declare no conflict of interest.

References

1. Peng, J.; Pan, Y.; Liu, Y.; Zhao, H.; Wang, Y. Linking ecological degradation risk to identify ecological security patterns in a rapidly urbanizing landscape. *Habitat Int.* **2018**, *71*, 110–124. [CrossRef]
2. Peng, J.; Yang, Y.; Liu, Y.; Du, Y.; Meersmans, J.; Qiu, S. Linking ecosystem services and circuit theory to identify ecological security patterns. *Sci. Total Environ.* **2018**, *644*, 781–790. [CrossRef] [PubMed]
3. Guo, R.; Wu, T.; Liu, M.; Huang, M.; Stendardo, L.; Zhang, Y. The Construction and Optimization of Ecological Security Pattern in the Harbin-Changchun Urban Agglomeration, China. *Int. J. Environ. Res. Public Health* **2019**, *16*, 1190. [CrossRef] [PubMed]
4. Fu, Y.; Shi, X.; He, J.; Yuan, Y.; Qu, L. Identification and optimization strategy of county ecological security pattern: A case study in the Loess Plateau, China. *Ecol. Indic.* **2020**, *112*, 106030. [CrossRef]
5. Li, J.L.; Xu, J.G.; Chu, J.L. The Construction of a Regional Ecological Security Pattern Based on Circuit Theory. *Sustainability* **2019**, *11*, 6343. [CrossRef]
6. Guo, S.S.; Wang, Y.H. Ecological Security Assessment Based on Ecological Footprint Approach in Hulunbeir Grassland, China. *Int. J. Environ. Res. Public Health* **2019**, *16*, 4805. [CrossRef]
7. Wang, Y.; Pan, J. Building ecological security patterns based on ecosystem services value reconstruction in an arid inland basin: A case study in Ganzhou District, NW China. *J. Clean. Prod.* **2019**, *241*, 118337. [CrossRef]
8. Hu, M.; Li, Z.; Yuan, M.; Fan, C.; Xia, B. Spatial differentiation of ecological security and differentiated management of ecological conservation in the Pearl River Delta, China. *Ecol. Indic.* **2019**, *104*, 439–448. [CrossRef]
9. Hodson, M.; Marvin, S. 'Urban Ecological Security': A New Urban Paradigm? *Int. J. Urban Reg. Res.* **2009**, *33*, 193–215. [CrossRef]
10. Peng, J.; Zhao, H.; Liu, Y. Urban ecological corridors construction: A review. *Acta Ecol. Sin.* **2017**, *37*, 23–30. [CrossRef]
11. Liu, D.; Chang, Q. Ecological security research progress in China. *Acta Ecol. Sin.* **2015**, *35*, 111–121. [CrossRef]
12. Wang, C.; Yu, C.; Chen, T.; Feng, Z.; Hu, Y.; Wu, K. Can the establishment of ecological security patterns improve ecological protection? An example of Nanchang, China. *Sci. Total Environ.* **2020**, *740*, 140051. [CrossRef] [PubMed]
13. Ezeonu, I.C.; Ezeonu, F.C. The environment and global security. *Environmentalist* **2000**, *20*, 41–48. [CrossRef]

14. Pirages, D. Ecological security: Micro-threats to human well-being. In *People and Their Planet*; Springer: Berlin/Heidelberg, Germany, 1999; pp. 284–298.
15. Lu, S.; Li, J.; Guan, X.; Gao, X.; Gu, Y.; Zhang, D.; Mi, F.; Li, D. The evaluation of forestry ecological security in China: Developing a decision support system. *Ecol. Indic.* **2018**, *91*, 664–678. [CrossRef]
16. Yu, G.; Zhang, S.; Yu, Q.; Fan, Y.; Zeng, Q.; Wu, L.; Zhou, R.; Nan, N.; Zhao, P. Assessing ecological security at the watershed scale based on RS/GIS: A case study from the Hanjiang River Basin. *Stoch. Environ. Res. Risk Assess.* **2014**, *28*, 307–318. [CrossRef]
17. Lu, S.; Tang, X.; Guan, X.; Qin, F.; Liu, X.; Zhang, D. The assessment of forest ecological security and its determining indicators: A case study of the Yangtze River Economic Belt in China. *J. Environ. Manag.* **2020**, *258*, 110048. [CrossRef]
18. Wu, X.; Liu, S.; Sun, Y.; An, Y.; Dong, S.; Liu, G. Ecological security evaluation based on entropy matter-element model: A case study of Kunming city, southwest China. *Ecol. Indic.* **2019**, *102*, 469–478. [CrossRef]
19. Xu, J.Y.; Fan, F.; Liu, Y.; Dong, J.; Chen, J. Construction of Ecological Security Patterns in Nature Reserves Based on Ecosystem Services and Circuit Theory: A Case Study in Wenchuan, China. *Int. J. Environ. Res. Public Health* **2019**, *16*, 3220. [CrossRef]
20. Gebler, D.; Wiegleb, G.; Szoszkiewicz, K. Integrating river hydromorphology and water quality into ecological status modelling by artificial neural networks. *Water Res.* **2018**, *139*, 395–405. [CrossRef]
21. Wang, Q.; Li, S.; He, G.; Li, R.; Wang, X. Evaluating sustainability of water-energy-food (WEF) nexus using an improved matter-element extension model: A case study of China. *J. Clean. Prod.* **2018**, *202*, 1097–1106. [CrossRef]
22. Gao, P.P.; Li, Y.P.; Sun, J.; Li, H.W. Coupling fuzzy multiple attribute decision-making with analytic hierarchy process to evaluate urban ecological security: A case study of Guangzhou, China. *Ecol. Complex.* **2018**, *34*, 23–34. [CrossRef]
23. Zhou, Y.; Varquez, A.C.G.; Kanda, M. High-resolution global urban growth projection based on multiple applications of the SLEUTH urban growth model. *Sci. Data* **2019**, *6*, 34. [CrossRef] [PubMed]
24. Feng, H.H.; Liu, H.P.; Lu, Y. Scenario Prediction and Analysis of Urban Growth Using SLEUTH Model. *Pedosphere* **2012**, *22*, 206–216. [CrossRef]
25. Chaudhuri, G.; Clarke, K.C. Modeling an Indian megalopolis-A case study on adapting SLEUTH urban growth model. *Comput. Environ. Urban Syst.* **2019**, *77*, 101358. [CrossRef]
26. Chaudhuri, G.; Clarke, K. The SLEUTH land use change model: A review. *Environ. Resour. Res.* **2013**, *1*, 88–105.
27. Jantz, C.A.; Goetz, S.J.; Shelley, M.K. Using the Sleuth Urban Growth Model to Simulate the Impacts of Future Policy Scenarios on Urban Land Use in the Baltimore-Washington Metropolitan Area. *Environ. Plan. B Plan. Des.* **2004**, *31*, 251–271. [CrossRef]
28. Saxena, A.; Jat, M.K. Capturing heterogeneous urban growth using SLEUTH model. *Remote Sens. Appl. Soc. Environ.* **2019**, *13*, 426–434. [CrossRef]
29. Saxena, A.; Jat, M.K. Land suitability and urban growth modeling: Development of SLEUTH-Suitability. *Comput. Environ. Urban Syst.* **2020**, *81*, 101475. [CrossRef]
30. Fujian Provincial Bureau of Statistics. *Fujian Statistical Yearbook 2019*; Fujian Provincial Bureau of Statistics: Fujian, China, 2019.
31. Xiao, B.; Wang, Q.-H.; Fan, J.; Han, F.-P.; Dai, Q.-H. Application of the SCS-CN Model to Runoff Estimation in a Small Watershed with High Spatial Heterogeneity. *Pedosphere* **2011**, *21*, 738–749. [CrossRef]
32. Liu, J.X.; Zhang, G.; Zhuang, Z.; Cheng, Q.; Gao, Y.; Chen, T.; Huang, Q.; Xu, L.; Chen, D. A new perspective for urban development boundary delineation based on SLEUTH-InVEST model. *Habitat Int.* **2017**, *70*, 13–23. [CrossRef]
33. Terando, A.J.; Costanza, J.; Belyea, C.; Dunn, R.R.; McKerrow, A.; Collazo, J.A. The Southern Megalopolis: Using the Past to Predict the Future of Urban Sprawl in the Southeast U.S. *PLoS ONE* **2014**, *9*, e102261. [CrossRef] [PubMed]
34. Kong, F.; Yin, H.; Nakagoshi, N.; Zong, Y. Urban green space network development for biodiversity conservation: Identification based on graph theory and gravity modeling. *Landsc. Urban Plan.* **2010**, *95*, 16–27. [CrossRef]

35. Zhang, K.H.; Song, S. Rural–urban migration and urbanization in China: Evidence from time-series and cross-section analyses. *China Econ. Rev.* **2003**, *14*, 386–400. [CrossRef]
36. Yu, Z.; Fryd, O.; Sun, R.; Jørgensen, G.; Yang, G.; Özdil, N.C.; Vejre, H. Where and how to cool? An idealized urban thermal security pattern model. *Landsc. Ecol.* **2020**. [CrossRef]

International Journal of
Environmental Research and Public Health

Article

The Construction and Validation of a Sustainable Tourism Development Evaluation Model

Han-Shen Chen [1,2]

1 Department of Health Diet and Industry Management, Chung Shan Medical University, No. 110, Sec. 1, Jianguo N. Rd., Taichung City 40201, Taiwan; allen975@csmu.edu.tw; Tel.: +886-4-2473-0022 (ext. 12225); Fax: +886-4-2324-8188

2 Department of Medical Management, Chung Shan Medical University Hospital, No. 110, Sec. 1, Jianguo N. Rd., Taichung City 40201, Taiwan

Received: 25 August 2020; Accepted: 5 October 2020; Published: 7 October 2020

Abstract: As climate change, food crises, sustainable development, and ecological conservation gain traction, the revival of traditional fishing villages has become an important governmental policy for Taiwan. To reduce cognitive bias, the choice experiment method was applied to construct an attribute function in fishing village tourism coupled with virtual reality headsets. Conditional logit and random parameter logit models were employed to estimate tourism utility functions. Moreover, a latent class model was employed to determine whether hetxerogeneous preferences regarding fishing village travel existed. The sampling sites were distributed across the Dongshi area. In total, 612 tourists and 170 local residents were interviewed. After incomplete questionnaires were removed, 816 valid questionnaires remained, representing 95.83% of the total questionnaires. Older residents and residents with shorter histories of education were inclined to increase land development and utilization by reducing natural landscapes; tourists preferred preserving landscapes and preventing land development. Residents with more education believed that local landscape imagery was essential. Tourists who were more educated, with high incomes, and those who were older believed that a selling platform incorporating local industries and products within the villages would be attractive for other tourists.

Keywords: climate change; environmental impact; environmental planning; willingness to pay; virtual reality

---

1. Introduction

Taiwan is a typical island country with 224 fishing ports. Fishing villages clustered around the ports as corresponding industries have flourished. However, due to depleting fishery resources, the pollution of the marine ecological environment, and the enforcement of the United Nations Convention on the Law of the Sea, the fishery economy witnessed a decline. The prosperity of these villages has also contracted because of the dearth of employment opportunities, thus further engendering the emigration of young residents and leaving behind a largely aging population in the fishing communities. The rural and remote nature of these villages, coupled with dated public facilities and constructions, only further exacerbates the problem. To increase their value, agricultural and fishing villages across the globe actively explore solutions for the socioeconomic problems that they encounter. These villages propose these solutions to induce industrial transformation and enhancement, to strike a balance between the development and preservation of resources, and to ensure the sustainable development of agricultural and fishing villages [1–4]. Keeping in mind the increasing significance of issues such as global climate change, food crises, sustainable development, and ecological conservation, the revival of fishing villages with traditional features becomes an important policy for the government.

The authorities introduced new plans, such as "Building Rich and Beautiful Agricultural and Fishing Villages," "New Style in Agricultural Villages," and "Agricultural Villages Revival 2.0," with the hope of accelerating agricultural and fishery transformation by combining local industries and culture in a bid to solve the existing problems. With special environmental attributes, such as natural landscapes, precious wildlife, and unique settlement cultures [5,6], tourism is often considered to provide a significant impetus for the development of agricultural and fishing villages [7,8]. It has been suggested that as a new form of tourism with multiple features, fishery tourism will become one of the main practices helming the transformation of coastal areas [9]. Fishery tourism can accelerate economic growth and resolve the livelihood problems of local residents as well as, more importantly, maintain a balance between the development of fishery tourism and the preservation of ocean resources while allowing for sustainable development.

The tourism industry contributed 8.8 trillion USD, i.e., nearly 10.40% of the global gross domestic product (GDP), in 2018, and it has created 319 million jobs across the world [10]. Furthermore, according to the Tourism Bureau Survey of Travel by R.O.C. Citizens (2018) [11], from 2001 to 2018, domestic travel by citizens increased by 75.58%, from 97.45 million to 171.09 million. Instead of seeking comfort in travel, travel by citizens looking to be close to nature increased from 64.70% to 75.60%. Other tourism destinations may include forest recreation areas and leisure farms. Compared to them, fishery tourism, which contributes to ecotourism by combining local natural resources and cultural celebrations as well as experiences in fishery activities and village folk customs, is not limited to providing leisure and entertainment to the public; rather, fishery tourism also provides economic benefits for villagers. Therefore, this serves as a critical juncture for fishing village development.

Overseas fishing villages proactively promote transformation and upgrade at the same time. Misali Island in the United Republic of Tanzania established a marine protected area and developed ecotourism to preserve ocean resources and engage in sustainable development [12]. Further, Andros Island of the Bahamas, housing one-third of the world's precious wildlife resources, vigorously promoted ecotourism, and its tourism income indicated an annual growth rate of 78% [13]. Similarly, Sebuyau, a fishing village in Malaysia that relied on agriculture, fishery, and handicraft during earlier times, promoted travel experiences to accelerate the local economy, resulting in a tourism income as high as 65 billion Ringgit (17 billion USD) in 2012 [14]. Padín et al. noted that when Galicia, a community in Spain famed for its seafood throughout Europe, encountered economic hardships due to fishing quotas, it transformed its primary economic source to industries that included exquisite handicraft, small-scale fishing, and tourism [15]. Tourists can fish on boats, spend nights in the villages, and visit fishing museums as a means of experiencing the production of the village, village life, and eco-industry. The authorities also collaborate with the tourism industry to promote water activities, such as glass-bottom boat tours and snorkeling, and thus, this has sharpened the industry's competitive edge through ecotourism. Additionally, Palau generated 40% of its jobs and 75% of its GDP from ecotourism [16]. Furthermore, Japan recently developed a Satoumi policy to promote active maintenance efforts among coastal fishing villagers and increase socioeconomic value across the coastal community. Nagasaki Prefecture, for example, used its ocean resources and added a novel value to its fishing villages by selling unique heart-shaped oysters to achieve sustainable economic and environmental development [17].

With reference to the aforementioned literature, we understand that whether local or overseas, fishing villages encounter various difficulties and changes in relation to the environment. To ensure continued developments, villages use their unique features to engender transformations and upgrades, promote leisure traveling, and design numerous fishery-related activities, thereby adding value to the villages. Jiménez de Madariaga and García del Hoyo suggested that the development of tourism in fishing villages could redefine fishery culture and usher in multiple new approaches into the economy [18]; nonetheless, cooperation and planning between the government and local communities remains essential for transformation. Mease et al. echoed the aforementioned aspect by recommending the inclusion of relevant stakeholders and the public in addition to governmental guidance to ensure that policies for the development and management of such villages are both effective and continuous [17,19].

Dongshi fishing village, the target research area, suffered from problems such as emigration, aging population, labor shortages, decline in fishery economy and productivity, and long-term ecological damage. Owing to guidance from and planning of the government, the Dongshi fishing village has gradually revived and transformed itself in recent years. By selling valuable fishery products and holding related cultural celebrations, Dongshi has earned a reputation in supplying fish specialties and fetches economic revenue for its surroundings. Experiences on the sea (e.g., oyster farm visiting above water), land–water interface zones (e.g., bird watching), special cultural activities on land (e.g., the fishery cultural season), and experiences related to ecology (e.g., oyster picking) has together made the village a new tourist hotspot. However, this transformation eventually affects the natural environment. Cars and motorcycles produce air and noise pollution, while human activities on the sea damage the habitats of animals and plants. The growing number of tourists negatively affects the local quality of life due to the leftover rubbish, polluted water, and ensuing security problems. Sustainability is threatened because natural resources are prone to being exhausted, thus leaving nothing for sightseeing. Owing to the vulnerability of the ecological environment in agricultural and fishing villages, especially due to limited transport, land, and economy, development in such locations could lead to positive or negative effects on the local environment and economy [20,21]. Mura and Ključnikov suggested that although human activities increase the construction of fishing villages and promote tourism industries and fish farming, the activities also deplete the ecological system's productivity, worsen the quality of the environment, change the landscape, and endanger ecological systems and biodiversity [22]. There is a crucial lesson in fishery tourism, that is, to maintain village characteristics and views while simultaneously ensuring the benefit of the ecological system under the principle of sustainability. Therefore, the present study took the Dongshi fishing village of Taiwan as an example to discuss the utility of sustainable fishery tourism development.

Sustainable tourism usually develops after precise planning and cautious consideration of revenue and environmental impact in areas housing highly sensitive environments. Consequently, evaluating sustainable tourism through an economic analysis entails managing units with respect to decisions in the planning of the usage or continuous functioning of local ecological resources [16,19]. In the field of recreational efficiency evaluation, several studies have used the travel cost method (TCM) and contingent valuation method (CVM) to assess ecotourism's utility [23–25]. However, because TCM and CVM have some limitations in their application that make it difficult to evaluate products with multiple attributes and standards, the choice experiment method (CEM) has the advantage of being used to explore product preferences with multiple attributes and levels.

CEM can assess use value and non-use value simultaneously as well as define a hypothetical market through questionnaire surveys to investigate public preferences in landscape preservation and natural development, which, in turn, reflects the value of environmental goods or services. As CEM can evaluate multiple attributes and levels, different combinations of alternative plans can be raised according to the important characteristics of non-market goods or services. Respondents can identify appropriate alternative plans in accordance with their preferences and avoid bias in assessment through choice sets in different scenarios [26]. Given these advantages, CEM has been widely applied for the assessment of non-market value in recent years, including studies on wildlife conservation [27,28], wetland restoration [29,30], ecotourism preferences [31,32], coastal areas [33–36], national parks [37,38], and ecosystem services [39–41]. Further, Diafas et al. employed CEM to examine ecosystem changes in rural communities and the efficiency of ecosystem preservation [42]. Ferretti and Gandino used CEM to study the revival of abandoned rural buildings at world heritage sites and estimated the benefits they could generate; the results have been provided to authorities as a reference for exploring new policies [43].

In empirical modeling, conditional logit (CL) models can estimate tourists' average preferences on multi-attributes of ecotourism and estimate the marginal willingness to pay (MWTP) of ecotourism attribute degree [32,44]. To examine the heterogeneous preferences of respondents and changes in their willingness to pay (WTP), given different attribute levels (attributes include experience in folk customs and culture or ecological activities), a random parameter logit (RPL) model is applied to demonstrate divergence among respondents from diverse backgrounds while facing various specific attribute preferences [45–47]. To segment target markets, a latent class model (LCM) is applied to segregate respondents into various groups as well as to investigate and compare the differences between groups, for example, the preferences and attitudes of interviewed tourists toward ecotourism along with their socioeconomic background [48]. From the aforementioned studies, we understand that CL, RPL, LCM, and other empirical models in CEM can be applied for the examination and evaluation of preference in relation to multi-attributes in ecotourism destinations with proven results.

Virtual reality (VR) is a high-technology, simulated system manufactured through computer assistance. A sense of space is created through computing, while substances in reality and digital data are turned into a simulated three-dimensional environment, which can be seen or even touched. Users can choose between various input and output equipment, be submerged into a three-dimensional space, and become direct participants in the virtual scene. Interactive, immersive, and real-time features create a vast difference between VR and traditional 3D animation. VR is now extensively applied in research encompassing virtual tours [49,50], shopping [51–53], education [54,55], virtual appreciation [56], visual and audio experiences [57], medical treatment [58,59], and disaster prevention [60,61]. A scene simulated by VR reflects more of the real world and diminishes cognitive bias compared to a description using words [62]. Innocenti suggested that in addition to reducing cognitive bias, contingency variables are strictly under control using VR [63]. In view of this, scholars have started combining applications of VR and CEM. Matthews et al. demonstrated various situations of coastal erosion using VR and employed CEM to exemplify the efficiency of coastal protection by tourist inflow [40]. Rid et al. used VR to illustrate a view of houses and their building styles while applying CEM to estimate consumers' preferences and weigh future development directions [64].

Extant research examining the economic value of non-market resources primarily focuses on forests, coastal areas, national parks, and naturally protected areas. Few studies examine the value of fishery tourism; no research has ever applied VR to the topic. This study combined CEM and VR as a cross-disciplinary approach to broaden the depth of research on sustainable tourism in fishing villages. The construction of this methodology is based on representativeness and research originality and fills a gap in the research efforts on sustainable fishery tourism development that are lacking. Studies on and contributions to the abovementioned problems will serve as references, thus benefiting researchers who aim to evaluate models for sustainable fishery tourism development and those who practice management and administration policies in fishery tourism.

2. Materials and Methods

2.1. Description of the Study Area

The Dongshi fishing village is located in western Taiwan, facing the Taiwan Strait. The village has an area of approximately 820,000 square kilometers (as illustrated in Figure 1) and a population of approximately 30,000. Some agricultural lands became unsuitable for farming because of land subsidence and soil salinization, and therefore, they were transformed into fish ponds for fish farming. The 14-km-long coastline is vital for the village's livelihood, with fish ponds covering around 2140 hectares and offshore fish farming area spanning 1800 hectares. The village contains the largest oyster farm in Taiwan, producing more than one-third of the oysters in Taiwan. Fish farming consumes substantial time and effort but provides limited vacancies and salaries. Consequently, many young adults are driven outside the village for their personal development. The agriculture and fishery industries were further damaged when the Taiwanese market was opened to imports of agricultural and fish products after the country enlisted with the World Trade Organization. In recent years, to solve the economic and social problems in the village, the government collaborated with local organizations to plan and develop fishery tourism proactively and demonstrate the unique style of fishing villages. For example, the largest wetland in Taiwan, Aogu Wetland, which covers 1500 hectares, has become an important habitat for waders and a hotspot for bird watchers. The largest shoal of the country, Waisandingzou, is a major oyster farming area. With considerable promotion in terms of culture, history, travel, and the integration of ecology education, the Dongshi fishing village is now a new tourist destination for ecological experiences, cultural education, and leisure.

Figure 1. Map of the study area.

2.2. Construction of an Utility Model for Fishing Village Travel Preference Attributes

The study applied CEM in the construction of a utility model for fishing village travel preference attributes. The study also employed CL and RPL models to estimate the indirect utility functions of fishery tourism. Using data on socioeconomic backgrounds as well as the cognitive and behavioral perspectives of interviewed tourists, differences in the MWTP of various attributes were examined, while an LCM was employed to determine whether heterogeneous preferences for fishing village travel exist among respondents. Finally, the results from the aforementioned empirical analysis were used to estimate the economic utility of fishery tourism.

CEM is a typical random utility model (RUM) for examining MWTP, given various attributes and levels [65]. Therefore, in a binary model, the utility for the nth respondent was assumed to be the different options (U_{ni}) that they meet, and the options would maximize utility as shown in Equation (1):

$$U_{ni} = V_{ni} + \varepsilon_{ni} \quad (1)$$

where U_{ni} is the utility of attribute n for person i, V_{ni} is the observable utility function, and ε_{ni} is an error that cannot be observed.

This study expects to examine different preferences and the MWTP of respondents from diverse socioeconomic backgrounds, given various attributes and levels; therefore, an RPL model and an LCM were employed for analysis. The overall utility in the LCM is as follows:

$$U_{ni} = V_{ni}(X_{ni}, S_n) + \varepsilon_{ni} \quad (2)$$

where V_{ni} is the utility coefficient of observable variable X_{ni} and respondents' characteristic S_n and represents the respondent's preference, while ε_{ni} is the error.

The estimation method of the LCM assumes that $f(\beta)$ distribution is separated, while β is the differentiate value of a finite set. It is assumed that under β, possible values of option k are represented by b_1, ... ,b_k, probability is represented by S_k, while $\beta = b_k$. Therefore, the option probability is as follows:

$$P_{ni} = \sum_{k-1}^{k} S_{nk}L(i|k) = S_{nk}L(i|k) \quad (3)$$

In the above equation, $L(i|k)$ is the CL probability by group, and S_{nk} is the group probability. Under the LCM, an individual can be incorporated into behavioral groups to analyze their preferences and WTP. In this manner, assumptions of homogeneity and heterogeneity within groups can be satisfied, while individual characteristics can be used to calculate group probability. However, to estimate the relative importance of product attributes on value, assuming that the degrees of attribute in alternative plan j remain the same, the marginal change in WTP of the kth attribute can be given by Equation (4):

$$\text{WTP} = \frac{-\beta_k}{\beta_c} \quad (4)$$

2.3. Fishery Tourism Attribute and Level Integration Plan

By referring to related survey reports and studies and interviewing professionals and scholars from different disciplines, this study set "land use planning," "cultural experience," "local imagery landscape and architecture," "product and industry promotion," and "willingness to pay/leisure attractiveness" as the five attributes. Travel attributes and the view of the fishing village were completely shown using VR to reduce cognitive differences due to words or a single photo, which might have an impact on research results. In an evaluation of efficiency, different measurements are applied to different groups of respondents. The investigation focus for the locals is set as the "willingness to pay," which concerns maintenance and construction with a foundation's support. For the tourists, the focus is on "leisure attractiveness," that is, to weigh their preferences regarding various attributes by assessing how much additional time they would spend on sightseeing, following the inclusion of new features. Table 1 lists the setting and details of these five attributes.

Table 1. Attributes and attribute levels of fishing village tourist attractions.

Attributes	Levels	Variable	Number of Levels
Land use planning(LUP)	1. Maintaining the status quo 2. Increase land use and planning 3. Maintenance of natural landscape	LUP± LUP+ LUP−	3
Cultural experience(CE)	1. Maintaining the status quo 2. Provides two cultural experiences 3. Provides three cultural experiences	CE CE+ CE++	3
Landscape architecture(LA)	1. Maintaining the status quo 2. Increase landscape architecture	LA± LA+	2
Product and industry promotion(PIP)	1. Maintaining the status quo 2. Product and industry promotion	PIP PIP+	2
Evaluation attributes Visitor: Extra travel time (min) Local residents: Tourist Site Maintenance Fund	Visitor/Local residents 1. Maintaining the status quo 2. 100 min/100 NTD per month 3. 150 min/150 NTD per month 4. 200 min/200 NTD per month 5. 250 min/250 NTD per month		5

The superscript ± describes the attribute level included in the basic alternative. The superscript + (++) indicates an increase (strong) compared with the basic alternative and the superscript (−) indicates a reduction.

2.3.1. Land Use Planning

Carlsson et al. observed that the maintenance of the natural landscape would significantly and positively affect tourism [66]. Ridding et al. emphasized that keeping and restoring forest land

and natural landscapes promote sustainable development in an ecological system [67]. Chen and Chen indicated that tourists and local residents were divided on how to use tourism resources, such as whether to maintain the status quo or develop new land uses [68]. This study took the present conditions as the base level, adding "maintenance of natural landscape" and "land utilization and planning" as two new levels for respondents to choose from.

2.3.2. Fishery Cultural Experience

Demarco suggested that a cultural experience would deepen tourists' understanding of local culture and history [69]. Chaminuka et al. examined tourist preferences regarding ecotourism development and their MWTP, and the studied experiences included rural accommodation, rural cultural travel, and rural handicraft [70]. While studying cultural travel to temples in Taiwan, Wu et al. discovered that most tourists like interactive experiences, such as understanding the history of temples by watching animations [71]. Increasing fishery cultural experience was thus listed as a level to estimate respondent preferences on cultural experience.

2.3.3. Local Landscape and Architecture Imagery

This explains how producers project compelling site meanings and how tourists decode these meanings through the lens of an imagined community. Oh et al. argue that tourists primarily seek and consume engaging experiences accompanied by the goods and/or service components of the destinations [72]. Thus, by designing and building local landscape and architecture imagery, an understanding of fishing villages via a viewing experience is enhanced. Farrelly used three memorial venues of different imageries to examine the experiences of tourists [73]. In this study, building new local landscape and architecture imagery was listed as a level for respondents to decide on.

2.3.4. Local Products and Industries Promotion

Ferretti and Gandino studied the efficiency of reviving and using abandoned rural buildings at world heritage sites and found that most respondents preferred to turn abandoned buildings into bicycle shops followed by traditional product shops [43]. Grebitus et al. examined the correlation between consumer behavior and agricultural development and concluded that consumers were attracted to fresher and higher-quality local products, which promoted the economy and incentivized industrial development [74]. Kastenholz et al. stated that economic sustainability requires the generation of economic benefits for local communities by providing value to local assets and competences, job creation, and the promotion of local products [75]. Thus, the study included whether to promote local products and industries as a level for respondents.

2.3.5. Willingness to Pay/Leisure Attractiveness

Agimass et al. determined the leisure attractiveness of forests to tourists based on the distance of forests from the respondents [76]. Further, Bakhtiari et al. suggested that the attributes of leisure environment and adding travel services could be examined with distance as a variable, as distance could indicate the degree of attraction that respondents have to a leisure location [77]. However, a minority of respondents had an unclear cognition of distance; consequently, extra travel time was employed instead as an evaluation attribute for tourists to understand the differences in attractiveness caused by changes in tourism factors. Tourist sites help the development of the region and the economy, while residents in the region also have the responsibility to maintain the development of the tourist sites. The study uses the tourism management and maintenance perspective to explore residents' preferences for the maintenance of tourist attractions through the maintenance fund.

2.4. Construction of Evaluation Model for Fishery Tourism Attributes

Orthogonal design is often used as a tool for scaling down experimental programs and can achieve statistical accuracy in the case of scaling down. The study reduced 180 ($2^2 \times 3^2 \times 5$) combinations to 18 alternative combinations through orthogonal design, and one current solution. After you include the current scenario in each group's selection set, each scenario contains two scenarios with random alternatives and one current scenario. Each questionnaire contains five selection sets drawn from it, with 34 versions. The combination of the design process and the above selection sets improves the statistical efficiency of the selection set design.

In addition, the model herein was built by panorama photos. Collections of photos taken by a panoramic camera were processed and combined using software to produce panoramic photos in 720 degrees; subsequently, the photos could be turned into VR spaces. The advantages of this building style include speed and a high resemblance to reality, as the user can experience a 720-degree reality view that is all-dimensional. Interactive features are also added to allow for a well-rounded understanding of the destinations' essence. This can reduce evaluation bias due to information differences in comparison with previous studies that used traditional pictures and descriptions involving words (as illustrated in Figure 2).

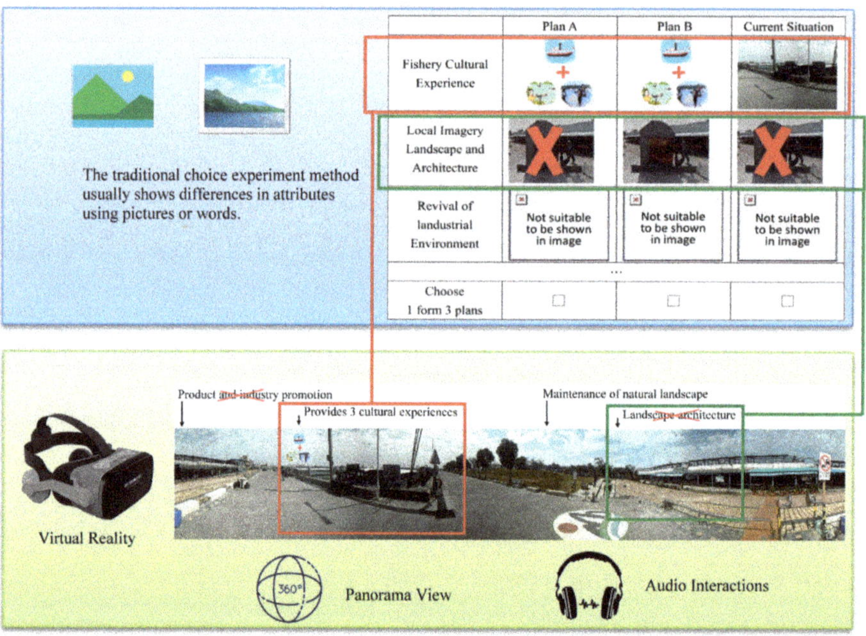

Figure 2. Observed differences between the traditional choice experiment method and after the use of virtual reality tools.

2.5. Survey Design and Respondents

The questionnaire was divided into three sections: "tourist behavior," "fishing village attribute preference," and "basic information of tourists." The details of the different sections are the following.

(1) Tourist behavior: this section was primarily designed to understand the traveling behavior of tourists, including the frequency of travel, types of activities, travel time, and satisfaction regarding the activities, in order to decipher a basic picture of tourists' situations and how they graded different conditions of tourism attributes.

(2) Fishing village attribute preference: this section involved applying CEM to examine respondents' preferences in fishery tourism, including "natural landscape protection," "fishery cultural experience," "local imagery landscape and architecture," "revival of industrial environment," and "leisure distance."

(3) Basic information of tourists: this section aimed to identify the socioeconomic background of respondents, including their gender, age, education level, occupation, residence location, and average monthly individual income.

Local residents and tourists were interviewed in early January 2019 using an initial questionnaire. This questionnaire was subsequently modified based on the conservation and management status, expert opinion and advice, and the initial results. Interviews using the final questionnaire were conducted from April to August 2019. Respondents were randomly selected and received an individual on-site interview. Respondents could experience the targeted destinations through VR and fully understood the research attributes in concrete expressions. This prevented information asymmetry confusion due to differences in understanding and cognition. The sampling sites were distributed across the Dongshi area. The respondents were divided into two groups: local residents and tourists. In total, 612 tourists and 170 local residents were interviewed. After incomplete questionnaires were removed, 816 valid questionnaires remained, representing 95.83% of the total questionnaires. (Table 2).

Table 2. Sociodemographic and economic characteristics of the respondents.

Description		Visitors		Local Residents	
		Number	%	Number	%
Gender	Male	293	47.9	88	51.8
	Female	319	52.1	82	48.2
Marital status	Single	101	16.5	21	12.4
	Married	511	83.5	149	87.6
Education	High school	54	8.8	69	40.6
	University	392	64.1	84	49.4
	Master's	166	27.1	17	10.0
Age (years)	20–29	158	25.8	14	8.2
	30–39	241	39.4	47	27.6
	40–49	131	21.4	47	27.6
	50–59	66	10.8	49	28.8
	≥60	16	2.6	13	7.6
Monthly income (NTD) [a]	<25,000	127	20.8	89	52.4
	25,001–50,000	377	61.6	76	44.7
	≥50,001	108	17.6	5	2.9

[a] NTD: new Taiwan dollar (1 NTD = 0.033 USD).

Of all the respondents, 401 respondents (51.30%) were female. Most respondents were in the age group of 31–40 years, with 288 (36.80%) in this group, while 178 respondents (22.80%) were aged 41–50 years. In terms of education, 476 (60.90%) respondents had a bachelor's degree, and 183 (23.40%) had a master's degree. Finally, 453 (57.90%) respondents had an average monthly individual income of 25,001–50,000 NTD, while 216 (27.60%) earned under 25,000 NTD.

3. Results

3.1. Analysis of the Preferences and Benefits of Fishing Village Environmental Resource Attributes

This study adopted CL and RPL models to estimate fishing village travel utility functions (Tables 3 and 4). The CL model showed that tourists and local residents had noticeable preferences in the designated tourism attributes. To further examine the differences in preferences between the groups of respondents, this study employed an RPL model to analyze the differences in fishing village attribute preferences of tourists and local residents and estimate related essences affecting each attribute.

These results can provide implications for the government and concerned parties when planning and implementing environmental policies in the future.

Table 3. Results of the conditional logit model.

Variables and Levels	Visitors		Local Residents	
	Coeff.	t-Statistic	Coeff.	t-Statistic
ASC	2.183	6.22 ***	1.719	2.82 ***
LUP+	−0.886	−11.59 ***	0.251	1.93 *
LUP−	1.57	15.48 ***	0.417	2.67 ***
CE+	1.589	18.99 ***	1.095	7.90 ***
CE++	2.208	21.10 ***	1.383	8.05 ***
LA+	0.232	6.69 ***	0.555	8.70 ***
PIP+	0.457	12.83 ***	0.621	9.75 ***
Willingness to Pay/Leisure Attractiveness	−0.009	−13.93 ***	−0.004	−4.25 ***
Number of choice sets	3060		850	
Log-likelihood ratio	−1791.69847		−573.27	

* $p < 0.1$; *** $p < 0.01$; alternative specific constant (ASC) for the status quo.

Table 4. Results of random parameter logit model.

Variables and Levels	Visitors					Local Residents				
	Coeff.	t-Statistic	Coeff. Std	t-Statistic	ETT	Coeff.	t-Statistic	Coeff. Std	t-Statistic	WTP
ASC	1.968	3.75 ***	0.996	1.74 *	-	1.28	1.08	1.11	1.98 **	-
LUP+	−1.365	−10.20 ***	1.384	9.26 ***	−109.71	0.735	2.31 **	1.996	5.65 ***	72.07
LUP−	2.291	12.78 ***	1.098	7.39 ***	184.16	1.059	2.66 ***	2.796	4.47 ***	103.81
CE+	2.259	16.04 ***	0.424	1.98 **	181.61	2.44	6.72 ***	0.657	2.10 **	239.25
CE++	3.164	16.22 ***	0.661	3.46 ***	254.32	3.104	6.07 ***	1.461	3.92 ***	304.35
LA+	0.299	5.73 ***	0.501	5.33 ***	24.06	1.133	6.07 ***	1.081	4.92 ***	111.05
PIP+	0.673	10.44 ***	0.652	5.97 ***	54.12	1.333	6.40 ***	1.187	4.58 ***	130.71
Willingness to Pay/Leisure Attractiveness	−0.012	−12.30 ***	–	–	–	−0.01	−4.21 ***	–	–	–
Number of choice sets	3060					850				
Log-likelihood ratio	−1686.84844 ***					−512.88136 ***				
Chi-square	3349.81032					841.87816				

* $p < 0.1$; ** $p < 0.05$; *** $p < 0.01$. ETT: extra travel time (minutes); WTP: willingness to pay; ASC: alternative specific constant.

The creation of three kinds of cultural experiences would increase attractiveness to visitors to the highest degree (254 min), followed by natural landscape preservation (184 min), the creation of two kinds of cultural experiences (181 min), product and industry promotion (54 min), and local landscape and architecture imagery (24 min). Overdevelopment of land would decrease the willingness to travel. From the perspective of tourists, the attractiveness of tourism attributes such as cultural experiences, natural landscape preservation, and product and industry promotion was much higher than that of local landscape and architecture imagery, which means that the attractiveness of local landscape and architecture imagery is weak.

Local residents' WTP was the highest (304 NTD) for the creation of three kinds of cultural experiences, followed by the creation of two kinds of cultural experiences (239 NTD), product and industry promotion (130 NTD), local imagery landscape and architecture (111 NTD), natural landscape preservation (103 NTD), and land development and utilization (72 NTD). From the perspective of local residents, preferences are higher for increasing cultural experiences, product and industry promotion, and creation of local imagery landscape and architecture. Notably, land development and utilization and natural landscape preservation both bore positive results, reflecting a division of opinion in land use within the residents.

3.2. Analysis of Fishery Tourism's Attractive Essence Differentiation

Respondents had different preferences regarding tourism development resources because of their identities and standpoints. Through an examination and comparison via a cross-analysis on their socioeconomic backgrounds, Tables 5 and 6 illustrate the correlation between respondents' backgrounds and different resources attributes.

The cross-analysis on tourists' socioeconomic backgrounds and attribute preferences indicated the following. (1) The two attributes, natural landscape preservation and lowering development and utilization of land, were highly attractive to more educated and younger respondents. (2) Increasing cultural experiences was highly attractive to married respondents under the age of 50 with bachelor's degrees. (3) Local landscape and architecture imagery was more attractive to older respondents. (4) The incorporation of local products and industries into the fishing village could effectively attract individuals who are more educated and with a higher income over the age of 50.

The cross-analysis on local residents' socioeconomic backgrounds and attribute preferences showed the following. (1) More educated and younger residents were willing to reduce land development and preserve natural landscapes. (2) Less educated and older residents believed natural landscapes should be reduced while increasing development and utilization of land. (3) More educated residents believed that the local landscape and architecture imagery were essential. (4) Residents from different age groups had divided opinions on incorporating local products and industries into the fishing village.

3.3. Analysis on Potential Categories

Through the LCM, this study further analyzed tourism environment preferences of different tourist categories. From Table 7, two potential categories and the difference in tourism environment preferences were recognized. Concerning the attractiveness of tourism environment, the first category of tourists was inclined toward an "increase in cultural experience," "natural resources preservation," and "product and industry promotion," while "imagery architecture" was the least attractive. It is noteworthy that "land development and utilization" created push factors in travel. The second category of tourists was inclined toward "natural resources preservation," "product and industry promotion," "imagery architecture," and "increase in cultural experience," while their sense of decreasing "land development and utilization" was stronger than that of the first category. The first category comprised 73.10% of the sample. They were clearly inclined toward cultural experiences and product and industry promotion and could be called "tourists for depth." Contrastingly, the second category expressed higher preferences only for "natural resources preservation," while expressing indifference to "cultural experiences," "product and industry promotion," and "imagery architecture." This category of tourists, which constituted 26.90% of the sample, can be named "general tourists." Comparing the socioeconomic backgrounds and travel behaviors of the two categories, "tourists for depth" were mostly married and under the age of 49. In terms of travel behavior, tourists are attracted to more entertainment facilities, more cultural experiences, natural landscape preservation, and the revival of local industries, while an increase in architecture imagery would push them away.

Table 5. Cross-analysis of tourist socioeconomic background and environment attractiveness.

Visitor		LUP+	LUP−		CE+		CE++		LA+		PIP+	
Social Characteristic	obs	Mean	*t*-Statistic/ F-Test	Mean	*t*-Statistic/ F-Test	Mean	*t*-Statistic/ F-Test	Mean	*t*-Statistic/ F-Test	Mean	*t*-Statistic/ F-Test	Mean
Single	101	−119.81	−1.573	178.66	−0.742	180.38	−1.187	238.07	−9.088 ***	26.86	1.547	52.28
Married	511	−106.52		182.89		181.98		256.63		23.82		52.85
High school	54	−85.83		163.07		181.36		254.59		23.78		52.37
University	392	−103.48	12.44 ***	182.49	6.027 ***	182.64	5.153 ***	254.72	3.196 **	24.15	0.123	49.79
Master's	166	−128.52		187.71		179.65		250.51		24.89		59.88
20–29 years old	158	−127.83		186.36		180.66		255.15		21.65		54.63
30–39 years old	241	−110.17		185.66		182.89		251.8		23.52		48.12
40–49 years old	131	−104.3	10.957 ***	179.39	2.91 **	181.28	1.443	252.7	3.704 ***	25.65	3.599 ***	52.3
50–59 years old	66	−82.77		170.86		180.76		259.87		28.24		63.54
More than 60 years old	16	−40.96		158.54		181.89		245.65		35.78		63.26
Less than 25,000 NTD	127	−103.37		180.88		183.43		252.52		21.54		42.53
25,001–50,000 NTD	377	−110.61	0.574	182.29	0.09	181.67	3.613 **	253.51	0.539	24.98	1.909	54.34
More than 50,001 NTD	108	−108.37		183.4		179.88		254.99		25.3		59.25
North	157	−102.2		179.22		180.78		267.69		22.04		53.06
East	76	−122.08	1.591	183.24	0.447	175.27	18.954 ***	259.44	16.358 ***	23.88	1.327	56.32
West	266	−109.41		184.25		184.48		249.6		25.07		53.49
South	113	−107.12		180.79		180.85		239.33		26.01		48.21

** $p < 0.05$; *** $p < 0.01$.

Table 6. Cross-analysis of resident socioeconomic background and willingness to pay for environment attributes.

Local Residents	LUP+			LUP−		CE+		CE++		LA+		PIP+	
Social Characteristic	obs	Mean	t-Statistic/ F-Test	Mean	t-Statistic/ F-Test	Mean	t-Statistic/ F-Test	Mean	t-Statistic/ F-Test	Mean	t-Statistic/ F-Test	Mean	t-Statistic/ F-Test
Single	21	−40.17	−3.124 ***	98	−0.094	235.13	−1.046	318.24	1.156	96.46	−1.011	112.49	−1.453
Married	149	62.57		102.83		244.7		300.1		112.69		138.81	
High school	69	128.58	28.938 ***	62.44	6.598 ***	248.11	1.83	311.02	1.943	97.45	2.699 *	125.82	1.057
University	84	15.69		109.91		240.9		299.29		116.64		140.29	
Master's	17	−100.63		225.8		237.82		282.12		135.01		151.7	
20–29 years old	14	−44.78	12.559 ***	262.57	6.986 ***	248.82	0.885	307.53	1.576	117.84	0.218	111.72	3.685 ***
30–39 years old	47	−29.72		157.79		240.5		312.33		108.44		159.97	
40–49 years old	47	51.84		68.91		241.49		291.81		116.99		119.77	
50–59 years old	49	133.9		54.83		244		300.51		106.34		146.89	
More than 60 years old	13	115.79		27.84		254.25		305.57		104.72		87.3	
Less than 25,000 NTD	89	68.41	1.868	89.43	0.587	242.61	1.284	305.39	0.442	107.76	0.45	144.1	1.198
25,001–50,000 NTD	76	32.81		114.28		245.68		300.35		112.43		125.32	
More than 50,001 NTD	5	−20.44		147.01		226.82		278.22		136.42		139.17	

* $p < 0.1$; *** $p < 0.01$.

Table 7. Empirical projection results on potential category model.

Attributes and Levels Parameters	Category 1 (73.10%)			Category 2 (26.90%)		
	Coefficient	t-Value	WTP	Coefficient	t-Value	WTP
Constant	−25.1	0	-	−12.54	−4.23	-
LUP+	−0.34	−3.38 ***	−52.19	−7.89	−4.79 ***	−145.93
LUP-	1.49	10.76 ***	228.84	3.39	6.59 ***	62.69
CE+	2.17	16.51 ***	333.62	0.22	0.82	-
CE++	2.73	17.29 ***	420.3	0.91	2.36 **	16.85
LA+	0.21	4.44 ***	31.6	1.25	4.32 ***	23.18
PIP+	0.63	12.11 ***	97.55	1.26	4.34 ***	23.36
FUND	0.01	−7.9 ***	-	−0.05	−5.31 ***	-
Constant	−0.42	−1.06				
Rich natural landscape	0.69	3.24 ***				
More entertainment facilities	2.38	8.28 ***				
More cultural experience	1.83	8.61 ***				
Increase in local architecture imagery	−0.43	−1.8 *				
Incorporation of local industry and product promotion	0.6	2.82 ***				
Married	0.49	2.16 **				
AGE ≤ 49	−0.66	−1.96 *				
N of choice sets	3060					
Log-likelihood ratio	−1562.247					
Chi squared (degree of freedom)	3599.012 [24]					

* $p < 0.1$; ** $p < 0.05$; *** $p < 0.01$.

4. Discussion

Sustainable tourism is usually regarded as a solution to the environmental impact of tourism, which emphasizes the appeal of environmental conservation [78], and environmentally responsible behavior [79]. Kumar (2002) believes that, in a sensitive and fragile ecosystem, tourism may not come without incurring costs [80]. Although sustainable tourism offers a basis on which many community tourism projects are founded, there is nevertheless an extensive amount of literature documenting the negative impacts of tourism—not only on natural and cultural resources but also on the local community itself, as well as on relations between residents and tourists [81,82]. Tourists and local residents are of paramount importance in tourism development [83], and their preferences and perceptions are of paramount importance among tourism stakeholders. Fishing villages are bounded by limited natural resources, a degree of vulnerability, restricted hazard resilience, and resultant economic dependence. To develop tourism and provide leisure services, Fishing villages must bear the negative effects on the ecological environment and social culture. Thus, the development of tourism resources with careful consideration of the ecology, economy, and social as well as sustainable development while cushioning the impact of leisure is significant.

From the abovementioned results, tourists and residents evidently had different cognitions concerning several tourism environment attributes. Factors for the tourists' environmental preferences included the following: (1) concerning the development and preservation of natural resources, tourists preferred reducing land development and utilization and supported preserving landscapes; (2) more cultural experiences would attract more visitors; (3) although the creation of local landscape and architecture imagery was still moderately attractive to tourists, their attractiveness was lower than other attributes; and (4) product and industry promotion led to evident positive results, suggesting that when a destination provides more local products, the willingness of tourists to visit increases. The environmental preferences influencing the factors of local residents included the following: (1) concerning the development and preservation of natural resources, some local residents believed in more development and utilization of land, while others supported enhancing landscape preservation; (2) residents agreed to increase cultural experiences to draw in more visitors; (3) residents generally believed that the creation of local landscape and architecture imagery attracts more visitors; and

(4) product and industry promotion brought clear positive results, indicating that residents believed that tourist destinations should open more spaces to sell and display local agricultural and fishery products.

The above analysis shows that tourists and local residents have great differences in their environmental preferences. Tourists hope to reduce land development and utilization and supported preserving landscapes, whereas residents hope to reinforce the development and utilization of land. The results of this analysis are same as those of Chen and Chen (2019) [68], which indicated that tourists and local residents were divided on how to use tourism resources. In the future, planners of fishing village development should consider the opinions of tourists and local residents and include their knowledge regarding the environmental resources of travel destinations as well as their needs to construct a more accurate foundation for decision-making [84] such as whether to maintain the status quo or develop new land uses [68]. Stefanica and Butnaru [85] argued that responsibility for the environmental impacts of tourism development should be shared by tourism industry operators and tourists alike. Several studies have reported that tourism creates new opportunities for residents such as new shopping and recreation opportunities [86,87].

Second, some residents believed that deficiencies exist in the current fishing villages tourism environment, which should be overcome by increasing land development and using hardware facilities to expand tourism capacity. They believed that local landscape and architecture imagery can attract visitors but were divided on incorporating local products and industries into the fishing village. However, the factors that further attracted tourists included experiencing the village life, the maintenance of the natural rural landscape, and the incorporation and promotion of local products and industries in the fishing village. Local landscape and architecture imagery was the least attractive attribute to tourists. Previous studies have indicated that a cultural experience would deepen tourists' understanding of local culture and history [69], and by designing and building local landscape and architecture imagery, an understanding of fishing villages via a viewing experience is enhanced [72]. Furthermore, consider economic sustainability requires the generation of economic benefits for local communities by job creation and the promotion of local products [74,75]. The findings of this study are consistent with the majority of studies especially when it comes to the perceived economic benefits of tourism [88,89]. This is because people firmly believe that tourism is the catalyst for local economic development.

In this study, only five environmental resource attributes are illustrated because of the limitation in expression via panoramic images. However, there are many more environmental resource attributes that can be discussed with regard to fishing villages, such as the maintenance of the leisure environment, car park spaces, the design of shaded walking trails, and guided tours. Through a related evaluation of environment attributes, we can learn more about tourists' actual traveling preferences, and the results can act as references for the government and concerned parties when planning and implementing policies. Furthermore, concerning the socioeconomic status of respondents, the survey only focused on gender, age, education level, occupation, income, and residence location. Future research may include more survey items, such as status of travel companions, the number of travel companions, the age of the companions, and the major aim of travel (historical heritage, natural landscape, local art and literature, etc.), to further examine if there are more aspects that affect respondents' preferences regarding environmental resources.

5. Conclusions

This study performed an extended application of CEM using VR panoramic images and audio to deliver attributes evaluated and to avoid cognitive differences caused by traditional expressions via pictures and words. The preferences of local residents and tourists on travel attributes were also further examined. Perspectives on land resources (land development and utilization and the maintenance of natural landscape), cultural experiences (fish farming experience and ecological experiments), history and humanities (local landscape and architecture imagery), and industry (local product and industry promotion) were included in the evaluation model. The results demonstrated that tourists and local residents had different preferences regarding the allocation of fishing villages tourism resources. Sociodemographic and economic characteristics such as age, education, and income were also found to significantly influence tourists and local residents' perceptions towards the environmental impacts of fishery tourism. Cognitive differences exist between tourists and local residents on various topics, including land development and utilization, local landscape and architecture imagery, as well as the incorporation of local industries and product development into villages. Less educated and older residents are inclined to increase land development and utilization by reducing natural landscapes, while tourists preferred preserving landscapes and preventing land developments. More educated residents believed that local landscape imagery was essential, but its attraction to tourists was the lowest of all the attributes. The opinions of local residents were divided on incorporating local industries and product development into villages, while tourists who were more educated, older and higher earners believed that a selling platform combining local industries and products within the villages would be attractive for tourists with similar backgrounds as them. In conclusion, to ensure the sustainable development of a tourism destination, we must incorporate stakeholders' ideas as well as consider tourists' needs and preferences in tourism environments together with the present situation of leisure resources, environmental characteristics, and the management of the destination.

In the future, planners of fishing village development should consider the opinions of tourists and local residents. Additionally, when there are plans to adjust the allocation of tourism environment resources, it is important to avoid a result that is merely beneficial to a minority of residents. Thus, local residents should be invited to discuss the adjustments. In such discussions, planners should explain the reasons for the reallocation of environmental resources and the benefits of the adjustments to resolve disagreements with residents and reduce any impediments in implementing policies.

Funding: This work was supported by the Ministry of Science and Technology (Taiwan) [Grant Number MOST 108−2410-H−040−011]. The funder had no role in study design; in the collection, analysis and interpretation of data; in writing the report; and in the decision to submit the article for publication.

Acknowledgments: First of all, I would like to express my deepest gratitude to Chu-Wei Chen, who provided the statistics used in my thesis. Second, I would like to express my heartfelt thanks to all the experts who have taken the time to review this article and provide valuable comments.

Conflicts of Interest: The author declares that there is no conflict of interest.

References

1. Kheiri, J.; Nasihatkon, B. The Effects of Rural Tourism on Sustainable Livelihoods (Case Study: Lavij Rural, Iran). *Mod. Appl. Sci.* **2016**, *10*, 10. [CrossRef]
2. Muresan, I.C.; Oroian, C.F.; Harun, R.; Arion, F.H.; Porutiu, A.; Chiciudean, G.O.; Todea, A.; Lile, R. Local Residents' Attitude toward Sustainable Rural Tourism Development. *Sustainability* **2016**, *8*, 100. [CrossRef]
3. Su, M.M.; Wall, G.; Jin, M. Island livelihoods: Tourism and fishing at Long Islands, Shandong Province, China. *Ocean. Coast. Manag.* **2016**, *122*, 20–29. [CrossRef]
4. Su, M.M.; Wall, G.; Wang, S. Yujiale fishing tourism and island development in Changshan Archipelago, Changdao, China. *Isl. Stud. J.* **2017**, *12*, 127–142. [CrossRef]
5. Mitchell, C.J.; Shannon, M. Exploring cultural heritage tourism in rural Newfoundland through the lens of the evolutionary economic geographer. *J. Rural Stud.* **2018**, *59*, 21–34. [CrossRef]
6. Petroman, C.; Mirea, A.; Lozici, A.; Constantin, E.C.; Marin, D.; Merce, I. The Rural Educational Tourism at the Farm. *Procedia Econ. Financ.* **2016**, *39*, 88–93. [CrossRef]
7. Chen, B.; Qiu, Z.; Usio, N.; Nakamura, K. Tourism's Impacts on Rural Livelihood in the Sustainability of an Aging Community in Japan. *Sustainability* **2018**, *10*, 2896. [CrossRef]
8. Wang, L.; Yotsumoto, Y. Conflict in tourism development in rural China. *Tour. Manag.* **2019**, *70*, 188–200. [CrossRef]
9. Porter, B.A.; Orams, M.B.; Lück, M. Surf-riding tourism in coastal fishing communities: A comparative case study of two projects from the Philippines. *Ocean. Coast. Manag.* **2015**, *116*, 169–176. [CrossRef]

10. World Travel and Tourism Council. Travel and Tourism Economic Impact. 2019. Available online: https://www.wttc.org/-/media/files/reports/economic-impact-research/regions-2019/world2019.pdf (accessed on 18 November 2019).
11. Tourism Bureau. Visitors to Principal Tourist Spots in Taiwan by Month. 2018. Available online: http://admin.taiwan.net.tw/English/index.aspx (accessed on 20 August 2019).
12. Chernela, J.M.; Ahmad, A.; Khalid, F.; Sinnamon, V.; Jaireth, H. Innovative governance of fisheries and ecotourism in community-based protected areas. *Parks* **2002**, *12*, 28–41.
13. Hayes, M.C.; Peterson, M.N.; Heinen-Kay, J.L.; Langerhans, R.B. Tourism-related drivers of support for protection of fisheries resources on Andros Island, The Bahamas. *Ocean. Coast. Manag.* **2015**, *106*, 118–123. [CrossRef]
14. Ahmad, J.A.; Abdurahman, A.Z.A.; Ali, J.K.; Khedif, L.Y.B.; Bohari, Z.; Kibat, S.A. Social entrepreneurship in ecotourism: An opportunity for fishing village of Sebuyau, Sarawak Borneo. *Tour. Leis. Glob. Chang.* **2014**, *1*, 38–48.
15. Padin, C.; Lima, C.; Pardellas, X.X. A market analysis for improving fishing tourism management in Galicia (Spain). *Ocean. Coast. Manag.* **2016**, *130*, 172–178. [CrossRef]
16. Wabnitz, C.C.; Cisneros-Montemayor, A.; Hanich, Q.; Ota, Y. Ecotourism, climate change and reef fish consumption in Palau: Benefits, trade-offs and adaptation strategies. *Mar. Policy* **2018**, *88*, 323–332. [CrossRef]
17. Mizuta, D.D.; Vlachopoulou, E.I. Satoumi concept illustrated by sustainable bottom-up initiatives of Japanese Fisheries Cooperative Associations. *Mar. Policy* **2017**, *78*, 143–149. [CrossRef]
18. De Madariaga, C.J.; García-Del-Hoyo, J.J. Enhancing of the cultural fishing heritage and the development of tourism: A case study in Isla Cristina (Spain). *Ocean. Coast. Manag.* **2019**, *168*, 1–11. [CrossRef]
19. Mease, L.A.; Erickson, A.; Hicks, C.C. Engagement takes a (fishing) village to manage a resource: Principles and practice of effective stakeholder engagement. *J. Environ. Manag.* **2018**, *212*, 248–257. [CrossRef]
20. Randelli, F.; Romei, P.; Tortora, M. An evolutionary approach to the study of rural tourism: The case of Tuscany. *Land Use Policy* **2014**, *38*, 276–281. [CrossRef]
21. Su, M.M.; Wall, G.; Wang, Y.; Jin, M. Livelihood sustainability in a rural tourism destination—Hetu Town, Anhui Province, China. *Tour. Manag.* **2019**, *71*, 272–281. [CrossRef]
22. Mura, L.; Ključnikov, A. Small Businesses in Rural Tourism and Agrotourism: Study from Slovakia. *Econ. Sociol.* **2018**, *11*, 286–300. [CrossRef]
23. Bertram, C.; Larondelle, N. Going to the Woods Is Going Home: Recreational Benefits of a Larger Urban Forest Site—A Travel Cost Analysis for Berlin, Germany. *Ecol. Econ.* **2017**, *132*, 255–263. [CrossRef]
24. He, J.; Huang, A.; Xu, L. Spatial heterogeneity and transboundary pollution: A contingent valuation (CV) study on the Xijiang River drainage basin in south China. *China Econ. Rev.* **2015**, *36*, 101–130. [CrossRef]
25. Pérez-Verdín, G.; Sanjurjo-Rivera, E.; Galicia, L.; Hernandez-Diaz, J.C.; Hernández-Trejo, V.; Linares, M.A.M. Economic valuation of ecosystem services in Mexico: Current status and trends. *Ecosyst. Serv.* **2016**, *21*, 6–19. [CrossRef]
26. Hoyos, D. The state of the art of environmental valuation with discrete choice experiments. *Ecol. Econ.* **2010**, *69*, 1595–1603. [CrossRef]
27. Lee, D.E.; Du Preez, M. Determining visitor preferences for rhinoceros conservation management at private, ecotourism game reserves in the Eastern Cape Province, South Africa: A choice modeling experiment. *Ecol. Econ.* **2016**, *130*, 106–116. [CrossRef]
28. Lew, D.K.; Wallmo, K. Temporal stability of stated preferences for endangered species protection from choice experiments. *Ecol. Econ.* **2017**, *131*, 87–97. [CrossRef]
29. Dias, V.; Belcher, K. Value and provision of ecosystem services from prairie wetlands: A choice experiment approach. *Ecosyst. Serv.* **2015**, *15*, 35–44. [CrossRef]
30. Franzén, F.; Dinnétz, P.; Hammer, M. Factors affecting farmers' willingness to participate in eutrophication mitigation—A case study of preferences for wetland creation in Sweden. *Ecol. Econ.* **2016**, *130*, 8–15. [CrossRef]
31. Romão, J.; Neuts, B.; Nijkamp, P.; Shikida, A. Determinants of trip choice, satisfaction and loyalty in an eco-tourism destination: A modelling study on the Shiretoko Peninsula, Japan. *Ecol. Econ.* **2014**, *107*, 195–205. [CrossRef]
32. Torres, C.; Faccioli, M.; Font, A.R. Waiting or acting now? The effect on willingness-to-pay of delivering inherent uncertainty information in choice experiments. *Ecol. Econ.* **2017**, *131*, 231–240. [CrossRef]

33. Cazabon-Mannette, M.; Schuhmann, P.W.; Hailey, A.; Horrocks, J. Estimates of the non-market value of sea turtles in Tobago using stated preference techniques. *J. Environ. Manag.* **2017**, *192*, 281–291. [CrossRef] [PubMed]
34. Peng, M.; Oleson, K.L. Beach Recreationalists' Willingness to Pay and Economic Implications of Coastal Water Quality Problems in Hawaii. *Ecol. Econ.* **2017**, *136*, 41–52. [CrossRef]
35. Schuhmann, P.W.; Bass, B.E.; Casey, J.F.; Gill, D.A. Visitor preferences and willingness to pay for coastal attributes in Barbados. *Ocean. Coast. Manag.* **2016**, *134*, 240–250. [CrossRef]
36. Xuan, B.B.; Sandorf, E.D.; Aanesen, M. Informing management strategies for a reserve: Results from a discrete choice experiment survey. *Ocean. Coast. Manag.* **2017**, *145*, 35–43. [CrossRef]
37. Ek, K.; Persson, L. Wind farms—Where and how to place them? A choice experiment approach to measure consumer preferences for characteristics of wind farm establishments in Sweden. *Ecol. Econ.* **2014**, *105*, 193–203. [CrossRef]
38. León, C.J.; Ledesma, J.D.L.; Araña, J.E.; González, M.M. Tourists' preferences for congestion, residents' welfare and the ecosystems in a national park. *Ecol. Econ.* **2015**, *118*, 21–29. [CrossRef]
39. Jaung, W.; Putzel, L.; Bull, G.Q.; Guariguata, M.R.; Sumaila, U.R. Estimating demand for certification of forest ecosystem services: A choice experiment with Forest Stewardship Council certificate holders. *Ecosyst. Serv.* **2016**, *22*, 193–201. [CrossRef]
40. Matthews, Y.; Scarpa, R.; Marsh, D. Using virtual environments to improve the realism of choice experiments: A case study about coastal erosion management. *J. Environ. Econ. Manag.* **2017**, *81*, 193–208. [CrossRef]
41. Sangha, K.K.; Jeremy, R.-S.; Morrison, S.C.; Costanza, R.; Edwards, A. Challenges for valuing ecosystem services from an Indigenous estate in northern Australia. *Ecosyst. Serv.* **2017**, *25*, 167–178. [CrossRef]
42. Diafas, I.; Barkmann, J.; Mburu, J. Measurement of Bequest Value Using a Non-monetary Payment in a Choice Experiment—The Case of Improving Forest Ecosystem Services for the Benefit of Local Communities in Rural Kenya. *Ecol. Econ.* **2017**, *140*, 157–165. [CrossRef]
43. Ferretti, V.; Gandino, E. Co-designing the solution space for rural regeneration in a new World Heritage site: A Choice Experiments approach. *Eur. J. Oper. Res.* **2018**, *268*, 1077–1091. [CrossRef]
44. Mejía, C.V.; Brandt, S. Managing tourism in the Galapagos Islands through price incentives: A choice experiment approach. *Ecol. Econ.* **2015**, *117*, 1–11. [CrossRef]
45. Fujino, M.; Kuriyama, K.; Yoshida, K. An evaluation of the natural environment ecosystem preservation policies in Japan. *J. For. Econ.* **2017**, *29*, 62–67. [CrossRef]
46. Lyu, S.O. Which accessible travel products are people with disabilities willing to pay more? A choice experiment. *Tour. Manag.* **2017**, *59*, 404–412. [CrossRef]
47. Randrianarison, H.; Wätzold, F. Are buyers of forest ecosystem services willing to consider distributional impacts of payments to local suppliers? Results from a choice experiment in Antananarivo, Madagascar. *Environ. Conserv.* **2016**, *44*, 74–81. [CrossRef]
48. Juutinen, A.; Mitani, Y.; Mäntymaa, E.; Shoji, Y.; Siikamäki, P.; Svento, R. Combining ecological and recreational aspects in national park management: A choice experiment application. *Ecol. Econ.* **2011**, *70*, 1231–1239. [CrossRef]
49. Carrozzino, M.; Bergamasco, M. Beyond virtual museums: Experiencing immersive virtual reality in real museums. *J. Cult. Herit.* **2010**, *11*, 452–458. [CrossRef]
50. Younes, G.; Kahil, R.; Jallad, M.; Asmar, D.; Elhajj, I.; Turkiyyah, G.; Al-Harithy, H. Virtual and augmented reality for rich interaction with cultural heritage sites: A case study from the Roman Theater at Byblos. *Digit. Appl. Archaeol. Cult. Herit.* **2017**, *5*, 1–9. [CrossRef]
51. Ouellet, É.; Boller, B.; Corriveau-Lecavalier, N.; Cloutier, S.; Belleville, S.; Belleville, S. The Virtual Shop: A new immersive virtual reality environment and scenario for the assessment of everyday memory. *J. Neurosci. Methods* **2018**, *303*, 126–135. [CrossRef]
52. Siegrist, M.; Ung, C.-Y.; Zank, M.; Marinello, M.; Kunz, A.; Hartmann, C.; Menozzi, M. Consumers' food selection behaviors in three-dimensional (3D) virtual reality. *Food Res. Int.* **2019**, *117*, 50–59. [CrossRef]
53. Van Kerrebroeck, H.; Brengman, M.; Willems, K. Escaping the crowd: An experimental study on the impact of a Virtual Reality experience in a shopping mall. *Comput. Hum. Behav.* **2017**, *77*, 437–450. [CrossRef]
54. Dávideková, M.; Mjartan, M.; Greguš, M. Utilization of Virtual Reality in Education of Employees in Slovakia. *Procedia Comput. Sci.* **2017**, *113*, 253–260. [CrossRef]

55. Feng, Z.; González, V.A.; Amor, R.; Lovreglio, R.; Cabrera-Guerrero, G. Immersive virtual reality serious games for evacuation training and research: A systematic literature review. *Comput. Educ.* **2018**, *127*, 252–266. [CrossRef]
56. Koller, S.; Ebert, L.; Martinez, R.M.; Sieberth, T. Using virtual reality for forensic examinations of injuries. *Forensic Sci. Int.* **2019**, *295*, 30–35. [CrossRef]
57. Flavián, C.; Ibáñez-Sánchez, S.; Orús, C. The impact of virtual, augmented and mixed reality technologies on the customer experience. *J. Bus. Res.* **2019**, *100*, 547–560. [CrossRef]
58. Du Sert, O.P.; Potvin, S.; Lipp, O.; Dellazizzo, L.; Laurelli, M.; Breton, R.; LaLonde, P.; Phraxayavong, K.; O'Connor, K.; Pelletier, J.-F.; et al. Virtual reality therapy for refractory auditory verbal hallucinations in schizophrenia: A pilot clinical trial. *Schizophr. Res.* **2018**, *197*, 176–181. [CrossRef]
59. Menelas, B.-A.J.; Haidon, C.; Ecrepont, A.; Girard, B. Use of virtual reality technologies as an Action-Cue Exposure Therapy for truck drivers suffering from Post-Traumatic Stress Disorder. *Entertain. Comput.* **2018**, *24*, 1–9. [CrossRef]
60. Lovreglio, R.; Gonzalez, V.; Feng, Z.; Amor, R.; Spearpoint, M.; Thomas, J.; Trotter, M.; Sacks, R. Prototyping virtual reality serious games for building earthquake preparedness: The Auckland City Hospital case study. *Adv. Eng. Inform.* **2018**, *38*, 670–682. [CrossRef]
61. Ronchi, E.; Kinateder, M.; Müller, M.; Jost, M.; Nehfischer, M.; Pauli, P.; Mühlberger, A. Evacuation travel paths in virtual reality experiments for tunnel safety analysis. *Fire Saf. J.* **2015**, *71*, 257–267. [CrossRef]
62. Patterson, Z.; Darbani, J.M.; Rezaei, A.; Zacharias, J.; Yazdizadeh, A. Comparing text-only and virtual reality discrete choice experiments of neighbourhood choice. *Landsc. Urban. Plan.* **2017**, *157*, 63–74. [CrossRef]
63. Innocenti, A. Virtual reality experiments in economics. *J. Behav. Exp. Econ.* **2017**, *69*, 71–77. [CrossRef]
64. Rid, W.; Haider, W.; Ryffel, A.; Beardmore, B. Visualisations in Choice Experiments: Comparing 3D Film-sequences and Still-images to Analyse Housing Development Alternatives. *Ecol. Econ.* **2018**, *146*, 203–217. [CrossRef]
65. Shoyama, K.; Managi, S.; Yamagata, Y. Public preferences for biodiversity conservation and climate-change mitigation: A choice experiment using ecosystem services indicators. *Land Use Policy* **2013**, *34*, 282–293. [CrossRef]
66. Carlsson, F.; Frykblom, P.; Liljenstolpe, C. Valuing wetland attributes: An application of choice experiments. *Ecol. Econ.* **2003**, *47*, 95–103. [CrossRef]
67. Ridding, L.E.; Redhead, J.W.; Oliver, T.H.; Schmucki, R.; McGinlay, J.; Graves, A.R.; Morris, J.; Bradbury, R.B.; King, H.; Bullock, J.M. The importance of landscape characteristics for the delivery of cultural ecosystem services. *J. Environ. Manag.* **2018**, *206*, 1145–1154. [CrossRef]
68. Chen, H.-S.; Chen, C.-W. Economic Valuation of Green Island, Taiwan: A Choice Experiment Method. *Sustainability* **2019**, *11*, 403. [CrossRef]
69. Demarco, D. Sustainable Urban Development Perspectives in the Era of Tourism Experience. *Procedia Soc. Behav. Sci.* **2016**, *223*, 335–341. [CrossRef]
70. Chaminuka, P.; Groeneveld, R.; Selomane, A.; Van Ierland, E. Tourist preferences for ecotourism in rural communities adjacent to Kruger National Park: A choice experiment approach. *Tour. Manag.* **2012**, *33*, 168–176. [CrossRef]
71. Wu, Y.-C.; Lin, S.-W.; Wang, Y.-H. Cultural tourism and temples: Content construction and interactivity design. *Tour. Manag.* **2020**, *76*, 103972. [CrossRef]
72. Oh, H.; Fiore, A.M.; Jeoung, M. Measuring Experience Economy Concepts: Tourism Applications. *J. Travel Res.* **2007**, *46*, 119–132. [CrossRef]
73. Farrelly, F. Revealing the memorial experience through the tourist-led construction of imagined communities. *Tour. Manag.* **2019**, *75*, 13–21. [CrossRef]
74. Grebitus, C.; Printezis, I.; Printezis, A. Relationship between Consumer Behavior and Success of Urban Agriculture. *Ecol. Econ.* **2017**, *136*, 189–200. [CrossRef]
75. Kastenholz, E.; Eusébio, C.; Carneiro, M.J. Segmenting the rural tourist market by sustainable travel behaviour: Insights from village visitors in Portugal. *J. Destin. Mark. Manag.* **2018**, *10*, 132–142. [CrossRef]
76. Agimass, F.; Lundhede, T.H.; Panduro, T.E.; Jacobsen, J.B. The choice of forest site for recreation: A revealed preference analysis using spatial data. *Ecosyst. Serv.* **2018**, *31*, 445–454. [CrossRef]
77. Bakhtiari, F.; Jacobsen, J.B.; Jensen, F.S. Willingness to travel to avoid recreation conflicts in Danish forests. *Urban. For. Urban. Green.* **2014**, *13*, 662–671. [CrossRef]

78. Björk, P. Ecotourism from a Conceptual Perspective, an Extended Definition of a Unique Tourism Form. *Int. J. Tour. Res.* **2000**, *2*, 189–202.
79. Chiu, Y.-T.H.; Lee, W.-I.; Chen, T.-H. Environmentally responsible behavior in ecotourism: Antecedents and implications. *Tour. Manag.* **2014**, *40*, 321–329. [CrossRef]
80. Kumar, S. Wildlife Tourism in India: Need to Tread with Care, In, B. D. Sharma (Ed.), Indian Wildlife: Threats and Preservation. *New Delhi Anmol Publ.* **2002**, *2*, 72–94.
81. Jokinen, M.; Sippola, S. Social sustainability at tourist destinations—Local opinions on their development and future in northern Finland. University of Lapland. *Arct. Cent. Rep.* **2007**, *50*, 89–99.
82. Tsartas, P.A. Tourism Development in Greek Insular and Coastal Areas: Sociocultural Changes and Crucial Policy Issues. *J. Sustain. Tour.* **2003**, *11*, 116–132. [CrossRef]
83. Krippendorf, J. *The Holiday-Makers: Understanding the Impact of Travel and Tourism*; Heinemann Publishing: Oxford, UK, 1987.
84. Kate, D.; Fisher, K.; Dickson, M.E.; Thrush, S.F.; Le Heron, R. Improving ecosystem service frameworks to address wicked problems. *Ecol. Soc.* **2015**, *20*. [CrossRef]
85. Ştefănică, M.; Butnaru, G.I. Research on Tourists' Perception of the Relationship between Tourism and Environment. *Procedia Econ. Financ.* **2015**, *20*, 595–600. [CrossRef]
86. Gursoy, D.; Jurowski, C.; Uysal, M. Resident Attitudes: A Structural Modeling Approach. *Ann. Tour. Res.* **2002**, *29*, 79–105. [CrossRef]
87. Walpole, M.J.; Goodwin, H.J. Local economic impacts of dragon tourism in Indonesia. *Ann. Tour. Res.* **2000**, *27*, 559–576. [CrossRef]
88. Jalani, J. Local people's perception on the impacts and importance of eco-tourism in Sabang, Palawan, Philippines. *Procedia Soc. Behav. Sci.* **2012**, *57*, 247–254. [CrossRef]
89. Türker, N.; Özturk, S. Perceptions of residents towards the impacts of tourism in the Küre Mountains National Park, Turkey. *Int. J. Bus. Soc. Sci.* **2013**, *4*, 45–56.

International Journal of
Environmental Research and Public Health

MDPI

Article

The Nexus between Economic Complexity and Energy Consumption under the Context of Sustainable Environment: Evidence from the LMC Countries

Hongbo Liu, Shuanglu Liang * and Qingbo Cui

School of Economics, Yunnan University, Kunming 650106, China; hongboliu@ynu.edu.cn (H.L.); cqb_ynu@ynu.edu.cn (Q.C.)
* Correspondence: shlliang@ynu.edu.cn

Abstract: The wide application of various energy resources in economic development is allegedly responsible for deepening environmental deterioration in terms of increasing pollution emissions and other negative consequences including climate change. This current work investigates the interdependent correlation between energy consumption (both fossil fuel energy consumption and renewable energy consumption) and economic complexity among Lancang-Mekong Cooperation (hereafter LMC) countries, from 1991 to 2017. As for empirical analysis, a panel vector autoregression (PVAR) model was employed. Outcomes of this research confirm the existence of a unidirectional relationship between energy consumption and economic complexity index. It is verified that renewable energy usage is a possible alternative to traditional energy and is able to increase economic complexity. This current research proposed to contribute as a pioneering exploration on LMC countries by adding original observations into existing studies. Finally, we will discuss policy implications of this work.

Keywords: economic complexity indicator (ECI); energy consumption; sustainable environment; Lancang-Mekong Cooperation (LMC); Panel Vector Autoregression (PVAR)

Citation: Liu, H.; Liang, S.; Cui, Q. The Nexus between Economic Complexity and Energy Consumption under the Context of Sustainable Environment: Evidence from the LMC Countries. *Int. J. Environ. Res. Public Health* **2021**, *18*, 124. https://dx.doi.org/10.3390/ijerph18010124

Received: 18 November 2020
Accepted: 12 December 2020
Published: 27 December 2020

Publisher's Note: MDPI stays neutral with regard to jurisdictional claims in published maps and institutional affiliations.

1. Introduction

According to a report released by the United Nation on October, 2018, that the world could be on the brink of a climate change disaster if immediate actions are not made. Based on a recent prediction released by U.S. Energy Information Administration (EIA), with the prevailing energy consumption rate, the world is expecting a 50% increase in energy usage by 2050, led by growth in Asia. Energy is indispensable to the functioning of human activities worldwide, nonetheless, increasing energy consumption, especially the non-renewable energy consumption, which has led to severe environmental concerns. Compared to renewable energy resources, fossil fuels in the form of crude oil, coal, along with natural gas, are more commonly adopted for energy resources in developing countries. The production process of fossil fuels is more harmful to the environment which is deemed to increase CO_2 emissions and deteriorate the environment. One of the major methods to deal with such environmental damage is adopting renewable energy, for instance, solar energy, wind energy and hydropower etc., instead of relying too much on non-renewable energy such as natural gas or coal [1,2]. Unlike non-renewable energy, most renewable energy produces little or nearly zero greenhouse gas emissions.

The transformation from the traditional economy to the green economy is on the top of policymakers' agendas, and is proposed to evoke a transition of economic operations worldwide [3]. Past decades have witnessed the steady growth of the world's energy demand and consumption [4–6]. Developing countries and emerging markets are developing in an accelerating rate with rapid population growth and industrialization. For developing countries to progress in a track of sustainable path, it is imperative for them to adopt cleaner alternatives for energy consumption in order to reduce climate change effects as well as pollution emission [7].

As a traditional energy source, non-renewable energy served for the development of human beings with a relatively long history. On the contrary, in comparative to the non-renewable energy usage, renewable energy technologies are relatively new, which indicates that they are unable to serve the society at a cost-effective level, especially among developing countries [7–9]. All six members of LMC are developing countries, due to their specific geographical location, they possess abundant of natural resources. But as a result of technology or capital restriction, they could not exploit them and make a significant renewable energy contribution to the power system as developed countries do.

The Lancang-Mekong River is one of the world's major river, ranked second after Amazon in respect to biodiversity. China, Laos, Myanmar, Vietnam, Thailand and Cambodia are six countries alongside this river, with a population amounts to 72 million live within the Mekong basin (as is demonstrated in Figure 1). LMC as a transnational organization encompasses the aforementioned six countries alongside Lancang Mekong river. LMC proposes to contribute in the area of social and political issues, sustainable development among its member countries, along with culture communication [10]. Among the aforementioned objectives, sustainable development is of key significance. Countries of this region have witnessed enormous economic growth, while the side-effects of this increase such as inequality and environmental pollution have also evoked attention. As is demonstrated in Figure 1, countries alongside the Lancang-Mekong river have the opportunity to use renewable energy owing to its congenital rivers dropping variance. The demand for renewable energy has encouraged the construction of hydropower facilities in the Lancang-Mekong river valley [11]. During this process, trans-boundary cooperation among its members will yield a win-win outcome to all collaborators in terms of both economic development and environmental protection.

Figure 1. Member countries of LMC.

First introduced by Hausmann and Hidalgo in year 2009, Economic Complexity specified the idea of the multiplicity of advantageous intellect deposit in certain country [12,13]. Economic Complexity is therefore explained as the composition of the productive yield and the arrangements that emerge to absorb and associate proficiency in a country. Through calculating the mix of products that a certain country is capable of producing, it is possible to get access to its Economic Complexity Index [12,13]. First provoked as an enrichment of export diversity, ECI is designed to measure the capability of an economy indirectly by looking at the mix of products that nation exports. However, based on the researches of Eric Kemp-Benedict and Penny Mealy, ECI is orthogonal to export diversity [14,15], which confirms that ECI is not an impeccable design. On the other hand, it is argued that ECI captures significant information concerning the process of economic development in essence, and is capable of supporting the discussion we proposed to make.

The correlation between Economic Complexity and energy consumption can be summarized into four theorems: Neutrality hypothesis, Growth hypothesis, Conservation hypothesis, as well as Feedback hypothesis [1,8,16–18]. Among which, no causal relationship between economic complexity and energy consumption indicates neutrality hypothesis which means economic complexity and energy consumption are independent with each other [19]; Growth hypothesis leads to a uni-directional correlation running from energy consumption to economic complexity, implying that innovative sources of energy should be adopted; the existing of a uni-directional causality from economic complexity to energy consumption refers to conservative hypothesis; Feedback hypothesis implies the existence of a bi-directional causal relationship between economic complexity and energy consumption [20–25].

Existing studies in the empirical literature have been inconclusive in exploring the causality between economic complexity and energy consumption (neither renewable energy consumption nor non-renewable energy consumption). This current research proposed to fill a gap in existing literature with a concentration on the LMC member countries by using Panel Vector-Autoregressive models. What is more, it is expected to contribute to the existing literature in terms of scrutinizing the relationship of Economic Complexity and energy consumption among LMC member countries.

The rest of the paper is structured as follows: Section 2 refers to the literature review. Section 3 presents the methodology framework, data description can be observed in Section 4, Section 5 presents empirical analysis, while Section 6 demonstrates discussion and Section 7 presents relevant conclusion and policy implications.

2. Literature Review

The relationship between energy consumption and economic development as well as energy consumption and economic complexity has attracted comprehensive attention from scholars to implement empirical analysis. Analysis of the former literature reveals ambiguous and controversial empirical outcomes. Literature review of this work is grouped into two research strands which were examining the aforementioned topics of environmental economics: first, the indicators we are adopting, including energy consumption, economic complexity; second, the methodologies used in relevant research.

2.1. Literature on Economic Development and Energy Consumption

Researches concerning energy consumption and economic development have been explored by numerous studies in former literature. In some studies, the concept of energy consumption was investigated separately from renewable energy consumption aspect, along with non-renewable energy consumption perspective [5,9,26,27]. In precise, Kahia and his colleagues observed a long-term equilibrium correlation between economic growth and renewable energy use and non-renewable energy use among eleven MENA countries from 1080 to 2012 [28]. Furthermore, the development of energy consumption had strong causal relationship to the economic growth as well as the development condition in the latest literature [3], in which, energy consumption was divided into renewable energy consumption and non-renewable consumption. In addition, a study focused on MIST countries including Mexico, Indonesia, South Korea as well as Turkey, revealed the existence of a long-term causal relationship between renewable and non-renewable energy use [29].

This research suggested that nuclear energy a viable approach to enhance energy security and promote sustainable energy economy.

In respect of researches empirically explored the renewable energy consumption exclusively, Adrienne Ohler and Ian Fetters examined the causal relationship between electricity generated from renewable energy sources and economic growth across 20 OECD countries from 1990 to 2008, the results from a panel error correction framework demonstrated the existing of a bidirectional causal relationship between real GDP and aggregate renewable generated from electricity [30]. Chinese scholar Yiping Fang, studied the how economic welfare influences renewable energy consumption in China from 1978 to 2008, by using a multivariate OLS methodology. This research disclosed that an increase in renewable energy usage will contribute to economic welfare development in China [31]. Unlike the above-mentioned research which discovered a uni-directional relationship between concerned variables, Usama Al-mulali identified a bi-directional relationship in the long run between GDP growth and renewable energy usage [32]. Besides, a more up-to-date research originated by Emrah Kocak observed a bi-directional causal relation and confirmed the "Feedback Hypothesis" among Black Sea and Balkan countries [33]. Kocak examined the renewable energy consumption and economic development nexus in 9 concerning countries, by using a heterogeneous panel causality approach from 1990 to 2012. What is more, a study concentrated on six newly industrialized countries revealed a cointegrated relationship between real GDP and renewable energy consumption [34]. In this research, Destek examined the causal relationship between concerned variables by employing an asymmetric causality mechanism, time ranging from 1971 to 2011.

Different from the aforementioned researches, there exists studies applying Economic Complexity Index as a proxy of economic development condition. As an effective variable to explain fluctuations in country development and economic growth, Economic Complexity Index (ECI) had obtained significant focus among researchers and policy makers from all over the globe [12,14,35–40]. According to former literature, nations with a relatively higher ECI index demonstrates similar export baskets with those other countries with a high ECI, which tend to be countries with advanced development status and be able to export products that are relatively more technologically complicated [3].

2.2. Literature on Panel Vector Autoregressive Model

Md.Samsul Alam indicated in his research that through the methodology of VECM (Vector Error Correction Model) and robust panel cointegration tests framework, it is able to certify that a significant long-run equilibrium relationship among economic growth, oil consumption, finance, trade openness and CO_2 emissions can be observed in 18 developing countries [41]. Similar methodology was utilized by Hasan Ertygrul in analyzing the influence of trade openness on global CO_2 emissions for the top 10 emitters among developing countries [42]. As for developed country groups, Tsangyao Chang verified the existence of a bi-directional causal relationship between economic growth and renewable energy from 1990 to 2011 across G7 countries [7]. A heterogeneous panel cointegration test was employed to testify the relationship between renewable energy consumption, economic development (indexed by real GDP and real gross fixed capital formation), as well as the labor force in six Central American nations, time ranges from 1980 to 2006 [16].

Besides the analysis among country groups, there are some methods that are employed on single country analysis. A research concerning the United States revealed that CO_2 emissions levels are negatively related to renewable energy usage by adopting cointegration and Granger-causality test [43]. Another study concerning the relationship between economic development and electricity usage in China used VAR and VECM model for exploration [44]. A research using time-frequency analysis on France illustrated none robust causal relationship between greenhouse gas emissions and trade openness which confirms the 'neutral hypothesis' of the target country [45]. ARDL is another methodology available for single country analysis in the relevant research area, Eyup Dogan found that through this method, it is possible to support the existence of the growth hypothesis in Turkey [20] (see Table 1).

Table 1. Literature review summary.

Authors	Time Period	Country	Methodology	Variables	Empirical Findings	Reference
Dogan Eyup	1990–2012	Turkey	Granger causality test, VECM	Economic growth, renewable energy consumption, nonrenewable energy consumption	Feedback consumption between NRELC and GR	[20]
Syed Ali Raza, Nida Shah	1991–2016	G7 countries	FMOLS, DOLS	CO_2 emission, GDP, export, import, trade, renewable energy consumption	Support EKC hypothesis	[46]
Nicholas Apergis, James Payne	1980–2004	Central America	Panel cointegration	GDP, energy usage, labor force, capital formation	Support growth hypothesis	[21]
Chi Zhang, Kaile Zhou	1978–2016	China	ARDL, VAR, ECM, OLS	GDP, electricity consumption	Interaction between electricity consumption and economic growth	[44]
Mohammad Jaforullah	1965–2012	United States	Granger causality test	CO_2 emission, nuclear energy consumption, renewable energy consumption, real GDP, real price of energy	Renewable energy decreases CO_2 emission	[43]
Medhdi Ben Jebli, Slim Ben Youssef	1980–2010	OECD countries	FMOLS, DOLS	CO_2 emission, trade, renewable energy consumption	Renewable energy consumption imports	[47]
Mihai Mutascu	1960–2013	France	Wavelet tool	CO_2 emission, trade openness	Confirm neutral hypothesis	[45]
Pao, Hsiao-Tien	1971–2005	BRIC countries	Panel causality test	CO_2 emission, GDP, energy consumption	Bidirection causality between energy and emission	[48]

The correlation between economic development and energy consumption have attracted intensive attention in the past three decades, however, analysis of the former literature reveals ambiguous and controversial empirical outcomes. This current research proposed to fill a gap in existing literature with a concentration on the LMC member countries by using Panel Vector-Autoregressive models.

To conclude, according to the literature summary in Table 1, there is few literatures available concerning economic complexity and energy consumptions regime under the context of sustainable environment. Therefore, this current work is expected to fill the above-mentioned gap. The contributions of this current work to existing literature include: first, we adopt the structural equation methodology technique to investigate the significant relationship between energy consumption and economic complexity under the context of sustainable environment. Second, the deep exploration on such topic in terms of LMC countries was the first time, to the best of our knowledge. Lastly, the work proposed to fill a gap of existing single country researches, for their limitations of reducing the power of unit root and cointegration [26].

3. Methodology

To observe the nexus between Energy Consumption and Economic Complexity in terms of sustainable environment development, a Panel Data Vector Auto-regressive (hereafter PVAR) methodology was adopted. To the best of our knowledge, this specific methodology had not been applied to the subject of energy consumption and economic complexity among LMC countries by so far. PVAR is a scientific research approach entitled with such advantages as: it facilitates the combination of existing VAR method with the panel data approach, to be more specific, it considers all concerning variables in the equation as endogenous factors, which facilitates unobserved individual heterogeneity [49,50]. Furthermore, through PVAR approach, we are able to conquer the issues generated by using granger causality analysis or Vector Error Correction model individually [51,52]. PVAR model enables all variables being considered to be treated as interdependent and endogenous, besides, it is able to model how shocks are transmitted among different countries [53,54].

A general PVAR model can be illustrated as the following equation:

$$Y_{it} = Y_{it-1}A_1 + Y_{it-2}A_2 + \cdots + Y_{it-p+1}A_{p-1} + Y_{it-p}A_p + X_{it}B + \mu_{it} + \varepsilon_{it} \quad (1)$$

$$i \in \{1,2,\ldots,N\},\ t \in \{1,2,\ldots,\ T_i\} \quad (2)$$

The above equation is a i-variate PVAR model of order t, with panel-specific fixed effects, where, Y_{it} is a (1 × i) vector of dependent variables; X_{it} is a (1 × l) vector of exogenous covariates; μ_{it} and ε_{it} are (1 × i) vectors of dependent variable-specific fixed-effect and idiosyncratic errors, respectively. The (i × i) matrices (A_1, A_2 … , A_{p-1}, A_p) and the (l × i) matrix B are parameters to be estimated. It is assumed that the equations have such characteristics as [55]:

$$E[e_{it}] = 0,\ E\left[e'_{it}e_{it}\right] = \textstyle\sum,\ E\left[e'_{it}e_{it}\right] = 0 \text{ for all } t > \text{s} \quad (3)$$

It is possible to estimate the above parameters through fixed effects or, alternately, independently of the fixed effects after some transformation, using ordinary least squares (OLS) equation-by-equation.

To accomplish those research objectives, this current study designed a second order panel VAR model as follows:

$$Z_{it} = \Gamma_0 + \Gamma_1 Z_{it-1} + \Gamma_2 Z_{it-2} + \mu_i + d_{c,t} + \varepsilon_t \quad (4)$$

where Z_{it} is a four-variable vector (lnEC, lnECI, lnEXP, lnTRADE), using i to index countries and t to index time, Γ is the parameters and ε is white noise the error term. EC means energy consumption (both renewable energy consumption and fossil fuel consumption will be considered), ECI refers to Economic Complexity Index, EXP is export diversification, TRADE means trade margins which will be represented by extensive trade margins and intensive trade margins, respectively.

In order to utilize a VAR model into panel data analysis, it is proposed to impose restrictions, to make sure that the specific econometric designations are in accordance with each cross-sectional units, in this current case, member countries of LMC cooperation [56]. Therefore, diagnostic investigations such as normality, functional form serial correlation as well as heteroscedasticity analysis are performed to guarantee the reliability of this current study [57]. Maddala and Wu test is proposed to examine the unit root of this research, additionally, in order to decide the lag-order selection, both general Coefficient Determination (CD) operation, and Hasen's J statistic (J) procedure are conducted. The comprehensive Coefficient Determination (CD), Hansen's J statistics (J), p-value, MBIC, MAIC, as well as MQIC are calculated to determine lag-order.

4. Data

To analyze the nexus between energy consumption and Economic Complexity under the background of sustainable environment, six countries alongside Lancang Mekong river were considered, namely China, Laos, Mymmar, Thailand, Cambodia, Vietnam. Time period ranged from 1991 to 2017, variables include Renewable energy consumption (% of total final energy consumption), Fossil energy consumption, Economic complexity. Renewable energy consumption represents the presentation of renewable energy consumption in total final energy consumption. ECI indicates the knowledge intensity embedded in one economy, it can be measured through considering the knowledge intensity of the products it exports, a higher value of ECI represents an economy with more sophisticated and knowledge intensive production. The value of ECI is calculated through trade data from the UN Comtrade Database, the data of economic complexity comes from Penn World Table version 9.1 [3,12,13,40].

According to Table 2, the total set of data table in our research comprised a sample of 932 observations, which indicated the suitability for adopting Panel Vector Auto-regression model. Besides, the above data summary table demonstrates that the standard deviation is smaller than the mean value, which is suitable for further data analysis [37,58,59].

Table 2. Data summary.

Variable	Observations	Mean	Standard Deviation	Minimum	Maximum
Renewable energy consumption	160	54.29	26.54	11.70	91.12
Fossil energy consumption	135	53.72	26.17	13.81	88.90
Economic complexity	160	−0.47	0.75	−1.48	1.16
Export diversification	159	3.20	1.01	1.86	4.85
Extensive margin	159	0.19	0.27	0.002	1.36
Intensive margin	159	3.03	0.97	1.73	4.80

5. Empirical Analysis

We adopted the maximum available data for energy consumption and economic complexity covering from 1991 to 2017, and panel vector autoregression (PVAR) models are used to testify whether the interactions between the concerning variables are bidirectional empirically.

5.1. Model Specification

$$\begin{bmatrix} EC_t \\ ECI_t \\ EXP_t \\ TRADE_t \end{bmatrix} = \begin{bmatrix} \alpha_1 \\ \alpha_2 \\ \alpha_3 \\ \alpha_4 \end{bmatrix} + \begin{bmatrix} A_{11,1} & A_{12,1} & A_{13,1} & A_{14,1} \\ A_{21,1} & A_{22,1} & A_{23,1} & A_{24,1} \\ A_{31,1} & A_{32,1} & A_{33,1} & A_{34,1} \\ A_{41,1} & A_{42,1} & A_{43,1} & A_{44,1} \end{bmatrix} * \begin{bmatrix} EC_{t-1} \\ ECI_{t-1} \\ EXP_{t-1} \\ TRADE_{t-1} \end{bmatrix} + \ldots$$
$$+ \begin{bmatrix} A_{11,w} & A_{12,w} & A_{13,w} & A_{14,w} \\ A_{21,w} & A_{22,w} & A_{23,w} & A_{24,w} \\ A_{31,w} & A_{32,w} & A_{33,w} & A_{34,w} \\ A_{41,w} & A_{42,w} & A_{43,w} & A_{44,w} \end{bmatrix} * \begin{bmatrix} EC_{t-w} \\ ECI_{t-w} \\ EXP_{t-w} \\ TRADE_{t-w} \end{bmatrix} + \begin{bmatrix} \varepsilon_{1t} \\ \varepsilon_{2t} \\ \varepsilon_{3t} \\ \varepsilon_{4t} \end{bmatrix}$$
$$+ \begin{bmatrix} A_{11,k} & A_{12,k} & A_{13,k} & A_{14,k} \\ A_{21,k} & A_{22,k} & A_{23,k} & A_{24,k} \\ A_{31,k} & A_{32,k} & A_{33,k} & A_{34,k} \\ A_{41,k} & A_{42,k} & A_{43,k} & A_{44,k} \end{bmatrix} * \begin{bmatrix} EC_{t-k} \\ ECI_{t-k} \\ EXP_{t-k} \\ TRADE_{t-k} \end{bmatrix}$$

where EC means energy consumption (both renewable energy usage and traditional energy consumption will be studied), ECI refers to Economic Complexity Index, EXP is export diversification, TRADE means trade margins which will be represented by extensive trade margins and intensive trade margins accordingly. $A_{i,j}$ are polynomials in the lag operator, ε_{it} are error-correction terms which are assumed to be random and uncorrelated with mean zero. The following H_{01} to H_{03} are the assumptions the research is focusing on:

H_{01}: $A_{12,1} = A_{12,2} = \ldots = A_{12,k} = 0$, meaning energy consumption is unable to Granger cause economic complexity.

H_{02}: $A_{21,1} = A_{21,2} = \ldots = A_{21,k} = 0$, referring to export diversification does not Granger cause energy consumption.

H_{03}: $A_{13,1} = A_{13,2} = \ldots = A_{13,k} = 0$, indicating trade margins does not Granger cause Energy consumption.

The similar is true for other variables.

5.2. Unit Root Test

In order to examine the stationarity of the concerning variables, it is proposed to perform unit root tests before we proceed to panel data estimation [60]. Dickey-Fuller (DF), Augmented Dickey-Fuller (ADF), and Phillips-Perron (PP) are the most popular methods adopted for unit root tests among scholars [16,61–64]. Based on our data condition, it is suitable to use Maddala and Wu test to examine the existing of unit root among variables.

The unit root test we performed above (Table 3) indicated that the variables in logarithms and the first difference with both non-trend and trend are I(1), therefore, these variables are stationary. Furthermore, the comprehensive Coefficient Determination (CD), Hansen's J statistics (J), p-value, MBIC, MAIC, as well as MQIC were calculated to determine lag-order [62–65].

Table 3. Unit root test.

	Maddala and Wu-Test			
Variables	**Non-TREND**		**TREND**	
	Zt-Bar	***p*-Value**	**Zt-Bar**	***p*-Value**
L.EC	1.050	0.902	3.564	0.468
L.ECI	1.230	0.873	4.466	0.347
L.EXP	10.288	0.036	3.481	0.481
L.TRADE	17.171	0.002 **	12.945	0.012
ΔEC	28.540	0.001 **	19.432	0.035
ΔECI	99.965	0.000 ***	78.068	0.000 ***
ΔEXP	32.169	0.000 ***	33.718	0.000 ***
ΔTRADE	82.270	0.000 ***	65.915	0.000 ***

Notes: ***, ** denote statistical significance levels of 1% and 5%. The lag length (1) was used. EC = energy consumption. ECI = Economic Complexity Index. EXP = export diversification. TRADE = trade margins.

5.3. Lag Optimum Test

In order to decide the lag-order selection, both general coefficient determination (CD) operation, and Hasen's J statistic (J) procedure were conducted. A maximum of four lag was used, including 160 observations, with six panels and an average number T of 17.000, the results are as follows in Table 4.

Table 4. Results of lag length selection procedure.

Lag	CD	J	J *p*-Value	MBIC	MAIC	MQIC
1	0.9991	25.56778	0.0604	−45.515	−6.432	−22.152
2	0.9985	8.077455	0.7842	45.300	−15.989	−27.779
3	0.9988	2.027557	0.9802	−33.514	−13.972	−21.832
4	0.9931	7,897,810	0	7,897,774	7,897,794	7,897,786

According to the result of lag length selection model, concerning equations do not include more than a single lag of energy consumption and Economic Complexity Index since the first-order lag demonstrated the smallest criteria.

5.4. Results

This section provides the outcomes of panel vector autoregression model, Eigenvalue stability condition, and the analysis of Granger causality Wald test, which can be observed in Table 5.

Table 5. Results of PVAR analysis.

Response of	Response to			
	Ln_ECI	**Ln_EXP**	**Ln_Renew**	**Ln_Fossil**
Ln_ECI	0.683 ** (4.72)	−1.786 (−0.97)	−0.296 (−0.57)	1.426 (0.37)
Ln_EXP	−0.00664 (−0.41)	0.704 ** (3.22)	0.0836 (1.50)	0.350 (0.84)
Ln_Renew	0.0599 (1.66)	−0.827 (−1.80)	0.916 *** (8.71)	−1.110 (−1.30)
Ln_Fossil	−0.00772 (1.35)	0.159 * (2.04)	0.00391 (0.26)	1.000 *** (7.48)

t statistics in parentheses indicate * $p < 0.05$, ** $p < 0.01$, *** $p < 0.001$. ECI = Economic Complexity Index. EXP = export diversification.

The outcomes of PVAR analysis revealed that the increasing economic complexity leads to lower renewable energy consumption while increases fossil fuel energy consumption. On the contrary, renewable energy consumption is expected to boost economic complexity but fossil fuel energy consumption demonstrates the opposite function. Besides, export margin exhibited positive connections between both renewable energy usage and fossil fuel usage.

Afterward, Eigenvalue stability test (Table 6) was performed in order to scrutinize the stability condition of the PVAR model we examined above.

Table 6. Eigenvalue stability test.

Eigenvalue			Graph
Real	**Imaginary**	**Modulus**	
0.979	0.247	1.009994	
0.979	−0.247	1.009994	
0.798	0	0.798	
0.547	0	0.547	

Note: the dots in the above graph indicate eigenvalue.

The Eigenvalue test performed above demonstrated that the PVAR model we built in this research is stable, because all four eigenvalues are inside or on the edge of the unity circle as shown in Table 7.

Table 7. Panel Granger test results.

Equation/Excluded		Chi2	DF	Prob > Chi2
Ln_ECI	Ln_EXP	0.932	1	0.334
	Ln_Renew	0.327	1	0.567
	Ln_Fossil	0.139	1	0.709
	ALL	5.687	3	0.128
Ln_EXP	Ln_ECI	0.166	1	0.683
	Ln_Renew	2.246	1	0.134
	Ln_Fossil	0.702	1	0.402
	ALL	4.898	3	0.179
Ln_Renew	Ln_ECI	2.763	1	0.096
	Ln_EXP	3.250	1	0.071
	Ln_Fossil	1.677	1	0.195
	ALL	8.486	3	0.037
Ln_Fossil	Ln_ECI	1.818	1	0.178
	Ln_EXP	4.173	1	0.041
	Ln_Renew	0.067	1	0.795
	ALL	11.131	3	0.011

ECI = Economic Complexity Index. EXP = export diversification.

As the VAR model belongs to the regime of an atheoretical model, which means it is unable to interpret the examined parameters [66–70]. Consequently, it is suggested to concentrate on analyzing impulse response functions and causality investigation. Generally speaking, the Impulse Response Functions (IRF) are capable of evaluating the influence of a certain variable's shock on the current and future values of endogenous variables while keeping irrelevant shocks mute. However, this technique exerts possible issues of correlation between the residuals in the system. Therefore, in order to remove the possible obstacles of correlation, it is suggested to adopt a shock orthogonalization by using the Cholesky decomposition to isolate the prevailing elements from residuals.

The impulse-response function (demonstrated above in Figure 2) illustrated the causal effects among variables in the short run, medium term, and long term. The Cholesky procedure was used to compute the impulse-response function, the procedure was performed repeatedly for 1000 times to calculate the 5th and 95th percentiles of the impulse responses.

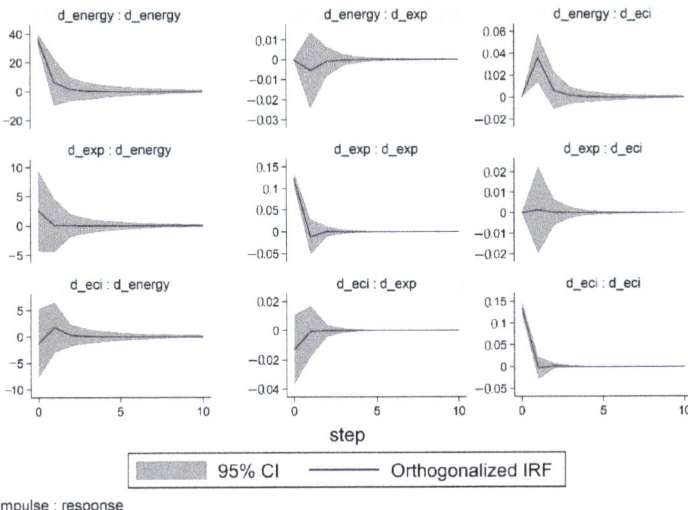

Figure 2. Impulse-response function.

The issues causing by fossil fuel adoption (including environmental deterioration as well as increasing greenhouse gas emission) are evoking serious social attention in LMC countries. To overcome the above-mentioned debatable issue, it is urgent to introduce renewable energy usage and encourage sustainable economic growth by developing complex economic growth.

6. Discussions

The wide adoption of various energy resources in economic development are deepening environmental deterioration [71]. Extensive comprehension of the relationship between energy consumption and economic development are significant to policy makers in order to make effective environmental policies [21]. The current work investigates the interdependent relationship between energy consumption (both fossil fuel energy consumption and renewable energy consumption) and Economic Complexity among Lancang-Mekong Cooperation (LMC) countries, time ranges from 1991 to 2017. To achieve this purpose, a Panel Vector Auto-regression (PVAR) model was introduced. Outcomes of empirical analysis confirms the presence of a uni-directional relationship between energy consumption and economic complexity index. It is verified that renewable energy usage is a possible alternative to traditional energy and be able to increase economic complexity at the same time contributes to green development. However, in comparative to the non-renewable energy usage, renewable energy technologies are relatively new, which indicates that they are unable to serve the society at a cost-effective level [5,8,20,72], especially among developing countries such as LMC member countries. This current research proposed to contribute as a pioneering exploration on LMC countries by adding original observations to existing studies.

7. Conclusions and Policy Implications

According to the empirical analysis presented in this current work, this paper provides substantial values to policy makers: the existence of an interdependent relationship between economic complexity and renewable energy consumption and nonrenewable encourage the usage of renewable energy more widely, and application of energy conservation policies among LMC countries. Furthermore, readjustment of industrial structure is proposed to increase economic complexity and ensure sustainable economic development. Besides, since hydroelectric power is one of the major energy sources in this region, it is suggested to attract investment to develop technologies that facilitate cleaner energy technologies such as hydroelectric schemes and installation to optimize energy usage efficiency, and exploit more environmentally friendly alternative energy resources. At last, the percentage of other renewable energy constitutions such as wind, nuclear energy, as well as solar should be considered to increase the renewable energy mix in LMC member countries. To conclude, it is suggested for LMC countries to further their commitment as well as cooperate in renewable energy technologies to achieve sustainable environment development.

Through an innovative and data driven approach, the current research shed new light on controlling environmental degradation and green industrialization among LMC countries and has several distinct implications for the development of sustainable industrial strategy to exert beneficial effect on the environmental quality in these countries. Further researches are encouraged to study the correlation between economic complexity and renewable energy usage by the types of energy, for example, wind energy, solar energy and nuclear energy usage. Besides, due to the limitation of data availability, this current work failed to achieve a more robust result. Therefore, it is rewarding for future researches to do profound analysis by adopting up-to-dated data from various sources in relevant academic realm.

Author Contributions: Conceptualization, methodology, software, data curation, writing, original draft preparation, formal analysis, H.L.; Writing—reviewing and funding acquisition, S.L.; reviewing and editing and funding acquisition, Q.C. All authors have read and agreed to the published version of the manuscript.

Funding: This research is sponsored by Yunnan University with fund number: c176220100058 and c176240103, we also received support from Yunnan Education Bureau with a fund number 2020J0038.

Institutional Review Board Statement: Not applicable.

Informed Consent Statement: Not applicable.

Data Availability Statement: Restrictions apply to the availability of these data. Data was obtained from UN Comtrade Data-base and Penn World. and are available https://growthlab.cid.harvard.edu/files/growthlab/files/atlas_2013_part1.pdf with the permission of UN Comtrade Database and Penn World.

Conflicts of Interest: The authors declare no conflict of interest.

References

1. Acheampong, A.O. Economic growth, CO_2 emissions and energy consumption: What causes what and where? *Energy Econ.* **2018**, *74*, 677–692. [CrossRef]
2. Ahmed, A.; Uddin, G.S.; Sohag, K. Biomass energy, technological progress and the environmental Kuznets curve: Evidence from selected European countries. *Biomass Bioenerg.* **2016**, *90*, 202–208. [CrossRef]
3. Mealy, P.; Teytelboym, A. Economic complexity and the green economy. *Res. Policy* **2020**, in press. [CrossRef]
4. Suganthi, L.; Samuel, A.A. Energy models for demand forecasting—A review. *Renew. Sustain. Energy Rev.* **2012**, *16*, 1223–1240. [CrossRef]
5. Al-mulali, U.; Fereidouni, H.G.; Lee, J.Y.M. Electricity consumption from renewable and non-renewable sources and economic growth: Evidence from Latin American countries. *Renew. Sustain. Energy Rev.* **2014**, *30*, 290–298. [CrossRef]
6. Brizga, J.; Feng, K.; Hubacek, K. Household carbon footprints in the Baltic States: A global multi-regional input–output analysis from 1995 to 2011. *Appl. Energy* **2017**, *189*, 780–788. [CrossRef]
7. Chang, T.; Gupta, R.; Inglesi-Lotz, R.; Simo-Kengne, B.; Smithers, D.; Trembling, A. Renewable energy and growth: Evidence from heterogeneous panel of G7 countries using Granger causality. *Renew. Sustain. Energy Rev.* **2015**, *52*, 1405–1412. [CrossRef]
8. Adewuyi, A.O.; Awodumi, O.B. Renewable and non-renewable energy-growth-emissions linkages: Review of emerging trends with policy implications. *Renew. Sustain. Energy Rev.* **2017**, *69*, 275–291. [CrossRef]
9. Destek, M.A.; Aslan, A. Renewable and non-renewable energy consumption and economic growth in emerging economies: Evidence from bootstrap panel causality. *Renew. Energy* **2017**, *111*, 757–763. [CrossRef]
10. Liu, H.; Liang, S. The Nexus between Energy Consumption, Biodiversity, and Economic Growth in Lancang-Mekong Cooperation (LMC): Evidence from Cointegration and Granger Causality Tests. *IJERPH* **2019**, *16*, 3269. [CrossRef]
11. Gu, Z.; Fan, H.; Wang, Y. Dynamic characteristics of sandbar evolution in the lower Lancang-Mekong River between 1993 and 2012 in the context of hydropower development. *Estuar. Coast. Mar. Sci.* **2020**, *237*, 106678. [CrossRef]

12. Hausmann, R.; Hidalgo, C. *The Atlas of Economic Complexity: Mapping Paths to Prosperity*; MIT Press Books: Cambridge, MA, USA, 2014; p. 1. Available online: https://growthlab.cid.harvard.edu/files/growthlab/files/atlas_2013_part1.pdf (accessed on 16 December 2020).
13. Hidalgo, C.A.; Hausmann, R. The building blocks of economic complexity. *Proc. Natl. Acad. Sci. USA* **2009**, *106*, 10570–10575. [CrossRef] [PubMed]
14. A New Interpretation of the Economic Complexity Index. Available online: https://www.inet.ox.ac.uk/files/main_feb4.pdf (accessed on 16 December 2020).
15. Kemp-Benedict, E. An Interpretation and Critique of the Method of Reflections. MPRA Paper. Available online: https://mpra.ub.uni-muenchen.de/60705/1/MPRA_paper_60705.pdf (accessed on 16 December 2020).
16. Apergis, N.; Payne, J.E. The renewable energy consumption–growth nexus in Central America. *Appl. Energy* **2011**, *88*, 343–347. [CrossRef]
17. Khan, M.K.; Teng, J.-Z.; Khan, M.I. Effect of energy consumption and economic growth on carbon dioxide emissions in Pakistan with dynamic ARDL simulations approach. *ESPR* **2019**, *26*, 23480–23490. [CrossRef]
18. Wolde-Rufael, Y. Electricity consumption and economic growth: A time series experience for 17 African countries. *Energy Policy* **2006**, *34*, 1106–1114. [CrossRef]
19. Shahbaz, M.; Hye, Q.M.A.; Tiwari, A.K.; Leitão, N.C. Economic growth, energy consumption, financial development, international trade and CO_2 emissions in Indonesia. *Renew. Sustain. Energy Rev.* **2013**, *25*, 109–121. [CrossRef]
20. Dogan, E. The relationship between economic growth and electricity consumption from renewable and non-renewable sources: A study of Turkey. *Renew. Sustain. Energy Rev.* **2015**, *52*, 534–546. [CrossRef]
21. Apergis, N.; Payne, J.E. Energy consumption and economic growth in Central America: Evidence from a panel cointegration and error correction model. *Energy Econ.* **2009**, *31*, 211–216. [CrossRef]
22. Apergis, N.; Payne, J.E. Renewable energy consumption and economic growth: Evidence from a panel of OECD countries. *Energy Policy* **2010**, *38*, 656–660. [CrossRef]
23. Shahbaz, M.; Lean, H.H. The dynamics of electricity consumption and economic growth: A revisit study of their causality in Pakistan. *Energy* **2012**, *39*, 146–153. [CrossRef]
24. Shahbaz, M.; Lean, H.H.; Shabbir, M.S. Environmental Kuznets Curve hypothesis in Pakistan: Cointegration and Granger causality. *Renew. Sustain. Energy Rev.* **2012**, *16*, 2947–2953. [CrossRef]
25. Ang, J.B. CO_2 emissions, energy consumption, and output in France. *Energy Policy* **2007**, *35*, 4772–4778. [CrossRef]
26. Maji, I.K.; Sulaiman, C.; Abdul-Rahim, A.S. Renewable energy consumption and economic growth nexus: A fresh evidence from West Africa. *Energy Reports* **2019**, *5*, 384–392. [CrossRef]
27. Doytch, N.; Narayan, S. Does FDI influence renewable energy consumption? An analysis of sectoral FDI impact on renewable and non-renewable industrial energy consumption. *Energy Econ.* **2016**, *54*, 291–301. [CrossRef]
28. Kahia, M.; Aïssa, M.S.B.; Lanouar, C. Renewable and non-renewable energy use - economic growth nexus: The case of MENA Net Oil Importing Countries. *Renew. Sustain. Energy Rev.* **2017**, *71*, 127–140. [CrossRef]
29. Pao, H.-T.; Li, Y.-Y.; Hsin-Chia, F. Clean energy, non-clean energy, and economic growth in the MIST countries. *Energy Policy* **2014**, *67*, 932–942. [CrossRef]
30. Ohler, A.; Fetters, I. The causal relationship between renewable electricity generation and GDP growth: A study of energy sources. *Energy Econ.* **2014**, *43*, 125–139. [CrossRef]
31. Fang, Y. Economic welfare impacts from renewable energy consumption: The China experience. *Renew. Sustain. Energy Rev.* **2011**, *15*, 5120–5128. [CrossRef]
32. Al-mulali, U.; Fereidouni, H.G.; Lee, J.Y.; Sab, C.N.B.C. Examining the bi-directional long run relationship between renewable energy consumption and GDP growth. *Renew. Sustain. Energy Rev.* **2013**, *22*, 209–222. [CrossRef]
33. Koçak, E.; Şarkgüneşi, A. The renewable energy and economic growth nexus in Black Sea and Balkan countries. *Energy Policy* **2017**, *100*, 51–57. [CrossRef]
34. Destek, M.A. Renewable energy consumption and economic growth in newly industrialized countries: Evidence from asymmetric causality test. *Renew. Energy* **2016**, *95*, 478–484. [CrossRef]
35. Hausmann, R.; Hwang, J.; Rodrik, D. What you export matters. *J. Econ. Growth* **2007**, *12*, 1–25. [CrossRef]
36. Ozcan, B. The nexus between carbon emissions, energy consumption and economic growth in Middle East countries: A panel data analysis. *Energy Policy* **2013**, *62*, 1138–1147. [CrossRef]
37. Shahzad, U.; Fareed, Z.; Shahzad, F.; Shahzad, K. Investigating the nexus between economic complexity, energy consumption and ecological footprint for the United States: New insights from quantile methods. *J. Clean. Prod.* **2021**, *279*, 123806. [CrossRef]
38. Can, M.; Gozgor, G. Dynamic relationships among CO_2 emissions, energy consumption, economic growth, and economic complexity in France. *SSRN* **2016**. [CrossRef]
39. Economic Complexity and Environmental Performance: Evidence from a World Sample. Available online: https://mpra.ub.uni-muenchen.de/92833/1/MPRA_paper_92833.pdf (accessed on 16 December 2020).
40. Tacchella, A.; Cristelli, M.; Caldarelli, G.; Gabrielli, A.; Pietronero, L. A new metrics for countries' fitness and products' complexity. *Sci. Rep.* **2012**, *2*, 723. [CrossRef]
41. Alam, M.S.; Paramati, S.R. Do oil consumption and economic growth intensify environmental degradation? Evidence from developing economies. *Appl. Econ.* **2015**, *47*, 5186–5203. [CrossRef]

42. Ertugrul, H.M.; Cetin, M.; Seker, F.; Dogan, E. The impact of trade openness on global carbon dioxide emissions: Evidence from the top ten emitters among developing countries. *Ecol. Indic.* **2016**, *67*, 543–555. [CrossRef]
43. Jaforullah, M.; King, A. Does the use of renewable energy sources mitigate CO_2 emissions? A reassessment of the US evidence. *Energy Econ.* **2015**. [CrossRef]
44. Zhang, C.; Lam, P.-L. On electricity consumption and economic growth in China. *Renew. Sustain. Energy Rev.* **2017**, *76*, 353–368. [CrossRef]
45. Mutascu, M. A time-frequency analysis of trade openness and CO_2 emissions in France. *Energy Policy* **2018**, *115*, 443–455. [CrossRef]
46. Raza, S.A.; Shah, N. Testing environmental Kuznets curve hypothesis in G7 countries: The role of renewable energy consumption and trade. *Environ. Sci. Pollut. Res.* **2018**, *25*, 26965–26977. [CrossRef] [PubMed]
47. Ben Jebli, M.; Ben, S.Y.; Ozturk, I. Testing environmental Kuznets curve hypothesis: The role of renewable and non-renewable energy consumption and trade in OECD countries. *Ecol. Indic.* **2016**, *60*, 824–831. [CrossRef]
48. Pao, H.-T.; Tsai, C.-M. CO_2 emissions, energy consumption and economic growth in BRIC countries. *Energy Policy* **2010**, *38*, 7850–7860. [CrossRef]
49. Liu, H.; Kim, H. Ecological Footprint, Foreign Direct Investment, and Gross Domestic Production: Evidence of Belt & Road Initiative Countries. *Sustainability* **2018**, *10*, 3527.
50. Tiwari, A.K. Comparative performance of renewable and nonrenewable energy source on economic growth and CO_2 emissions of Europe and Eurasian countries: A PVAR approach. *Econ. Bull.* **2011**, *31*, 2356–2372.
51. Shahiduzzaman, M.; Alam, K. Information Technology and Its Changing Roles to Economic Growth and Productivity in Australia. *Telecomm. Policy* **2014**, *38*, 125–135. [CrossRef]
52. Khayyat, N.T.; Lee, J.D. A measure of technological capabilities for developing countries. *Technol. Forecast. Soc. Change* **2014**, *92*, 210–223. [CrossRef]
53. Koop, G.; Korobilis, D. Model uncertainty in Panel Vector Autoregressive models. *Eur. Econ. Rev.* **2016**, *81*, 115–131. [CrossRef]
54. Koop, G.; Korobilis, D. *Forecasting with High-Dimensional Panel VARs*; Working Paper; 2018. Available online: https://econpapers.repec.org/paper/glaglaewp/2015_5f25.htm (accessed on 16 December 2020).
55. Estimation of Panel Vector Autoregression in StatA: A Package of Programs. Available online: https://www.economics.hawaii.edu/research/workingpapers/WP_16-02.pdf (accessed on 16 December 2020).
56. Love, I.; Zicchino, L. Financial development and dynamic investment behavior: Evidence from panel VAR. *QREF* **2006**, *46*, 190–210. [CrossRef]
57. Bölük, G.; Mert, M. The renewable energy, growth and environmental Kuznets curve in Turkey: An ARDL approach. *Renew. Sustain. Energy Rev.* **2015**, *52*, 587–595. [CrossRef]
58. Chu, Z. Logistics and economic growth: A panel data approach. *Ann. Reg. Sci.* **2012**, *49*, 87–102. [CrossRef]
59. Vásconez Rodríguez, A. *Economic Growth and Gender Inequality: An Analysis of Panel Data for Five Latin American Countries*; CEPAL Review; 2017. Available online: https://repositorio.cepal.org/handle/11362/42660 (accessed on 16 December 2020).
60. Liu, H.; Kim, H.; Choe, J. Export diversification, CO_2 emissions and EKC: Panel data analysis of 125 countries. *APJRS* **2019**, *3*, 361–393. [CrossRef]
61. Choi, I. Unit root tests for panel data. *J. Int. Money Financ.* **2001**, *20*, 249–272. [CrossRef]
62. Dinda, S.; Coondoo, D. Income and emission: A panel data-based cointegration analysis. *Ecol. Econ.* **2006**, *57*, 167–181. [CrossRef]
63. Elliot, B.; Rothenberg, T.; Stock, J. Efficient tests of the unit root hypothesis. *Econometrica* **1996**, *64*, 13–36.
64. Engle, R.F.; Granger, C.W.J. Co-Integration and Error Correction: Representation, Estimation, and Testing. *Econometrica* **1987**, *55*, 251–276. [CrossRef]
65. Andreoni, J.; Levinson, A. The simple analytics of the environmental Kuznets curve. *J Public Econ.* **2001**, *80*, 269–286. [CrossRef]
66. Levin, A.; Lin, C.-F.; James, C. –S. Unit root tests in panel data: Asymptotic and finite-sample properties. *J. Econom.* **2002**, *108*, 1–24. [CrossRef]
67. Hovhannisyan, N.; Sedgley, N. Using panel VAR to analyze international knowledge spillovers. *Int. Econ. Rev.* **2019**, *27*, 1633–1660. [CrossRef]
68. Topcu, E.; Altinoz, B.; Aslan, A. Global evidence from the link between economic growth, natural resources, energy consumption, and gross capital formation. *Resources Policy* **2020**, *66*, 101622. [CrossRef]
69. Sassi, S.; Gasmi, A. The Dynamic Relationship Between Corruption–Inflation: Evidence from Panel Vector Autoregression. *Jpn. Econ. Rev.* **2017**, *68*, 458–469. [CrossRef]
70. Johansen, S. The Interpretation of Cointegrating Coefficients in the Cointegrated Vector Autoregressive Model. *Oxf. Bull. Econ. Stat.* **2005**, *67*, 93–104. [CrossRef]
71. Ozturk, I.; Al-Mulali, U. Investigating the validity of the environmental Kuznets curve hypothesis in Cambodia. *Ecol. Indic.* **2015**, *57*, 324–330. [CrossRef]
72. Belaid, F.; Youssef, M. Environmental degradation, renewable and non-renewable electricity consumption, and economic growth: Assessing the evidence from Algeria. *Energy Policy* **2017**, *102*, 277–287. [CrossRef]

International Journal of
Environmental Research and Public Health

MDPI

Article

The Variability of Nitrogen Forms in Soils Due to Traditional and Precision Agriculture: Case Studies in Poland

Anna Podlasek [1,*], Eugeniusz Koda [1] and Magdalena Daria Vaverková [1,2]

[1] Department of Revitalization and Architecture, Institute of Civil Engineering, Warsaw University of Life Sciences, Nowoursynowska 159 St., 02-776 Warsaw, Poland; eugeniusz_koda@sggw.edu.pl (E.K.); magda.vaverkova@uake.cz (M.D.V.)

[2] Department of Applied and Landscape Ecology, Faculty of AgriSciences, Mendel University in Brno, Zemědělská 1, 613 00 Brno, Czech Republic

* Correspondence: anna_podlasek@sggw.edu.pl

Abstract: The soil and human health issues are closely linked. Properly managed nitrogen (N) does not endanger human health and increases crop production, nevertheless when overused and uncontrolled, can contribute to side effects. This research was intended to highlight that there is a need for carrying out monitoring studies in agricultural areas in order to expand the available knowledge on the content of N forms in agricultural lands and proper management in farming practice. The impact of two types of fertilization, concerning spatially variable (VRA) and uniform (UNI) N dose, on the distribution of N forms in soils was analyzed. The analysis was performed on the basis of soil monitoring data from agricultural fields located in three different experimental sites in Poland. The analyses performed at selected sites were supported by statistical evaluation and recognition of spatial diversification of N forms in soil. It was revealed that the movement of unused N forms to deeper parts of the soil, and therefore to the groundwater system, is more limited due to VRA fertilization. Finally, it was also concluded that the management in agricultural practice should be based on the prediction of spatial variability of soil properties that allow to ensure proper application of N fertilizers, resulting in the reduction of possible N losses.

Keywords: ammonium; nitrate; migration of nitrogen forms; fertilization

Citation: Podlasek, A.; Koda, E.; Vaverková, M.D. The Variability of Nitrogen Forms in Soils Due to Traditional and Precision Agriculture: Case Studies in Poland. *Int. J. Environ. Res. Public Health* **2021**, *18*, 465. https://doi.org/10.3390/ijerph18020465

Received: 11 November 2020
Accepted: 6 January 2021
Published: 8 January 2021

Publisher's Note: MDPI stays neutral with regard to jurisdictional claims in published maps and institutional affiliations.

1. Introduction

Excess nitrogen (N) in the soil, aquatic and atmospheric environments is a global problem, resulting mostly from human activities, which have a great impact on the N cycle at all environmental scales [1]. Available N from soil, fertilizers, and manure sources, when inefficiently used in crop production systems, can move from agricultural fields and contaminate surface and groundwater resources and also contribute to greenhouse gas emissions [2].

In recent years, numerous technologies and scientific attempts have been employed to improve the efficiency of N use. New technologies are especially intended to predict and control the N impact on the natural environment. They are also crucial for accurate measurement of spatial variability of crop yields and N availability in soils. Several authors reported that the most significant factors that determine N content in soils are, i.e., bulk density, soil clay content, organic matter, pH, climate, vegetation, terrain topography, and human activity [3,4]. The amount of N released into the soil–water environment is also significantly influenced by factors like the dose and type of N fertilizers, time and frequency of their application, efficiency of the use of N by plants, depth of the plant root system, and soil permeability [5–8]. According to Mosier et al. [9], in Europe, where N fertilizers are used the most frequently, leaching and denitrification are the main processes responsible for N losses to the environment.

The nitrate and ammonium ions released in the solubilization process are used by plants and affect crop yield. However, their excess (especially nitrate) may penetrate into groundwater. Ammonium ions supplied in mineral fertilizers (e.g., ammonium nitrate), despite the fact that they are not susceptible to leaching, like nitrate ions, may contribute directly to soil acidification. Increasing the content of nitrite (NO_2^-) and nitrate (NO_3^-) in plants is also treated as an undesirable phenomenon due to the harmfulness of N compounds to animal and human organisms. For example, N compounds, including nitrate and nitrite, were classified along with lead (Pb), cadmium (Cd), and sulfur (S), among the most dangerous factors that can have a harmful effect on human and animal health. The ammonium can also undergo a nitrification process carried by *Nitrosomonas* and *Nitrobacter*, contributing then to the accumulation and further loss of nitrate.

N losses associated with its leaching can vary significantly even over short distances of the agricultural lands, due to the variability of the soil profiles and landscape topography. These differences may also be visible in specific regions where different fertilization practices are applied. It is also possible to distinguish temporary changes in the amount of leached N forms, which is related to the climatic conditions. Gray [10] reported that in favorable environmental conditions, about 50–70% N supplied in fertilizers could be taken up by plants, 2–20% of the delivered dose is lost to the atmosphere, 15–25% is retained on organic particles and clay minerals, and 10% is leached to groundwater and surface waters. Leaching of inorganic N forms from arable fields takes place by gradual displacement of nitrate to deeper layers of the soil. Füleky [11] assessed that, on average, nitrate leaching losses were equal to 30–40 kg ha^{-1} from sandy soils and 20–30 kg N ha^{-1} from loamy soils, whereas the N dose applied during the treatment was in the range 112–280 kg N ha^{-1} year^{-1}.

Therefore, in relation to the above, it is necessary to take actions to reduce the concentration of nitrate in groundwater and surface waters in arable lands as well as to monitor the N cycle in the soil and aquifer in order to assess the impact of fertilization on the quality of the soil–water environment, and consequently on human health [12–14].

The common agricultural policy (CAP) was launched in the European Union (EU) as a partnership between agriculture and society, as well as between Europe and its farmers. The CAP is intended to: support farmers and improve agricultural productivity; to make reasonable living for EU farmers; to help tackle climate change and the sustainable management of natural resources; to maintain rural areas and landscapes across the EU; and to keep the rural economy alive by promoting jobs in farming and associated sectors. The second pillar of CAP is rural development, which aim is providing member states with an envelope of EU funding to manage nationally or regionally under multiannual, co-funded programs. The Polish Rural Development Program (RDP) is aimed at all six rural development priorities with the main priority being farm viability and competitiveness. The RDP focuses on the support of physical investments as well as on farms located in nitrate vulnerable zones and Natura 2000 areas. In addition, the support is reserved for environmental and climate friendly services and practices aimed at enhancing biodiversity, high nature value farming, improving water management, and preventing soil erosion. The focus of adopted priorities concerns: (1) Knowledge transfer and advisory services in agriculture; (2) Competitiveness of agricultural sector; (3) Food chain organization, including processing and marketing of agricultural products, (4) Restoring, preserving, and enhancing ecosystems; (5) Low carbon and climate resilience economy in the agriculture and forestry sector; (6) Social inclusion and local development.

In Poland, within 14.5 million hectares of utilized agricultural area, 74.7% is arable land and 22.4% is permanent grassland and meadows. The total population is 38.5 million and there is a relatively high share of the population working in agriculture (12%, compared to the EU average of 5%) due to the socio-economic structure of Polish agriculture which is dominated by small family farms (out of 1.5 million farms, c.a. 55% are below 5 ha). Moreover, low soil quality, combined with frequent rainfall shortages, also have a negative impact on agricultural productivity. Due to that, the 62.5% of agricultural land is classified as areas with natural constraints (ANC) [15]. Approximately 19.4% of arable lands in Poland face various environmental challenges. It concerns areas particularly endangered by water and/or wind erosion (8.2%), problems with low humus levels (3.6%), and nitrate vulnerable zones (7.4%) [15]. The relatively intensive use of arable land is putting pressure on ecosystems and naturally valuable areas. Moreover, the relatively poor water quality and the high eutrophication of Polish lakes, waterways, and the Baltic Sea make it necessary to reduce nitrogen, phosphorus, pesticide, and herbicide emissions.

In EU countries, significant changes are observed with regard to reducing the use of fertilizers and thereby reducing the N balance surplus, leading to a more efficient use of this nutrient in agriculture. In the 1990s, the annual N surplus reached over 200 kg N ha^{-1} for agricultural lands in Belgium, Malta, and Cyprus, and even up to 300 kg N ha^{-1} for agricultural lands in the Netherlands. Currently, the largest annual N surplus, exceeding 100 kg N ha^{-1}, is still observed in the abovementioned countries. Poland, with the annual surplus of 40 kg N ha^{-1}, belongs to the group of EU countries with the lowest gross N balance (Figure 1).

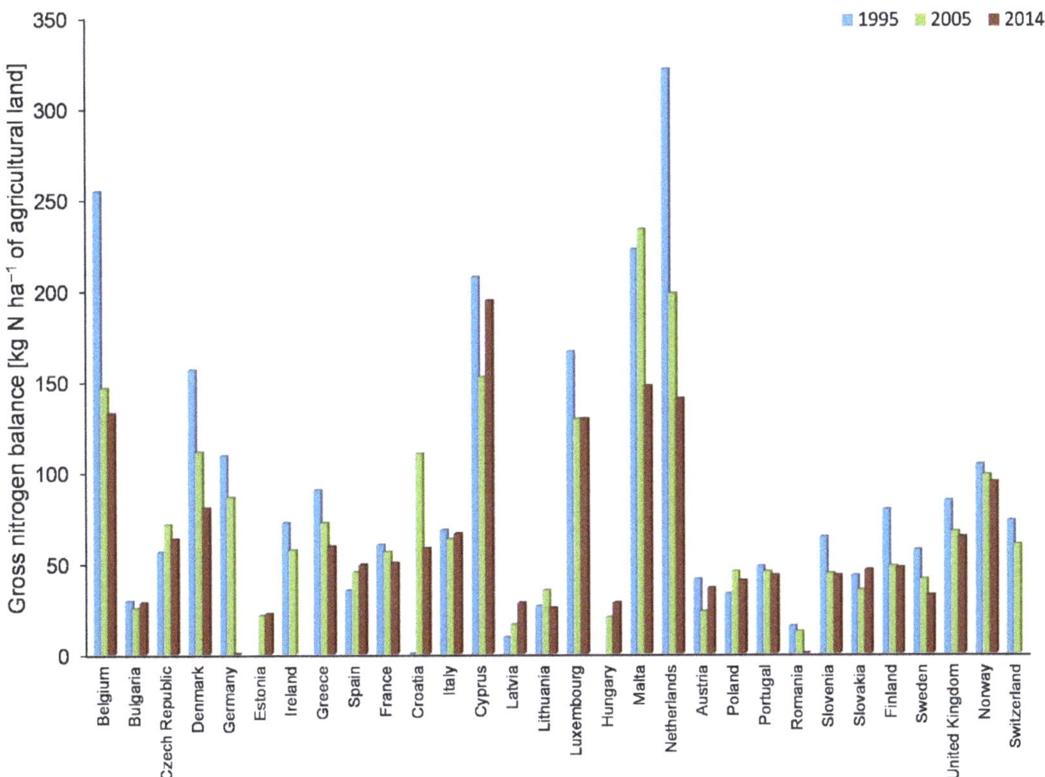

Figure 1. Annual gross N balance (kg N ha^{-1} of agricultural land) in selected EU countries (own study based on Eurostat data [15].

The losses of fertilizer components (i.e., N forms) from agriculture can be reduced by using modern fertilization techniques within precision agriculture (Table 1). The principles of precision fertilization are most often associated with limiting the excessive application of mineral fertilizers and reducing the risk of losses of unused fertilizer components. The most important approach is the optimization of fertilization rates, regarding the method and time of fertilizer application, adapted to the current needs of plants.

Table 1. Summary of selected precision agriculture research carried out in Poland in 2015–2020.

No.	Objective	Key Results	Ref.
1	Presentation of the existing methods of micronutrient fertilization.	Precise fertilization techniques based on low-solubility fertilizers, coated fertilizers, bio-based, and nanofertilizers are a new trend in modern agriculture.	[16]
2	Identifying the diversity of soil profile cohesion on the basis of non-invasive measurement of the electrical conductivity.	Maps of spatial differentiation of electrical conductivity within the field for their further use in precision agriculture.	[17]
3	Concept of a circularly polarized antenna with partially reflecting surface (PRS) has been adopted for precision farming applications.	Designed antennas employed for point to point communication in systems of mobile devices or vehicles used under precision farming.	[18]
4	Creation of independent, multi-criteria models for the prediction of winter rapeseed yield.	Forecasting winter rapeseed yields using artificial neural networks makes it possible to obtain an accurate yield forecast before harvesting. The concept of neural modeling may contribute to sustainability by reducing the doses of mineral fertilizers.	[19]
5	Study on precision agriculture concept and application.	Obtaining data concerning spatial variability of soil and plants, discussion on remote sensing, and advanced digital technology application in precision agriculture.	[20]
6	Study on the use of remote sensing in precision agriculture.	Discussion on precision agriculture in aspect of steering of farm machinery, monitoring of biomass and crop yields, soil collection, doses of mineral fertilization.	[21]
7	Analysis of use of machine vision in modern agriculture.	Examples of the use of the CloverCam system, the WeedSeeker system, Robot RoniBob Amazone Bosch, autonomous robot Agrobob.	[22]
8	Presentation of the latest trends related to the digitization of agricultural processes.	Due to the process of digitization of agriculture, in the near future, resource management will be more effective, which will reduce the impact of farming and crops on the environment, supporting sustainable agriculture.	[23]
9	Estimation of potato yields.	Examples of use of remote sensing, vegetation indices, forecasting models, artificial neural networks, and image analysis methods in yields prediction.	[24]
10	Analysis of the application of a high precision positioning system ASG-EUPOS and its service NAWGEO for agricultural machines positioning.	Field tests show usefulness of the ASG-EUPOS network and its VRS NAWGEO service for precise positioning of agricultural machinery in dynamic conditions. The obtained data can be used to create numerical models of fields on-line, for example, in selective cereals harvesting technology.	[25]
11	Identification of soil properties in different weather conditions during the growing season and mapping soil properties and crop yields using inverse distance weighting.	Geostatistical analysis is a useful tool to determine spatial interrelationships of crop yield and soil properties in the scale of agricultural field.	[26]
12	Analysis of vegetation indices used to carry out a precise and non-invasive assessment of plants condition.	Creation of vegetation indices maps (NDVI, GNDVI, SAVI). It was concluded that proper interpretation of the obtained indicators will allow for the preparation of fertilizer applications.	[27]
13	Analysis of precision agriculture methods and application.	The application of the principles of precision farming has a positive effect on reducing contamination. Discussion on variable rate application of fertilizers.	[28]
14	Evaluation of the sensitivity of sensor-based N-rate prescriptions for winter wheat to selection of sample strips for AOS calibration.	The choice of a sample strip for AOS calibration could significantly affect sensor variable N rates prescribed for winter wheat.	[29]
15	Presentation of innovative solutions for plant production in Poland.	Innovative technologies in agricultural production may reduce the negative impact of climate change.	[30]

Table 1. *Cont.*

No.	Objective	Key Results	Ref.
16	Evaluation of the precision agriculture technology on the territory of Podlaskie Voivodeship in Poland.	Only 10% of farmers use the positioning system and only 8% of the surveyed farmers apply the system for guiding agricultural machines. In addition, 14% of the investigated farmers use the system of parallel guiding.	[31]
17	Analysis of techniques for photographing and scanning crops from drones and creating field maps.	Information was obtained that could be read by the automatic control systems of machines used for fertilization and plant protection, as well as for harvesting crops. It was concluded that the use of drones in agriculture contributes to economic results.	[32]
18	Evaluation of the performance of active optical sensor (AOS) by determination of grain yield, N fertilizer use, grain protein content, N use efficiency, and N balance, utilizing a built-in algorithm for variable N rate fertilization of winter wheat.	Implementation of AOS for variable N application would minimize N surplus in areas of low productivity and improve the sustainability of N management.	[33]
19	Analysis of the use of remote sensing data in crop yield forecasting, assessing nutritional requirements of plants and nutrient content in soil, determining plant water demand and weed control.	Use of remote sensing to determine fertilization needs of plants based on the nutrient content of crops and soils helps to increase yields and improve the crop profitability.	[34]
20	Presentation of the evolutionary transition of conventional systems of agricultural activity to environmentally sustainable systems, integrated with the rural environment.	The concept of the organization of the agricultural precision production system in selected (certified) ecological farms was presented.	[35]
21	Evaluation of the soil texture prediction accuracy and the main criteria by which prediction accuracy is estimated.	All soil texture fractions were predicted with similar accuracy, using inverse distance weighting, radial basis function, ordinary kriging, and ordinary cokriging.	[36]

Moreover, the application of spatially variable (VRA) rates of fertilizers improves soil productivity and the efficiency of fertilization [37]. It is also advisable to divide the total N dose into several splits applied on several dates, during the period of plant growth, due to the fact that a high single dose of easily soluble mineral fertilizers causes the accumulation of N in the soil and as a result can lead to groundwater pollution. As an example, Kabala et al. [38] reported that a single application of the entire fertilizer dose (90 and 180 kg ha^{-1}) resulted in a significantly higher concentration of ammonium and nitrate in soil compared to a split dose. The authors also confirmed that the concentrations of nitrate and ammonium in soil were the highest with standard urea fertilization and the lowest in variants fertilized with slow-release urea. It was also pointed out that higher concentrations of both N forms were noted at the fertilizer dose of 180 kg ha^{-1}. Findings presented in the cited study indicated that the concentrations of N forms in soil are influenced by the type of fertilizer, fertilizer dose, and its division into parts.

The determination of soil mineral N patterns and variability is extremely important for agricultural management and planning [39,40]. Knowledge about the N content in the soil is also important for hydrogeological modeling and predictive analysis of N fate and transport [41,42]. Continuous development of modeling techniques and the widespread use of computational programs make it possible to solve most of the environmental tasks, including estimation of the extent of pollutant or nutrients migration. Nevertheless, it involves the risk connected with the lack of sufficient data concerning parameters occurring in mathematical equations describing the processes of pollutant transport. Typically, the prediction of contaminant migration parameters is based on the data found in the literature without considering under what conditions they were measured. Hence, the basis for creating each model of migration of N forms is always a proper database with the parameters of the environment in which the migration or transformation takes place [43].

Our study was intended to highlight that there is a need for carrying out monitoring studies in agricultural areas in order to expand the available knowledge on the content of N forms (the range and spatial diversification) in agricultural lands of selected experimental sites in Poland. We hypothesized that the concentration and distribution of N forms in soil is impacted by the fertilization method as well as the content of N forms in soil is spatially diversified and correlated with the content of clay, silt, and sand fractions. Due to the above, the objectives of this study were: (1) to revise the literature findings in aspect of N forms content in soils and factors influencing them, and (2) to evaluate the effect of UNI and VRA N application as well as soil texture on the variability of N forms in soils.

2. Materials and Methods

2.1. Study Site Description and Soil Sampling

Soil samples were collected from three agricultural fields in Poland, located in the Lower Silesian (A—50°48′58.38″ N, 17°05′17.85″ E), Pomeranian (B—54°31′12.43″ N, 17°18′33.77″ E), and Masovian Voivodships (C—52°04′31.04″ N, 21°10′54.93″ E), respectively (Figure 2).

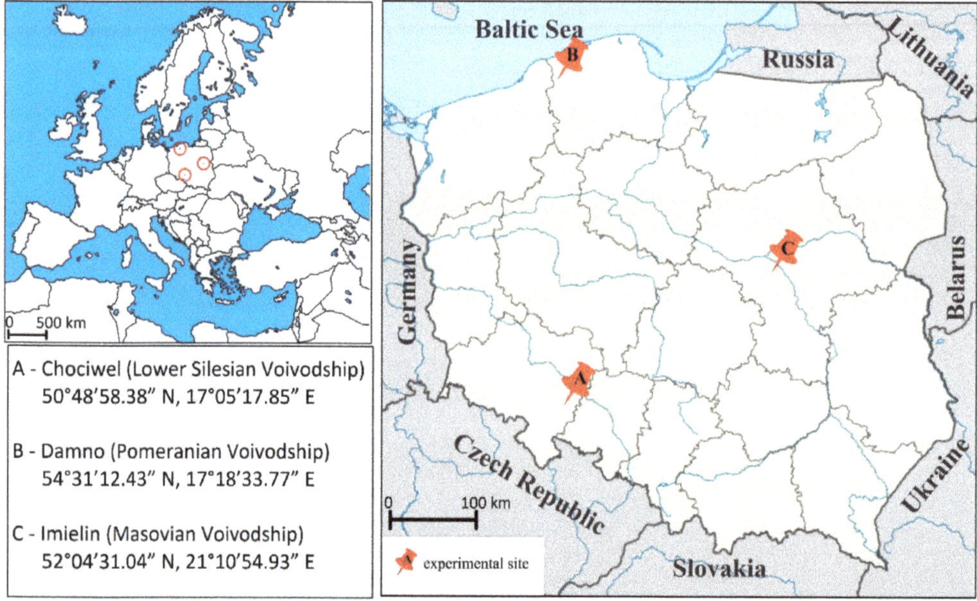

Figure 2. Locations of experimental sites.

The selected study sites are comparable by means of meteorological conditions (Figure 3), described by the data obtained from meteorological stations located nearby the experimental sites.

The average concentrations of N forms in soil before the application of N fertilizers were as follows:

- A experimental site: 1.1 mg kg^{-1} (NH_4-N), 4.3 mg kg^{-1} (NO_3-N), 0.4 g kg^{-1} ($N_{Kjeldahl}$);
- B experimental site: 0.9 mg kg^{-1} (NH_4-N), 10.2 mg kg^{-1} (NO_3-N), 0.8 g kg^{-1} ($N_{Kjeldahl}$);
- C experimental site: 1.9 mg kg^{-1} (NH_4-N), 15.1 mg kg^{-1} (NO_3-N), 0.6 g kg^{-1} ($N_{Kjeldahl}$).

In this research, the analyzed sites were fertilized with two different method (Table 2).

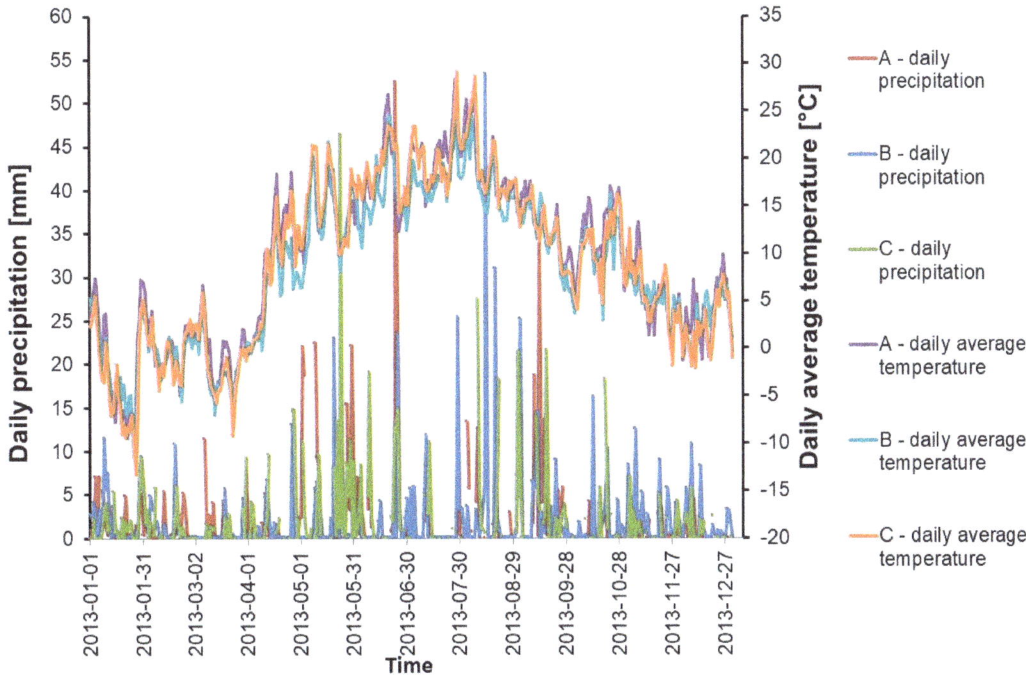

Figure 3. Meteorological data for the A, B, C experimental sites in the year of the research: A—Chociwel in Lower Silesian Voivodship; B—Damno in Pomeranian Voivodship, C—Imielin in Masovian Voivodship.

Table 2. Characteristics of the study sites and fertilization.

Characteristics	Experimental Site		
	A	B	C
Area	20 km^2	40 km^2	20 km^2
Climate classification [1]	Cfb	Dfb	Cfb
Average annual temperature	9.5 °C	8.2 °C	9.0 °C
Total annual precipitation	646	721	612
First sampling/fertilization type	May/UNI	May/UNI	May/UNI
Days from UNI fertilization to soil sampling	46	33	25
Second sampling/fertilization	September/VRA	November/VRA	September/VR
Days from VRA fertilization to soil sampling	136	159	126
Fertilizer used—UNI	32% ammonium nitrate	32% urea ammonium nitrate solution	24% N and 15% S Sulfan
Fertilizer used—VRA	32% ammonium nitrate	32% ammonium nitrate	34% ammonium nitrate
UNI fertilization doses	74 kg N ha^{-1}	80 kg N ha^{-1}	60 kg N ha^{-1}
VRA fertilization doses	30–70 kg N ha^{-1}	40–90 kg N ha^{-1}	55–105 kg N ha^{-1}

[1] according to the Köppen–Geiger system [44].

The first one considered traditional fertilization with the use of uniform N dose of fertilizer for the entire field. The second one concerned precision fertilization with the use of spatially variable rate of N fertilizer. In the case of precise fertilization, the N rate was adjusted to plant demands on the basis of measurement performed by the use of Veris mapping system which allows for the rapid determination of pH, electrical conductivity, and the content of organic matter in soil.

During the monitoring period, winter wheat was cultivated at tested sites. The locations of the sampling points were designated with the density of approximately one soil sample per one hectare. The co-ordinates of sampling points were set in reference to the PUWG 1992 geodetic system using the GPS equipment. The number of soil samples collected for laboratory analyses of N form concentrations varied from 18 to 39. To track the content of N forms in the topsoil, the soil samples were taken from the depths of: 0.00–0.30 m; 0.30–0.60 m, and 0.60–0.90 m. The particle size analysis of soil samples was performed using the sieve and areometric methods according to PN-B-04481:1988 [45]. The basis of a sieve analysis was to allow the non-cohesive soil to pass through a series of sieves of progressively smaller mesh size and weighting the amount of material that was stopped by each sieve as a fraction of the whole mass. Then, the percentage of grains remaining on the sieves was calculated in relation to the total mass of the tested sample. The areometric method was based on determination of the falling speed of soil particles in water. During the examination, the change in suspension density over time was evaluated with the use of the areometer. By means of areometric analysis, the actual dimensions of the soil particles were not determined, but the equivalent diameters, i.e., the diameters of the particles with the same specific density of the soil skeleton as the tested soil and falling in water at the same speed as real particles.

Soils were classified in accordance with the U.S. Department of Agriculture [46]. Soil particle-size distribution was described in relation to the percentages of sand (Sa) (0.05–2.0 mm), silt (Si) (0.002–0.05 mm), and clay (Cl) (<0.002 mm).

2.2. Laboratory Analysis

The ammonium content in soil was determined by a colorimetric method with Nessler's reagent, whereas the nitrate content was determined by a colorimetric method with phenol disulfonic acid, using a UV-VIS DR 6000 spectrophotometer. In this method, ammonium and nitrate ions were extracted from the soil using 1% K_2SO_4 solution. The sum of ammonium and organic N ($N_{Kjeldahl}$) in soil was determined in accordance with application note ASN 3313 [47].

2.3. Statistical Calculations

The results of laboratory tests were subjected to statistical analysis using Statistica 12 software (StatSoft Inc., Tulsa, OK, USA). The data set was tested for normality using the Shapiro–Wilk test. In case of a lack of normality, the Mann–Whitney U test [48] was applied to check the differences between measured variables (N concentrations). The homogeneity of variances was checked using the Levene test. When the data revealed the normality, the Student *t*-test was used [49]. The significance of differences in ammonium, nitrate, and the sum of ammonium and organic N were checked for both uniform and spatially variable (UNI and VRA) fertilization systems. It was assumed that the average concentrations of N forms are the same after UNI and VRA variable fertilization ($c_1 = c_2$, where c_1—average concentration of N form after UNI fertilization, c_2—average concentration of N form after VRA fertilization). The differences in the concentrations of N forms at the examined sites after UNI and VRA fertilization were also compared. The hypothesis was based on the assumption that the concentrations of N forms in soils after UNI or VRA fertilization do not differ significantly between the analyzed fields. A comparative analysis of the N content at different sites after different types of fertilization was also performed following the Kruskal–Wallis test [50]. To check whether our statistical testing was performed properly, we have also used the post-hoc tests. The post-hoc tests were performed to refine the differences detected by analysis of variance. This allowed the separation of homogeneous groups. Two tests were used: Duncan's test and Fisher's NIR test, which are usually used as verification for other tests. Performed tests allowed us to distinguish one or two homogenous groups. The Spearman correlation coefficients were calculated to determine the relationships between the concentration of N in soils and various soil properties. Moreover, the distribution maps of the content of N forms in the soil were prepared using Surfer 10 software. The soil data were interpolated using the kriging method [51], which is commonly used in order to estimate the spatial distributions of pollutants, nutrients, or other soil properties.

3. Results

3.1. Distribution of Soil Types

The performed investigations revealed that silt loams dominate in the soil at the A experimental site (Chociwel in Lower Silesian Voivodship) (Figure 4).

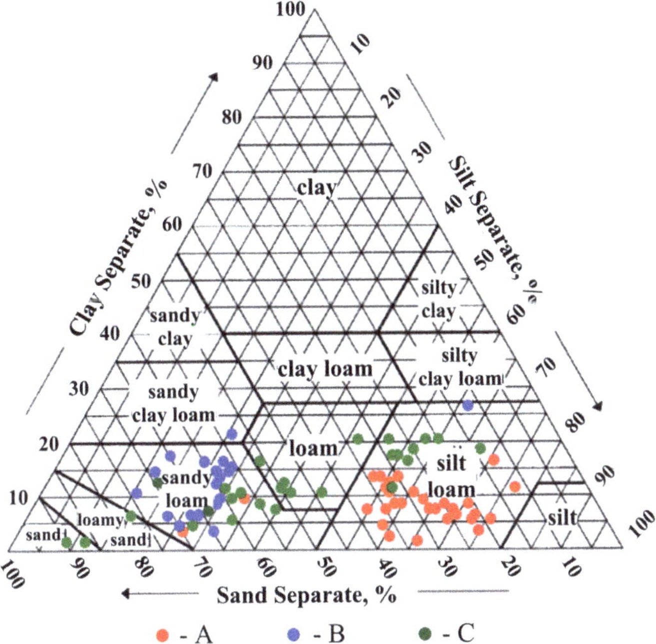

Figure 4. Classification of soils collected from the analyzed experimental sites A, B, C.

The average content of sand, silt, and clay fractions in soils collected from this experimental field were: 32%, 60%, and 8%, respectively. Soils collected from the B experimental site (Damno in Pomeranian Voivodship) were represented mainly by sandy loams and contain on average 63% sand, 26% silt, and 11% clay fractions. Soils from the C experimental site (Imielin in Masovian Voivodship) were characterized by the greatest variability. In majority, they were represented by sandy loams, loams, and silt loams. The average contents of sand, silt, and clay fractions in soils collected from this site were: 48%, 40%, and 12%, respectively.

3.2. Statistical Testing of Differences between N Forms Content in Soil

Based on the results of the Mann–Whitney U test (Table 3), the concentrations of ammonium after UNI and VRA fertilization were different at experimental sites A and B.

Table 3. Summary of the statistical analysis of differences in the measured content of N forms in soils after uniform (UNI) and spatially variable (VRA) fertilization.

N Form	Field	Test	*p* Value	Mean Concentration [1] ± SD [2]	
				UNI	VRA
NH_4-N	A	Mann–Whitney U	0.000867 [3]	1.49 ± 1.38	4.59 ± 4.46
	B	Mann–Whitney U	0.018582 [3]	1.10 ± 0.22	1.29 ± 0.35
	C	Mann–Whitney U	0.083671	1.23 ± 0.44	1.96 ± 1.33
NO_3-N	A	Mann–Whitney U	0.002866 [3]	5.49 ± 5.40	18.78 ± 18.23
	B	Student t	0.027702 [3]	15.46 ± 5.88	19.34 ± 8.53
	C	Mann–Whitney U	0.079757	13.81 ± 10.03	18.08 ± 10.02
$N_{Kjeldahl}$	A	Mann–Whitney U	0.000000 [3]	0.79 ± 0.28	1.54 ± 0.36
	B	Mann–Whitney U	0.068934	1.49 ± 0.54	1.30 ± 0.38
	C	Student t	0.000140 [3]	1.39 ± 0.49	0.88 ± 0.48

[1] expressed in mg kg^{-1} for NH_4-N and NO_3-N and g kg^{-1} for $N_{Kjeldahl}$, [2] standard deviation, [3] values indicate concentrations statistically different.

Like the concentration of ammonium in soils, the concentration of nitrate was different between both types of fertilization at the experimental site A. The homogeneity of variances was checked using the Levene test (p = 0.119959), and the obtained significance level meant that the variances were homogeneous. Moreover, the $N_{Kjeldahl}$ was found to be significantly different for both fertilization strategies at the A and C experimental sites.

The calculations made by the Kruskal–Wallis test indicated that the experimental sites A, B, and C did not differ significantly (p = 0.0745) with regard to ammonium concentration when UNI fertilization was applied. Based on the results obtained, it was also revealed that experimental sites differ significantly (p = 0.0000) in terms of nitrate in soils when UNI fertilization was applied. The same conclusion can be drawn when comparing the differences of the sum of ammonium and organic N in soils taken from the experimental sites after UNI fertilization (p = 0.0000).

The analysis performed within the presented study showed that the sites differ significantly in terms of ammonium concentrations when VRA fertilization was adopted (p = 0.0002). In contrast to the above, the concentrations of nitrate and the sum of ammonium and organic N in soils did not differ significantly at the analyzed fields after VRA fertilization, which was proven by the Kruskal–Wallis test (p = 0.0654 for N-NO_3^- and p = 0.1432 for the sum of ammonium and organic N).

It was revealed by the post-hoc tests that the object C was different from the object A and B, by means of ammonium concentration after UNI fertilization. It was also indicated that the objects B and C are similar by means of ammonium concentration after VRA fertilization and nitrate concentration after UNI concentration. All of tested objects were similar by means of nitrate concentration and the sum of ammonium and organic N after VRA fertilization. The post-hoc tests allow us also to draw the conclusion that the sum of ammonium and organic N at the B experimental site is different from the concentrations measured at A and C experimental sites.

In this study, we based our approach, and then the interpretation of results, on findings presented in guideline adopted for Statistica software, reported by Rabiej [49] who presented several examples which indicate that the interpretation of data (and its significance) should come down to p value analysis (p greater or lower than α). Nevertheless, the researchers should be careful when interpreting their data based on only p values and all results should be interpreted in the context of the study design, including the nature of the samples, the concept of the experiment, validity of instruments, and rules with which the study was performed.

3.3. Content of N Forms in Soil vs. Soil Depth

The results show that the concentrations of N forms in the soil profile were negatively correlated with the soil depth for both types of fertilization (Table 4).

Table 4. Correlation between the content of N forms in soil profile and soil depth, calculated separately for A, B, C experimental sites.

	NH_4-N	NO_3-N	$N_{Kjeldahl}$	Fertilization	Experimental Site
Depth	−0.16	−0.32	−0.17	UNI	A
	−0.59 [1]	−0.47	−0.71 [1]	VRA	
	−0.24	−0.65 [1]	−0.52 [1]	UNI	B
	−0.38 [1]	−0.44 [1]	−0.46 [1]	VRA	
	−0.38 [1]	−0.58 [1]	−0.43 [1]	UNI	C
	−0.69 [1]	−0.85 [1]	−0.55 [1]	VRA	

[1] correlation significant at the level of $p < 0.05$.

It was also shown that greater concentrations occurred after VRA fertilization at A and B experimental sites (Figure 5).

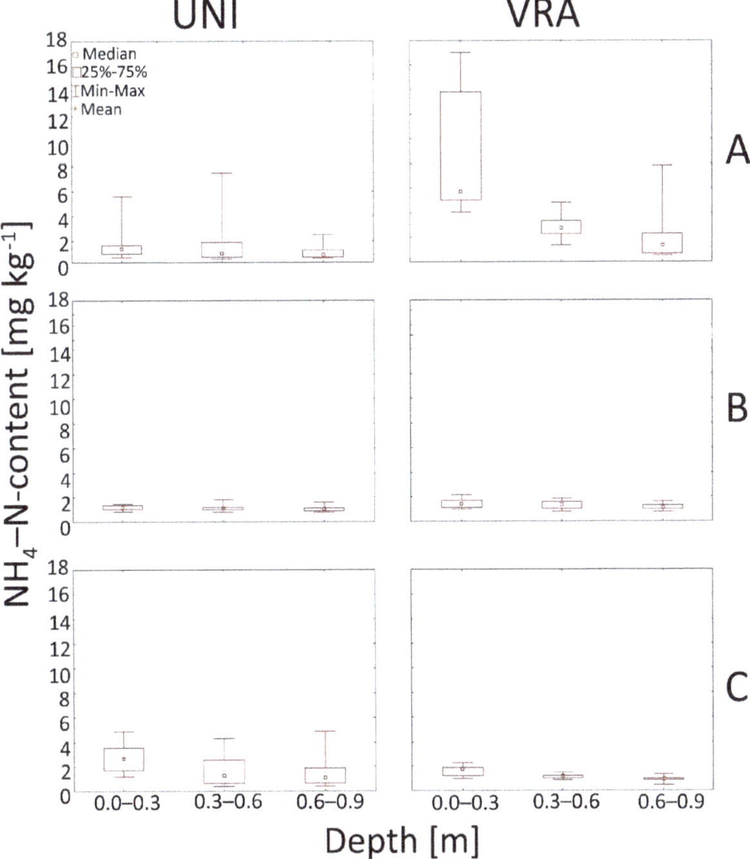

Figure 5. Box and whiskers plots presenting the ammonium content in the soil after UNI and VRA fertilization at the analyzed experimental sites A, B, and C. (**A–C**).

The highest concentrations of NH_4-N in soils were observed after VRA fertilization at the A experimental site. The average concentrations of ammonium in the soil at the B and C experimental site were slightly lower and did not exceed 3 mg kg^{-1}.

Our study also pointed out that nitrate as well as the content of the sum of ammonium and organic N in soils have a decreasing trend with depth both for UNI and VRA fertilization (Figures 6 and 7).

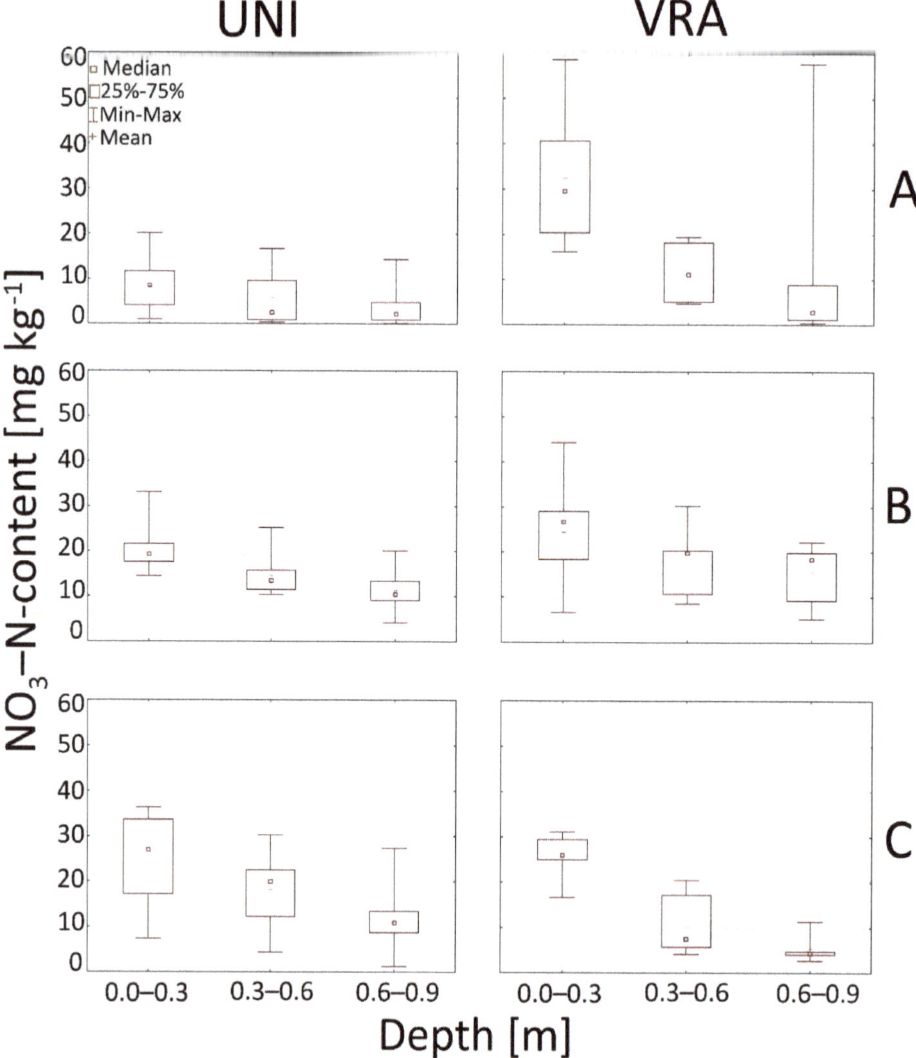

Figure 6. Box and whiskers plots presenting nitrate content in the soil after UNI and VRA fertilization at the analyzed experimental sites A, B, and C. (**A–C**).

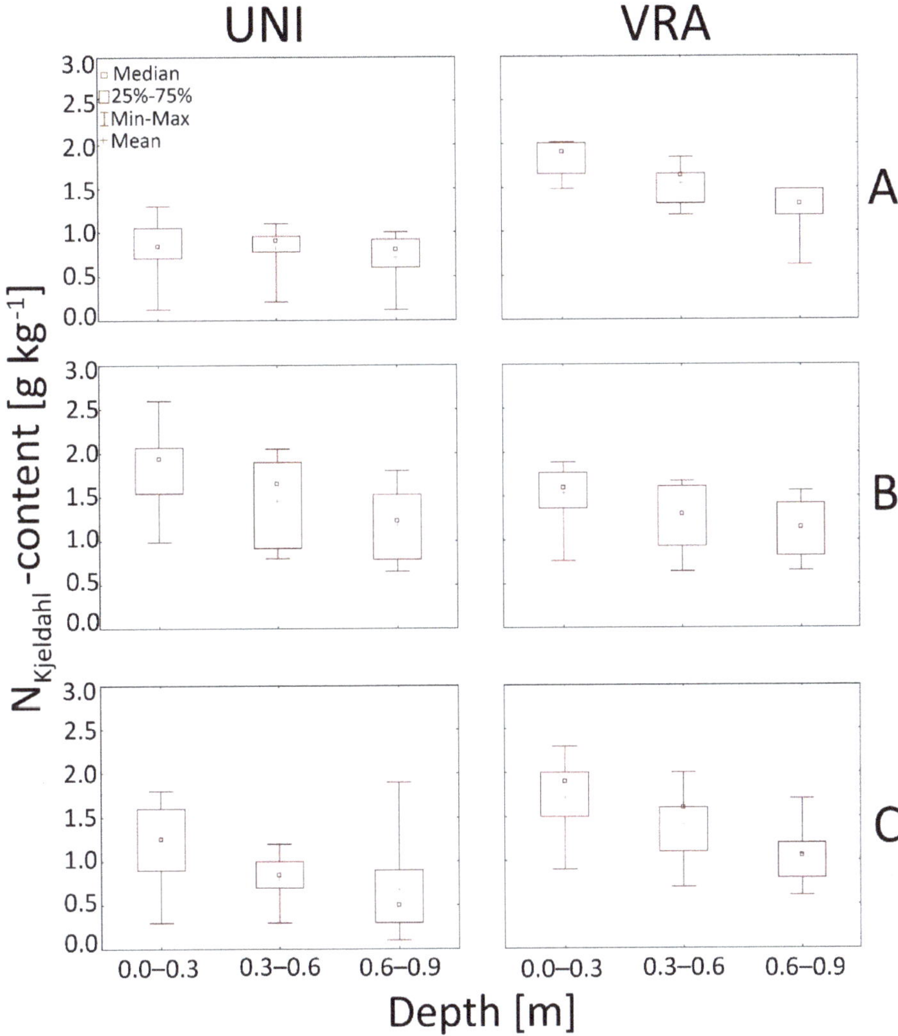

Figure 7. Box and whiskers plots presenting $N_{Kjeldahl}$ content in the soil after uniform UNI and VRA fertilization at analyzed experimental sites A, B, and C. (**A–C**).

The simple regression relationship between the concentration of specific N forms in soil and the soil depth were presented using mathematical equations (Table 5).

Assuming a linear relationship between the concentration of N forms and depth, the equations were presented on the basis of which it was possible to determine the concentration of a selected form of N at a selected depth. Using the regression equations, it was calculated at what depth below the surface the concentration of N forms from fertilizers would be present in soil in trace amounts (concentration almost equal to 0 mg kg^{-1} for NH_4-N and NO_3-N or 0 g kg^{-1} for $N_{Kjeldahl}$).

As it has been shown in Table 3, it was calculated that the ammonium in soil of selected agricultural areas after UNI fertilization could be observed in trace amounts at the maximum depth 2.11 m, whereas the concentration of this form at the same fields after VRA fertilization could be observed at the maximum depth 1.16 m. The same was revealed for nitrate which could be observed in trace amount in soil after UNI fertilization at the depth of 1.46 m and at the depth of 1.28 m after VRA fertilization. The differences in calculated depths let us suppose that when the precise fertilization (adjusted to plant demands) was applied, the amount of unused N would be smaller and hence, the smaller amount of N forms would be transported to deeper parts of soil profile, and therefore to the groundwater system.

Table 5. Concentration of N form as a function of soil depth.

Fertilization	N Form	Equation	Maximum Depth D [m]
UNI	NH_4-N	NH_4-N = 2.0787 − 0.9813 × D	2.11
UNI	NO_3-N	NO_3-N = 21.511 − 14.67 × D	1.46
UNI	$N_{Kjeldahl}$	$N_{Kjeldahl}$ = 1.4910 − 0.7103 × D	2.09
VRA	NH_4-N	NH_4-N = 4.4291 − 3.786 × D	1.16
VRA	NO_3-N	NO_3-N = 32.189 − 25.14 × D	1.28
VRA	$N_{Kjeldahl}$	$N_{Kjedahl}$ = 1.8387 − 0.7958 × D	2.31

3.4. Content of N Forms in Soil vs. Soil Fractions

From all the data obtained, we can see that regardless the soil depth, after UNI fertilization, a positive correlation existed between ammonium and nitrate ions in soil. The same can be attributed to $N_{Kjeldahl}$ and nitrate which showed a positive relationship. When comparing the relationships between specific soil fractions and the content of N forms, it can be seen that between sand fraction and nitrate as well as between clay fraction and nitrate, the positive correlation existed for all depths analyzed after UNI fertilization (Table 6).

Table 6. Spearman correlation between the content of N forms and soil fractions at different depths; after uniform fertilization (UNI).

Depth	Parameter	NH_4-N	N-NO_3	$N_{Kjeldahl}$
0.00–0.30 m b.s.l.	NH_4-N	1.00	0.35	−0.10
	NO_3-N	0.35	1.00	0.35 [1]
	$N_{Kjeldahl}$	−0.10	0.35 [1]	1.00
	Sa	−0.04	0.39 [1]	0.68 [1]
	Si	0.02	−0.47 [1]	−0.67 [1]
	Cl	0.45 [1]	0.26	−0.12
	Si + Cl	0.05	−0.39 [1]	−0.69 [1]
0.30–0.60 m b.s.l.	NH_4-N	1.00	0.31	0.21
	NO_3-N	0.31	1.00	0.16
	$N_{Kjeldahl}$	0.21	0.16	1.00
	Sa	0.00	0.36 [1]	0.57 [1]
	Si	−0.11	−0.44 [1]	−0.57 [1]
	Cl	0.24	0.29	−0.10
	Si + Cl	−0.01	−0.35	−0.57 [1]
0.60–0.90 m b.s.l.	NH_4-N	1.00	0.42 [1]	0.13
	NO_3-N	0.42 [1]	1.00	0.11
	$N_{Kjeldahl}$	0.13	0.11	1.00
	Sa	0.19	0.46 [1]	0.39 [1]
	Si	−0.15	−0.48 [1]	−0.41 [1]
	Cl	−0.04	0.15	0.00
	Si + Cl	−0.15	−0.38 [1]	−0.38 [1]

[1] correlation significant at the level of $p < 0.05$.

In case of VRA fertilization, it was observed that for all depths analyzed, the positive correlation existed between ammonium and $N_{Kjeldahl}$. Moreover, there were positive relationships obtained between $N_{Kjeldahl}$ and silt and clay fraction, respectively. It was also revealed that for each depth, the concentration of ammonium was positively correlated with the sum of silt fraction in soil (Table 7). That could be explained by the sorption of positively charged ion (ammonium) on negatively charged particles of soil.

Table 7. Spearman correlation between the content of N forms and soil fractions at different depths; after variable fertilization (VRA).

Depth	Parameter	N-NH4	N-NO3	$N_{Kjeldahl}$
0.00–0.30 m b.s.l.	NH_4-N	1.00	0.09	0.30
	NO_3-N	0.09	1.00	−0.21
	$N_{Kjeldahl}$	0.30	−0.21	1.00
	Sa	−0.56 [1]	−0.09	−0.35
	Si	0.58 [1]	0.15	0.32
	Cl	−0.09	−0.24	0.26
	Si + Cl	0.55 [1]	0.08	0.33
0.30–0.60 m b.s.l.	NH_4-N	1.00	−0.06	0.11
	NO_3-N	−0.06	1.00	−0.08
	$N_{Kjeldahl}$	0.11	−0.08	1.00
	Sa	−0.20	0.02	−0.32
	Si	0.16	−0.05	0.24
	Cl	−0.05	0.12	0.26
	Si + Cl	0.21	−0.01	0.30
0.60–0.90 m b.s.l.	NH_4-N	1.00	0.41 [1]	0.18
	NO_3-N	0.41 [1]	1.00	−0.05
	$N_{Kjeldahl}$	0.18	−0.05	1.00
	Sa	−0.19	0.16	−0.31
	Si	0.23	−0.04	0.31
	Cl	0.09	0.06	0.18
	Si + Cl	0.28	−0.05	0.28

[1] correlation significant at the level of $p < 0.05$.

3.5. Spatial Fluctuations of N Content in Soil

Our study revealed that concentrations of N forms show the spatial variability at each experimental site, after UNI and VRA fertilization. The highest concentrations observed at the A experimental site (Figures 8 and 9) was lower than the average concentrations of ammonium measured for Polish arable soils [52].

It was also observed that the highest concentrations of nitrate were accumulated at the south part of the A experimental site, in locations of the lowest altitude (Figures 8 and 9).

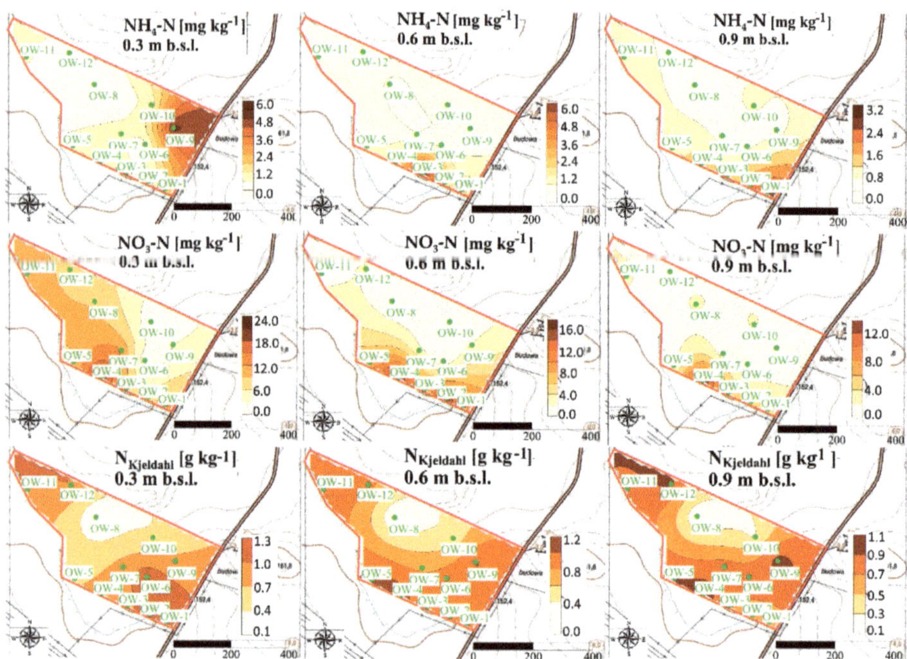

Figure 8. Spatial distribution of ammonium, nitrate, and $N_{Kjeldahl}$ in soil at the A experimental site after UNI fertilization.

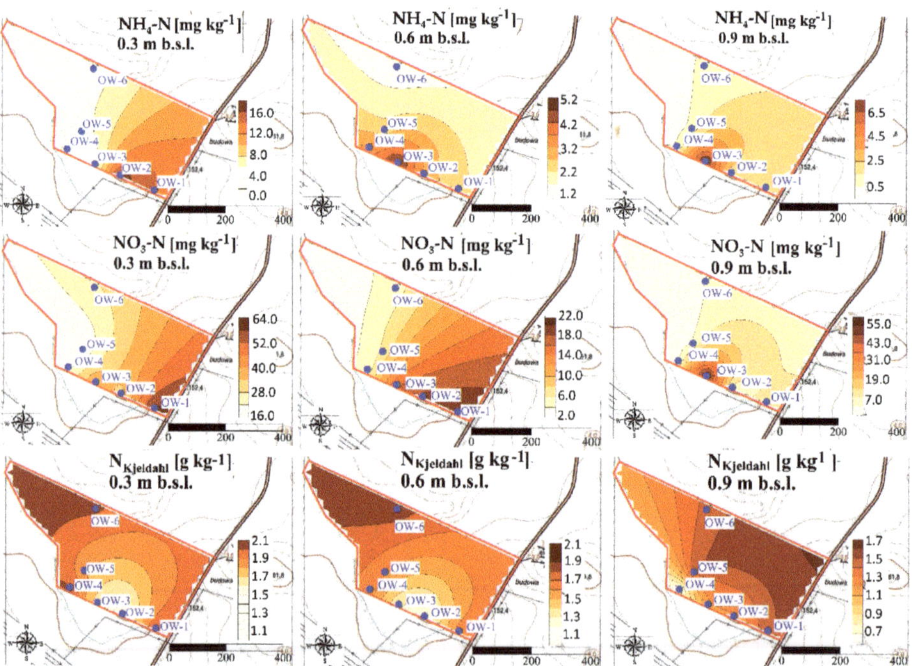

Figure 9. Spatial distribution of ammonium, nitrate, and $N_{Kjeldahl}$ in soil at the A experimental site after VRA fertilization.

The fluctuations of N forms are also clearly visible at the B experimental site (Figures 10 and 11).

The greatest concentrations of ammonium at that site were observed after UNI fertilization in the vicinity of the sampling point OW-3 at the depth of 0.0–0.30 m (Figures 10 and 11).

The accumulation of ammonium was observed at the depth of 0.30–0.60 m and 0.60–0.90 m in the vicinity of the sampling point OW-6.

It was also revealed that increased concentrations of nitrate after UNI fertilization were observed in the vicinity of sampling point OW-1, at each of the analyzed depths, which is slightly different after VRA fertilization (Figure 11). The content of the sum of ammonium and organic N and their hot-spots are distributed non-uniformly at the B experimental site. The greatest concentrations of ammonium at the C experimental site were observed in the vicinity of sampling point OW-7, after UNI fertilization (Figure 12).

No evidence for this tendency can be observed at this site after VRA fertilization. A local increase in nitrate concentrations is observed at the C experimental site, both after UNI and VRA fertilization. The greatest concentrations of the sum of ammonium and organic N were observed in the vicinity of the sample point OW-7 and OW-8, both after UNI and VRA fertilization (Figures 12 and 13). A deviation from this was observed at the depth of 0.60–0.90 m, after UNI fertilization, as the greatest $N_{Kjeldahl}$ concentration was measured in the vicinity of the sampling point OW-6 (north part of the study site). The occurrence of elevated N concentrations should be treated as incidental and can be explained by local fluctuations in weather conditions.

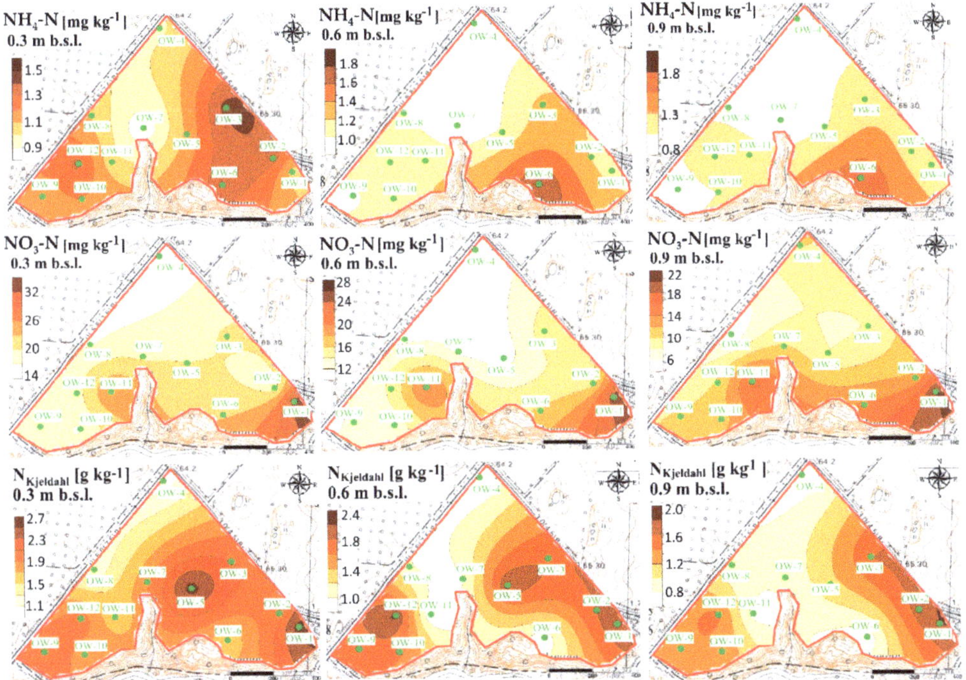

Figure 10. Spatial distribution of ammonium, nitrate, and $N_{Kjeldahl}$ in soil at the B experimental site after UNI fertilization.

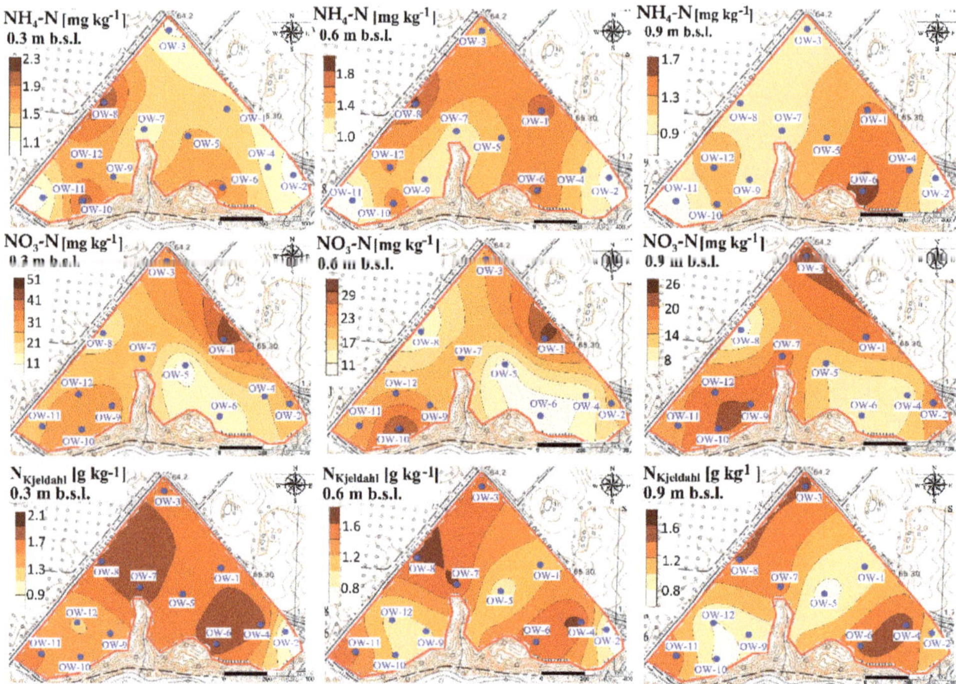

Figure 11. Spatial distribution of ammonium, nitrate, and $N_{Kjeldahl}$ in soil at the B experimental site after VRA fertilization.

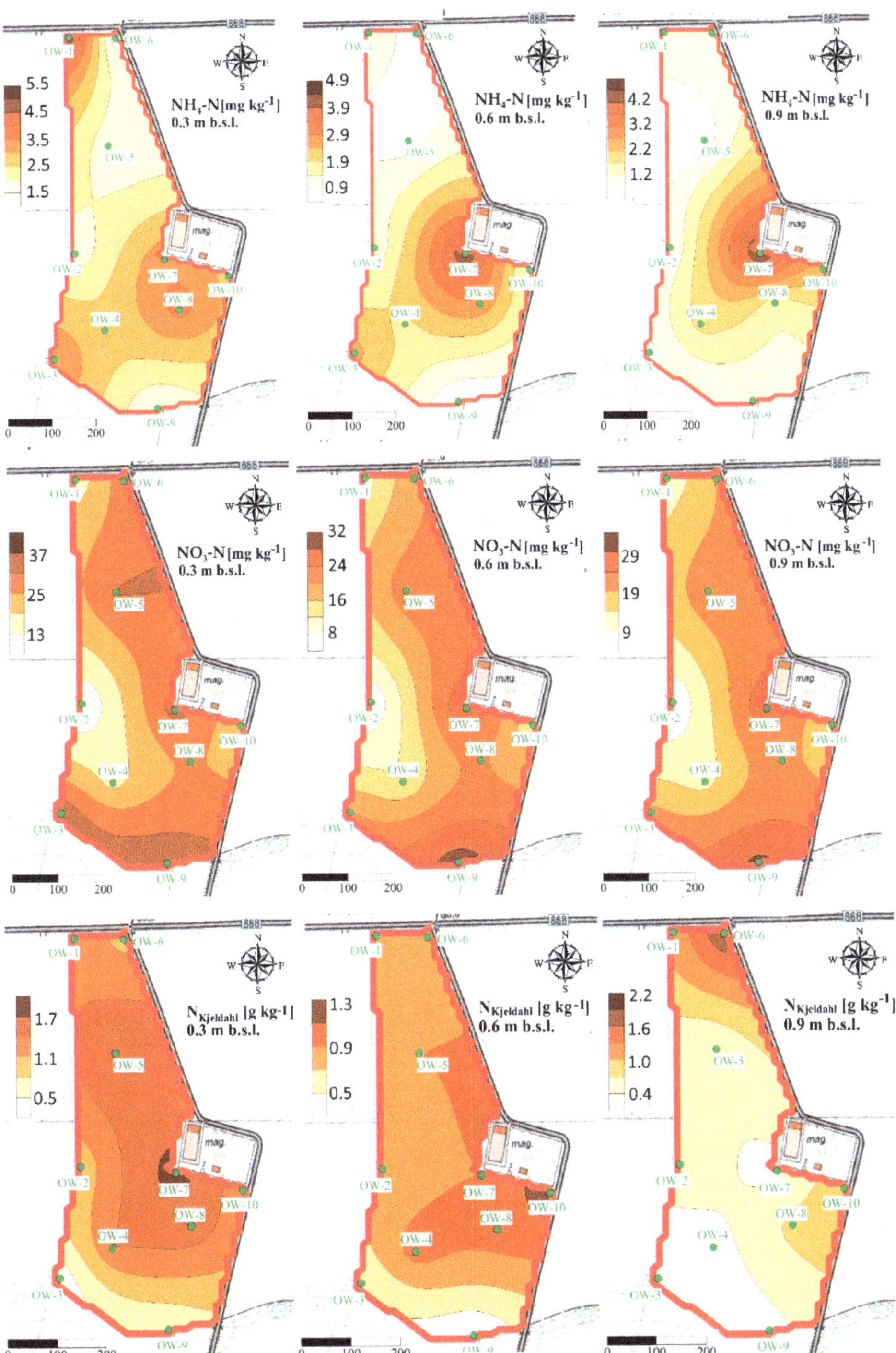

Figure 12. Spatial distribution of ammonium, nitrate, and $N_{Kjeldahl}$ in soil at the C experimental after UNI fertilization.

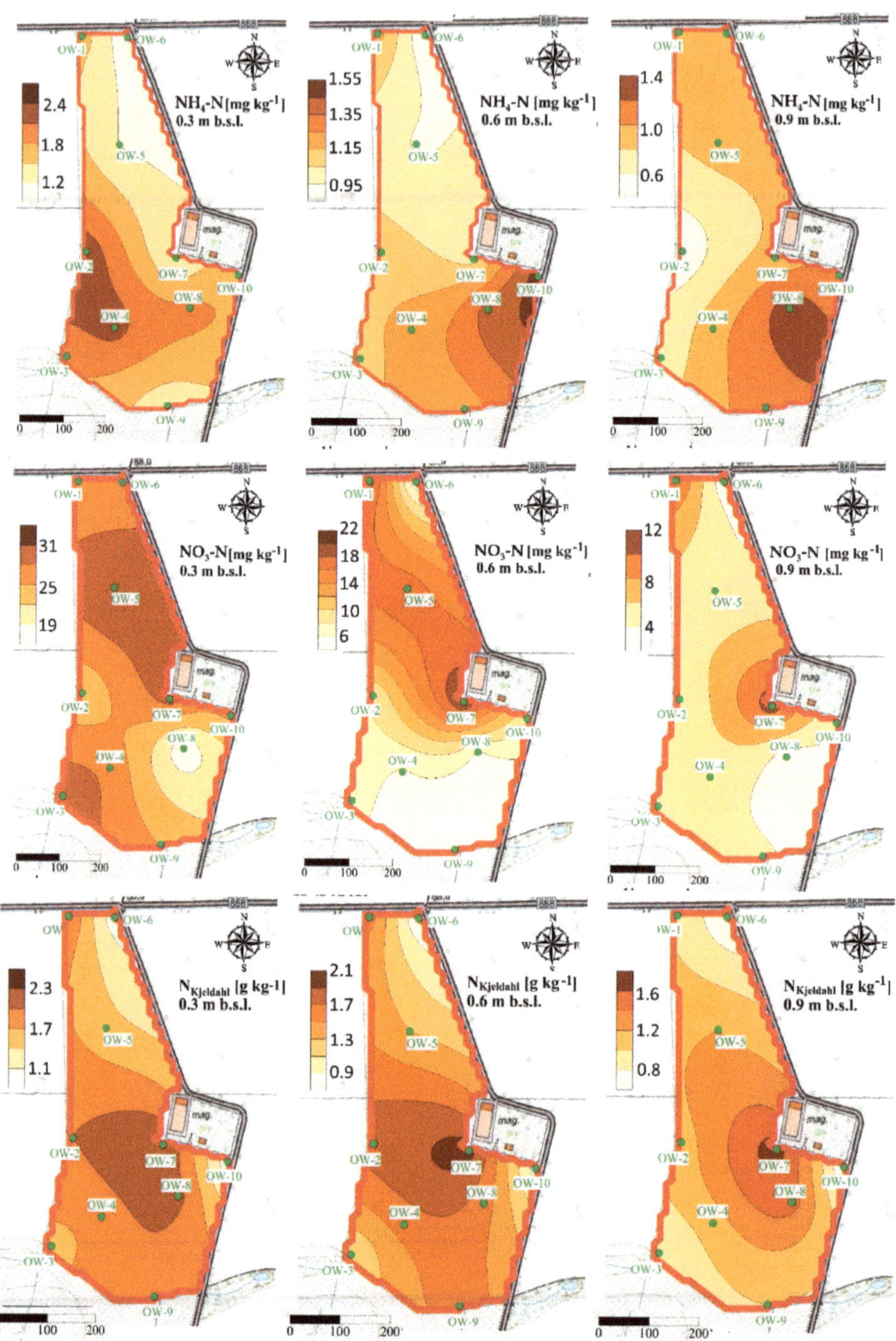

Figure 13. Spatial distribution of ammonium, nitrate, and $N_{Kjeldahl}$ in soil at the C experimental site after VRA fertilization.

4. Discussion

4.1. Typical Concentrations of N in Soil

According to the Monitoring of the Chemistry of Polish Arable Soils [52], the content of total N in soil is in the range of 400–4100 mg kg^{-1}, while its average content is 1100 mg kg^{-1}. In mineral soils, the N content is in the range of 200–4000 mg kg^{-1}, while in organic soils, it can be up to 14,000 mg kg^{-1} [53]. It was also reported that minimum concentrations of nitrate in arable soils are at a level lower than 1 mg kg^{-1}, whereas the maximum and median concentrations of nitrate in agricultural soils are equal to 110.58 and 5.42 mg kg^{-1}, respectively. For comparison, Lee et al. [54] reported that the median concentration of nitrate measured in agricultural soils was 3 mg kg^{-1}. The concentrations of nitrate measured at the analyzed sites were slightly higher than those measured for Polish arable soils. Monitoring data [55] showed that the minimum concentration of ammonium in agricultural soils was at the level of 0.43 mg kg^{-1}, whereas the maximum concentration of this component was at the level of 42.6 mg kg^{-1}. It can be stated that for agricultural sites presented in this study, the concentrations of ammonium were lower than the average concentrations measured for Polish agricultural soils. According to Lee et al. [54], the median concentration of ammonium in agricultural soils was equal to 4 mg kg^{-1}, whereas in Lithuanian agricultural lands, concentrations of ammonium were observed at the level of 1 mg kg^{-1} at the depth of 0.00–0.30 m; 0.84 mg kg^{-1} at the depth of 0.30–0.60 m; and 0.84 mg kg^{-1} at the depth of 0.60–0.90 m. Nitrate concentrations measured in Lithuanian soils were equal to 6.45 mg kg^{-1} at the depth of 0.00–0.30 m; 4.19 mg kg^{-1} at the depth of 0.30–0.60 m, and 2.86 mg kg^{-1} at the depth of 0.60–0.90 m [56]. For comparison, Długosz and Piotrowska-Długosz [57] revealed that the average concentrations of the total N, nitrate, and ammonium in the soils of northwestern Poland were 1.99 g kg^{-1}, 17.1 mg kg^{-1} and 11.1 mg kg^{-1}, respectively. Literature findings indicate that proportions between ammonium and nitrate in soils show a specific tendency. Yan et al. [58] presented, based on their field plot experiment, that the content of ammonium was markedly lower than nitrate in the soil solution. In our study, it was revealed that at the analyzed experimental sites, the concentrations of ammonium are lower than nitrates (from 2 to 17 times at corresponding depths). In contrast to our findings, Sądej and Przekwas [59] revealed that ammonium dominated over nitrate in soils. Similar results were presented by Arbačauskas et al. [56] who documented that concentrations of nitrate also have a decreasing tendency through depth. Similar tendencies were also presented in our study. The comparison of measured concentrations of N forms with ranges of N form concentrations observed in different regions of the world let us state that analyzed experimental sites are comparable, in terms of N forms content, with other agricultural regions fertilized with the use of N fertilizers.

Keeping in mind that at a given N dose and under identified soil conditions N forms were observable in the examined concentration range may be an indication whether the doses of fertilizers used in the traditional system are adequate and there was no excessive accumulation of N in soil, or whether the application was excessive and unused N forms were sorbed in the soil or leach out, which could be also indicated by increased nitrate concentrations at greater depths of the soil profile.

4.2. Effect of Soil Type on N Forms Migration

Several authors reported that clay fractions can retain ammonium because of negative charge on their particles [60,61]. Literature findings indicated also that coarser soils are characterized by smaller retention than finer soils, which allows for faster leaching of nitrate into groundwater [62]. Moreover, the positive relationships between silt-clay content in soil and total N was proved [63].

It is known that nitrate is an ion that can migrate with water flow. Water does not flow easily in clayey layers and, as a result, nitrate is not freely leached to the groundwater system. Some scientific research proved that N losses by leaching were in the range from less than 10 to 30%, but it was also emphasized that leaching of N forms from coarse-textured soils may be greater than 30% [16].

It was also reported that nitrate is lost from clayey soils mainly by denitrification, which has an impact on the reduction of nitrate amount that can be released to groundwater. In this aspect, Arbačauskas et al. [56] revealed that the concentrations of nitrate in light-textured soils were lower in comparison with heavier textured soils.

Ibrahim et al. [64] indicated that the clay had a positive direct effect on the N content, what can be attributed to the sorption phenomena. Then, it is also important that besides the sorption abilities, the retention of N forms in soil is also related to the form of clay fraction [65]. Generally, it was confirmed that organic matter in the fine clay fraction contained more N than coarse clay. According to Alamgir [66], the clay fraction has a high sorption capacity due to its specific surface area, chemical and mechanical stability, layered structure, and high cation exchange capacity. In contrast to clay soils, sandy soils are characterized by a lower sorption capacity and larger pore size, which makes them ineffective in retaining pollution. On this basis, Xu et al. [67] concluded that the total N in soil had a positive correlation with silt-clay fraction which has the greater ability to retain N ions migrating through the soil profile. What is also important, the proportion of clay and silt fraction in soils may have a great impact on the crop yield [68]. Due to this, it was revealed that the yield from soils represented by sandy loams was almost 30% higher than that from loamy sands, and the highest yield was reported for silty soils.

Regarding literature findings reported above, we can say that at analyzed experimental sites, the vertical movement of N forms through the soil profile was successively limited by the occurrence of cohesive soils in the top parts of analyzed soil profiles. Because of the dominance of silty loams, and loams, no evidence of leaching N forms (no excessive concentrations) was observed at tested agricultural fields. Moreover, the lack of excessive concentrations of ammonium in deeper parts of the soil profile pointed out that existing soil conditions are unfavorable for the biochemical reduction of nitrate to ammonium.

4.3. Accumulation of N in Soil

Some scientific studies revealed that the excessive application of N fertilizers has a significant impact on the residual nitrate in soil, which can be seen by the accumulation of this ion in deeper soil layers [69]. The same was proved by Yan et al. [58], who indicated that excessive N application may lead to increased nitrate accumulation in the lower soil layers. The greater accumulation of N forms in soils can be observed more frequently in heavy texture soils (loams and clays) than in light texture soils (sands, sandy loams) [56,70]. Due to these facts and based on our findings presented in this study, we can state that the lack of accumulation of nitrate ions in deeper parts of analyzed soil profiles indicates that the N is applied suitably to the current plant demands. It was also revealed that lower concentrations of ammonium at the B experimental site can be attributed to the occurrence of sandy loams which have lower possibilities to retain N ions than heavier soils occurring at the A and C experimental sites.

The differences between N concentrations and accumulation in soil can be also attributed to contamination. In this regard, Uwah et al. [71] stated that significant differences ($p < 0.05$) between nitrate concentrations in soil samples and their corresponding levels in the control samples might be attributed to possible pollution resulting from an excessive use of agro-chemicals, and wastewater soil irrigation, and also can be attributed to the environmental conditions pertinent in the analyzed sites.

4.4. Decrease of N Form Concentration through the Depth

In the majority of cases it is not surprising that the decrease of N form concentrations with soil depth is observed, but without any analysis performed, the negative correlation between concentration of all N forms and soil depth should not be taken for granted. Regardless of the results obtained in this study, it is worth noting that the negative correlation between all N forms and depth is not always valid because sometimes totally opposite trends may be observed and they were also given by nature.

We may say that the correlation between N forms and depth is rather related to the specific study area and may be different for different sites. These observations can be also confirmed by the scientific research presented in the literature. For example, Bahmani et al. [60] reported that the content of ammonium decreases with the depth, and the concentration of ammonium is much lower than nitrate. Specifically, the decrease of ammonium through the depth can be explained by the nitrification process in the upper layers in which N in the form of ammonium is converted to nitrate. The same was proved by Yan et al. [58] who stated that ammonium derived from soil and fertilizers is rapidly transformed into nitrate through nitrification, resulting in lower ammonium and higher nitrate concentrations in the soil profile. The higher content of nitrate than ammonium in the same soil layer was also observed by Sądej and Przekwas [59] and Xu et al. [67]. The mentioned authors also reported that the gradual decrease of the total N content with soil depth can be attributed to N in organic matter. The decreasing trend of changes in N forms concentrations was also revealed in our study and can be explained by mentioned phenomenon.

4.5. Spatial Variability of N Forms Concentrations

Our study revealed that the fluctuations of N forms concentrations are clearly visible, but no patterns in their variability can be assessed as well as no tendency in occurring hot-spots of N concentrations can be indicated globally for analyzed fields. The concentrations of ammonium and nitrate change spatially over time and hot-spots of N forms may occur irregularly in time, depending on precipitation or N uptake by plants [72]. The spatial independence of N forms in soils was also reported by Hope et al. [73] and Spohn et al. [74] who pointed out that the contents of N forms (ammonium and nitrate) show high dynamics and variability. The random spatial variations and distributions of N forms in soils can be also attributed to differences in physical, chemical, and biological processes taking place in soils [75]. The soil moisture can be also treated as the factor determining the spatial distribution of nitrate which can be accumulated in dry soils due to reduced leaching [76]. The differences in the total N, ammonium, and nitrate distributions in soils were also reported by Długosz and Piotrowska-Długosz [57] who pointed out that it is crucial to take into consideration the variability of soil characteristics (soil type, content of N forms) for the purpose of a proper soil management and better understanding of the N transformations in the ecosystem. Hence, the knowledge about spatio-temporal behavior and the variability of nutrients is also essential for the management within precision agriculture [77]. The knowledge related to the spatial distribution of N forms was also found to be significant when assessing the patterns of groundwater contamination in the vicinity of landfill sites [78].

Due to the fact that N forms are very susceptible to transformation, maps presenting their distribution should be prepared to track their fluctuations only in a very short-term period. The susceptibility of N forms to transformations should be also taken into consideration when preparing and analyzing maps of their spatial distribution [79]. In this aspect, the best practice would be to create N distribution maps for every season, before the formulation any guidelines concerning site management [79].

4.6. Premises of Sustainable Development Goals (SDGs) Implementation

The performed study addresses the Sustainable Development Goals (SDGs) adopted by the United Nations (UN). In most of the SDGs presented, soils play a crucial role as they are a link between the atmosphere–geosphere–hydrosphere–biosphere systems. The chemical and physical soil properties are also prerequisites for soil health. The performed study pointed out that proper agricultural management is required to ensure sustainable development and environmental safety. Soil vulnerability to anthropogenic threats should be studied as widespread contamination can threaten sustainable food production, clean water supply, climate change, and the health and quality of human life. To fulfill the SDGs, the integrated soil–crop system management is required. It would be possible by the application of the precision agriculture, which is a crucial way to maximize yield and minimize environmental impact by adapting agricultural management to local conditions and by optimizing nutrient application (N doses adjusted to real plant demands). Based on the performed research, we can conclude that the content of N forms in soils in connection with the soil data is relevant to optimize the application of N fertilizers in agricultural practice. Our outcomes strictly indicated that there are significant differences in N concentrations in soil, depending on the type of fertilization, soil properties, or soil depth. Significant differences were visible for the concentration of ammonium and nitrate, and the sum of ammonium and organic N ($N_{Kjedahl}$) almost at each study site. We may also state that according to the precise fertilization, N is applied suitably to plant demands, which was proved by no accumulation of nitrate ions in lower parts of soil profiles. In this aspect, the movement of unused N forms to deeper parts of the soil, and therefore, to the groundwater system, is more limited due to the VRA fertilization (Figure 14).

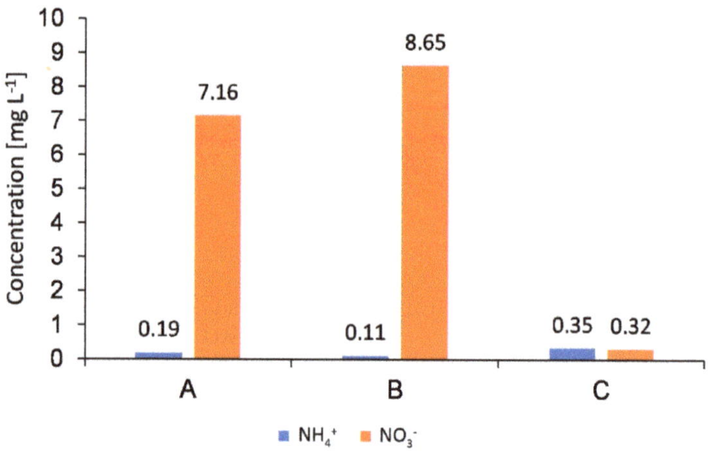

Figure 14. Concentrations of ammonium and nitrate ions in groundwater at the analyzed experimental sites: A, B, C [80,81].

It was confirmed that the concentrations of N ions observed in groundwater samples (both nitrate and ammonium) indicate good chemical statuses. The average concentrations of nitrate measured in groundwater samples collected from piezometers located within the agricultural areas did not exceed the limit value of 10 mg L^{-1}, which can be assigned to the first class of groundwater quality. Moreover, due to the precise fertilization, the average concentrations of ammonium in groundwater did not exceed the limit value of 0.5 mg L^{-1}, characteristic for the first class of groundwater quality [80].

The results of performed study indicate that the recognition of soil and its properties is required for proper production of plant biomass and ensuring food safety, as stated in SDG 2. Moreover, it was indicated that the sorption capacities of soils can play a filtering role against contaminants and therefore protect the groundwater, as outlined in SDG 6.

4.7. Future Research Needs

Future research should be performed as a holistic investigation to control N losses and reduce agricultural costs while maintaining productivity. The joint actions should be taken among researchers, land managers, and policy makers to implement improved agricultural management practices that optimize agricultural production, minimize adverse effects on human health, and reduce environmental contamination. The approaches should also be put forward to strengthen environmental and economic sustainability to improve understanding soil N processes and plant uptake.

The attempts should be made to improve the current knowledge related to the N use efficiency, especially with the special care to N recommendations, precision farming technologies, improving soil testing, and calculating nutrient budgets with advanced computer modeling tools and computer software.

The presented study should also be the starting point for future research focused on hydrogeological modeling aiming at the recognition of paths of N forms' migration to the soil–water environment, taking into consideration information about their spatial diversity and influencing factors.

5. Conclusions

The presented research is a comprehensive study on N forms distribution in soils, with special attention given to different N forms and factors determining their content. Wide-scale analyses performed at selected study sites in Poland, and supported by the statistical evaluation of differences in N forms concentrations in soil, constitute an original contribution to the soil science and environmental management in agricultural areas.

In accordance with the presented approach, management in agricultural practice should be based on the prediction of spatial variability of soil properties which allows to ensure proper application of fertilizers, resulting in the reduction of possible N losses. Nevertheless, the maps presenting spatial distribution of N forms in soils should be taken into consideration in the temporal aspect, concerning the fact that N forms are very susceptible to transformation due to physical, chemical, and biological processes.

Author Contributions: Conceptualization, A.P., E.K. and M.D.V.; methodology, A.P. and E.K.; software, A.P.; formal analysis, A.P., E.K. and M.D.V.; investigation, A.P., E.K. and M.D.V.; data curation, A.P.; writing—original draft preparation, A.P.; writing—review and editing, E.K. and M.D.V.; visualization, A.P. and M.D.V.; supervision, E.K. and M.D.V.; project administration, A.P.; funding acquisition, A.P. and E.K. All authors have read and agreed to the published version of the manuscript.

Funding: This research was supported by the National Science Centre, Poland, under a grant No. 2017/25/N/ST10/00909 ("Analysis of contaminant migration processes in the soil-water environment using laboratory tests and numerical modeling techniques"). Preliminary studies of the presented research were supported by the European Regional Development Fund under the Innovative Economy Operational Programme: "BIOPRODUCTS, innovative production technologies of pro-healthy bakery products and pasta with reduced caloric value" (POIG.01.03.01-14-041/12).

Institutional Review Board Statement: Not applicable.

Informed Consent Statement: Not applicable.

Data Availability Statement: Data sharing is not applicable to this article.

Acknowledgments: We thank the research team from the Faculty of Civil and Environmental Engineering of WULS-SGGW in Warsaw, involved in the realization of Task 3: "Monitoring of selected elements of the natural environment in cereal production using precision farming tools", for technical support and help in collecting soil samples.

Conflicts of Interest: The authors declare no conflict of interest.

References

1. Xia, X.; Zhang, S.; Li, S.; Zhang, J.; Wang, G.; Zhang, L.; Wang, J.; Li, Z. The cycle of nitrogen in river systems: Sources, transformation, and flux. *Environ. Sci. Process. Impacts* **2018**, *20*, 863–891. [CrossRef] [PubMed]
2. Pajewski, T. Water pollution as a negative effect of agricultural activity. *Sci. Ann. Assoc. Agric. Agribus. Econ.* **2016**, *18*, 191–195. (In Polish)
3. Bechmann, M. Nitrogen losses from agriculture in the Baltic Sea region. Long-term monitoring of nitrogen in surface and subsurface runoff from small agricultural dominated catchments in Norway. *Agric. Ecosyst. Environ.* **2014**. [CrossRef]
4. Vaverková, M.; Adamcová, D. Can Vegetation Indicate a Municipal Solid Waste Landfill's Impact on the Environment? *Polish J. Environ. Stud.* **2014**, *23*, 501–509.
5. Wick, K.; Heumesser, C.; Schmid, E. Groundwater nitrate contamination: Factors and indicators. *J. Environ. Manag.* **2012**, *111*, 178–186. [CrossRef] [PubMed]

6. Kyllmar, K.; Bechmann, M.; Deelstra, J.; Iital, A.; Blicher-Mathiesen, G.; Jansons, V.; Koskiaho, J.; Povilaitis, A. Long-term monitoring of nutrient losses from agricultural catchments in the Nordic-Baltic region: A discussion of methods, uncertainties and future needs. *Agric. Ecosyst. Environ.* **2014**, *198*, 4–12. [CrossRef]
7. Elbl, J.; Vaverková, M.; Adamcová, D.; Plošek, L.; Kintl, A.; Lošák, T.; Hynšt, J.; Kotovicová, J. Influence of fertilization on microbial activities, soil hydrophobicity and mineral nitrogen leaching. *Ecol. Chem. Eng. S* **2014**, *21*, 661–675. [CrossRef]
8. Gao, S.; Xu, P.; Zhou, F.; Yang, H.; Zheng, C.; Cao, W.; Tao, S.; Piao, S.; Zhao, Y.; Ji, X.; et al. Quantifying nitrogen leaching response to fertilizer additions in China's cropland. *Environ. Pollut.* **2016**, *211*, 241–251. [CrossRef] [PubMed]
9. Mosier, A.R.; Syers, J.K.; Freney, J.R. Agriculture and the Nitrogen Cycle. In *Assessing the Impacts of Fertilizer Use on Food Production and the Environment*; Island Press: Washington, DC, USA, 2004.
10. Gray, N.F. Drinking Water Quality. In *Problems and Solutions*, 2nd ed.; Cambridge University Press: Cambridge, UK, 2008.
11. Füloky, G. Accumulation and depletion of fertilizer originated nitrate-N and ammonium-N in deeper soil layers. *Columella J. Agric. Environ. Sci.* **2014**, *1*, 7–16. [CrossRef]
12. Jego, G.; Sanchez-Perez, J.M.; Justes, E. Predicting soil water and mineral nitro-gen contents with the STICS model for estimation nitrate leaching under agricultural fields. *Agric. Water Manag.* **2012**, *107*, 54–65. [CrossRef]
13. Sieczka, A.; Bujakowski, F.; Falkowski, T.; Koda, E. Morphogenesis of a Floodplain as a Criterion for Assessing the Susceptibility to Water Pollution in an Agriculturally Rich Valley of a Lowland River. *Water* **2018**, *10*, 399. [CrossRef]
14. Kalenik, M.; Chalecki, M. Investigations on the effectiveness of wastewater purification in medium sand with assisting clinoptilolite layer. *Environ. Prot. Eng.* **2019**, *45*, 117–126. [CrossRef]
15. Eurostat. Available online: https://ec.europa.eu/eurostat/home (accessed on 12 September 2018).
16. Mikula, K.; Izydorczyk, G.; Skrzypczak, D.; Mironiuk, M.; Moustakas, K.; Witek-Krowiak, A.; Chojnacka, K. Controlled release micronutrient fertilizers for precision agriculture—A Review. *Sci. Total Environ.* **2020**, *712*, 136365. [CrossRef] [PubMed]
17. Kiełbasa, P.; Zagórda, M.; Jabłoński, P.; Koronczok, J. Porównanie zróżnicowania przewodności elektrycznej gleby wykonanej urządzeniem Topsoil Mapper i jej charakterystyk penetrometrycznych. *Przegląd Elektrotechniczny* **2020**, *96*, 146–150. (In Polish)
18. Leszkowska, L.; Rzymowski, M.; Nyka, K.; Kulas, Ł. Simple Superstrate Antenna for Connectivity Improvement in Precision Farming Applications. In Proceedings of the 2020 IEEE International Symposium on Antennas and Propagation and North American Radio Science Meeting, Montréal, QC, Canada, 5–10 July 2020. [CrossRef]
19. Niedbała, G. Application of Artificial Neural Networks for Multi-Criteria Yield Prediction of Winter Rapeseed. *Sustainability* **2019**, *11*, 533. [CrossRef]
20. Barwicki, J.; Mazur, K.; Borek, K.; Wardal, W.J. Development of plant cultivation using precision agriculture. *Polish Tech. Rev.* **2019**, *2*, 15–20. [CrossRef]
21. Barwicki, J.; Borek, K. Remote sensing in farm machinery design—As fundamental improvement of development—Of precision agriculture phenomenon. *Polish Tech. Rev.* **2019**, *3*, 26–30.
22. Zawada, M.; Ciechanowski, M.; Szulc, T.; Szychta, M.; Smela, A.; Kamprowski, R. Systemy wizyjne we współczesnym rolnictwie. *Technika Ogrodnicza Rolnicza i Leśna* **2019**, *1*, 13–16.
23. Lorencowicz, E. Cyfrowe rolnictwo-cyfrowe zarządzanie. *Roczniki Naukowe Stowarzyszenia Ekonomistów Rolnictwa i Agrobiznesu* **2018**, *XX*, 104–110. (In Polish) [CrossRef]
24. Piekutowska, M.; Niedbała, G.; Adamski, M.; Czechlowski, M.; Wojciechowski, T.; Czechowska-Kosacka, A.; Wójcik Oliveira, K. Modeling methods of predicting potato yield—Examples and possibilities of application. *J. Res. Appl. Agric. Eng.* **2018**, *63*, 176–180.
25. Czechlowski, M.; Wojciechowski, T.; Adamski, M.; Niedbała, G.; Piekutowska, M. Application of ASG-EUPOS high precision positioning system for cereal harvester monitoring. *J. Res. Appl. Agric. Eng.* **2018**, *63*, 44–50.
26. Usowicz, B.; Lipiec, J. Spatial variability of soil properties and cereal yield in a cultivated field on sandy soil. *Soil Tillage Res.* **2017**, *174*, 241–250. [CrossRef]
27. Mazur, P.; Chojnacki, J. Wykorzystanie dronów do teledetekcji. *Technika Rolnicza Ogrodnicza Leśna* **2017**, *1*, 25–28.
28. Laskowska, P.J. Informatyka w rolnictwie, czyli rolnictwo precyzyjne. In *Materiały pokonferencyjne III Ogólnopolskiej Konferencji Interdyscyplinarnej "Współczesne Zastosowania Informatyki"*; Instytut Informatyki Uniwersytetu Przyrodniczo-Humanistycznego w Siedlcach: Siedlce, Poland, 2017. (In Polish)
29. Samborski, S.M.; Gozdowski, D.; Walsh, O.S.; Kyveryga, P.; Stępień, M. Sensitivity of sensor-based nitrogen rates to selection of within-field calibration strips in winter wheat. *Crop Pasture Sci.* **2017**, *68*, 101–114. [CrossRef]
30. Żak, A. Innovations in plant production as an opportunity for Polish agriculture. *J. Agribus. Rural Dev.* **2017**, *1*, 239–246. [CrossRef]
31. Borusiewicz, A.; Kapela, K.; Drożyner, P.; Marczuk, T. Application of precision agriculture technology in Podlaskie Voivodeship. *Agric. Eng.* **2016**, *20*, 5–11. [CrossRef]
32. Berner, B.; Chojnacki, J. Wykorzystanie dronów w rolnictwie precyzyjnym. *Technika Rolnicza Ogrodnicza Leśna* **2016**, *6*, 19–21.
33. Samborski, S.M.; Gozdowski, D.; Stępień, M.; Walsh, O.S.; Leszczyńska, E. On-farm evaluation of an active optical sensor performance for variable nitrogen application in winter wheat. *Eur. J. Agronomy* **2016**, *74*, 56–67. [CrossRef]
34. Wójtowicz, M.; Wójtowicz, A.; Piekarczyk, J. Application of remote sensing methods in agriculture. *Commun. Biometry Crop Sci.* **2016**, *11*, 31–50.
35. Wójcicki, Z.; Rudeńska, B. Systemy rolniczej produkcji ekologicznej i precyzyjnej (informacyjnej). *Problemy Inżynierii Rolniczej* **2015**, *2*, 5–15.

36. Gozdowski, D.; Stępień, M.; Samborski, S.; Dobers, E.S.; Szatyłowicz, J.; Chormański, J. Prediction accuracy of selected spatial interpolation methods for soil texture at farm field scale. *J. Soil Sci. Plant Nutr.* **2015**, *15*, 639–650. [CrossRef]
37. Borsato, E.; Galindo, A.; Tarolli, P.; Sartori, L.; Marinello, F. Evaluation of the Grey Water Footprint Comparing the Indirect Effects of Different Agricultural Practices. *Sustainability* **2018**, *10*, 3992. [CrossRef]
38. Kabala, C.; Karczewska, A.; Gałka, B.; Cuske, M.; Sowiński, J. Seasonal dynamics of nitrate and ammonium ion concentrations in soil solutions collected using MacroRhizon suction cups. *Environ. Monit. Assess.* **2017**, *189*, 304. [CrossRef] [PubMed]
39. Staszewski, Z. Nitrogen in soil and its impact on the environment. *Sci. Notebooks Civ. Eng. Environ. Improv.* **2011**, *4*, 50–58. (In Polish)
40. Pawlak, J. Precision farming, its role and economic efficiency. *Adv. Agric. Sci.* **2008**, *1*, 3–14. (In Polish)
41. Gao, T.; Pang, H.; Zhang, J.; Zhang, H.; Zhou, J. Migration and transformation rule of nitrogen in vadose zone and groundwater: A case of Hebei plain. In Proceedings of the 2011 International Conference on Electrical and Control Engineering, Yichang, China, 16–18 September 2011; pp. 3524–3527. [CrossRef]
42. Sieczka, A.; Bujakowski, F.; Koda, E. Modelling groundwater flow and nitrate transport: A case study of an area used for precision agriculture in the middle part of the Vistula River valley, Poland. *Geologos* **2018**, *24*, 225–235. [CrossRef]
43. Koda, E.; Sieczka, A.; Osiński, P. Ammonium Concentration and Migration in Groundwater in the Vicinity of Waste Management Site Located in the Neighborhood of Protected Areas of Warsaw, Poland. *Sustainability* **2016**, *8*, 1253. [CrossRef]
44. Kottek, M.; Grieser, J.; Beck, C.; Rudolf, B.; Rubel, F. World Map of the Köppen-Geiger climate classification updated. *Meteorol. Z.* **2006**, *15*, 259–263. [CrossRef]
45. PN-B-04481:1988. Soils. In *Soil Sample Testing*; Polish Committee for Standardization: Warsaw, Poland, 1988. (In Polish)
46. U.S. Department of Agriculture. Soil Conservation Service. In *Soil Survey Staff. Soil Survey Manual*; Agricultural Handbook No. 18; U.S. Department of Agriculture: Washington, DC, USA, 1951.
47. FOSS Tecator AB. *Application Sub Note 3313, The Determination of Nitrogen According to Kjedahl in Soil*; FOSS: Höganäs, Sweden, 2003.
48. Mann, H.B.; Whitney, D.R. On a test of whether one of two random variables is stochastically larger than the other. *Ann. Math. Stat.* **1947**, *18*, 50–60. [CrossRef]
49. Rabiej, M. *Statistical Analysis with Statistica and Excel*; Helion: Gliwice, Poland, 2018. (In Polish)
50. Kruskal, W.H.; Wallis, W.A. Use of ranks in one-criterion variance analysis. *J. Am. Stat. Assoc.* **1952**, *47*, 583–621. [CrossRef]
51. Oliver, M.A.; Webster, R. How geostatistics can help you? *Soil Use Manag.* **1991**, *7*, 206–217. [CrossRef]
52. Monitoring of Soil Chemistry in Poland. Available online: http://www.gios.gov.pl/chemizm_gleb/ (accessed on 20 August 2020).
53. District Chemical-Agricultural Station in Lublin (DCAS). Nitrogen in Soil. Available online: http://www.oschr.pl/index.php/aktualnoci/45-azot-w-glebie.html (accessed on 12 June 2020).
54. Lee, J.; Garland, G.M.; Viscarra Rossel, R.A. Continental soil drivers of ammonium and nitrate in Australia. *Soil* **2018**, *4*, 213–224. [CrossRef]
55. Environmental Protection Inspection (EPI). *Monitoring of Soil Chemistry in Poland in 2010–2012*; Environmental Monitoring Library: Warsaw, Poland, 2012. (In Polish)
56. Arbačauskas, J.; Masevičienė, A.; Žičkienė, L.; Staugaitis, G. Mineral nitrogen in soils of Lithuania's agricultural land: Comparison of oven-dried and field-moist samples. *Zemdirb. Agric.* **2018**, *105*, 99–104. [CrossRef]
57. Długosz, J.; Piotrowska-Długosz, A. Spatial variability of soil nitrogen forms and the activity of N-cycle enzymes. *Plant Soil Environ.* **2016**, *62*, 502–507. [CrossRef]
58. Yan, C.; Du, T.; Yan, S.; Dong, S.; Gong, Z.; Zhang, Z. Changes in the inorganic nitrogen content of the soil solution with rice straw retention in northeast China. *Desalin. Water Treat.* **2018**, *110*, 337–348. [CrossRef]
59. Sądej, W.; Przekwas, K. Fluctuations of nitrogen levels in soil profile under conditions of a long-term fertilization experiment. *Plant Soil Environ.* **2008**, *54*, 197–203. [CrossRef]
60. Bahmani, O.; Nasab, S.B.; Behzad, M.; Naseri, A.A. Assessment of nitrogen accumulation and movement in soil profile under different irrigation and fertilization regime. *Asian J. Agric. Res.* **2009**, *3*, 38–46. [CrossRef]
61. Sieczka, A.; Koda, E. Kinetic and Equilibrium Studies of Sorption of Ammonium in the Soil-Water Environment in Agricultural Areas of Central Poland. *Appl. Sci.* **2016**, *6*, 269. [CrossRef]
62. Witheetrirong, Y.; Tripathi, N.K.; Tipdecho, T.; Parkpian, P. Estimation of the effect of soil texture on nitrate-nitrogen content in groundwater using optical remote sensing. *Int. J. Environ. Res. Public Health* **2011**, *8*, 3416–3436. [CrossRef]
63. Gami, S.K.; Lauren, J.G.; Duxbury, J.M. Influence of soil texture and cultivation on carbon and nitrogen levels in soils of the eastern Indo-Gangetic Plains. *Geoderma* **2009**, *153*, 304–311. [CrossRef]
64. Ibrahim, H.; Hatira, A.; Gallali, T. Relationship between nitrogen and soil properties: Using multiple linear regressions and structural equation modeling. *Int. J. Res. Appl. Sci.* **2013**, *2*, 1–7.
65. Corinna, M.; Markus, K.; Reinhold, J. Soil organic matter stabilization pathways in clay sub-fractions from a time series of fertilizer deprivation. *Org. Geochem.* **2005**, *36*, 1311–1322. [CrossRef]
66. Alamgir, M. The effects of soil properties to the extent of soil contamination with metals. In *Environmental Remediation Technologies for Metal-Contaminated Soils*; Hasegawa, H., Rahman, I.M.M., Rahman, M.A., Eds.; Springer: Berlin/Heidelberg, Germany, 2016; pp. 1–19.

67. Xu, G.; Li, Z.; Li, P. Fractal features of soil particle-size distribution and total soil nitrogen distribution in a typical watershed in the source area of the middle Dan River, China. *Catena* **2013**, *101*, 17–23. [CrossRef]
68. Tkaczyk, P.; Bednarek, W.; Dresler, S.; Krzyszczak, J. The effect of some soil physicochemical properties and nitrogen fertilisation on winter wheat yield. *Acta Agrophys.* **2018**, *25*, 107–116. [CrossRef]
69. Lopez-Bellido, L.; Munoz-Romero, V.; Lopez-Bellido, R.J. Nitrate accumulation in the soil profile: Long-term effects of tillage, rotation and N rate in a Mediterranean Vertisol. *Soil Tillage Res.* **2013**, *130*, 18–23. [CrossRef]
70. Sieczka, A.; Koda, E. Identification of nitrogen compounds sorption parameters in the soil-water environment of a column experiment. *Ochr. Sr.* **2016**, *38*, 29–34. (In Polish)
71. Uwah, E.I.; Abah, J.; Ndahi, N.P.; Ogugbuaja, V.O. Concentration levels of nitrate and nitrite in soils and some leafy vegetables obtained in Maiduguri, Nigeria. *J. Appl. Sci. Environ. Sanit.* **2009**, *4*, 233–244.
72. Cain, M.L.; Subler, S.; Evans, J.P.; Fortin, M.J. Sampling spatial and temporal variation in soil nitrogen availability. *Oecologia* **1999**, *118*, 397–404. [CrossRef]
73. Hope, D.; Zhu, W.; Gries, C.; Oleson, J.; Kaye, J.; Baker, L.A. Spatial variation in soil inorganic nitrogen across an arid urban ecosystem. *Urban Ecosyst.* **2005**, *8*, 251–273. [CrossRef]
74. Spohn, M.; Novák, T.J.; Incze, J.; Giani, L. Dynamics of soil carbon, nitrogen, and phosphorus in calcareous soils after land-use abandonment—a chronosequence study. *Plant Soil* **2016**, *401*, 185–196. [CrossRef]
75. Liu, Y.; Gao, P.; Zhang, L.; Niu, X.; Wang, B. Spatial heterogeneity distribution of soil total nitrogen and total phosphorus in the Yaoxiang watershed in a hilly area of northern China based on geographic information system and geostatistics. *Ecol. Evol.* **2016**, *6*, 6807–6816. [CrossRef] [PubMed]
76. Tang, X.; Xia, M.; Guan, F.; Fan, S. Spatial Distribution of Soil Nitrogen, Phosphorus and Potassium Stocks in Moso Bamboo Forests in Subtropical China. *Forests* **2016**, *7*, 267. [CrossRef]
77. Morales, L.A.; Vázquez, E.V.; Paz-Ferreiro, J. Spatial distribution and temporal variability of ammonium-nitrogen, phosphorus, and potassium in a rice field in Corrientes, Argentina. *Sci. World J.* **2014**, *2014*, 135906. [CrossRef] [PubMed]
78. Koda, E.; Osiński, P.; Sieczka, A.; Wychowaniak, D. Areal Distribution of Ammonium Contamination of Soil-Water Environment in the Vicinity of Old Municipal Landfill Site with Vertical Barrier. *Water* **2015**, *7*, 2656–2672. [CrossRef]
79. Gallardo, A.; Parama, R.; Covelo, F. Differences between soil ammonium and nitrate spatial pattern in six plant communities. Simulated effect on plant populations. *Plant Soil* **2006**, *279*, 333–346. [CrossRef]
80. Podlasek, A.; Bujakowski, F.; Koda, E. The spread of nitrogen compounds in an active groundwater exchange zone within a valuable natural ecosystem. *Ecol. Eng.* **2020**, *146*, 105746. [CrossRef]
81. Sieczka, A. Migracja Związków Azotu Pochodzenia Nawozowego w Środowisku Gruntowo-Wodnym [Migration of Nitrogen Compounds from Fertilizers in the Soil-Water Environment]. Ph.D. Thesis, Warsaw University of Life Sciences, Warsaw, Poland, 2018. (In Polish).

International Journal of
Environmental Research and Public Health

MDPI

Article

Decoupling and Decomposition Analysis of Land Natural Capital Utilization and Economic Growth: A Case Study in Ningxia Hui Autonomous Region, China

Shanshan Guo [1], Yinghong Wang [2,*], Jiu Huang [2], Jihong Dong [2] and Jian Zhang [1]

1 School of Public and Management, China University of Mining and Technology, Xuzhou 221116, China; guoshanshan@cumt.edu.cn (S.G.); LB20160018@cumt.edu.cn (J.Z.)
2 School of Environment Science and Spatial Informatics, China University of Mining and Technology, Xuzhou 221116, China; jhuang@cumt.edu.cn (J.H.); dongjihong@cumt.edu.cn (J.D.)
* Correspondence: 1369@cumt.edu.cn

Abstract: In order to reduce the depletion of land natural capital and develop economy simultaneously, it is necessary to study how to achieve the strong decoupling relationship between them. However, so far such studies have been relatively limited. Thus, taking the case of Ningxia Hui Autonomous Region, China, this paper firstly analyzes the state of land natural capital utilization in 1999–2017 by using improved ecological footprint. Then, decoupling state is quantified by Tapio decoupling model. Last, major driving factors on the decoupling relationship are explored with combination of LMDI decomposition and Kaya identity equation. Results showed that: (1) Both natural capital flows and stock depletion of cultivated land decrease obviously during the transition to corn-based intensive ecological agriculture. Grassland and water are the most unsustainable development sectors among all land types with their stock depletion intensified. Forest land and construction land could basically meet the consumer demand, but the flow occupancy of construction land is the fastest-growing segment. (2) Decoupling relationship is in an alternating state between weak decoupling and strong decoupling in 1999–2017. Wherein, the cultivated land and forest land showed a preferred decoupling state, followed by grassland, while the water and construction land showed the unfavorable expansive negative decoupling and weak decoupling. (3) Decomposition results show that intensity effect is the major factor that promotes the decoupling while economic effect inhibits the decoupling, but this negative impact is weakening in the process of industrial transformation. The other three factors affect less on the decoupling. This study has a certain reference value to construct an ecological civilization in eco-fragile regions and formulate relevant policies on the increase of land natural capital efficiency.

Keywords: land; natural capital utilization; economic growth; decoupling analysis; decomposition

Citation: Guo, S.; Wang, Y.; Huang, J.; Dong, J.; Zhang, J. Decoupling and Decomposition Analysis of Land Natural Capital Utilization and Economic Growth: A Case Study in Ningxia Hui Autonomous Region, China. *Int. J. Environ. Res. Public Health* **2021**, *18*, 646. https://doi.org/10.3390/ijerph18020646

Received: 23 November 2020
Accepted: 8 January 2021
Published: 14 January 2021

1. Introduction

With the significant population growth, urban expansion and the extensive use of powerful scientific and technological means, global depletion of natural resources has reached an unprecedented scale and become largely unsustainable [1,2]. This puts forward a severe challenge of the rational use of natural resource to increase the supply of ecosystem goods and services [3] and to maintain biodiversity that is vital to ecosystem functionality [4]. In this process, the Sustainable Development Goals (SDGs) proposed by the United Nations [5] has increasing been the focus of attention in the world, which enriches the insights of economists and policymakers into economic growth's possible effects on the ecological environment, shifting their paradigms from the simple pursuit of economic growth to concern about ecologically friendly economic growth. The concept of "natural capital" originated from the idea of sustainable development, highlighting the efficient and sustainable use of natural resources [6]. Natural capital has replaced artificial capital as an important limiting factor for the well-being of human society today and natural capital management has increasingly become an important issue in the study of sustainable development [7–9]. Wherein, Land Natural Capital (LNC), as an important index and criterion, has played a significant role in achieving the effective management of resource and the economically, socially and environmentally sustainable development.

As the second-largest economy as well as most populous country in the world, China has been playing a key role in the process of global sustainable development and has formulated a set of policies to achieve this goal. However, the natural resource extraction and exploitation accelerate in parallel with economic growth that leads to decrease biocapacity while increasing the ecological footprint, which has posed a major barrier to China's sustainable development [10,11]. In striving to promote environmental quality and resource efficiency, China has initiated the Ecological Civilization Construction [12], emphasizing the need to build a Life Community of Mountains, Rivers, Forests, Lakes and Grasses. This highlights the urgency for well-formulated metrics that could help policymakers better understand the progress on resource sustainable use regionally. But on the whole, there were few studies on qualify the relationship between economic development with ecological quality and environmental quality at home and abroad, and lacked of relevant researches on the utilization of resources in different stages of economic development.

Ningxia Hui Autonomous Region (Ningxia) is located primarily in arid and semi-arid zones, belonging to the typical ecologically fragile region and one of the most desertified provinces in China. Beset by drought and lack of water resources as well as unreasonable resource utilization and long-term overgrazing and overcultivation, the fragile ecological environment is facing greater challenges, seriously restricting the sustainable development of society and economy in Ningxia. In order to better develop the pilot area of ecological civilization construction and industrial transformation in western region of China, National Development and Reform Commission (NDRC) of Ningxia issued the Comprehensive Management Plan for Key Ecological Areas in Western Region (2012–2020) and Action Plan of Boosting the Development and Transition of the Western Region (2013–2020) [13], highlighting that Ningxia should undertake important task of ecological management and ecological construction in the western region of China. Therefore, as a typical representative of the western provinces in both fragile environment and economic transformation, research on the characteristics of land natural capital utilization (LNCU), decoupling economic outputs from land occupancy, as well as the primary factors affecting their decoupling nexus, is conducive to achieving the goal of accelerating natural capital appreciation meanwhile providing guidance to the economic transition in ecologically fragile regions.

2. Literature Review

At present, various measurement and accounting methods of natural capital utilization have been explored by scholars in studies on sustainable development goals monitoring. Of the latter approaches, the ecological footprint (EF) is defined as a bio-productive land area that maintains human living needs while absorbing pollution caused by human activities [14] from the ecological perspective, which has been broadly used to reflect environmental degradation because it focuses on production and consumption activities on the environment both directly and indirectly [8]. The outstanding merits of this method are that it is easy to apply, repeatable, and simple to understand [15,16]. To date, the ecological footprint model has been improved and extended from one-dimension [17] to two-dimension [14,15] and then to three-dimensional model [18,19], and has been applied to researches at multiple scales of global [15], national [20], province [21], city [22], and enterprise [23]. Especially in ecologically sensitive areas [24,25], the ecological footprint model has fully demonstrated its advantages of ecological sustainability. In the latest applications of three-dimensional ecological footprint in Shandong Province [26], Jiangsu Province [27], Hainan Province [28], Shaanxi Province [29], Beijing-Tianjin-Hebei Metropolitan region [30], Pearl River Delta Urban Agglomerations [31], Yellow River Delta Region [32], Southern Qin Ling piedmont [33], Beijing city [34] and Guiyang city [8], scholars studied the dynamics of regional natural capital stock and flow, ecological sustainability and driving factors influencing the changes of ecological footprints. While this metric is appealing as a communication tool for showing human impact on the natural resources and environment, its methodology and usefulness have also been challenged. Previous researches mostly involve the "real state" ecological footprint (land footprint) and "virtual state" ecological footprint (energy footprint) into natural capital accounting for the purpose of incorporation of the key natural capitals that underpin human society as fully as possible. But the point is that energy footprints do not exist actual ecological capacity correspondingly [9], which results in an irrationally high of land footprint depth. Therefore, Fang [8] discussed defining the "real state" land ecological footprint from the perspective of production to account for the dynamic changes of natural capital utilization of urban and rural land in Guiyang city based on the improved three-dimensional ecological footprint. Above all, despite intense research based on the three-dimensional ecological footprint, there are less researches on the LNCU and its balanced relationship with regional economic in terms of "real state" land footprints.

The core process of human-land relationship is mainly embodied in the relationship between the pressure of human activities and the pressure capacity of resources and environment [35]. At present, various theoretical approaches, such as bivariate causality, correlation analysis, multivariate cointegration, simple regressions, variance decomposition, have widely applied to research the causal relationship between land resource utilization and economic growth [36–38]. However, whether an economy is becoming less dependent on land capital consumption has become another important issue that was less on information. The decoupling analysis has become an important method to study that problem. This theory, originated from the concept of physics, was originally applied to explain the decrease or absence of interrelationships between physical quantities [39]. Later, the Organization for Economic Cooperation and Development (OECD) [40] firstly introduced the theory to analysis the fields of resources and environment in the early 21st century. However, the decoupling index defined by OECD varies with the selection of base period and is classified decoupling as relative or absolute, which cannot accurately reflect the decoupling nexus between those two indicators. To deal with this problem, Tapio [41] defined the decoupling index based on the idea of elastic coefficient and further divided the decoupling state into eight logical possibilities, i.e., strong decoupling (SD), weak decoupling (WD), recessive decoupling (RD), expansive coupling (EC), recessive coupling (RC), expansive negative decoupling (END), weak negative decoupling (WND) and strong negative decoupling (SND). Then numerous studies are available in analyzing the decoupling nexus between economic growth and energy consumption [42,43], economic growth and urban expansion [44] as well as GDP and environmental and resource interrelationships [45,46]. For example, Wang et al. [47] compared both the carbon emissions and the decoupling performance between China and the United States from 2000 to 2014, and found that China has experienced expansive coupling and weak decoupling, while the U.S. experienced mostly weak and strong decoupling. Song [48] examined the decoupling effects of economic growth on cultivated land decreases in China, and found that cultivated land was separately occupied by the expansion of urban areas (1986–1995), rural settlements (1995–2000) and "other construction" land use category (2000–2005) at different stages of China's economic development.

Nevertheless, decoupling indicator method alone cannot evaluate the effects of environmental externalities and obtain the genuine feedback for improvement [49]. To overcome this shortcoming, another group of scholars have concentrated their efforts on investigating the inner mechanism of decoupling by integrating decoupling index with the decomposition approach. At present, the Structural Decomposition Analysis (SDA), the Index Decomposition Analysis (IDA), and the Production-theoretical Decomposition Analysis (PDA) are the three primary types of decomposition approaches [47]. Among which, the Logarithmic Mean Division Index method (LMDI), as the most representative of the IDA decomposition method, performs best because of the advantage of decomposition without residuals [50]. Therefore, it is regarded as the most accurate and practical method in the current decomposition system [51]. Zhang et al. [10] studied the decoupling relationship between coal consumption and economic growth in China, uncovering that the energy intensity effect played the crucial role in decreasing coal consumption, while the economic activity effect and population effect propelled the continual increase of coal consumption in 1991–2013. Liu et al. [52] adopted the LMDI method to decompose the changes of rural residential land area into rural residential land intensive use effect, urban-rural population structure effect, urbanization effect and total population effect. They found that the urban-rural population structure effect was the strongest driver in China's rural residential land changes.

The preceding literatures lay important theoretical foundations and provide feasible and effective methods to study the situation of natural capital utilization as well as the decoupling relationship between resource depletion and economic growth. However, there is little research on the LNCU and its nexus with economic growth. The contributions of this article are focused on the following three aspects: (1) The status of LNCU is analyzed based on the improved EF method in Ningxia during 1999–2017; (2) The Tapio model is introduced to examine the decoupling nexus between LNCU and economic growth; (3) Instead of analyzing the influencing factors of LNCU, this paper systematically studies the major factors affecting the decoupling relationship with combination of the Logarithmic Mean Divisia Index (LMDI) method and Kaya identity equation.

3. Methods and Data Sources

3.1. Study Area and Data Resource

The Ningxia Hui Autonomous Region (104°17′–107°39′ E, 35°14′–149 39°23′ N) is located at the middle and upper reaches of the Yellow River, and the transition zone between Loess Plateau and desert in western China, belonging to the typical ecologically fragile area in western China, with the average annual precipitation and evaporation about 305 mm and 1800 mm separately. Therefore, most area of the region is now in a state of extreme drought and prone to a variety of natural disasters. The provincial area is 66,400 km^2, including five prefectures (Figure 1). By 2017, the area of cropland, grassland and forest land respectively accounted for 19.5%, 31.4% and 11.6% of the total area [53]. The grassland is the dominate land use type. The per capita GDP of Ningxia was 5.1×10^4 RMB in 2017, which only accounted for 85% of China's average per capita GDP [54], facing the double pressure of environmental protection and economic growth. In recent years, Ningxia is undergoing the critical period of economic transformation and territorial space planning. Easing the pressure on the environment, reducing economic dependence on land resource and improving the efficiency of resource utilization are of great importance to achieve environmental protection and economic transition in Ningxia Province.

In this paper, considering that the time-series data in Ningxia started in 1999 and the latest data available to us was 2018, we selected the research period from 1999 to 2017 to avoid inaccuracies caused by inconsistent statistical coverage before 1999. The biological resource production data and socioeconomic data used herein are collected from China Statistical Yearbook (2000–2018) and Ningxia Statistical Yearbook (2000–2018). Wherein, the GDP data take the constant price in 2000 as the base year, and further obtained the actual GDP value. Additionally, we introduced the Liu's [55] research on yield factors: the cultivated land and construction land are 0.94, the grassland is 2.25, the forest land is 0.85 and the water area is 2.25. The equivalence factor is adopted from the National Ecological Footprint Accounting Guidance: the cultivated and construction land are 2.52, the grassland is 0.46, the forest land is 1.29, and the water area is 0.37.

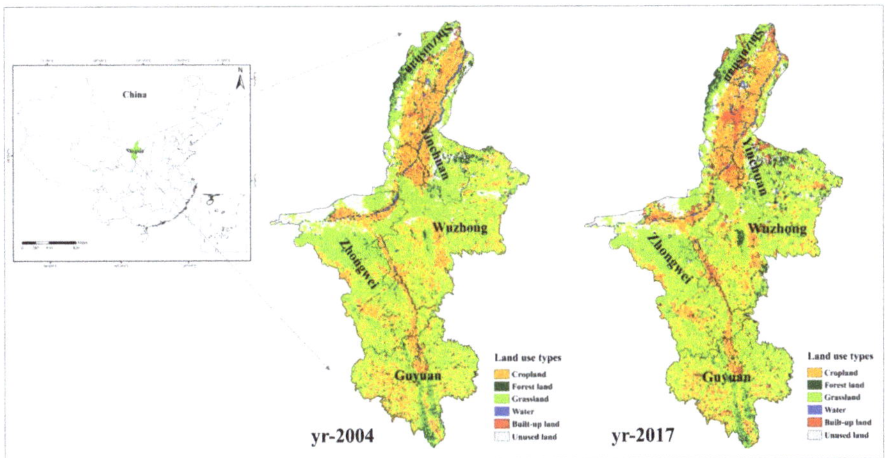

Figure 1. Study area.

3.2. Improved Ecological Footprint Method

In this paper, we applied Fang's [8] improved ecological footprint method to calculate the "real state" land EF from a productive perspective, that is, the land EF is defined as the area of biologically productive land (cropland, forest land, grassland, water and construction land) needed to support the production of regional biological products (agricultural products, forest products, livestock products and aquatic products) and the expansion of construction land [9,56]. The corresponding land capacity is defined as the area of biologically productive land actually available in a region. The calculation formulas for land footprint and land ecological capacity are shown below:

$$LF = \sum_{j=1}^{n}\sum_{i=1}^{m}(\frac{P_{ij}}{AP_{w,ij}} \times r_j) \tag{1}$$

$$LC = 0.88 \times \sum_{j=1}^{n}(A_j \times r_j \times y_j) \tag{2}$$

where LF represents the total land footprint (gha) in a certain year, and LC indicates the total land ecological capacity(gha). P_{ij} is the amount of product i harvested from the land use type j (t yr^{-1}), $AP_{w,ij}$ is the global average yield of product i in land use type j (t wha^{-1} yr^{-1}), A_j represents the area of the biologically productive land required, r_j and y_j are the equivalence factor and the yield factor for a given land use type, respectively, n and m are the number of land use types and product items, respectively. According to the requirements of the World Commission on Environment and Development (WCED), the area of biologically productive land is required to reduce by 12% for biodiversity conservation [57].

The $LF_{siz,reg}$ is defined as the annual occupation area of bio-productive land within the limit of land capacity, which can represent the scale of human utilization of natural capital flow [8,18]. $LF_{dep,reg}$ is the portion of land natural capital occupation that exceeds the range of the land capacity. The formulas are as follows:

$$LF_{siz,reg} = \sum_{j=1}^{n}\min\{LF_j, LC_j\} \tag{3}$$

$$LF_{dep,reg} = 1 + \frac{\sum_{j=1}^{n} \max\{LF_j - LC_j, 0\}}{\sum_{j=1}^{n} LC_j} \quad (4)$$

where LF_j and LC_j represent the land footprint and land ecological capacity for the jth land use type (gha), separately. $LF_{siz,reg}$ is the regional land footprint size of natural capital flow (gha), and $LF_{dep,reg}$ is the regional land footprint depth of natural capital stock (gha).

When the land capital flow of a certain region is not completely occupied, the $LF_{dep,reg}$ equals 1. Utilization efficiency of capital flows (UE_{flo}) is introduced to represent the actual efficiency of human utilization of land capital flow. While when the land capital flow is completely occupied, the use ratio of land capital stocks to flows (UR^{sto}_{flo}) is introduced to represent the extent to which the land capital stock exceeds the capital flow, following the formulas:

$$UE_{flo} = \frac{LF_{size}}{LC} \times 100\% \ (LF \leq LC) \quad (5)$$

$$UR^{sto}_{flo} = \frac{LF - LF_{size}}{LF_{size}} = LF_{size} - 1(LF > LC) \quad (6)$$

3.3. Decoupling Indicator

Tapio decoupling theory breaks the linkage of resource and environmental pressure from economic benefits [41] by comparing the change rates between variables. Therefore, based on this theory, the decoupling elasticity index between LNCU and economic growth can be illustrated as:

$$\beta_t = \frac{\delta LF_t}{\delta G_t} = \frac{\Delta LF}{LF} / \frac{\Delta G}{G} = \Delta LF \times G / LF \times \Delta G \quad (7)$$

where β_t denotes decoupling elasticity index, δLF_t and δG_t represent the growth rate of land capital consumption and gross domestic product (GDP) from a base year 0 to a target year t, respectively. ΔLF and ΔG represent the change value of land capital utilization and economic growth.

Table 1 presents eight grades corresponding to β_t. The growth rate of GDP and the indicator of LF can be coupled, decoupled or negatively decoupled [41,58]. Strong decoupling means that the land natural capital consumption declines while GDP grows. And it's the preferred state for development. Achieving a strong decoupling is also the key point in double intensive of LNCU and economic growth. On the contrary, the strong negative decoupling is the most unfavorable state for the economic development, in which the land natural capital consumption increases and GDP declines. Coupling donates that there is synchronization between these two.

3.4. Decomposition Model

Since the Tapio decoupling model merely considers the decoupling relationship between LNCU and GDP, but does not explore the reasons behind the changes. Therefore, in this paper, several influencing factors of LNCU is decomposed by combination of Kaya identity, LMDI decomposition and Tapio decoupling model. Based on extended Kaya identity [59], the influencing factors can be divided into five parts: structure effect, technical effect, economic effect, labor force effect and population effect, which are defined as follows:

$$FL_t = \sum_{i=1}^{5} LF_{it} = \sum_{i=1}^{5} \frac{LF_{it}}{LF_t} \times \frac{LF_t}{Y_t} \times \frac{Y_t}{L_t} \times \frac{L_t}{P_t} \times P_t = \sum_{ij} S^{LF}_{it} \times I^{LF}_t \times M^{LF}_t \times N^{LF}_t \times P^{LF}_t \quad (8)$$

where LF_{it} represents the footprint of the ith land use type in year t. Y_t, L_t and P_t are the gross domestic product (GDP), the number of working population and resident population respectively in year t. S^{LF}_{it} is the share of the ith land use type footprint to the total land footprint in year t, which represents structure effect; I_t^{LF} is the land footprint depletion caused by ten thousand yuan GDP in year t, which represents technical effect, M_t^{LF} is the output per unit of labor in year t, which represents economic effect; N_t^{LF} is the size of labor force in year t, which represents labor force effect; P_t^{LF} is the number of resident population in year t, which represents population effect.

Table 1. Six decoupling states.

Decoupling State		δLF_t	δG_t	β_t	Decoupling Type	Meaning
Decoupling	Strong decoupling (SD)	−	+	$(-\infty, 0)$	I	Economic grows while land natural capital depletion decreases
	Weak decoupling (WD)	+	+	$[0, 0.8)$	II	Economic grows while land natural capital depletion decreases slowly
	Recessive decoupling (RD)	−	−	$(1.2, +\infty)$	III	Economic grows slowly while land natural capital depletion decreases significantly
Coupling	Expansive coupling (EC)	+	+	$[0.8, 1.2]$	IV	Economic grows and land natural capital depletion increases moderately
	Recessive coupling (RC)	−	−	$[0.8, 1.2]$	V	Economic declines and land natural capital depletion increases moderately
Negative decoupling	Weak negative decoupling (WND)	−	−	$[0, 0.8)$	VI	Economic declines and land natural capital depletion decreases slowly
	Expansive negative decoupling (END)	+	+	$[1.2, +\infty)$	VII	Economic grows and land natural capital depletion increases significantly
	Strong negative decoupling (SND)	+	−	$(-\infty, 0)$	VIII	Economic declines while land natural capital depletion increases

According to LMDI decomposition and Kaya identity, the change of LF from a base year 0 and a target year t, represented by ΔLF, can be decomposed into five effects as follows: (i) the changes in the structure effect (represented by ΔLF_S); (ii) the changes in the intensity effect (represented by ΔLF_I); (iii) the changes in the economic effect (represented by ΔLF_M); (iv) the change in the labor force effect (represented by ΔLF_N), and (v) the changes in the population effect (represented by ΔLF_P), as shown in Equation (9):

$$\Delta LF = LF_t - LF_0 = LF_S + LF_I + LF_M + LF_N + LF_P \tag{9}$$

Each effect in the right-hand side of Equation (9) can be expressed as follows:

$$\Delta LF_S = \sum_{i=1}^{5} \frac{LF_{it} - LF_{i0}}{\ln(LF_{it}) - \ln(LF_{i0})} \times \ln\left(\frac{S_{it}^{LF}}{S_{i0}^{LF}}\right) \quad \Delta LF_I = \sum_{i=1}^{5} \frac{LF_{it} - LF_{i0}}{\ln(LF_{it}) - \ln(LF_{i0})} \times \ln\left(\frac{S_t^{LF}}{S_0^{LF}}\right)$$

$$\Delta LF_M = \sum_{i=1}^{5} \frac{LF_{it} - LF_{i0}}{\ln(LF_{it}) - \ln(LF_{i0})} \times \ln\left(\frac{M_t^{LF}}{M_0^{LF}}\right) \quad \Delta LF_N = \sum_{i=1}^{5} \frac{LF_{it} - LF_{i0}}{\ln(LF_{it}) - \ln(LF_{i0})} \times \ln\left(\frac{N_t^{LF}}{N_0^{LF}}\right)$$

$$\Delta LF_P = \sum_{i=1}^{5} \frac{LF_{it} - LF_{i0}}{\ln(LF_{it}) - \ln(LF_{i0})} \times \ln\left(\frac{P_t^{LF}}{P_0^{LF}}\right)$$

Combining Equations (7) and (9), a decoupling model between LF and GDP is established based on the LMDI method, which can decompose the decoupling elasticity index for the relationship between LNCU and GDP into corresponding decoupling elasticity indexes.

$$\begin{aligned} \beta_t &= \frac{\Delta LF}{LF} / \frac{\Delta G}{G} = \Delta LF \times G / LF \times \Delta G \\ &= (\Delta LF_S + \Delta LF_I + \Delta LF_M + \Delta LF_N + \Delta LF_P) \times G / LF \times \Delta G \\ &= \beta_S + \beta_I + \beta_M + \beta_N + \beta_P \end{aligned} \tag{10}$$

4. Results and Discussion

4.1. Analysis of the Land Natural Capital Utilization

Based on the method illustrated in Section 3.2, the land footprint and land capacity, land footprint size and land footprint depth, as well as the land footprint utilization efficiency of land capital flows and the use ratio of land capital stocks to flows can be obtained.

4.1.1. Natural Capital Utilization of Cultivated Land

Cultivated land is the main component of Ningxia's ecological footprint. As shown in Figure 2a, the cultivated land footprint increased rapidly from 1999 to 2012, and then leveled off during the following years at around 4.13 gha. Specifically, the footprints of rice and wheat showed a significant downward trend, and their proportions of total land footprint decreased by 23.23% and 17.02% to 0.092 gha and 0.505 gha respectively over the past 19 years. Oppositely, the proportion of corn increased obviously to 77.02%, followed by oil plants and vegetables (Figure 2b). This shows that Ningxia is constantly optimizing the agricultural planting structure to realize the transformation from extensive to intensive, decentralized to large-scale specialized modern agriculture. The footprint depth increased first and then decreased significantly in 2013, resulting in the large reduction of UR_{flo}^{sto} (Figure 3). But the values of footprint depth were greater than 1, which meant the capital stock consumption of cultivated land was still relatively serious, and food supply pressure would continue to exist with the decrease of cultivated land per capita.

4.1.2. Natural Capital Utilization of Grassland

The livestock products produced from grassland provide a large amount of animal protein and are one of the most important sources of food consumption for Ningxia residents. Figure 4a showed that the grassland footprint increased from 0.321 gha in 1999 to 0.803 gha in 2017, showing an increase of 1.50 times. Specifically, the footprint of pork decreased from 10.05% in 1999 to 8.11% in 2017, while footprints of beef, mutton and milk increased obviously from 7.65%, 6.48% and 3.45% to 22.29%, 20.24% and 21.52% separately (Figure 4b), showing the changes in people's diet structure. In 1999, the grassland footprint depth was 3.247. By 2017, the footprint depth reached 12.699, suggesting that almost 13 times the current grassland area was required to support the livestock consumption in Ningxia and that this situation was likely to continue well into the future as urban and rural residents demanding more meat, eggs and dairy product. Based on the model hypothesis, the land capital flow is assumed to be used first, followed by capital stock [8]. However, during the process of using actual land natural capital, the footprint size (capacity) of grassland was rather low and had fallen by 56.28% since 1999, which exacerbated the contradiction between supply and demand of grassland resources.

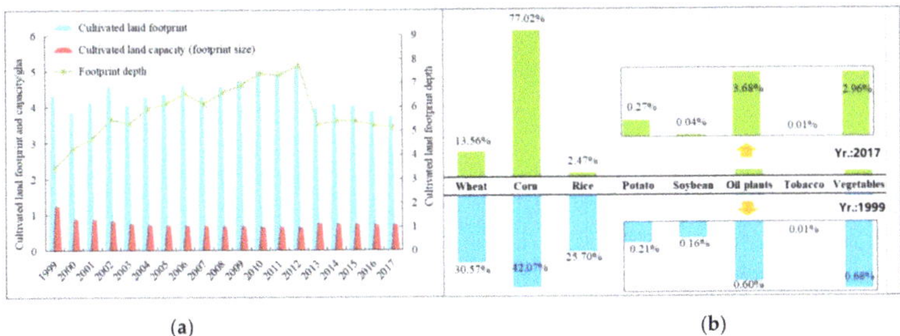

Figure 2. Changes in cultivated land's natural capital use (**a**) and compositions of cultivated land footprint (**b**) in Ningxia from 1999 to 2017.

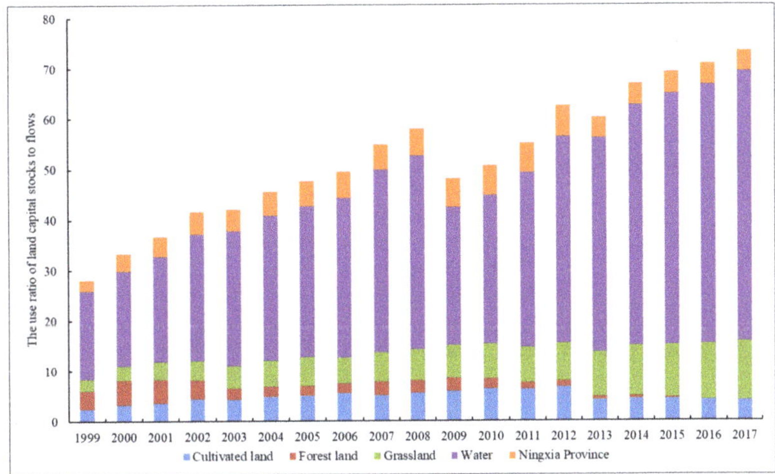

Figure 3. Changes in the UR_{flo}^{sto} of different land use types in Ningxia from 1999 to 2017.

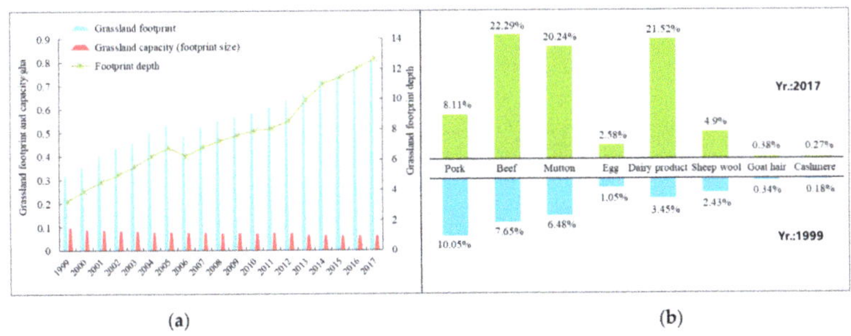

Figure 4. Changes in grassland's natural capital use (**a**) and compositions of grassland footprint (**b**) in Ningxia from 1999 to 2017.

4.1.3. Natural Capital Utilization of Forest Land

Affected by human activity, the forest land footprint in Ningxia fluctuated more dramatically. From 1999 to 2006, forest land footprint firstly showed a mild downward trend but was generally sustained at around 0.311 gha, then it showed a rapid increase during 2007–2009 and peaked at 0.379 gha in 2007, and finally it declined rapidly to 0.101 gha in 2017 (Figure 5a). Forest land capacity changed similarly with footprint size from 1999 to 2015. But since then, forest land capacity remained stable at around 0.118 gha while the footprint size declined again for the reason that Ningxia enforced effective measures like Grain for Green Project and Three North Shelterbelt Forest Program to strictly control ecological land occupation during the early stage of the study period because of serious soil erosion. However, as the accelerating urbanization and implementation of the strictest farmland protection system, construction land occupied a part of forest land. Therefore, footprint depth of forest land showed a rapid volatility drop until it leveled off in 2015 at 1, meaning existing forest land could basically meet the current consumption demand.

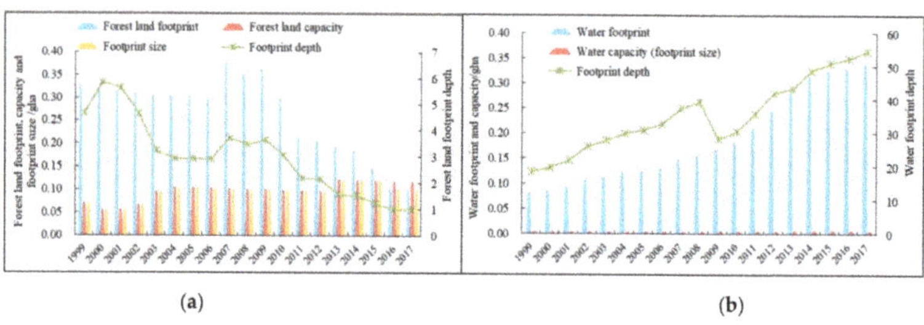

(a) (b)

Figure 5. Changes in the natural capital use of forest land (**a**) and water (**b**) in Ningxia from 1999 to 2017.

It is noted that if the energy footprint was incorporated into the forestland to calculate the footprint depth according to the practice of Niccolucci [19], the forestland footprint depth of Ningxia in 2017 was up to 12.69 [60], which is obviously too pessimistic to reflect the actual situation of the local forest ecosystem.

4.1.4. Natural Capital Utilization of Water

The water footprint in Ningxia increased from 0.082 gha in 1999 to 0.339 gha in 2017, up more than tripled (Figure 5b). The water capacity (footprint size) was slowly increasing year by year at an annual rate of 1.89%, from 0.004 gha in 1999 to 0.006 gha in 2017. The explosive growth of water deficit has not met the people's greater demand of aquatic products. The water footprint depth experienced a substantial upswing with a change process of "N" shape over the past 19 years, resulting in the maximum level of UR^{sto}_{flo} in five land use types. Then, the water footprint depth soared again from 28.337 in 2009 to 54.578 in 2017 after a brief decline in 2009, much higher than the water footprint in eastern China (2.67 in Zhejiang Province) [61], suggesting that water is the most unsustainable development sector among all land types and almost 55 times the current area is needed to sustain the water resource consumption. Moreover, as a severe shortage of water resources and the water capacity that can be provided in Ningxia is observed, the imbalance between water supply and demand will continue to deteriorate under the existing consumption patterns. Therefore, it is the fundamental way to adjust the aquaculture structure and increase the output per unit of aquatic product while protecting the lake and wetland.

4.1.5. Natural Capital Utilization of Construction Land

The construction land includes urban land, industrial and mining land, transportation land and other built-up land types, which is most closely related to people's daily production and life. As showed in Figure 6a, construction land footprint (footprint size) increased rapidly from 0.006 gha in 1999 to 0.023 gha in 2017, representing a 7.59-fold increase. Although construction land occupied the smallest area of biological productive lands in Ningxia, it was the fastest-growing segment. The construction land capacity remained relatively stable in 1999–2008, and then began to increase after a slight drop in 2009, suggesting that with the expansion of urban scale in Ningxia, urban construction land use pattern is gradually optimized. Over the past 19 years, the footprint depth maintained at 1, and the utilization efficiency of capital flows increased from 4.639% to 21.228%, with an average annual growth rate of 8.82% (Figure 6b), showing that there was still a large ecological surplus space and the land capital flow could meet the demand of urbanization development.

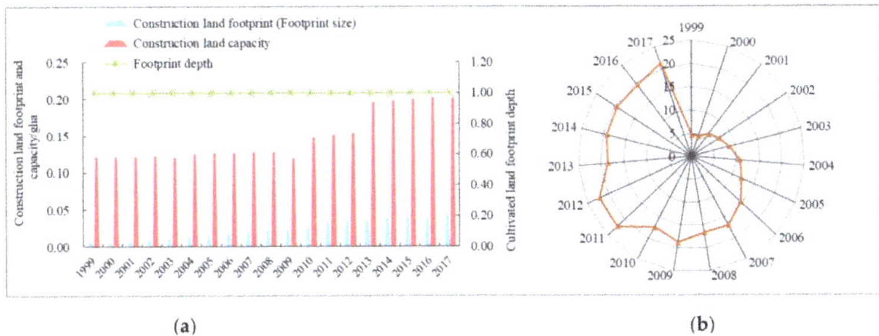

Figure 6. Changes in construction land's natural capital use (**a**) and utilization efficiency of capital flows (**b**) in Ningxia.

4.2. Analysis of the Decoupling Relationship

Based on the method illustrated in Section 3.3, the decoupling status between economic growth and LNCU during 1999–2017 can be easily verified, listed in Table 2. Overall, as the growth rate of economy was positive throughout, only three statuses occurred in the eight decoupling states, that is, expansive negative decoupling, weak decoupling and strong decoupling.

Table 2. Decoupling state between land natural capital utilization and economic growth from 1999 to 2017.

Year	δLF_t	δG_t	D_t	Decoupling	Year	δLF_t	δG_t	D_t	Decoupling
1999–2000	−0.088	0.102	−0.860	I	2008–2009	0.037	0.119	0.307	II
2000–2001	0.076	0.101	0.755	II	2009–2010	0.041	0.135	0.305	II
2001–2002	0.098	0.102	0.959	IV	2010–2011	−0.019	0.121	−0.160	I
2002–2003	−0.095	0.127	−0.750	I	2011–2012	0.043	0.115	0.372	II
2003–2004	0.060	0.112	0.532	II	2012–2013	−0.172	0.098	−1.754	I
2004–2005	0.023	0.109	0.211	II	2013–2014	0.036	0.080	0.449	II
2005–2006	0.038	0.127	0.299	II	2014–2015	−0.008	0.080	−0.104	I
2006–2007	−0.033	0.127	−0.263	I	2015–2016	−0.032	0.081	−0.391	I
2007–2008	0.051	0.126	0.403	II	2016–2017	−0.016	0.078	−0.208	I

4.2.1. Decoupling State of Economic Growth and Land Capital Utilization

In Table 2, there was a strong decoupling in 1999–2000, which is the optimal decoupling state, indicating the land natural capital occupation declined while the economic growth increased. But soon after, decoupling state gradually deteriorated and showed the weak decoupling and expansive coupling in 2000–2002, which meant that the growth trend of land capital consumption was almost the same as that of economy. During the year of 2003–2010, such decoupling state had improved: economy kept stable growth, while land capital consumption maintained slow growth, and even showed negative growth in 2006–2007. Therefore, this period was basically stable in the state of weak decoupling except for the strong decoupling in 2006–2007. While, such relatively stable state was broken in 2000–2014 due to the changeable rate of land capital consumption, and there became alternative between weak decoupling and strong decoupling. Since then, decoupling state between land capital utilization and economic growth has been remained preferred strong decoupling, suggesting that marginal effect of economic growth is smaller than that of land natural capital utilization.

4.2.2. Decoupling State of Economic Growth and Each Bioproductive Land Type

Table 3 summarizes the decoupling status between five bioproductive land types and economic growth. Overall, both cultivated land and forest land showed strong decoupling from economic growth, while grassland, water and construction land experienced weak decoupling from economic growth. These clear periodical characteristics illustrated the differences in the priorities of regional development policies in different periods.

Table 3. The decoupling index of different land capital use and GDP growth.

Year	Cultivated Land	Forest Land	Grassland	Water	Construction Land
1999–2000	−1.081	−0.195	1.121	0.354	−0.027
2000–2001	0.772	−0.156	1.140	0.976	3.167
2001–2002	1.047	−0.143	0.757	1.558	1.733
2002–2003	−0.936	−0.117	0.280	0.358	1.470
2003–2004	0.534	−0.049	0.662	0.739	2.560
2004–2005	0.173	0.030	0.532	0.154	1.149
2005–2006	0.450	−0.115	−0.905	0.338	1.821
2006–2007	−0.553	2.069	0.596	1.034	1.291
2007–2008	0.498	−0.594	0.163	0.427	−0.096
2008–2009	0.293	0.295	0.747	0.640	0.465
2009–2010	0.429	−1.282	0.084	0.642	1.108
2010–2011	−0.157	−2.377	−0.055	1.314	2.577
2011–2012	0.356	−0.328	0.388	1.373	0.098
2012–2013	−2.247	−0.614	0.414	1.639	0.612
2013–2014	0.315	−0.540	0.253	1.368	0.696
2014–2015	−0.114	−2.561	0.530	0.430	0.313
2015–2016	−0.479	−3.795	0.529	0.216	−0.009
2016–2017	−0.417	−0.182	0.423	0.328	1.178

Specifically, the decoupling statuses between cultivated land capital use and economic growth experienced a positive process from expansive coupling to strong decoupling. In 1999–2003, the decoupling status fluctuated a lot and showed an obvious circleprocess of SD-WD-RD-SD. And then, there was a long period of weak decoupling in 2003–2010 except for 2006–2007 and 2010–2011 in which it showed the preferred strong decoupling. With continuous decrease of cultivated land footprint during the period 2012–2017, the decoupling relationship at this stage was generally stable at strong decoupling except for weak decoupling in 2013–2014. Due to strict ecological protection measures, natural capital consumption of forest land continued to decrease and showed the optimal decoupling state. In 1999–2017, the decoupling index of forest land maintained a preferred strong decoupling except for weak decoupling in 2004–2005 and 2008–2009, as well as expansive negative decoupling in 2006–2007.

During the year of 1999–2001, the decoupling index of grassland capital use and GDP growth showed expansive negative decoupling. In the following period this situation got better and appeared a long weak decoupling, even some strong decoupling status occurred intermittently in 2005–2006 and 2010–2011. Decoupling statuses of water capital use and GDP growth experienced a deteriorated trend from weak decoupling in 1999–2000 to expansive coupling in 2000–2001, and further to expansive negative decoupling in 2001–2002. Then, this situation improved and appeared a seven-year weak decoupling in 2002–2006 and 2007–2010. However, during the year of 2010–2014 the decoupling state deteriorated again and showed the unfavorable expansive negative decoupling throughout. Until recent years, the growth rate of water footprint slowed down significantly and decoupling state returned back to weak decoupling. Compared with other decoupling indexes of productive land types, the decoupling state between construction land and economic growth was poor and fluctuated a lot. In 1999–2007, the decoupling status of construction land undergone the transition from the strong decoupling to the expansive negative decoupling. Then after a brief preferred decoupling in 2007–2009, it returned to expansive negative decoupling again in 2010–2011. While, during the following years, the decoupling state stepwise improved to strong decoupling along with construction land control and spatial layout optimization.

4.3. Analysis of the Driving Factors on Decoupling Relationship

According to the decomposition Equation (10) presented in Section 3.4, land natural capital utilization was jointly determined by five main factors. Figure 7 depicts the decomposition results listed in Table 4 to illustrate the contribution of each index to the decoupling state in every year from 1999–2017. As the growth rate of GDP is positive throughout, the negative value of decoupling index indicates a promoting effect to the decoupling between LNCU and economic growth. On the contrary, it performs a negative effect on the decoupling state.

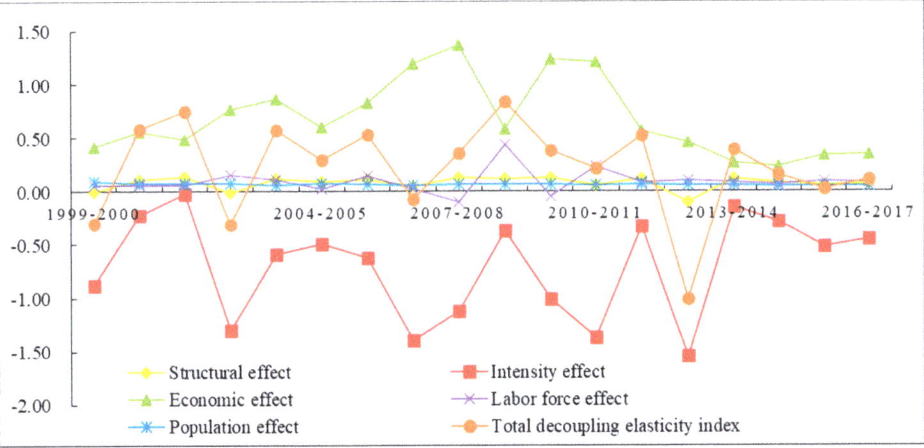

Figure 7. Changes in the total decoupling index and the sub-decoupling index in Ningxia from 1999 to 2017.

Table 4. The decoupling index of different land capital use and GDP growth.

Year	β_S	β_I	β_M	β_N	β_P	β_T
1999–2000	−0.001	−0.873	0.422	0.055	0.095	−0.301
2000–2001	0.117	−0.220	0.556	0.059	0.075	0.587
2001–2002	0.139	−0.015	0.493	0.062	0.075	0.753
2002–2003	−0.008	−1.299	0.767	0.159	0.076	−0.305
2003–2004	0.118	−0.585	0.863	0.112	0.064	0.572
2004–2005	0.094	−0.496	0.602	0.026	0.074	0.300
2005–2006	0.109	−0.622	0.832	0.148	0.066	0.534
2006–2007	0.049	−1.393	1.194	0.024	0.057	−0.070
2007–2008	0.130	−1.119	1.375	−0.097	0.065	0.353
2008–2009	0.124	−0.366	0.584	0.434	0.067	0.843
2009–2010	0.127	−1.005	1.238	−0.044	0.071	0.387
2010–2011	0.062	−1.364	1.218	0.239	0.060	0.213
2011–2012	0.121	−0.322	0.566	0.084	0.071	0.520
2012–2013	−0.097	−1.541	0.465	0.106	0.059	−1.007
2013–2014	0.118	−0.140	0.274	0.083	0.056	0.391
2014–2015	0.085	−0.280	0.236	0.070	0.048	0.159
2015–2016	0.057	−0.515	0.339	0.094	0.052	0.027
2016–2017	0.072	−0.451	0.351	0.086	0.049	0.108

4.3.1. Structural Effect

Figure 7 presents the impact of structural effect changes on decoupling for the period from 1999 to 2017. Overall, the impact of this factor was small and in slight fluctuations. However, this negative impact had weakened with respect to the zero point of decoupling index, and even appeared positive impacts on the decoupling in the year 1999–2000, 2002–2003 and 2012–2013 due mostly to the large reduction of natural capital consumption of cultivated land. It indicated that the adjustment and optimization of land use structure could play a role in decoupling economic growth and LNCU to a certain degree.

4.3.2. Intensity Effect

The intensity effect reflects the impact of technological progress on the decoupling. As shown in Table 4, the decoupling index of intensity effect was significantly less than 0 throughout but fluctuated a lot, indicating that the improvement of technology played a dominant role in decoupling. In recent years, with establishment of Ecological Agriculture Demonstration Area, the High-tech Development Zone and Inland Opening-up Pilot Economic Zone in Ningxia, scientific and technological progress have greatly improved regional capital utilization efficiency and supply capacity. While, during the year of 2013–2017, this positive impact decreased a bit. For instance, the intensity effect weakened and its decoupling index showed a downward trend due to the marginal utility (Figure 7).

4.3.3. Economic Effect

The results presented that the decoupling indexes of economic output were positive throughout, with a mean value of 0.687, which was much greater than that of structural effect, labor force effect and population effect. It means that economic profits in Ningxia mainly rely on the consumption of natural resource and result in the most obvious negative impact on the decoupling. As shown in Figure 7, the curve of economic effect can be divided into three stages: a rapid-increase stage from 1999 to 2007, a great fluctuation period from 2009 to 2011, and a steadily decrease phase from 2012 to 2017, reaching the minimum 0.236 in 2014–2015. This change can explain the transition of economic pattern from resource-dependent development to innovation-independent economy in Ningxia.

4.3.4. Effects of Labor Force and Population

It can be observed from Table 4 that during 1999–2017, the effects of labor force and population showed an opposite change characteristic and a negative impact on the decoupling as a whole. Specially, population effect fluctuated marginally around its mean value of 0.099 during the research period, indicating that with the continuous development of the economy of Ningxia, the status quo of the populous province has not been significantly changed. On the contrary, the labor force appeared a rapid fluctuation in the year of 2007–2011. The reason is that due to economic slowdown in Ningxia, large numbers of labor forces in the secondary and tertiary industries have shifted to the agricultural labor force. During this period, the decoupling index between GDP growth and LNCU increased rapidly. Nevertheless, from 2011 onwards the influence of this factor on the decoupling has weakened and gradually stabilized around 0.1 owing to the concentration of urban population in the accelerating process of industrialization and urbanization.

5. Conclusions and Policy Implications

Natural capital utilization in typical ecologically sensitive and fragile areas has attracted wide attention. Although existing studies have contributed theoretical foundations and various methods to study the dynamic change of natural capital utilization from multiple scales, the "real state" land footprint analysis of LNCU research is still insufficient and there are less studies concentrated on the decoupling relationship between natural capital depletion and economic growth. Therefore, in this paper, we established the improved EF model to analysis the states of LNCU in Ningxia Province, introduced Tapio decoupling theory was to explore the nexus between LNCU and economic growth, and decomposed the major factors affecting this relationship in combination with the Kaya identity and LMDI model, which further expands the existing research. However, despite our promising results, there are still some limitations in this study: first of all, we mainly focused on the perspective of production to study the changes in the use of land natural capital, which may produce different results from the perspective of consumption. Second, our study adopts a top-down approach, and most data used are from the Ningxia Statistical Yearbook and the China Statistical Yearbook. When calculating the ecological footprint of the city and county scale, the accuracy of the data is insufficient, and therefore the further research on a small scale should be carried out in combination with field investigation. The primary conclusions are as follows:

(1) From the analysis of land natural capital utilization in Ningxia, it can be observed that the natural capital stock utilization of cultivated land decreased obviously, resulting in the declining trend of the UR_{flo}^{sto} in recent years. Similarly, in forest land, it decreased constantly and the flow occupation of natural capital could basically meet consumer demand since 2015. While, the UR_{flo}^{sto} of grassland and water increased rapidly and performed as the most unsustainable sectors among all land types. Moreover, the footprint of construction land occupied the smallest area of biological productive lands in Ningxia, but was the fastest-growing segment.
(2) The decoupling analysis showed that the pressure of economic development on the sustainable use of land natural capital always exists. Overall, the decoupling state was preferred and dominated by strong decoupling and weak decoupling. The two stages appeared almost at the same frequency and only in 2001–2002 there was an expansive coupling. Specially, the cultivated land and forest land showed a preferred decoupling state in recent years, followed by grassland, while water and construction land showed the unfavorable expansive negative decoupling and weak decoupling.
(3) In terms of the decomposition results, it can be obtained that economic effect is the biggest unfavorable factor limiting the strong decoupling, while intensity effect is the most favorable factor to promote the decoupling of land capital occupation from economic growth in Ningxia Province. Moreover, the impacts of structural effect, labor force effect and population effect on the decoupling are relatively weak.

Given the above findings and regional characteristics of Ningxia, we propose the following policies to promote the rational use of land natural capital and the decoupling of land capital occupation from economic growth in Ningxia based on the partition perspective:

(1) In the central and southern mountains with fragile ecological environment (Guyuan city, South of Wuzhong City and Zhongwei city), considering the reality that grassland resource is abundant but overutilized seriously, measures related to the grassland and forest land protection, such as the Grain for Green Project, the Region-Wide Grazing Ban and the Three North Shelterbelt Forest Program, should be strictly implemented to improve ecological carrying capacity. Apart from that, due to the rapid expansion of urban construction land, later monitoring and management are crucial for the maintenance of ecological restoration achievements.
(2) In the northern Yellow River irrigation area where is featured with higher economic development and better ecological environment (Yinchuan city and Shizuishan city), it is imperative to promote the transformation and upgrading of traditional industries and advance scientific and technological innovation for the purpose of reducing the impediment of economic effect on the decoupling relationship, and improving the intensity effect by improving the efficiency of resource utilization. At the same time, reasonably controlling the boundary of urban space expansion and optimizing the population structure to give full play to the powerful driving effect of talents on the economy instead of the overuse of natural resources.
(3) Moreover, in the Ningxia Plain along the Yellow River, it is suggested to vigorously develop ecological, water-saving agriculture, reduce the use of fertilizer and control the loss of water resources, to alleviate the pressure of the extreme shortage of water resources.

Author Contributions: Conceptualization, S.G. and Y.W.; data collection process, S.G. and J.Z.; methodology, S.G. and J.D.; data analysis, S.G.; writing—original draft preparation, S.G.; writing—review and editing, Y.W., J.H., and J.D.; funding acquisition, Y.W. and J.D. All authors have read and agreed to the published version of the manuscript.

Funding: This research was funded by the National Key Research and Development Program of China (2019YFC1904303), and the National Natural Science Foundation of China (U1710120, 51874306).

Institutional Review Board Statement: Not applicable.

Informed Consent Statement: Informed consent was obtained from all subjects involved in the study.

Data Availability Statement: No new data were created or analyzed in this study. Data sharing is not applicable to this article.

Conflicts of Interest: The authors declare no conflict of interest.

References

1. Kontokosta, C.E.; Jain, R.K. Modeling the determinants of large-scale building water use: Implications for data-driven urban sustainability policy. *Sustain. Cities Soc.* **2015**, *18*, 44–55. [CrossRef]
2. Wang, Z.; Yang, L.; Yin, J.; Zhang, B. Assessment and prediction of environmental sustainability in China based on a modified ecological footprint model. *Resour. Conserv. Recycl.* **2018**, *132*, 301–313. [CrossRef]
3. Blignaut, J.; Aronson, J. Getting serious about maintaining biodiversity. *Conserv. Lett.* **2008**, *1*, 12–17. [CrossRef]
4. Benayas, J.M.R.; Newton, A.C.; Diaz, A.; Bullock, J.M. Enhancement of Biodiversity and Ecosystem Services by Ecological Restoration: A Meta-Analysis. *Science* **2009**, *325*, 1121–1124. [CrossRef]
5. United Nations (UN). Transforming Our World: The 2030 Agenda for Sustainable Development. 2015. Available online: https://sustainabledevelopment.un.org/content/documents/21252030%20Agenda%20for%20Sustainable%20Development%20web.pdf (accessed on 5 December 2020).
6. Fairbrass, A.; Mace, G.; Ekins, P.; Milligan, B. The natural capital indicator framework (NCIF) for improved national natural capital reporting. *Ecosyst. Serv.* **2020**, *46*, 101198. [CrossRef]
7. Costanza, R.; Fioramonti, L.; Kubiszewski, I. The UN Sustainable Development Goals and the dynamics of well-being. *Front. Ecol. Environ.* **2016**, *14*, 59. [CrossRef]
8. Fang, K.; Zhang, Q.; Yu, H.; Wang, Y.; Dong, L.; Shi, L. Sustainability of the use of natural capital in a city: Measuring the size and depth of urban ecological and water footprints. *Sci. Total Environ.* **2018**, *631–632*, 476–484. [CrossRef]
9. Zhang, Z.; Hu, B.; Shi, K.; Su, K.; Yang, Q. Exploring the dynamic, forecast and decoupling effect of land natural capital utilization in the hinterland of the Three Gorges Reservoir area, China. *Sci. Total Environ.* **2019**, *718*, 134832. [CrossRef]
10. Zhang, M.; Bai, C.; Zhou, M. Decomposition analysis for assessing the progress in decoupling relationship between coal consumption and economic growth in China. *Resour. Conserv. Recycl.* **2016**, *129*, 454–462. [CrossRef]
11. Wang, Q.; Jiang, R.; Li, R. Decoupling analysis of economic growth from water use in City: A case study of Beijing, Shanghai, and Guangzhou of China. *Sustain. Cities Soc.* **2018**, *41*, 86–94. [CrossRef]
12. Zhang, X.; Wang, Y.; Qi, Y.; Wu, J.; Liao, W.; Shui, W.; Zhang, Y.; Deng, S.; Peng, H.; Yu, X.; et al. Reprint of: Evaluating the trends of China's ecological civilization construction using a novel indicator system. *J. Clean. Prod.* **2017**, *163*, S338–S351. [CrossRef]
13. Guan, Y.; Kang, L.; Shao, C.; Wang, P.; Ju, M. Measuring county-level heterogeneity of CO_2 emissions attributed to energy consumption: A case study in Ningxia Hui Autonomous Region, China. *J. Clean. Prod.* **2017**, *142*, 3471–3481. [CrossRef]
14. Wackermagel, M.; Rees, W.E. *Our Ecological Footprint, Reducing Human Impact on the Earth*; New Society Publishers: Gabriela Island, PA, USA, 1996.
15. Wackernagel, M.; Onisto, L.; Bello, P.; Callejas Linares, A.; Ina, S.L.F.; Mendez Garcia, J.; Ana, I.S.G.; Guadalupe Suarez Guerrero, M. National natural capital accounting with the ecological footprint concept. *Ecol. Econ.* **1999**, *29*, 375–390. [CrossRef]
16. Li, J.; Chen, Y.; Xu, C.; Li, Z. Evaluation and analysis of ecological security in arid areas of Central Asia based on the emergy ecological footprint (EEF) model. *J. Clean. Prod.* **2019**, *235*, 664–677. [CrossRef]
17. Rees, W.E. Ecological Footprints and appropriated carrying capacity: What urban economics leaves out. *Environ. Urban.* **1992**, *4*, 121–130. [CrossRef]
18. Niccolucci, V.; Bastianoni, S.; Tiezzi, E.B.P.; Wackernagel, M.; Marchettini, N. How deep is the footprint? A 3D representation. *Ecol. Model.* **2009**, *220*, 2819–2823. [CrossRef]
19. Niccolucci, V.; Galli, A.; Reed, A.; Neri, E.; Wackernagel, M.; Bastianoni, S. Towards a 3D National Ecological Footprint Geography. *Ecol. Model.* **2011**, *222*, 2939–2944. [CrossRef]
20. Erb, K. Actual land demand of Austria 1926-2000: A variation on Ecological Footprint assessments. *Land Use Policy* **2004**, *21*, 247–259. [CrossRef]
21. Miao, C.; Sun, L.; Yang, L. The studies of ecological environmental quality assessment in Anhui Province based on ecological footprint. *Ecol. Indic.* **2016**, *60*, 879–883. [CrossRef]
22. Isman, M.; Archambault, M.; Racette, P.; Konga, C.N.; Llaque, R.M.; Lin, D.; Iha, K.; Ouellet-Plamondon, C.M. Ecological Footprint assessment for targeting climate change mitigation in cities: A case study of 15 Canadian cities according to census metropolitan areas. *J. Clean. Prod.* **2018**, *174*, 1032–1043. [CrossRef]
23. Fan, Y.; Qiao, Q.; Xian, C.; Xiao, Y.; Fang, L. A modified ecological footprint method to evaluate environmental impacts of industrial parks. *Resour. Conserv. Recycl.* **2017**, *125*, 293–299. [CrossRef]

24. Guo, S.; Wang, Y. Ecological Security Assessment Based on Ecological Footprint Approach in Hulunbeir Grassland, China. *Int. J. Environ. Res. Public Health* **2019**, *16*, 4805. [CrossRef]
25. Li, X.; Tian, M.; Wang, H.; Wang, H.; Yu, J. Development of an ecological security evaluation method based on the ecological footprint and application to a typical steppe region in China. *Ecol. Indic.* **2014**, *39*, 153–159. [CrossRef]
26. Xun, F.; Hu, Y. Evaluation of ecological sustainability based on a revised three-dimensional ecological footprint model in Shandong Province, China. *Sci. Total Environ.* **2019**, *649*, 582–591. [CrossRef]
27. He, J.; Wan, Y.; Feng, L.; Ai, J.; Wang, Y. An integrated data envelopment analysis and emergy-based ecological footprint methodology in evaluating sustainable development, a case study of Jiangsu Province, China. *Ecol. Indic.* **2016**, *70*, 23–34. [CrossRef]
28. Dong, H.; Li, P.; Feng, Z.; Yang, Y.; You, Z.; Li, Q. Natural capital utilization on an international tourism island based on a three-dimensional ecological footprint model: A case study of Hainan Province, China. *Ecol. Indic.* **2019**, *104*, 479–488. [CrossRef]
29. Yang, Y.; Hu, D. Natural capital utilization based on a three-dimensional ecological footprint model: A case study in northern Shaanxi, China. *Ecol. Indic.* **2018**, *87*, 178–188. [CrossRef]
30. Du, Y.; Peng, J.; Gao, Y.; Zhao, H. Sustainability evaluation of natural capital utilization based on a three-dimensional ecological footprint model: A case study of the Beijing-Tianjin-Hebei Metropolitan region. *Prog. Geogr.* **2016**, *35*, 1186–1196.
31. Zhang, X.; Zeng, H. Dynamic of three-dimensional ecological footprint in the Pearl River Delta and its driving factors. *Acta Sci. Circum.* **2017**, *37*, 771–778.
32. Cheng, Y.; Yin, J.; Wang, J. Research on natural capital evolution and influencing factors in the Yellow River Delta Region. *China Popul. Resour. Environ.* **2019**, *29*, 127–136.
33. Yang, Y.; Ling, S.; Zhang, T.; Yao, C. Three-dimensional ecological footprint assessment for ecologically sensitive areas: A case study of the Southern Qin Ling piedmont in Shaanxi, China. *J. Clean. Prod.* **2018**, *194*, 540–553. [CrossRef]
34. Peng, J.; Du, Y.; Ma, J.; Liu, Z.; Liu, Y.; Wei, H. Sustainability evaluation of natural capital utilization based on 3DEF model: A case study in Beijing City, China. *Ecol. Indic.* **2015**, *58*, 254–266. [CrossRef]
35. Li, X.; Yang, Y.; Liu, Y. Research progress in man-land relationship evolution and its resource-environment in China. *Acta Geogr. Sin.* **2016**, *71*, 2067–2088. [CrossRef]
36. Handavu, F.; Chirwa, P.W.C.; Syampungani, S. Socio-economic factors influencing land-use and land-cover changes in the miombo woodlands of the Copperbelt province in Zambia. *Forest Policy Econ.* **2019**, *100*, 75–94. [CrossRef]
37. Skonhoft, A.; Solem, H. Economic growth and land-use changes: The declining amount of wilderness land in Norway. *Ecol. Econ.* **2001**, *37*, 289–301. [CrossRef]
38. Yu, J.; Zhou, K.; Yang, S. Land use efficiency and influencing factors of urban agglomerations in China. *Land Use Policy* **2019**, *88*, 104143. [CrossRef]
39. Ning, Y.; Zhang, B.; Ding, T.; Zhang, M. Analysis of regional decoupling relationship between energy-related CO_2 emission and economic growth in China. *Nat. Hazards* **2017**, *87*, 867–883. [CrossRef]
40. Organization for Economic Co-operation and Development (OECD). Sustainable Development: Indicators to measure Decoupling of Environmental Pressures from Economic Growth. 2002. Available online: http://www.oecd.org/officialdocuments/publicdisplaydocumentpdf/?cote=sg/sd(2002)1/final&doclanguage=en (accessed on 15 October 2002).
41. Tapio, P. Towards a theory of decoupling: Degrees of decoupling in the EU and the case of road traffic in Finland between 1970 and 2001. *Transp. Policy* **2005**, *12*, 137–151. [CrossRef]
42. Kovanda, J.; Hak, T. What are the possibilities for graphical presentation of decoupling? An example of economy-wide material flow indicators in the Czech Republic. *Ecol. Indic.* **2007**, *7*, 123–132. [CrossRef]
43. Song, Y.; Sun, J.; Zhang, M.; Su, B. Using the Tapio-Z decoupling model to evaluate the decoupling status of China's CO2 emissions at provincial level and its dynamic trend. *Struct. Chang. Econ. Dyn.* **2020**, *52*, 120–129. [CrossRef]
44. Li, M.; Hao, J.; Chen, L.; Gu, T.; Chen, A. Decoupling of urban and rural construction land and population change in China at the prefectural level. *Resour. Sci.* **2019**, *41*, 1897–1910. [CrossRef]
45. Mattila, T. Any sustainable decoupling in the Finnish economy? A comparison of the pathways and sensitivities of GDP and ecological footprint 2002–2005. *Ecol. Indic.* **2012**, *16*, 128–134. [CrossRef]
46. Zhao, X.; Pan, Y.; Zhao, B.; He, R.; Liu, S.; Yang, X.; Li, H. Temporal-spatial evolution of the relationship between resource-environment and economic development in China: A method based on decoupling. *Prog. Geogr.* **2011**, *30*, 706–714.
47. Wang, Q.; Zhao, M.; Li, R.; Su, M. Decomposition and decoupling analysis of carbon emissions from economic growth: A comparative study of China and the United States. *J. Clean. Prod.* **2018**, *197*, 178–184. [CrossRef]
48. Song, W. Decoupling cultivated land loss by construction occupation from economic growth in Beijing. *Habitat Int.* **2014**, *43*, 198–205. [CrossRef]
49. Wu, Y.; Tam, V.W.Y.; Shuai, C.; Shen, L.; Zhang, Y.; Liao, S. Decoupling China's economic growth from carbon emissions: Empirical studies from 30 Chinese provinces (2001–2015). *Sci. Total Environ.* **2019**, *656*, 576–588. [CrossRef]
50. Ang, B.W. Decomposition analysis for policymaking in energy: Which is the preferred method? *Energy Policy* **2004**, *32*, 1131–1139. [CrossRef]
51. Ren, S.; Yin, H.; Chen, X.H. Using LMDI to analyze the decoupling of carbon dioxide emissions by China's manufacturing industry. *Environ. Dev.* **2013**, *9*, 61–75. [CrossRef]

52. Liu, S.; Ye, Y.; Lin, Y. Evolution characteristics and decomposition of driving factors on rural residential land based on decoupling theory and LMDI model. *Trans. Chin. Soc. Agric. Eng.* **2019**, *35*, 272–280.
53. Cadavid Restrepo, A.M.; Yang, Y.R.; Hamm, N.A.S.; Gray, D.J.; Barnes, T.S.; Williams, G.M.; Soares, M.R.J.; McManus, D.P.; Guo, D.; Clements, A.C.A. Land cover change during a period of extensive landscape restoration in Ningxia Hui Autonomous Region, China. *Sci. Total Environ.* **2017**, *598*, 669–679. [CrossRef]
54. Ningxia Statistics Bureau. *Ningxia Statistical Yearbooks 2018*; China Statistical Press: Beijing, China, 2018.
55. Liu, M.; Zhang, D.; Min, Q.; Xie, G.; Su, N. The calculation of productivity factor for ecological footprints in China: A methodological note. *Ecol. Indic.* **2014**, *38*, 124–129. [CrossRef]
56. Yang, L.; Yang, Y. Evaluation of eco-efficiency in China from 1978 to 2016: Based on a modified ecological footprint model. *Sci. Total Environ.* **2019**, *662*, 581–590. [CrossRef] [PubMed]
57. Wackernagel, M.; Schulz, N.B.; Deumling, D.; Linares, A.C.; Jenkins, M.; Kapos, V.; Monfreda, C.; Loh, J.; Myers, N.; Norgaard, R. Tracking the ecological overshoot of the human economy. *Proc. Natl Acad. Sci. USA* **2002**, *99*, 9266–9271. [CrossRef] [PubMed]
58. Xie, P.; Gao, S.; Sun, F. An analysis of the decoupling relationship between CO_2 emission in power industry and GDP in China based on LMDI method. *J. Clean. Prod.* **2019**, *211*, 598–606. [CrossRef]
59. Ma, M.; Cai, W. What drives the carbon mitigation in Chinese commercial building sector? Evidence from decomposing an extended Kaya identity. *Sci. Total Environ.* **2018**, *634*, 884–899. [CrossRef] [PubMed]
60. Guo, S.; Wang, Y.; Hou, H.; Wu, C.; Yang, J.; He, W.; Xiang, L. Natural Capital Evolution and Driving Forces in Energy-Rich and Ecologically Fragile Regions: A Case Study of Ningxia Province, China. *Sustainability* **2020**, *12*, 562. [CrossRef]
61. Jin, X.; Li, X.; Zhang, H.; Zhang, N.; Zhe, F. Dynamic evaluation of natural capital in Zhejiang Province based on the three-dimensional ecological footprint extension model. *Chin. J. Ecol.* **2019**, *38*, 2177–2183.

International Journal of
Environmental Research and Public Health

Article

Romanian Students' Environment-Related Routines during COVID-19 Home Confinement: Water, Plastic, and Paper Consumption

Vasile Gherheș *, Mariana Cernicova-Buca *, Marcela Alina Fărcașiu and Adina Palea

Department of Communication and Foreign Languages, Politehnica University of Timisoara, 300006 Timisoara, Romania; marcela.farcasiu@upt.ro (M.A.F.); Adina.palea@upt.ro (A.P.)
* Correspondence: vasile.gherhes@upt.ro (V.G.); mariana.cernicova@upt.ro (M.C.-B.)

Abstract: The disruptive force of the COVID-19 pandemic is lessening in power and plans are being made for the postcrisis period, among which increasing the sustainability of higher education is of significant importance. The study aims at establishing students' existing environment-related routines during their home confinement, as a basis for applying green measures to campus living once academic life is resumed with the physical presence of students. The collected data rely on self-reported information provided by 816 students from Politehnica University of Timisoara (Romania), collected via an online, anonymous survey. The novelty of the approach is that household environment-related routines are investigated during a crisis period, with the possibility to build upon the results to implement tailored measures to encourage or diminish environmentally relevant consumption by young, highly skilled individuals. The students display a moderate awareness of environmental issues and indicate consumption routines that may be steered towards an increased sustainability-conscious campus life, through the combined intervention of the university, city administration, and stakeholder involvement. The findings are used to explore the possible directions for action towards increasing or contributing to the territorial sustainability in the socio-ecological context of Timisoara, the largest university city in the western part of Romania via educational, managerial and policy interventions.

Keywords: student campus; post-COVID-19 higher education; socio-ecological system; territorial sustainability; environmental policy; green university; environment-related routine; water consumption; plastic consumption; paper consumption

Citation: Gherheș, V.; Cernicova-Buca, M.; Fărcașiu, M.A.; Palea, A. Romanian Students' Environment-Related Routines during COVID-19 Home Confinement: Water, Plastic, and Paper Consumption. *Int. J. Environ. Res. Public Health* **2021**, *18*, 8209. https://doi.org/10.3390/ijerph18158209

Academic Editors: Salvador García-Ayllón Veintimilla and Antonio Espín Tomás

Received: 29 June 2021
Accepted: 30 July 2021
Published: 3 August 2021

1. Introduction

Students' daily routines changed dramatically with the outbreak of the COVID-19 pandemic. Instead of commuting to college, meeting with peers, drinking morning and break coffee, bringing along plastic bottles of water or refreshing drinks to school, taking notes on paper and/or handing in home assignments, final projects or exam work on paper, they had to stay indoors, many of them—instead of in campus dormitories—in their parents' homes, drastically limit face-to-face interactions and outdoor activities, adopt social distancing rules, use digitally supported solutions for their school life and other activities, and submit electronic versions of their academic-related work. The impact of the pandemic on students' lives is a topic of great interest, analyzed from a variety of angles, such as the resilience of students and their capacity to continue their professional trajectories [1], the existence of digital infrastructure and the (digital) habits for solving academic and life-related issues, or the students' satisfaction with the support offered by universities [2]. Equally important are research items that investigate the probability that the post-pandemic life is a digitally more inclusive one [3,4] and, at the same time, a more sustainable, environmentally conscious one [5,6], where young people are expected to display a solid range of sustainability competences, from green lifestyles and green economy to circular economy and environmentally friendly mindsets [7,8].

Prior to the COVID-19 pandemic outbreak in 2020, an event that confined to their homes more than 1.37 billion students of all levels of education, according to the statistics provided by UNESCO [9], studies on students' environment-related behaviors dealt with either their behavior on campus [10], at college or as part of their generation, in the household, from multiple angles—as part of a lifestyle [11], as a consciously adopted choice [12], as perceptions related to sustainability actions in green universities [10], up to behaviors and attitudes on campus [13,14]. Universities also attracted their share of attention, mainly in the form of calls to action under the framework of the United Nations Environment Programme for going green and joining the Green Universities Network (UNEP) [15], or as practices developed on regional or local levels [10,16–18]. The health crisis interrupted much of the habitual academic life and forced universities to focus on their capacity to deliver education and reduce the disruptive force of the pandemic on academic life, by accelerating the adoption of digitalization [19], a topic that was viewed mainly from a technological angle and with an impact on educational practice, but that has a spillover effect on the environmental concerns and greening, if the reduction of physical paper is taken under consideration.

If post-COVID-19 higher education is to be oriented towards adopting environmental innovations and steered towards increasing the sustainability commitment, as part of the action plan to increase the resilience of the university as the main organization relevant for preparing a highly skilled labor force and serving as beacon of knowledge, a profound rethinking of the *modus operandi* is necessary [20]. Additionally, it is important to keep in mind that, while universities can be viewed as agents of change and drivers of innovation in the regions where they operate, their activity is not isolated from the socio-economic, administrative, and political context of the territory [21,22], and their large-scale decisions are intertwined with the capabilities and prerequisites existing in that given territory [10,15]. Therefore, future planning at the university level should resonate and influence, wherever possible, the public policies in the geographical space where they operate.

Against this background, the present study aims to investigate the "ground zero" level for a specific university, to be taken into consideration when planning a sustainability plan for the student campus in the post-COVID-19 period, to identify the students' potential to change their environment-related routines, due to EU policy changes and/or due to the university's regulations, and to explore the possible directions for action towards increasing or contributing to the territorial sustainability in the socio-ecological context of Timisoara, the largest university city in the western part of Romania. Home confinement in 2020 and the first half of 2021 kept students away both from the physical academic facilities and, relevant for this study, from the campus. Therefore, the research team selected as topics that students could pinpoint from practice (and not from recollection) only those environment-related elements of consumption that they experienced in everyday life: water, plastic, and paper, leaving aside other elements such as mobility, clothing, or food consumption outside the household. The following research objectives are set for the study:

- (RO_1) To determine the students' water consumption routines (shower and drinking);
- (RO_2) To identify the students' environment-related routines with plastic and paper (avoiding waste generation and reducing plastic and paper consumption);
- (RO_3) To investigate gender-bound preferences in environment-related routines in the household;
- (RO_4) To identify the students' potential to change their environment-related routines, due to EU policy changes and/or due to the university's decision to go green in the post-pandemic period.

2. Literature Review

2.1. University as Part of the City's Socio-Ecological System

The university is a constantly evolving institution, transformative for the generations of students, coming to seek knowledge and paths for their futures, but also capable of transforming itself, due to its tremendous, never-ending capacity to adapt to new levels of knowledge and societal expectations [23]. Despite the well-known metaphor of the "ivory tower" that is often applied to higher education institutions, throughout their history universities have always chosen to have some form of relationship with each other and with the city in which they are located, which led to the town vs. gown debate [24–26]. In the 21st century universities are seen as centers of learning and research, but also as agents of change—catalysts for social and political action towards sustainable development [10,24]. Their contribution to the sustainable development of the territory in which they operate is conducted both internally, by focusing on the curricula taught to students, and by ensuring that sustainability policies govern research, campus life and internal operations, and externally, via the university's performance in the region [10]. There are voices that strongly speak for universities to step up to the role of being forerunners in facilitating the transition to a sustainable future [22]. In his seminal book, *Being a University*, Ronald Barnett bluntly states that "the time is such that the ecological university can be glimpsed, both as an idea and in its institutional form. The time is such also that the world needs its universities to be ecological" [23] (p. 141). However, his idea of ecology goes along the concerns regarding the university's impact on the environment, and towards the philosophy behind the ecological project, an almost utopian enterprise [23,27].

Socio-ecological systems thinking, in which nature and society are understood as coupled and mutually produced, attracts the interest of an even larger audience [28]. Social-ecological systems are defined as "integrated complex adaptive systems in which social and ecological subsystems are coupled and interdependent, each a function of the other, expressed in a series of mutual feedback relationships" [29] (p. 3). The university fits into this interpretation, due to its continuous action toward creating links between generations of highly skilled intellectuals, the labor market, the regional and global arena, the ideal of the knowledge-based economy and the capability of providing models and guidance in promoting new solutions to the concerns of the society at large. On an action plan, many universities all over the world align themselves to the sustainability and the environmental aspect of their scholarship and operations through declarations, projects, and initiatives [15], the greening of universities being a phenomenon specific to the beginning of the 21st century [10], with various levels of success. As Renata Dagiliūtė states, the idea of green universities, so popular in Western practice, is less salient in post-socialist countries. In these countries, although ample transformative processes have taken place, higher education institutions wanting to join the green university model still have to invest massively in campus sustainability "by encompassing the programs of energy saving, waste management, food services, etc., especially behavioral shaping, that reaches students directly and enables them to act sustainably" [10].

In their role as institutions promoting societal change, being confronted with growing expectations regarding compelling contributions to sustainable changes, universities play a role also in educating young adults to contribute to sustainability goals, including the choices they make for their (future) households. Young people acquire "stuff" and organize their lives in a specific social and spatial–temporal context, thus reducing or increasing the environmental burden on their place of residence [30,31]. Especially as students, young adults face the thrills and challenges of becoming independent, "a process in life's journey when practices are altered or become entrenched, for better or worse" [30]. Studies show that educational institutions (and governments) need to work towards reducing the gap between objective and subjective knowledge of environmental issues, so that students can make the right decisions based on their actual knowledge [32]. University-led initiatives toward the greening of institutional and the campus life are perceived as beneficial on multiple levels, such as offering vivid examples of social responsibility, saving money by reducing its resource consumption, or responding to the students' activist spirit through events such as the Fridays for Futures climate strikes [15]. Ultimately, such initiatives contribute not only to demonstrating the university's commitment to sustainability, but also to increasing the sustainability of the geographic territory in which they operate. Out of the many dimensions pertaining to the sustainable consumption, this study focuses only on water, plastic, and paper, as self-reported by students during the disruptive time of COVID-19.

2.2. Water Consumption

Besides the fact that the human body consists of 60–75% water and that a mere 15% loss of that amount can be fatal [33], water is necessary to prepare food, to wash, and to live in safe and healthy conditions.

The United Nation's Renewable Development Goal 6 (SDG 6) aims at providing clean water supply for the world population by 2030 since billions of people are still deprived of clean drinking water. In other words, SDG 6 plans on ensuring the availability and sustainable management of water and sanitation for all people by 2030 in alignment with the fundamental human rights [34].

According to the SDG 6 indicators, in 2017, 2.2 billion people (i.e., 29% of the world's population) did not have access to a safe drinking water source, 4.5 billion people did not have access to safe sanitation services and 3 billion people (i.e., 40% of the world population) did not have the possibility of washing themselves with water and soap at home [34].

Population growth and socio-economic development, as well as changes in consumption habits, have generated water stress. Water consumption has increased globally by approximately 1% a year since 1980, and it is estimated that, by 2050, it is going to increase by almost 20–30% over the current consumption level, with the domestic sector contributing to this trend as well [35].

Although, globally, the agricultural sector is still the biggest freshwater consumer, the statistics show that the water demand in the domestic sector has grown by 600%, at an accelerated pace, during 1960–2014 (Appendix A Figure A1) [36].

The European Union is no stranger to these problems, also being faced with concerns regarding water scarcity. Climate change (drought, less precipitation), pollution, urbanization, overconsumption, and mass tourism are the main contributors to water scarcity. Roughly 88.2% of Europe's freshwater supply comes from rivers and groundwater, whereas the rest comes from reservoirs (10.3%) and lakes (1.5%), all these sources being extremely vulnerable to the above-mentioned threats [37].

In the General Comment Nr. 15, the UN's Committee on Economic, Social and Cultural Rights defines the right to water as "sufficient, safe, acceptable, physically accessible and affordable water for personal and domestic uses". The same comment stipulates the basic human need for water, pointing out that, "for personal and domestic use", resources must include "drinking, personal sanitation, washing of clothes, food preparation, personal and household hygiene". According to the World Health Organization, between 50 and 100 L of water per person, per day, are necessary to ensure basic water needs [38], which raises a serious question about how and if the needs of each individual can be covered at the current status quo.

As far as the Europeans are concerned, the European Union data of 2016, for the period 2014–2015, show that the Italians used approximately 243 L per day, ranking first, whereas Romanians ranked last with only 74 L a day, based on data calculated by the providers of water services (Appendix A Figure A2) [39].

Romania also finds itself among the last countries in the 2018 ranking related to the drinking water consumption from public or individual sources within the EU, with 26 m^3 per capita (Appendix A Figure A3) [40]. At the same time, a decrease in the drinking water use can also be noticed, from 30% in 2012 to 26.4% in 2018 (Appendix A Figure A4) [40].

The reason why water consumption is so low in Romania can be linked to the water price increase as a result of Romania's alignment with the European directives, Romanian service providers being forced to increase their prices as a result of the investments carried out in the water treatment plants and of the works related to the extension of the water networks. The effect of the increase in water prices on household water consumption during 2002–2010 was studied by Ciomoș et al. [41].

Another reason why, in Romania, water consumption is so low could also stem from the percentage of households that are connected to the public water supply. According to the National Institute of Statistics, the population connected to the public water supply system represented 70.9% (13,728,144 people) of the resident population in Romania in 2019 (19,370,448 people), which means that 5,642,304 people do not have access to the public water supply. The lowest percentage of households connected to the public water supply is in the north-east of Romania (50.1%), followed by Oltenia in the south-west (59.3%). The Bucharest–Ilfov region has the highest percentage of the population connected to the water supply (87.7%). At the same time, in 2019, the highest percentage of water (43.4%) was distributed to the population (Appendix A Figure A5) [42].

The reasons behind the different domestic water consumption behaviors are varied (the number of inhabitants, socio-economic factors, water saving devices, etc.) and have been studied by many researchers [43–49], some of them even suggesting social and economic models based on water consumption behaviors.

Corral-Verdugo et al. [43] devised a social model based on Mexicans' water consumption behaviors. The authors noted that the perception that other people did not save water also reduced the other inhabitants' desire to save water and led to increased water consumption; in other words, the citizens replicated others' water consumption behaviors. Another social model, put forth by Gregory and Di Leo [44] and tested on Australian citizens, suggested that the annual water consumption was influenced by awareness of the environment, personal involvement, habits and other demographic characteristics (income, age, education, etc.) that influence water use (e.g., the use of a washing machine). An economic model, developed by Kenney et al. [47] on the inhabitants of Colorado, USA, showed that domestic water consumption depends on several parameters, such as price, weather as well as the restrictions imposed during severe climate conditions, such as drought.

Jorgensen. et al. [48] proposed an integrated socio-economic model, based on the already mentioned studies, which suggested that the demographic parameters, the household's size, and the number of inhabitants have a direct influence on the water consumption. At the same time, the authors developed the Theory of Planned Behavior, stipulating that past behavior toward water consumption, trust in other persons (whether others save water or not) and in institutions, attitudes toward restrictions and the price of water during shortages influence the way in which people consume water.

Narrowing down the investigation to the students' showering routines, British researchers highlighted the impact that administrative measures and persuasive campaigns had on the water consumption in Bristol [50], but the specificity of the local context makes it difficult to replicate the research in a different cultural and administrative context. The gap in researching students' awareness of their water consumption habits, their attitudes toward saving water, or the issue of water management was highlighted in the scientific literature [51].

The above-mentioned studies do not fully succeed in explaining domestic water consumption based on the contextual parameters of the investigated populations. In Romania, just as in the case of electricity [8], there have been no studies to shed light on the domestic water consumption, be it sustainable or not. Such studies would be important since they could provide vital information for understanding a group of people's attitudes towards water consumption in order for the authorities to develop a sustainable managerial plan and also in order to observe the necessity for educational resources on the sustainable water consumption.

Besides all these variables influencing the domestic water consumption, at the beginning of 2020, a new factor entered the equation, i.e., the lockdown following the COVID-19 pandemic. Therefore, because of schools closing and people working from home, water consumption began to increase in the households. In order to better their services for the population and to understand the rules of this new type of water consumption, the Water utility Stadtwerke Karlsruhe (SWKA) carried out an analysis on water consumption pattern changes in the city of Karlsruhe, Germany, which showed that the lockdown changes impact when and how much water people consume [52]. Even if official statistics regarding this topic are practically non-existent, researchers have studied these societal changes in different parts of the world, e.g., England [53,54], Germany [55], Uganda [56], Brazil [57]. In the case of Brazil, for instance, the showering routine increased by 32.4% during the lockdown, in comparison to the pre-COVID-19 period [58]. All these studies show a clear increase in domestic water consumption during lockdown and, thus, on people's incomes, and were undertaken with the purpose of creating forecast models on household water consumption for better water supply during emergency times. Romanian studies on domestic water consumption during lockdown, as well as on its effects on the Romanians' revenues have not been performed so far.

2.3. Paper and Plastic Use

As has already been pointed out, the COVID-19 pandemic has severely impacted the routines of university students and may very well lead to permanent changes and new recommendations worldwide. In Romania, the context prior to the lockdown points to an acknowledgement of man's impact on the environment, as results of surveys published by Statista [59,60] show that Romanians perceive as the main cause of global warming the increasing number of vehicles together with car traffic (31%), with the second most common responses being people's actions (12%) and improper disposal of waste (11%), therefore setting the ground for changes in behavior and legislation (Appendix A Figure A6) [59]. Similarly to other Europeans, people in Romania seem to have a fair understanding of what can and cannot be achieved in the short term and opt to set realistic goals for sustainable development—such as combating plastic waste—rather than "aim of achieving zero pollution" (Appendix A Figure A7) [60].

It is worth noting that, since plastics first became popular more than half a century ago, the annual production has increased significantly. In 1950, the global plastic production amounted to 1.5 million metric tons. In comparison, 359 million metric tons were produced in 2018, of which 61.8 million metric tons were produced in Europe, for the packaging sector. Managing plastic waste is crucial to sustainable environmental planning, and even if Eurostat shows that the recycling rate of municipal waste has been rising constantly [61], it has to be noted that Romania (11.5%) is far behind other European countries such as Austria (58.2%), Slovenia (59.2%), or France (46.3%).

An interesting fact unveiled by the literature review is that the amount of plastic wastes generated worldwide since the outbreak has increased, and it is estimated at 1.6 million tons/day. Most of it was caused by medical needs, with approximately 3.4 billion single-use facemasks/face shields estimated to be discarded daily because of the COVID-19 pandemic, globally [62].

Similar information on paper waste management could not be found, as the majority of articles focus on estimating the changes that the COVID-19 outbreak has imposed on businesses [63,64]. Nevertheless, there is an abundance of data concerning global paper production and use, until 2019. For example, Appendix A Figure A8 [65] highlights the differences in paper consumption per region. The numbers are directly related to the current life habits of the inhabitants of the regions and are consistent with the findings of other research (Appendix A Figure A9) [65] which show that 55% of paper consumption is generated by wrapping and packaging, and 26% by printing and writing.

Taking into account the fact that most paper is produced from forestry products, usually trees, and that paper production is one of the largest polluters to air, water, and land, it becomes obvious why it is important to find solutions and develop environmental policies that will encourage the sustainable consumption of paper. These concerns, along with cost savings, are among the reasons for the shift towards going paperless, a trend in accordance with the needs of a globalized and digitalized world. On the other hand, cardboard is considered by many to be a comparatively renewable and sustainable option compared to plastic packaging, which is why we are witnessing a steady increase in the production of packaging paper and board.

However, if this paper were to be recycled, the positive impact on the environment would be enormous [66] because recycled paper production results in 40% fewer greenhouse gases, recycled paper requires 26% less energy to produce, and the production of the recycled paper creates 43% less wastewater.

One positive aspect of the COVID-19 home confinement could reside in the way educational institutions were obliged to adapt to a digital environment. For example, if a major constraint in reducing paper consumption was the fact that some lecturers provided hard-copy course documents and rejected electronically submitted assignments, preferring instead to have assignments submitted in the form of double-spaced or single-sided hard copies [67], the pandemic forced academics to adopt new recommendations and accept the context.

With increased awareness of these impacts and the rise and prevalence of digital technologies, there have been increased efforts among higher education institutions to go paperless [68]. While the paper industry works to minimize the impact of paper manufacturing, reducing human paper use from a consumer's standpoint has its own set of unique challenges. Universities have worked to develop policies and incentives to decrease on-campus paper consumption including initiating printing quotas, restricting printing access for students, and requiring double-sided printing [69].

Even more, environmental sustainability has become a priority for many tertiary institutions and different initiatives have been formulated through declarations to foster sustainable development guidelines on how to incorporate sustainability into the university system [18,70].

3. Materials and Methods

3.1. Local Context

Timisoara, the third largest city in Romania, became a university city in 1920, with the foundation of the (nowadays) Politehnica University. Currently, it is the educational choice for around 45,000 students, attending one of the four public universities or two private ones. Out of the total number of students, almost 40% are locals, and the rest come to the city during the academic year, which creates an increase in the number of residents of 10% from September through June, with short winter and spring breaks. The four public universities offer between them, institutionally, 13,000 places on the student campus (half of which belong to Politehnica University) (Figure 1), with the rest of the incoming students privately renting rooms or apartments in the city.

The outbreak of the COVID-19 pandemic in March 2020 sent the student population home and practically closed the campus, situated close to the city downtown area. According to local media, the local economy felt a loss of EUR 150 million due to the absence of students during the state of emergency and state of alert periods (2020—first half of 2021) [71]. The empty student campus, with almost zero consumption of municipal utilities and services, also impacted the overall administrative planning and activity during the period, this unprecedented crisis highlighting the ties between the city and the campus. On the other hand, even during the pandemic, Timisoara City Hall continued implementing, at the initiative of university-led projects, environment-related processes fitting into the European Green Deal [72], thus signaling its interest in promoting an active environmental policy for the development of the city and offering solutions to increase the sustainability of households and institutional residences. In its projections regarding the economic, social and environment planning for the city, Timisoara City Hall explicitly refers to UN Agenda

2030 for Sustainable Development and commits itself to take steps towards implementing SDG goals [73].

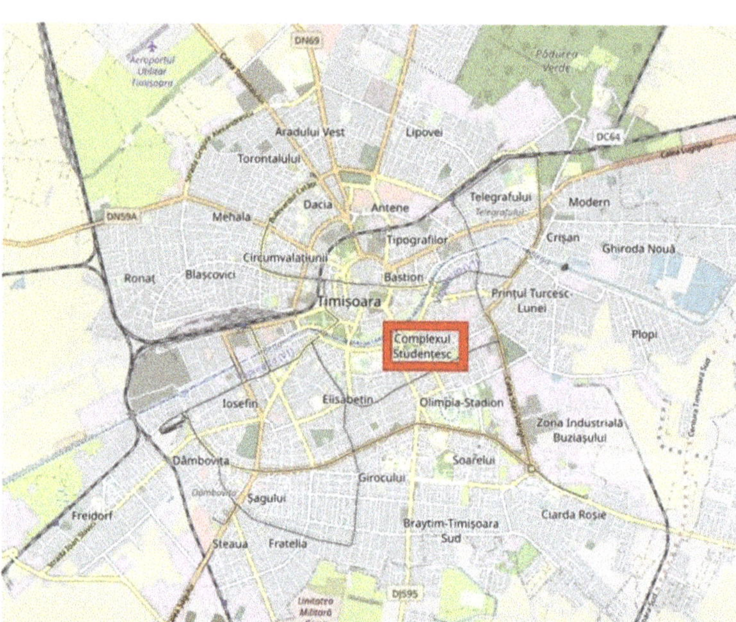

Figure 1. Map of Timisoara, with the student campus highlighted in the box.

3.2. Sample

The study is a snapshot of students' environment-related routines during the home confinement caused by the measures undertaken at state and local levels to contain the COVID-19 pandemic. No difference was made between students who live with their parents during their studies, those who resided on student campus in the pre-COVID-19 period or those who privately rent a residence during their studies. The survey was distributed to students of Politehnica University of Timisoara, the sample being one of convenience. In total, 816 subjects took part in the survey from all the study years. As the university's student body counts around 13,000 students, the calculated margin of error is of $\pm 3.3\%$. Their average age, according to the recorded results, is 20.37 years old. Prior studies indicate that gender is a significant factor influencing environment-related choices in students [32]. The research team decided to investigate whether the gender variable is relevant for the Romanian case, and analyzed the sample through these lenses, with 409 female and 407 male subjects. Participation in the study was optional. No incentives were used to stimulate participation and students could opt out of the study at any time of the data collection. Additionally, no personal data were collected, to ensure students' privacy and anonymous participation. The data were collected between March and April 2021.

3.3. Questionnaire

To collect the data, a non-standardized questionnaire was used, its content being validated through the following steps: assessment by experts (sociologists), followed by its qualitative and quantitative pretesting. To build the questionnaire, the everyday sustainable water, plastic, and paper consumption was referenced on specialized websites. Thus, a list of recommendations aimed at adopting sustainable behavior regarding the above-mentioned aspects was made. Finally, they were transformed into 5-point Likert-type scale questions that were included in the questionnaire.

The questionnaire was written in Romanian and was completed in 15 minutes on average. The statements were phrased in a neutral language, to collect students' reflections on the matter without indicating the social desirability of environment-related routines, which could affect the reliability of the data. Additionally, the research team built the questionnaire having in mind the aim of reducing the social desirability biases, although the subjective character of responses is acknowledged [74].

Out of the three sustainable household elements of capability: (1) household practices (e.g., recycling and water conservation); (2) household structure (e.g., income, employment, dwelling type and composition); (3) household sustainability judgements (e.g., knowledge, awareness, concern towards climate change) [75], the research team focused only on the first one. For the purpose of the study, preeminence was given to the students' environment-related routines, understood as familiar action patterns that involve regularity [76,77], and which are likely to be performed on a daily basis in the home confinement period caused by the COVID-19 pandemic.

3.4. Method

The instrument used for the data collection was the anonymous, self-administered online survey, posted on the Isondaje.ro platform (a Romanian free online survey service). The data were analyzed using SPSS (IBM SPSS Statistics variant 25).

At the end of their online classes, the students received from their teachers the link to the online questionnaire and the details needed to fill in the form. The average time required to answer the questionnaire was 15 min and the recorded response rate was of approximately 50%.

The analysis of the corpus of data relies on the self-reported information provided by the respondents and due to the variety of situations in the students' housing arrangements (living alone, with peers, parents or grandparents, renting, in Timisoara or in other cities, etc.), they cannot be aggregated with actual data of their consumption of water, plastic, and paper, such a study being possible, at best, in the post-COVID 19 period, with an investigation of such parameters related to student campus life. To verify the gender-related difference in the obtained responses a chi-square Pearson test was applied.

4. Results

A first point of interest of the study was finding out the level of awareness of the respondents regarding environmental protection. As can be noted in Figure 2, the highest percentages were recorded for "to a moderate extent" variant that was chosen by a little over half of the respondents (52.7%) and for "to a high extent" variant (31.4%). The percentage of those who consider themselves informed about environmental protection "to a small extent" and "to a very small extent" cumulates to 10.5%.

The next aspect researched by the study was determining the water consumption behaviors among students. Five sets of statements were created in the questionnaire, collected from specialized websites. For a better view and interpretation, the results were grouped according to the "often" and "always" as well as "rarely" and "never" response variants. The results can be seen in the figure below (Figure 3).

The statement that accumulated the highest percentage of responses and which shows that a sustainable water consumption behavior exists in households was "You repair the sink faucet if it leaks". The "often" and "always" variants in this case cumulated 72.4%. "You do not turn on the washing machine or the dishwasher until it is fully loaded" cumulated a total of 70.3%. Although the statement "You let the water run while brushing your teeth" recorded a high percentage for the "rarely" and "never" variants (60.9%), they actually represent a positive and sustainable behavior regarding water consumption. The last variant is "Don't know/Don't want to answer" (DNK/DWA).

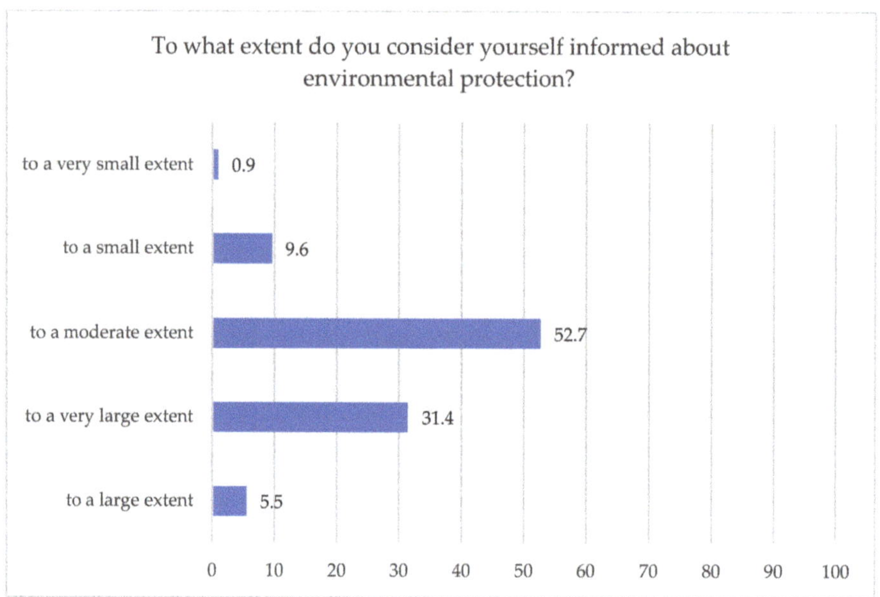

Figure 2. Respondents' awareness of environmental protection.

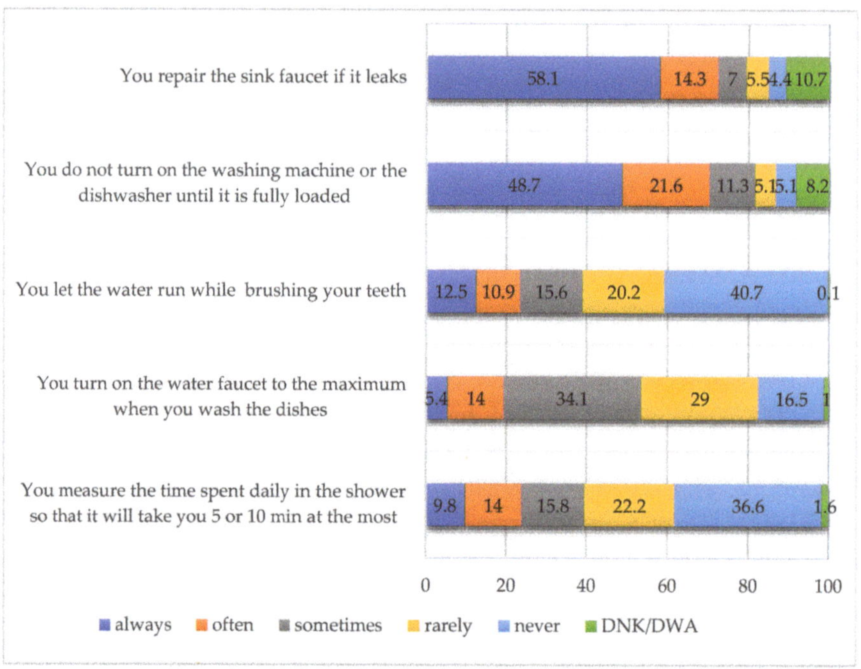

Figure 3. Respondents' behavior towards water consumption in the household.

The same situation is recorded for "You turn on the water faucet to the maximum when you wash the dishes" where 45.6% of the responses suggest the absence of such a behavior. Still, in this case, this statement has recorded the highest percentage of responses for the "sometimes" variant, which leads to the presumption that the respondents' behavior toward this household activity fluctuates.

An example of behavior that reveals the lack of awareness towards high water consumption has to do with personal hygiene and daily showers. "You measure the time spent daily in the shower so that it will take you 5 or 10 min at the most" recorded the highest percentages for the "rarely" and "never" variants, which together accumulated to 58.8%. In other words, it is likely that the level of awareness regarding the water waste in the household might be rather low.

A secondary analysis was carried out to verify whether there are significant differences between the female and male respondents. After applying the χ^2 test, the results show that there are differences between males and females only in the case of the research statement "You measure the time spent daily in the shower so that it will take you 5 or 10 min at the most". For this statement a value of $\chi^2 = 28.615$ and a value of $p = 0.00$ ($p < 0.05$) were recorded. The results pinpoint the fact that there are notable differences based on gender, namely, that this type of behavior is more specific to females than to males.

Another objective of this study was to capture the respondents' behaviors regarding the use of plastic, and thus five statements referring to this aspect were included in the questionnaire. As can be seen in the figure below (Figure 4), the statement that accumulated the highest percentages for the "often" and "always" variants was "You use a mug/glass and not disposable glasses when you drink". They scored 16.9% and 72.9%, respectively. Taking these results into account, it can be concluded that the respondents do not use plastic containers.

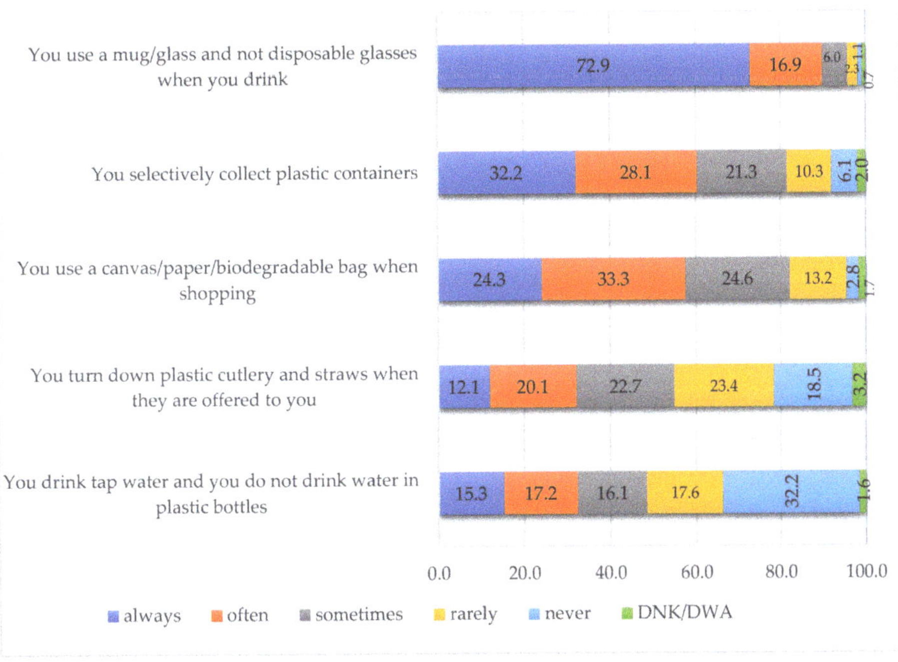

Figure 4. Respondents' behavior towards plastic consumption in the household.

This statement is followed at quite a distance, as far as the percentages are concerned, by "You selectively collect plastic containers", almost a third of the respondents declaring that they do this "often" (28.1%) and "always" (32.2%). By looking at the results obtained from those who answered "never" and "rarely" to the question, it can be observed that their percentage is relatively low, of 6.2% and 10.3%, respectively. "You use a canvas/paper/biodegradable bag when shopping" obtained almost a quarter of the responses for "always" (24.3%) and almost a third for "often" (33.3%). Therefore, as for the previous statements, this behavior can be found in more than half of the respondents.

At the other extreme, there are two statements for which the percentages obtained rather indicate the existence of a behavior of acceptance and use of plastic and not its refusal. "You turn down plastic cutlery and straws when they are offered to you" obtained a total score of 41.9% by cumulating the "rarely" and "never" response variants. This is the share of the respondents who accept their use. The last statement included in the study highlights again an acceptance of the use of plastic in everyday life. "You drink tap water and you do not drink water in plastic bottles" is the statement for which the highest values were recorded for the "never" variant (32.2%). The scores for the "rarely" variant, which obtained 17.6%, can be added to these values.

These statements were also checked to see whether significant differences between males and females exist. Following the application of the χ^2 test, no significant differences related to the respondents' gender were recorded. By analyzing the results obtained from the association tables, small differences in "You turn down plastic cutlery and straws when they are offered to you" can be noticed: female respondents accounted for 21.8% responses for "often" variant, while male respondents chose the same option only in 18.4% cases. For the "always" variant, female respondents chose the option in a proportion of 13.4% by comparison to male respondents with 10.8%.

The last objective of this study was to identify those behaviors that can lead to the avoidance of waste generation and to paper consumption reduction. As mentioned in the introductory part of this article, paper is one of the most common wastes in almost all fields of activity. Efficient paper management was captured in this study through the seven statements in the questionnaire targeting this behavior. To better view the results below, they were presented hierarchically, based on the results obtained for the "always" and "often" variants. As can be seen in the figure below (Figure 5), by accumulating the percentages for the above-mentioned variants, percentages higher than 50% were obtained for 3 of the 7 statements. Although "You print your e-mails" obtained the highest percentages for the variants situated at the other extreme, this result places it among the statements that lead to the idea of the existence of a behavior of avoidance of waste generation and of waste consumption reduction. By cumulating the results received for the "never" and "rarely" variants, a total of 83.7% is obtained, which confirms what was stated above. In the order of the percentages recorded for the variants that denote the existence of an ecological behavior, "You use cloth napkins in the kitchen to reduce the use of the paper ones" is the statement that follows, for which 63.8% of the respondents chose the "always" and "often" variants (31.7% and 32.1%, respectively). The same happens for "You selectively collect paper whenever possible", which accumulated values that exceeded 50% for the "always" and "often" variants (57.8%), being followed by "You reduce paper consumption by choosing to pay your bills online", with a total of 53.9%.

"You print on both sides of a single A4 paper" scored exactly 50%. Lower percentages were obtained for "You set the printer to automatically print on both sides of the paper and in black and white" and "You borrow books from the library anytime you can in order not to buy them", where the behavior of avoiding waste generation and of reducing paper consumption seems to be lacking.

These statements were also subjected to a second analysis to see whether there are significant differences between the male and the female population. Following the application of the χ^2 test, differences between the two genders were identified for "You borrow books from the library anytime you can in order not to buy them", which scored a value of $\chi^2 = 35.182$ and a value of $p = 0.00$ ($p < 0.05$). Therefore, these results show that there are significant differences related to gender in this case as this behavior seems to be more present in males and less present in females.

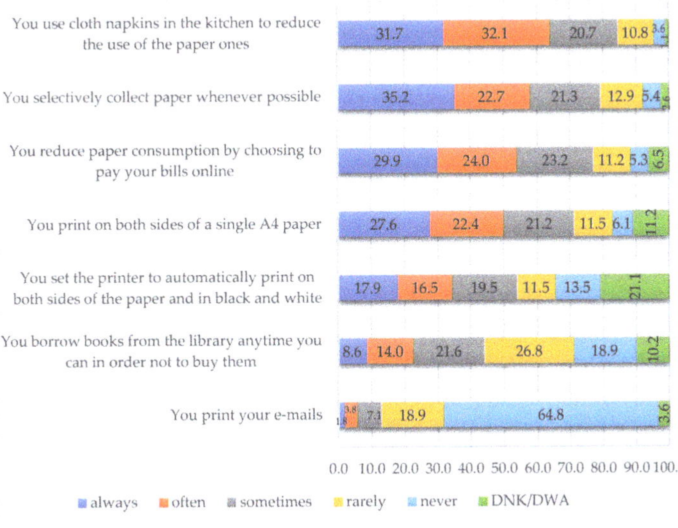

Figure 5. Respondents' behavior towards paper consumption in the household.

5. Discussion

The survey enabled an evaluation of the environmental awareness of students at a given moment in time, in 2021, when home confinement due to measures undertaken to contain the COVID-19 pandemic limited their mobility and face to face interactions outside their households. The findings indicate a general inclination of respondents of practicing routines that spare resources (water and paper), control and reduce pollution (paper and plastic waste), resonating with those highlighted by Carducci et al. [78], who identified students as displaying sensitivity towards environmental issues. These authors also proposed that during the development of interventions to promote pro-environmental behaviors, the target population (i.e., students) should be studied including the health-related aspects, determinants, and obstacles in overcoming the gap between attitudes and behaviors in sustainability projects [78].

Research has shown that individual choices and lifestyles have a spillover effect and ultimately impact the environment, the natural resources, and the quality of (urban) living [11]. However, the complex socio-ecological system in which an individual evolves represents to a considerable extent the 'sustenance base' of their daily domestic (and work) routines [11]. The physical substratum to the domestic life is of consequence and in the case of environment-oriented individuals, striving for more sustainable lifestyles and patterns of domestic consumption, the possibilities existing in the socio-ecological systems are of strategic importance. Therefore, universities wanting to adopt a greening strategy for campus life need to understand the patterns of consumption specific for the generation that, after a period of home confinement, will return to the campus life and rejuvenates, at the same time, the city. The ecological modernization of production—-consumption cycles will be easier to implement after the disruptive times of the COVID-19 pandemic and should aggregate the individuals affected by change (students), the university as a transformative institution, and also the city administration, as a provider of services that shape the domestic/living routines. Such environmental innovations can be accomplished via new techniques, but also via procedures, financial arrangements, etc. [11].

A sustainability-oriented public policy, such as the one declared by the Timisoara City Hall [72], should take into consideration not only the institutional efforts, but also the individual contributions to a better quality of life, a healthier environment, and a more prosperous economy. For this purpose, both at the individual and societal level, changes need to be made in the way resources are used, for their sustainable use and for the minimization of the environmental impact. The university campus can have significant effects on the environment both in terms of the resources it consumes and in terms of the problems it generates (pollution, waste, etc.). It should be the place where students, who represent an educated category of the population, can absorb the values related to sustainable development and can become the promoters of ecological behaviors both within their own households and at their future workplaces. It is a real laboratory where they can learn about sustainability, where models and opportunities can be created to change students' behaviors, with the purpose of embracing values that can lead to societal transformation [15].

As mentioned above, this research does not aim at explaining students' behaviors or at providing information about their triggers, but it is rather an assessment of these habits, capturing the reality during the COVID-19 pandemic. As D'Alessandro et al. rightfully state, "the ongoing pandemic of COVID-19 is a strong reminder that the lockdown period has changed the way that people and communities live, work, and interact" [79], and both outdoor and indoor spaces need to be re-examined from the point of view of their resilience and capacity to ensure the wellbeing of citizens. In the case of Romania, the absence of such studies at a local, regional, and even national level and the lack of information regarding the way water, plastic and paper are consumed are solid arguments in favor of the need for this analysis. Even though the study was carried out on a distinct category of respondents, the students of Politehnica University of Timisoara, these results are useful in planning policies that could lead to the adoption of sustainable behaviors both by this category of consumers and by other categories of the population as well. They are also necessary to carry out campaigns meant to inform the population about sustainable consumption, which cannot be conducted without being first aware of these habits. Previous research warns that tailoring information and persuasion campaigns can prove not to be cost-effective in many cases and that at different levels of aggregation (such as the community, municipal level, regional, or even country levels) a thorough knowledge of the differentials in the target population is necessary [80].

One aspect that can be drawn from the results of the study is that awareness campaigns should be conducted among students regarding general environmental problems, since about half of them say that they are informed about them "to a moderate extent" and about a third of them consider themselves informed of such problems to a large extent and "to a very large extent".

As far as water consumption is concerned, for more than half of the respondents, habits that show the existence of sustainable consumption behaviors, such as "You repair the sink faucet if it leaks", "You do not turn on the washing machine or the dishwasher until it is fully loaded" and "You do not let the water run while brushing your teeth", have been identified.

Another aspect highlighted by the study is the fact that actions should be taken primarily to raise awareness of the high water consumption in daily showers, a situation in which more than half of the respondents said that they do not "measure the time spent daily in the shower so that it will take them 5 or 10 min at the most" and in kitchen activities when they "turn on the water faucet to the maximum when they wash the dishes". The secondary analysis carried out on the corpus has also pinpointed a notable difference based on gender, i.e., a certain type of water consumption behavior is more specific to females than to males, for the above statement, resonating with findings in previous research [32]. However, although the presence of sustainable consumption behaviors has been identified for more than half of the respondents, there is still a significant share of the student population that still needs to be made aware of these problems and of these behaviors. Additionally, the fact that the results resonate with the findings of Vicente-Molina that gender roles might be decreasing in importance in some environmental tasks, especially in the student population, is an issue that requires an in-depth analysis [32].

Awareness campaigns for sustainable water consumption could be carried out by involving student organizations, which can act as "pipelines" through which information could be passed on to the students. These campaigns could also be carried out by involving the university management to identify some technical measures to optimize water consumption on campus (water faucet timers, economical shower heads, dual toilet tanks, etc.) and the dormitory management to help implement these solutions [50]. Another possibility is to introduce disciplines or new chapters in the existing curricula that target sustainable development. Finally, environmental associations, water suppliers and local authorities could also be involved; through mutual efforts, they could create a snowball effect with significant results in embracing sustainable water consumption behaviors. Studies show that when pro-active measures are undertaken, students respond positively, reflect on the possibility of adopting sustainable patterns of resource consumption and feel empowered to act responsively on campus [81].

Reducing the negative impact of plastic on human health and on the environment can be achieved by limiting its everyday use through reuse, or selective collection with the purpose of recycling. Analyzing students' environment-related routines has revealed interesting aspects regarding plastic consumption, which may indicate that further efforts should be made to educate students on how to reduce their carbon footprint. For example, the results showed that the majority of students selectively collect plastic containers (60.3%) and use "a canvas/paper/biodegradable bag when shopping" (57.6%), but 41.9% of them do not "turn down plastic cutlery and straws when they are offered" to them. Behavioral differences based on gender were merely evident, with green-oriented behavior being slightly more present in female respondents than in male respondents.

As mentioned in the introductory part of this study, paper is one of the most common wastes in almost all fields of activity, therefore understanding students' routines was key to evaluating the needs of future in-campus policies. The results lead to the idea of the existence of a behavior of avoidance of waste generation and of waste consumption reduction. A cumulated percent of 83.7% of students "never" or "rarely" print their emails, 57.8% of them "selectively collect paper whenever possible", and 53.9% "reduce paper consumption by choosing to pay your bills online". On the other hand, printing on both sides of the paper and in black and white is not yet the prevalent behavior. The research also identified a significant gender-based difference in the case of the statement "you borrow books from the library anytime you can in order not to buy them", this behavior being more present in males than in female students.

The sustainable household capability is an achievement, the outcome of a specific social and cultural context that offers sufficient possibilities to reduce emissions, as Gordon Waitt et al. convincingly demonstrate [75]. According to these authors, "some households may accept the environmental science that frames sustainability issues, including climate change, but their everyday social contexts, habits and routines may render pro-sustainability actions as unthinkable" [75]. As D'Alessandro et al. also convincingly point out, to increase the sustainability and resilience of households it is necessary to "pool the knowledge from the technical field and Public Health expertise", to identify exportable and scalable best practices and engage multisectoral responsibility, in order to overcome the vulnerabilities highlighted by the COVID-19 pandemic [79]. In the case of a university campus, as this study proposes, there are sufficient tiers of interventions that the university, jointly with the municipal administration, can undertake to encourage more vigorous pro-environmental behaviors, to increase the sustainability of the campus and, due to the spillover effect, of the administrative territory where the campus is operated.

The COVID-19 pandemic has obliged students and teachers to adapt to a virtual environment and find solutions to continue the educational process under new conditions. Therefore, with respect to environmental protection and sustainable development, the quick changes that had to be implemented can be considered a big step forward and universities could incorporate efforts toward resource preservation and the encouragement of green behavior as part of their recovery plan from the disruptive times of the pandemic, responding in a pro-active manner to the calls for a greener, more sustainable higher education [5–7]. Such actions can be implemented by all European universities, along the path already opened by the Green University model [15], with the cautious selection of measures that Renata Dagiliūtė et al. [10] stated for the post-socialist bloc that includes Romania. The present study focuses on a university that is perceived by its students as "trustworthy, effective, modern", and thus capable of leading the way for implementing innovation [82]. It is, therefore, to be expected that a greening action would be welcomed by students. Since students represent a quarter of all decision-making bodies in the university, their informed action is a precondition for all initiatives, environment-related ones included.

6. Limitations and Future Extension of the Study

The main limitation of this study acknowledged by the research team is that the results cannot be extrapolated to the whole student population in Timisoara, to other university students in Romania or to other geographical contexts. Researchers have shown that environmental choices are culture and context-bound [22,32]. These findings also need to be compared with data on students with a variety of contextual and situational factors (e.g., students who are working and studying simultaneously, who live with their parents or rent independently, outside the university campus, or chose to stay on campus for their studies and work). Such factors were not relevant during the COVID-19 pandemic, with students being practically home-confined for more than a year, but they need to be accounted for in future studies. Another limitation is that the study is based on self-reporting, and it would be useful to replicate the current research with objective behavior measures and aggregate the findings with data from relevant authorities (campus administration and/or municipal services). The broader survey methodology literature [32] highlights that inaccuracy in the results may stem from the subjective interpretation of the respondents, who attribute different meanings to the issues raised by the questionnaire or who, under the influence of the topic of the survey, attempt to overreport their pro-environmental behavior. Additionally, the research should be expanded with more survey instruments to model latent factors and measure variables such as motivations/value orientations, and environmental concern, which other researchers have considered to be influential.

The initial results are being used to help develop targeted strategies for minimizing water, plastic and paper consumption on campus and will be incorporated into the university's zero waste management plan that is currently under development. The effectiveness of such measures, the practical aspects of implementation and the resonance of this university-led initiative are also of research interest, especially since they are correlated with the larger context of the socio-ecological urban system and with the efforts undertaken by administrative bodies to increase Timisoara's sustainable feature.

7. Conclusions

The findings in the present study have theoretical, practical and managerial consequences, as well as policy implications. From a theoretical point of view, this research contributes to the body of knowledge regarding young people's behavior during emergency times when many of the habitual choices are canceled. While research on crises is extensive and well-documented, the COVID-19 pandemic puts its specific imprint and requires distinctive attention, with effects still to be evaluated when the health issued are solved.

Research on routines related to sustainable consumption should help practitioners understand how to motivate individuals (in this case—students, young, highly skilled adults) to engage in more sustainable consumption. From the perspective of institutions (universities) and administrative bodies, being confronted with sustainability expectations and/or trying to incorporate sustainability arguments for shaping more resilient policies, it is important to know which foci regarding the sustainability dimensions [12] resonate with the various segments of population, the student body in the case of Timisoara representing

10% of the city's residents. Insufficient motivation, environmental knowledge, and the lack of physical availability of sustainable solutions might be detrimental to pro-environmental behaviors [32]. Educational institutions, local administrations and governments should attempt to reduce the gap between objective and subjective knowledge of environmental issues, so that students can make informed decisions based on their actual knowledge and practice and feel empowered to act responsively on and off-campus.

Therefore, universities wanting to adopt a greening strategy for campus life need to focus on developing awareness campaigns, which, in accordance with the findings of other studies [10,13,50,51,68,78], is crucial to the success of implementing sustainable consumption habits. Universities alone cannot implement extensive greening measures without coordinating their efforts with the utility services providers (water services, waste services, etc.) and, further, with the authorities in charge of organizing such services. At the specific level of Timisoara, which has been analyzed in this study, the political will is present as expressed in the city administration's program that makes reference to the European Green Deal. The appropriate policy measures, however, still have to be undertaken. The timing is of essence and the disruption brought about by the COVID-19 pandemic can become the needed opportunity for universities, as transformative institutions, to be vectors of change.

Author Contributions: Conceptualization, V.G., M.C.-B., M.A.F., A.P.; methodology, V.G., M.C.-B.; software, V.G., M.C.-B.; validation, V.G., M.C.-B.; formal analysis, V.G., M.C.-B., M.A.F., A.P.; investigation, V.G., M.C.-B., M.A.F., A.P.; resources, V.G., M.C.-B., M.A.F., A.P.; writing—original draft preparation, V.G., M.C.-B., M.A.F., A.P.; writing—review and editing, V.G., M.C.-B., M.A.F., A.P. All authors have read and agreed to the published version of the manuscript.

Funding: This research received no external funding.

Institutional Review Board Statement: Not applicable.

Informed Consent Statement: Not applicable.

Data Availability Statement: The data presented in this study are available on request from the corresponding author.

Conflicts of Interest: The authors declare no conflict of interest.

Appendix A

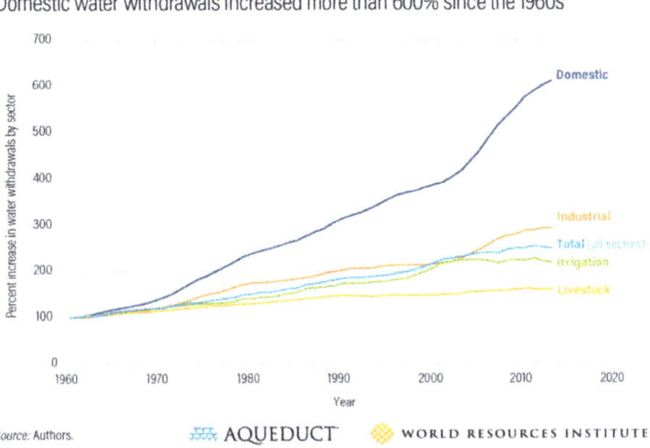

Figure A1. Global water use increase, 1960–2014. Source: https://www.wri.org/insights/domestic-water-use-grew-600-over-past-50 years (accessed on 17 June 2021).

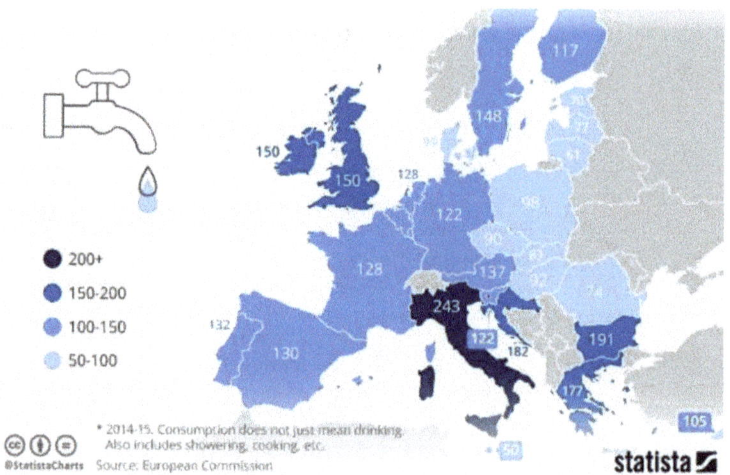

Figure A2. Average consumption of tap water per person in the EU, 2014–2015 (liters/day). Source: https://www.statista.com/chart/19591/average-consumption-of-tap-water-per-person-in-the-eu/ (accessed on 19 June 2021).

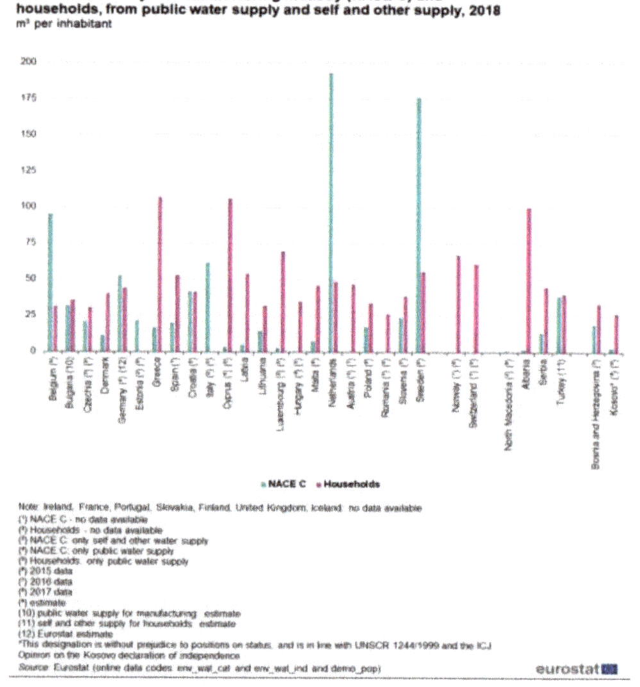

Figure A3. Drinking water consumption in the EU in 2018, m^3 per inhabitant. Source: Water statistics—Statistics Explained (europa.eu) (accessed on 20 May 2021).

Household water use from public water supply, 2008-2018

(m^3 per inhabitant)

	2008	2009	2010	2011	2012	2013	2014	2015	2016	2017	2018
Belgium	13.7	13.6			31.7	32.4	33.6	31.8			
Bulgaria	36.1	36.3	35.6	36.1	37.0	35.8	34.6	35.9	36.0	36.2	35.9
Czechia	32.4	31.9	30.9	31.0	30.8	30.2	30.1	30.2	30.5	30.7	30.9
Denmark (1)	42.6	42.6	42.3	42.4	42.5	43.1	40.5	37.4	37.2	39.5	40.6
Germany (2)			43.7	44.4	44.2	44.0	44.4	44.7	44.7	44.3	
Estonia											
Ireland											
Greece (4)	35.7			91.8	92.1	92.8	93.4	94.0	106.5	106.7	107.0
Spain	61.8	61.7	68.9	54.7	53.6	51.8	52.2	52.8	53.5		
France											
Croatia	42.6	42.6	44.0	42.6	43.1	45.5	40.6	42.5	41.6	41.0	41.6
Italy											
Cyprus	76.8	85.0	96.8	93.1	88.7	90.3	90.5	94.3	103.8	107.4	
Latvia	42.2	38.4	37.0	39.1	37.5	37.3	36.1	35.8	40.1	47.3	49.9
Lithuania	20.3	20.4	18.7	19.0	19.3	21.9	23.1	23.5	24.3	24.7	26.2
Luxembourg								74.9			
Hungary	36.0	35.9	34.1	34.0	34.4	33.4	33.0	34.0	34.1	34.8	35.0
Malta (3)	44.5	39.8	41.3	42.0	44.1	44.0	43.6	42.7	43.3	42.7	41.1
Netherlands	48.1	47.8	47.4	46.9	46.8	46.8	46.5	47.0	47.4	45.8	48.7
Austria (5)	45.7		45.6				44.9		43.6		
Poland	31.8	31.3	31.5	31.6	31.5	31.3	31.5	32.5	32.6	32.2	33.7
Portugal	55.7	58.6									
Romania					30.0	29.3	24.8	25.1	25.4	26.3	26.4
Slovenia (6)	40.3	39.9	39.6	39.5	41.3	38.2	38.1	38.1	38.0	38.2	38.4
Slovakia											
Finland											
Sweden			52.5					50.1			
Norway (7)	77.7	77.6						66.1	64.5	64.5	67.3
Switzerland	79.3	74.4	71.9	70.3	68.4	65.9	63.4	63.9	62.1	61.3	61.3
United Kingdom				46.0							
Albania										98.4	71.6
Serbia	47.2	46.4	45.2	44.1	44.8	45.1	43.2	44.6	43.4	45.0	45.2
Turkey	30.5		32.8		34.9		32.6		39.1		39.5
Bosnia and Herzegowina	29.8	30.3	30.6	29.9	30.2	28.5	28.1	28.8	29.6	30.3	28.1
Kosovo*			19.0	23.7	24.4	24.3	26.0	28.1	29.5	27.6	26.9

(:) not available

(1) Data for 2008 and 2009: Eurostat estimate; 2010: break in series
(2) Data for 2011 : estimated; data for 2017: Eurostat estimate
(3) Data estimated
(4) Data for 2007 instead of 2008
(5) Data for 2008 and 2014: Eurostat estimate
(6) Data for 2008 - 2011: Eurostat estimate
(7) Data for 2008 and 2009: estimated
* This designation is without prejudice to positions on status, and is in line with UNSCR 1244/1999 and the ICJ Opinion on the Kosovo declaration of independence

Source: Eurostat (online data codes: env_wat_cat and demo_pjan)

eurostat

Figure A4. Household water use in the EU, 2008–2018, m^3/inhabitant. Source: Water statistics—Statistics Explained (europa.eu) (accessed on 20 May 2021).

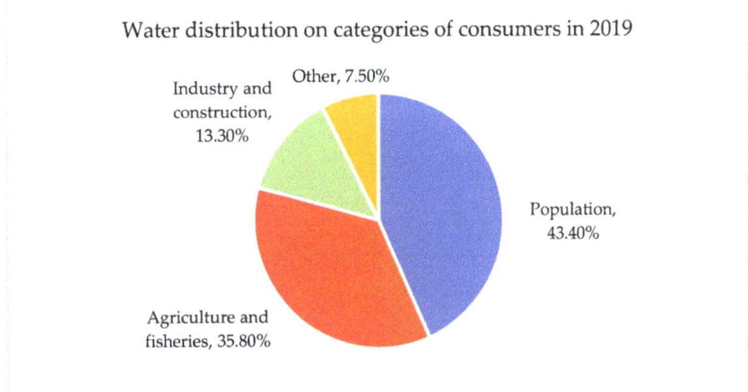

Figure A5. Water distribution based on categories of consumers in 2019, Romania. Source: https://insse.ro/cms/sites/default/files/com_presa/com_pdf/distributia_apei20r.pdf (accessed on 27 May 2021).

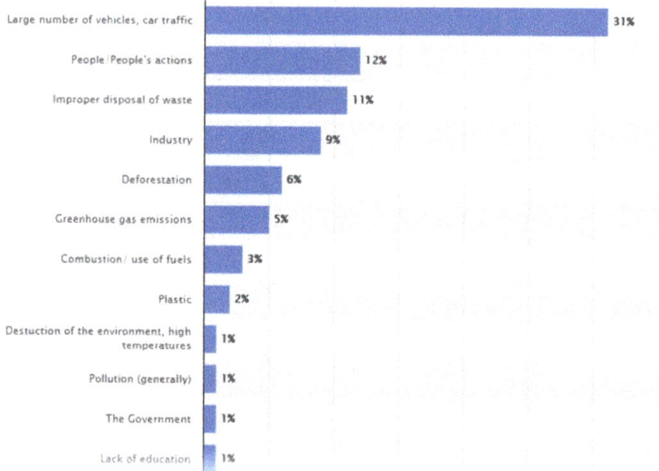

Figure A6. Main cause of global warming as perceived in Romania, 2019. Source: https://www.statista.com/statistics/1093956/perceived-causes-of-global-warming-romania/ (accessed on 8 June 2021).

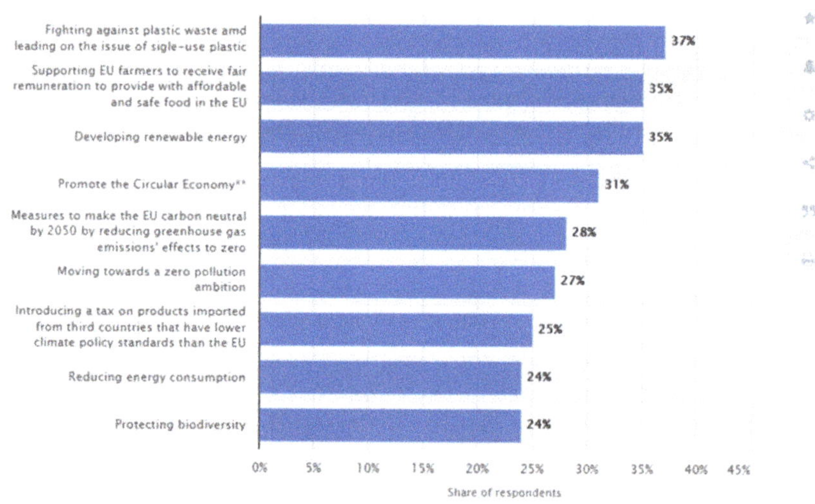

Figure A7. Romania: public opinion on the objectives in a European Green Deal 2019. Source: https://www.statista.com/statistics/1110123/european-green-deal-objectives-romania/ (accessed on 8 June 2021).

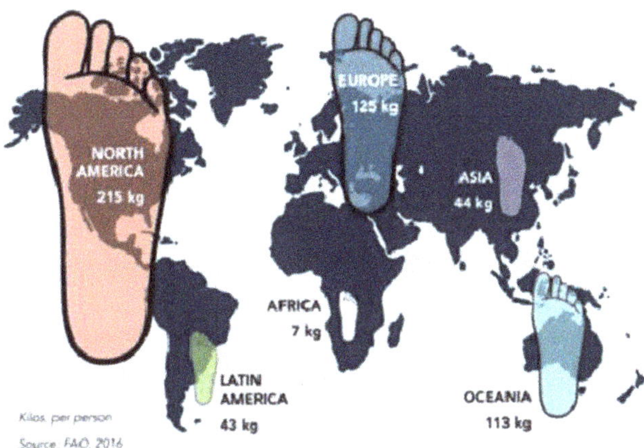

Figure A8. Per capita paper consumption, by region. Source: https://environmentalpaper.org/wp-content/uploads/2018/04/StateOfTheGlobalPaperIndustry2018_FullReport-Final-1.pdf (accessed on 8 June 2021).

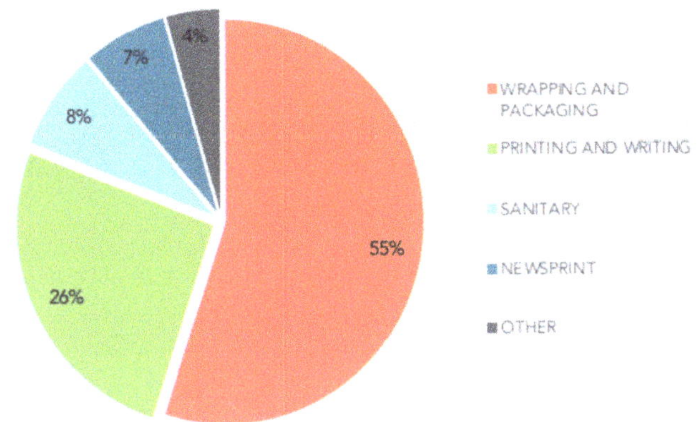

Figure A9. Global consumption by paper category, in tons. Source: https://environmentalpaper.org/wp-content/uploads/2018/04/StateOfTheGlobalPaperIndustry2018_FullReport-Final-1.pdf (accessed on 8 June 2021).

References

1. Student Life during the COVID-19 Pandemic Lockdown: Europe-Wide Insights. European Students' Union: Brussels. Available online: https://www.esu-online.org/wp-content/uploads/2021/04/0010-ESU-SIderalCovid19_WEB.pdf (accessed on 26 June 2021).
2. Aristovnik, A.; Keržič, D.; Ravšelj, D.; Tomaževič, N.; Umek, L. Impacts of the COVID-19 Pandemic on Life of Higher Education Students: A Global Perspective. *Sustainability* **2020**, *12*, 8438. [CrossRef]
3. Policy Brief: Education during COVID-19 and beyond. United Nations. Available online: https://www.un.org/development/desa/dspd/wp-content/uploads/sites/22/2020/08/sg_policy_brief_covid-19_and_education_august_2020.pdf (accessed on 15 June 2021).
4. Zhao, Y.; Watterston, J. The changes we need: Education post COVID-19. *J. Educ. Change* **2021**, *22*, 3–12. [CrossRef]
5. Youth and COVID-19: Response, Recovery and Resilience. OECD. Available online: https://www.oecd.org/coronavirus/policy-responses/youth-and-covid-19-response-recovery-and-resilience-c40e61c6/ (accessed on 17 June 2021).

6. Build Back Better: Education Must Change after COVID-19 to Meet the Climate Crisis. UNESCO. Available online: https://en.unesco.org/news/build-back-better-education-must-change-after-covid-19-meet-climate-crisis (accessed on 7 June 2021).
7. Bianchi, G. *Sustainability Competences, EUR 30555 EN*; Publications Office of the European Union: Luxembourg, 2020.
8. Gherheș, V.; Fărcașiu, M.A. Sustainable Behavior among Romanian Students. A Perspective on Electricity Consumption in Households. *Sustainability* **2021**, in press.
9. Education: From Disruption to Recovery. UNESCO. Available online: https://en.unesco.org/news/137-billion-students-now-home-covid-19-school-closures-expand-ministers-scale-multimedia (accessed on 27 May 2021).
10. Dagiliūtė, R.; Genovaitė Liobikienė, G.; Audronė Minelgaitė, A. Sustainability at universities: Students' perceptions from Green and Non-Green universities. *J. Clean. Prod.* **2018**, *181*, 473–482. [CrossRef]
11. Spaargaren, G.; Van Vliet, B. Lifestyles, consumption and the environment: The ecological modernization of domestic consumption. *Environ Pol.* **2000**, *9*, 50–76. [CrossRef]
12. Balderjahn, I.; Buerke, A.; Kirchgeorg, M.; Peyer, M.; Seegebarth, B.; Wiedmann, K.P. Consciousness for sustainable consumption: Scale development and new insights in the economic dimension of consumers' sustainability. *AMS Rev.* **2013**, *3*, 181–192. [CrossRef]
13. Fu, L.; Zhang, Y.; Xiong, X.; Bai, Y. Pro-Environmental Awareness and Behaviors on Campus: Evidence from Tianjin, China. *EURASIA J. Math. Sci. Tech. Ed.* **2018**, *14*, 427–445. [CrossRef]
14. Wang, R.; Jia, T.; Qi, R.; Cheng, J.; Zhang, K.; Wang, E.; Wang, X. Differentiated Impact of Politics- and Science-Oriented Education on Pro-Environmental Behavior: A Case Study of Chinese University Students. *Sustainability* **2021**, *13*, 616. [CrossRef]
15. United Nations Environment Programme, Transforming Universities into Green and Sustainable Campuses: A Toolkit for Implementers. Available online: https://www.unep.org/resources/report/greening-universities-toolkit-v20-transforming-universities-green-and-sustainable (accessed on 27 May 2021).
16. Martin, J.; Samels, J.E. *The Sustainable University: Green Goals and New Challenges for Higher Education Leaders*; Johns Hopkins University Press: Baltimore, MD, USA, 2012.
17. Badea, L.; Șerban-Oprescu, G.L.; Dedu, S.; Piroșcă, G.I. The Impact of Education for Sustainable Development on Romanian Economics and Business Students' Behavior. *Sustainability* **2020**, *12*, 8169. [CrossRef]
18. Șimon, S.; Stoian, C.E.; Gherheș, V. The Concept of *Sustainability* in the Romanian Top Universities' Strategic Plans. *Sustainability* **2020**, *12*, 2757. [CrossRef]
19. Cernicova-Buca, M.; Dragomir, G.-M. Romanian Students' Appraisal of the Emergency Remote Assessment due to the COVID-19 Pandemic. *Sustainability* **2021**, *13*, 6110. [CrossRef]
20. Le Grange, L.L.L. Sustainability and higher education: From arborescent to rhizomatic thinking. *Educ. Philos. Theory* **2011**, *43*, 742–754. [CrossRef]
21. Tezel, E.; Uğural, M.; Giritli, H. Pro-environmental Behavior of University Students: Influence of Cultural Differences. *Eur. J. Sust. Develop.* **2018**, *7*, 43–52. [CrossRef]
22. Orr, D.W. *The Nature of Design: Ecology, Culture and Human Intention*; Oxford University Press: Oxford, UK, 2002.
23. Barnett, R. *Being a University*; Routledge: Abingdon, UK, 2011.
24. Goddard, J.; Vallance, P. *The University and the City*; Routledge: London, UK, 2013.
25. O'Mara, M. Beyond town and gown: University economic engagement and the legacy of the urban crisis. *J. Techn. Transfer.* **2012**, *37*, 234–250. [CrossRef]
26. Gavazzi, S.M.; Fox, M.; Martin, J. Understanding Campus and Community Relationships through Marriage and Family Metaphors: A Town-Gown Typology. *Innov. High. Educ.* **2014**, *39*, 361–374. [CrossRef]
27. Guattari, F. *The Three Ecologies*; Continuum: London, UK, 2005.
28. Oestreicher, J.S.; Buse, C.; Brisbois, B.; Patrick, R.; Jenkins, A.; Kingsley, J.; Tevora, R.; Fatorelli, L. Where ecosystems, people and health meet: Academic traditions and emerging fields for research and practice. *Sustain. Debate* **2018**, *9*, 45–65.
29. Berkes, F. Environmental Governance for the Anthropocene? Social-Ecological Systems, Resilience, and Collaborative Learning. *Sustainability* **2017**, *9*, 1232. [CrossRef]
30. Stanes, E.; Klocker, N.; Gibson, C. Young adult households and domestic sustainabilities. *Geoforum* **2015**, *65*, 46–58. [CrossRef]
31. Ziesemer, F.; Hüttel, A.; Balderjahn, I. Young People as Drivers or Inhibitors of the Sustainability Movement: The Case of Anti-Consumption. *J. Consum. Policy* **2021**. [CrossRef]
32. Vicente-Molina, M.A.; Fernández-Sainz, A.; Izagirre-Olaizola, J. Does gender make a difference in pro-environmental behavior? *The case of the Basque Country University students J. Clean. Prod.* **2018**, *176*, 89–98. [CrossRef]
33. Biological Roles of Water: Why Is Water Necessary for Life? 2019. Available online: https://sitn.hms.harvard.edu/uncategorized/2019/biological-roles-of-water-why-is-water-necessary-for-life/ (accessed on 12 June 2021).
34. Summary Progress Update 2021: SDG 6—Water and Sanitation for All. UN-Water. Available online: https://www.unwater.org/publications/summary-progress-update-2021-sdg-6-water-and-sanitation-for-all/ (accessed on 15 June 2021).
35. UN World Water Development Report 2019. UN-Water. Available online: https://www.unwater.org/publications/world-water-development-report-2019/ (accessed on 15 June 2021).
36. Domestic Water Use Grew 600% over the Past 50 Years. World Resources Institute. Available online: https://www.wri.org/insights/domestic-water-use-grew-600-over-past-50-years (accessed on 17 June 2021).

37. Water Use in Europe—Quantity and Quality Face Big Challenges—European Environment Agency. Available online: https://www.eea.europa.eu/signals/signals-2018-content-list/articles/water-use-in-europe-2014 (accessed on 1 June 2021).
38. UN, International Decade for Action "Water for Life" 2005–2015. Available online: https://www.un.org/waterforlifedecade/human_right_to_water.shtml (accessed on 21 June 2021).
39. Where Europeans Consume the Most Tap Water. Statista. Available online: https://www.statista.com/chart/19591/average-consumption-of-tap-water-per-person-in-the-eu/ (accessed on 19 June 2021).
40. Eurostat-Water Statistics. 2018. Available online: https://ec.europa.eu/eurostat/statistics-explained/index.php?title=Water_statistics#Water_uses (accessed on 20 May 2021).
41. Ciomoş, V.I.; Ciatarâş, D.M.; Ciomoş, A.O. Price Elasticity of the Residential Water Demand. Case-Study: Investment Project in Cluj County, Romania. *Transylv. Rev. Adm. Sci.* **2012**, *8*, 67–76.
42. INS, Distributia Apei in 2019. Available online: https://insse.ro/cms/sites/default/files/com_presa/com_pdf/distributia_apei20r.pdf (accessed on 27 May 2021).
43. Corral-Verdugo, V.; Frías-Armenta, M.; Pérez-Urias, F.; Orduña-Cabrera, V.; Espinoza-Gallego, N. Residential Water Consumption, Motivation for Conserving Water and the Continuing Tragedy of the Commons. *Environ. Manag.* **2002**, *30*, 527–535. [CrossRef] [PubMed]
44. Gregory, G.D.; Di Leo, M. Repeated Behavior and Environmental Psychology: The Role of Personal Involvement and Habit Formation in Explaining Water Consumption. *J. Appl. Soc. Psychol.* **2003**, *33*, 1261–1296. [CrossRef]
45. Loh, M.; Coghlan, P. Domestic Water Use Study: In Perth, Western Australia 1998–2001. Water Corporation of Western Australia, 2003. Available online: https://www.water.wa.gov.au/__data/assets/pdf_file/0016/5029/42338.pdf (accessed on 21 June 2021).
46. Berk, R.A.; Schulman, D.; McKeever, M.; Freeman, H.E. Measuring the impact of water conservation campaigns in California. *Clim. Chang.* **1993**, *24*, 233–248. [CrossRef]
47. Kenney, D.S.; Goemans, C.; Klein, R.; Lowrey, J.; Reidy, K. Residential Water Demand Management: Lessons from Aurora, Colorado. *J. Am. Water Resour. Assoc.* **2008**, *44*, 192–207. [CrossRef]
48. Jorgensen, B.; Graymore, M.; O'Toole, K. Household Water Use Behavior: An Integrated Model. *J. Environ. Manag.* **2009**, *91*, 227–236. [CrossRef]
49. Wolters, E.A. Attitude-Behavior Consistency in Household Water Consumption. *Soc. Sci. J.* **2014**, *51*, 455–463. [CrossRef]
50. Simpson, K.; Staddon, C.; Ward, S. Challenges of Researching Showering Routines: From the Individual to the Socio-Material. *Urban Sci.* **2019**, *3*, 19. [CrossRef]
51. Ruiz-Garzón, F.; Olmos-Gómez, M.d.C.; Estrada Vidal, L.I. Perceptions of Teachers in Training on Water Issues and Their Relationship to the SDGs. *Sustainability* **2021**, *13*, 5043. [CrossRef]
52. Case Study: Data Links Covid-19 Lockdowns to Consumption Change. Available online: https://www.aquatechtrade.com/news/utilities/covid-19-lockdowns-impact-water-consumption/ (accessed on 23 June 2021).
53. Alda-Vidal, C.; Smith, R.; Lawson, R.; Browne, A.L. *Understanding Changes in Household Water Consumption Associated with Covid-19*; Artesia Consulting: Bristol, UK, 2020.
54. Abu-Bakar, H.; Williams, L.; Hallett, S.H. Quantifying the impact of the COVID-19 lockdown on household water consumption patterns in England. *NPJ Clean Water* **2021**, *4*, 13. [CrossRef]
55. Lüdtke, D.U.; Luetkemeier, R.; Schneemann, M.; Liehr, S. Increase in Daily Household Water Demand during the First Wave of the Covid-19 Pandemic in Germany. *Water* **2021**, *13*, 260. [CrossRef]
56. Sempewo, J.I.; Mushomi, J.; Tumutungire, M.D.; Ekyalimpa, R.; Kisaakye, P. The impact of COVID-19 on households' water use in Uganda. *Water Supply* **2021**. [CrossRef]
57. Kalbusch, A.; Henning, E.; Brikalski, M.P.; de Luca, F.V.; Konrath, A.C. Impact of coronavirus (COVID-19) spread-prevention actions on urban water consumption. *Resour. Conserv. Recycl.* **2020**, *163*, 105098. [CrossRef] [PubMed]
58. Campos, M.A.S.; Carvalho, S.L.; Melo, S.K.; Gonçalves, G.B.F.R.; dos Santos, J.R.; Barros, R.L.; Morgado, U.T.M.A.; da Silva Lopes, E.; Reis, R.P.A. Impact of the COVID-19 pandemic on water consumption behavior. *Water Supply* **2021**. [CrossRef]
59. Main Cause of Global Warming as Perceived in Romania 2019. Statista. Available online: https://www.statista.com/statistics/1093956/perceived-causes-of-global-warming-romania/ (accessed on 8 June 2021).
60. Which of the Following Objectives Should Be Given Top Priority in an European Green Deal? Statista. Available online: https://www.statista.com/statistics/1110123/european-green-deal-objectives-romania/ (accessed on 8 June 2021).
61. Recycling Rate of Municipal Waste. Available online: https://ec.europa.eu/eurostat/databrowser/view/sdg_11_60/default/table?lang=en (accessed on 9 June 2021).
62. Benson, N.U.; Bassey, D.E.; Palanisami, T. COVID pollution: Impact of COVID-19 pandemic on global plastic waste footprint. *Heliyon* **2021**, *7*, e06343. [CrossRef]
63. Sarcodie, S.A.; Owusu, P.A. Impact of COVID-19 pandemic on waste management. *Environ. Dev. Sustain.* **2021**, *23*, 7951–7960. [CrossRef]
64. Mancombu, S.R. How the paper industry has been affected by the 3 'C' factors. *BusinessLine*. 2021. Available online: https://www.thehindubusinessline.com/markets/commodities/how-the-paper-industry-has-been-affected-by-the-3-c-factors/article34131826.ece (accessed on 9 June 2021).

65. Martin, J.; Haggith, M. The State of the Global Paper Industry. *Environmental Paper Network*; 2018. Available online: https://environmentalpaper.org/wp-content/uploads/2018/04/StateOfTheGlobalPaperIndustry2018_FullReport-Final-1.pdf (accessed on 8 June 2021).
66. *Facts About Paper: How Paper Affects the Environment*; TonerBuzz, 2021. Available online: https://www.tonerbuzz.com/facts-about-paper/ (accessed on 9 June 2021).
67. Smyth, S.; Fredeen, A.; Booth, L. Reducing solid waste in higher education: The first step towards 'greening' a university campus. *Resour. Conserv. Recycl.* **2010**, *54*, 1007–1016. [CrossRef]
68. Reaz, M.; Hussain, S.; Kahdem, S. Multimedia University: A Paperless Environment to Take the Challenges for the 21st Century. *AACE J.* **2007**, *15*, 289–314.
69. Dripps, W.R.; Bartholomew, S. Student Paper Consumption at a Liberal Arts University. In Proceedings of the AAG International Conference, Tampa, FL, USA, 2014. Available online: https://www.researchgate.net/publication/262559543_Student_Paper_Consumption_at_a_Liberal_Arts_University (accessed on 22 June 2021).
70. Kapuka, M. Students' Attitudes to Paper Consumption in relation to Carbon Emissions and the Impact of Electronic Course Documents. *South. Afr. J. Environ. Educ.* **2017**, *33*, 84–98. [CrossRef]
71. Iszlai, E. Timișoara fără studenți: Pierderi de 150 de milioane de euro în economia locală. 2021. Available online: https://www.tion.ro/economic/timisoara-fara-studenti-pierderi-de-150-de-milioane-de-euro-in-economia-locala-1408205/ (accessed on 2 June 2021).
72. Proiectul "PadovaFIT Expanded"—"Expanding PadovaFIT! Home Solutions" ID 847143, Finanţat în Cadrul Programului Horizon 2020. Available online: https://www.primariatm.ro/proiecte-cu-finantare-europeana/activitatile-proiectului-padovafit-expanded-expanding-padovafit-home-solutions-id-847143-finantat-in-cadrul-programului-horizon-2020-continua-prin-forumuri-de/ (accessed on 2 June 2021).
73. Starea Economică, Socială Si de Mediu a Municipiului Timişoara 2020. Available online: https://www.primariatm.ro/wp-content/uploads/2021/04/Starea_economica_2020.pdf (accessed on 2 June 2021).
74. Kormos, C.; Gifford, R. The validity of self-report measures of proenvironmental behavior: A meta-analytic review. *J. Environ. Psychol.* **2014**, *40*, 359–371. [CrossRef]
75. Waitt, G.; Caputi, P.; Gibson, C.; Farbotko, C.; Head, L.; Gill, N.; Stanes, E. Sustainable Household Capability: Which households are doing the work of environmental sustainability? *Aust. Geogr.* **2012**, *43*, 51–74. [CrossRef]
76. Gallimore, R.; Lopez, E.M. Everyday Routines, Human Agency, and Ecocultural Context: Construction and Maintenance of Individual Habits. *OTJR Occup. Particip. Health* **2002**, *22*, 70S–77S. [CrossRef]
77. Ersche, K.D.; Tsen-Vei Lim, T.-V.; Ward, L.H.E.; Robbins, T.W.; Stochl, J. Creature of Habit: A self-report measure of habitual routines and automatic tendencies in everyday life. *Personal. Individ. Differ.* **2017**, *116*, 73–85. [CrossRef]
78. Carducci, A.; Fiore, M.; Azara, A.; Bonaccorsi, G.; Bortoletto, M.; Caggiano, G.; Calamusa, A.; De Donno, A.; De Giglio, O.; Dettori, M.; et al. Environmental Behaviors: Determinants and Obstacles among Italian University Students. *Int. J. Environ. Res. Public Health* **2021**, *18*, 3306. [CrossRef]
79. D'Alessandro, D.; Gola, M.; Appolloni, L.; Dettori, M.; Fara, G.M.; Rebecchi, A.; Settimo, G.; Capolongo, S. COVID-19 and Living Space Challenge. Well-Being and Public Health Recommendations for a Healthy, Safe, and Sustainable Housing. *Acta Bio-Med. Atenei Parm.* **2020**, *91*, 61–75. [CrossRef]
80. Martínez-Espiñeira, R.; García-Valiñas, M.A.; Nauges, C. Households' pro-environmental habits and investments in water and energy consumption: Determinants and relationships. *J. Environ. Manag.* **2014**, *133*, 174–183. [CrossRef]
81. Petersen, J.E.; Frantz, C.M.; Shammin, M.R.; Yanisch, T.M.; Tincknell, E.; Myers, N. Electricity and Water Conservation on College and University Campuses in Response to National Competitions among Dormitories: Quantifying Relationships between Behavior, Conservation Strategies and Psychological Metrics. *PLoS ONE* **2015**, *10*, e0144070. [CrossRef] [PubMed]
82. Cernicova, M.; Dragomir, M.; Palea, A. A students' and professors' view on the image of their university. Case study: Politehnica University of Timisoara. *Procedia Soc. Behav. Sci.* **2015**, *191*, 98–102. [CrossRef]

International Journal of
Environmental Research and Public Health

MDPI

Article

Identification of Policies Based on Assessment-Optimization Model to Confront Vulnerable Resources System with Large Population Scale in a Big City

Xueting Zeng [1,2,*], Hua Xiang [1], Jia Liu [3,*], Yong Xue [1], Jinxin Zhu [4] and Yuqian Xu [5]

1 School of Labor Economics, Capital University of Economics and Business, Beijing 100072, China; xiangh1203@cueb.edu.cn (H.X.); 12019205067@cueb.edu.cn (Y.X.)
2 Center for Population Development Research, Capital University of Economics and Business, Beijing 100072, China
3 College of Urban Economics and Public Administration, Capital University of Economics and Business, Beijing 100072, China
4 School of Geography and Planning, Sun Yat-Sen University, Guangzhou 510000, China; zhujx29@mail.sysu.edu.cn
5 Henry M. Gunn High School, Capital University of Economics and Business, 780 Arastradero Rd., Palo Alto, CA 94306, USA; yx46407@pausd.us
* Correspondence: zengxueting@cueb.edu.cn (X.Z.); liujia@cueb.edu.cn (J.L.); Tel.: +86-132-692-79739 (X.Z.)

Abstract: The conflict between excessive population development and vulnerable resource (including water, food, and energy resources) capacity influenced by multiple uncertainties can increase the difficulty of decision making in a big city with large population scale. In this study, an adaptive population and water–food–energy (WFE) management framework (APRF) incorporating vulnerability assessment, uncertainty analysis, and systemic optimization methods is developed for optimizing the relationship between population development and WFE management (P-WFE) under combined policies. In the APRF, the vulnerability of WFE was calculated by an entropy-based driver–pressure–state–response (E-DPSR) model to reflect the exposure, sensitivity, and adaptability caused by population growth, economic development, and resource governance. Meanwhile, a scenario-based dynamic fuzzy model with Hurwicz criterion (SDFH) is proposed for not only optimizing the relationship of P-WFE with uncertain information expressed as possibility and probability distributions, but also reflecting the risk preference of policymakers with an elected manner. The developed APRF is applied to a real case study of Beijing city, which has characteristics of a large population scale and resource deficit. The results of WFE shortages and population adjustments were obtained to identify an optimized P-WEF plan under various policies, to support the adjustment of the current policy in Beijing city. Meanwhile, the results associated with resource vulnerability and benefit analysis were analyzed for improving the robustness of policy generation.

Keywords: population–resource management framework; vulnerability assessment; two-stage dynamic fuzzy programming with Hurwicz criterion; policy scenario analysis

Citation: Zeng, X.; Xiang, H.; Liu, J.; Xue, Y.; Zhu, J.; Xu, Y. Identification of Policies Based on Assessment-Optimization Model to Confront Vulnerable Resources System with Large Population Scale in a Big City. *Int. J. Environ. Res. Public Health* **2021**, *18*, 13097. https://doi.org/10.3390/ijerph182413097

Academic Editors: Salvador García-Ayllón Veintimilla and Antonio Espín Tomás

Received: 25 October 2021
Accepted: 6 December 2021
Published: 11 December 2021

1. Introduction

Reliable water resources, safe gain production, and sustainable energy supply can be regarded as important factors to support urban development, which can provide basic power supply for urban operation and meet the material needs for human living [1]. However, under the background of accelerated urbanization, overpopulation growth, and rapid economic development, the demands of water–food–energy resources (WFE) are increased, which might surpass what natural or artificial systems can afford, leading to a resource crisis [2]. In particular, in some big cities such as Beijing (China) with a large population, the conflict between increasing resource demand and limited WFE carrying capacity can increase the risk of resource shortage. For instance, water shortage is a serious problem in Beijing city, where water resource per capita is one-seventh of the national average [3]. Meanwhile, the energy self-sufficiency (including coal resource, natural gas and electricity supply) of Beijing was 6% in 2017, resulting in energy supply being dependent on importation from other provinces [4]. Moreover, high urbanization under interregional coordination policy has accelerated agriculture transfer to adjacent regions (such as Hebei province), increasing its dependence level on agricultural resource importation. However, to implement water-saving and carbon emission reduction strategies in China, resource importation from other provinces may be restricted due to limited water rights and carbon emission permits. Thus, a fragile resource–supply (i.e., WFE supply) system under overpopulation growth and gathering can be deemed as an obstacle to synergistic economic gain and urban development in a big city [5]. Economic income can be obtained when the expected resource demands based on population scales are satisfied; otherwise, resource shortages generate penalties or losses. Therefore, an identification of the relationship between population development and resource management is required, attracting the attention of policymakers today.

As early as 1968, the contradiction between population growth and resource (e.g., land resource) scarcity was disclosed in a "tragedy of the commons", which was extended to other resources such as water, food, energy, space, and forest resources [6]. In the process of urbanization, the negative correlation between population development and natural resource supply has been verified by previous researchers [7]. Therefore, overpopulation due to urban development can raise the stress on the WFE supply system, enhancing its vulnerability due to artificial and natural driving influences [8]. Previously, various researchers paid attention to the vulnerability of resource (including WFE resources) supply capacity due to population development, whereby excessive population growth and gathering would result in resource shortage or unavailability in response to limited resource supply capacity [9–12]. For instance, Vorosmarty et al. (2000) built a hybrid numerical framework incorporating climate model outputs, water budgets, and socioeconomic information to reflect vulnerable water supply influenced by human impact, representing a potentially important facet of the larger global change question [13]. Nadine et al. (2013) used a vulnerability model to analyze two major industries on the Great Barrier Reef, reflecting the exposure level, sensitivity degree, and adaptive capacity of interaction between social components and the natural resource (including WFE) system [14]. Yang et al. (2015) developed an analytic hierarchy process combining set pair analysis model to display the vulnerability of water resources in response to population growth and climate change [15]. Chen et al. (2018) proposed a water–energy–food (WEF) condonation by PSR model, reflecting the vulnerability and coordination problems of the WEF system in northwest China from 2006 to 2015 for human survival and development. In general, previous researchers paid more attention to simulation and assessment analysis methods to reflect the relationship among population growth, economic development, and resource vulnerability [16]. In order to confront the above shortcomings, several studies associated with an adaptive management framework have been proposed. For instance, Young (2010) used resilience analysis, vulnerability assessment, and adaptation management to reflect the relationship between human lives and the ecological system, which is beneficial to prepare for brief windows of opportunity to make planned changes [17]. Li et al. (2018) combined system dynamic analysis (SD) and an optimal allocation model into a framework to support optimal water utilization according to the vulnerability of water resources, with the aim of supporting regional sustainable development in the context of population growth [18]. Xiang and Li (2020) used a functional model to assess the vulnerability of resources (including WFE) based on RAGA projection, achieving an adaptive resources management strategy to confront resource vulnerability due to excessive human activities [19].

In an adaptive management framework, the system optimization method is an effective approach to incorporate population–economy development and WFE management into a framework to coordinate various components and their relationships [20–26]. However, different uncertainties existing in subsystems and components of population–resource (including WFE) systems generate complex actions and reactions to other subsystems (such as population, economy, resource, and government subsystems) [27–29]. For example, spatiotemporal variations in natural WFE capacity caused by disparate natural features of topography and precipitation can be regarded as an important stochastic factor, leading to population fluctuation as a function of net system benefits. Meanwhile, dynamic socioeconomical development and governmental policies can result in different WFE consumption structures, which would bring about various resource stresses. Thus, a two-stage stochastic dynamic programming (TSDP) was introduced to deal with such objective or subjective randomness, which can build a link between regulated population policies and adjusted economic/resource policies under uncertainty [1]. Nevertheless, in a practical adaptive management issue, some fuzziness due to limited data acquirement (such as error estimation and lost data) can increase the difficulty of decision making, which is a challenge for TSDP [20,24]. Therefore, a credibility-based fuzzy programming (FCP) model was joined to improve the ability of tackling such precise values (e.g., vague inaccurate economic benefits or inexact losses of population adjustment) under weaker sources of information [22,25]. In addition, although random events and fuzzy data in a population–resource plan can be handled by TSDP and FCP methods, the fuzzy risk preference of policymakers would influence the robustness of the optimized population-WFE plan. Under these situations, a scenario analysis with consideration of risk preferences of policymakers based on Hurwicz criterion (SAL) can be designed to reflect the risk adaptation of policymakers in an eclectic optimistic and pessimistic manner [1,26]. However, previous studies focused little on incorporating hybrid methods (e.g., TSSP, FCP and HCA) into a framework to handle various uncertainties in a population and WFE issue.

Therefore, the objective of this study was to develop an adaptive population–WFE (P-WEF) management framework (APRF) to confront the conflict between population development and the WFE supply system. Vulnerability assessment, uncertainty analysis, and systemic optimization methods can be incorporated into this APRF as hybrid method manner to deal with multiple uncertainties, which can confront the complexity of P-WFE for policymakers. This hybrid method has the following advantages: (a) the vulnerability of WFE is calculated using an entropy-based driver–pressure–state–response (E-DPSR) model to reflect the exposure, sensitivity, and adaptability caused by population growth, economic development, and resource governance; (b) a population–resources (P-WFE) optimization can be conducted to identify various policies associated with population adjustment, resource regulation, and technique improvement; (c) a scenario-based dynamic fuzzy model with Hurwicz criterion (SDFH) is developed and embedded into P-WFE optimization to deal with uncertainties expressed as possibility and probability distributions. Meanwhile, the risk preferences of policymakers can also be reflected in an elected manner. The developed APRF was applied to a real case study of Beijing city, which has characteristics of a large population and resource deficit. The results of WFE shortages and population–economy adjustments under various scenarios were obtained to identify the optimized P-WEF under various policies. Furthermore, the obtained results associated with resource vulnerability in various scenarios and benefit analysis under various risk levels were analyzed, reflecting the tradeoff between population development and WEF management. The above results are beneficial for adjusting current policies in a risk-averse and effective manner.

2. Materials and Methods

2.1. Problem Statement

Beijing is the capital city of China with a total area of 16,412 square kilometers, located in a north temperate subhumid continental monsoon climate (located at 115.7°–117.4° E longitude, 39.4°–41.6° N latitude). It has average rainfall of 483.9 mm, which presents an uneven spatial and temporal distribution (approximately 65% of precipitation occurring in July and August; more than 77.59% of water availability distributed in Hebei province) (WRB, 2006–2017) [30]. Furthermore, although 67 types of mineral products have been found, Beijing is not a major mineral-producing area, whereby energy supply and coal products are imported from other provinces. Moreover, in the context of the coordination strategy for Beijing, Tianjin, and Hebei agglomeration (CBTH), agriculture was transferred from Beijing city to Hebei province; thus, food resources have relied on importation from circumjacent provinces in recent years.

As a result of the speed of urban agglomeration in recent decades, the population of Beijing presented an increasing tendency from 120 to 200 billion from 2010 to 2018 [31]. Furthermore, accelerated economic growth and developed social resource support (e.g., educational resources, public service facilities, cultural deposits, employment opportunities, and other aspects) have led to constant population gathering in the core area of Beijing, which requires high-quality resource support. However, the natural resources (land, water, energy, food, forest, etc.) of Beijing struggle to meet the demands of population agglomeration due to their own characteristics (such as limited quantity, uneven spatial distribution, and interrelation of different resources). The contradiction between population development and the water–food–energy (WFE) supply requires more effective approaches to deal with these urban effects.

Although a number of policies were introduced to relieve the WFE pressure in recent years, they brought about a number of challenges for policymakers. The polarization and siphon effects of a central city can lead to excessive population agglomeration. Excessive density of population increases the pressure on WFE in core regions [25,32]. Moreover, limited natural resources are deemed a bottleneck problem in urban development. For instance, Beijing has only one-seventh of the national average water resources per capita, which results in a serious water crisis; furthermore, other resources (such as food and mineral products) rely on importation from other provinces. In the context of the Beijing–Tianjin–Hebei coordinated development strategy, industry transformation facilitated agricultural and food product industries moving from Beijing to Heibei, reducing the self-sufficiency of food resources. Urbanization requires a high quality of energy supply (such as coal resources and corresponding derivatives), which cannot be met by an energy-deficient region dependent on importation from other provinces. However, the water-saving and carbon emission reduction strategies in China can restrict resource importation from other provinces due to limited water rights and carbon emission permits. Thus, the vulnerability of the WFE supply system has increased due to natural and artificial factors in recent decades. Although various strategies such as population adjustment based on industrial transformation, resource-saving techniques, and resource importation planning have been advocated, how to identify the interaction between population development and WFE management is a first point for remitting the resource crisis in Beijing city. Therefore, an adaptive population—WFE management framework (APRF) is required to confront the above challenges.

2.2. The Framework of APRF under Combined Policies

Figure 1 presents an adaptive population–WFE management framework (APRF) under combined policies to confront the conflict between rapid population development and the vulnerable WFE supply system in Beijing city. In this framework, population development (including population growth, gathering, and employment structure) and resource (including WFE) supply management can be incorporated into a system, where the population based on industrial layout (PIL) can be deemed as an important indicator to support economic growth, WFE consumption, and urban development. A proper PIL can support economic development in various industrial sectors at regional levels, which can also minimize the damage to WFE as a function of regional resource carrying capacity. Otherwise, this can lead to the vulnerability of the natural resource supply system, which may be expressed as resource shortages, economic loss, and system failure. With the agglomeration of population, the per capita possession of WFE in Beijing has decreased, and the utilization of WFE has intensified, resulting in increasing pressure on WFE vulnerability. Under these situations, the policymakers of Beijing have to adopt a series of policies and measures (such as population adjustment, industrial reallocation, technique improvement, and resource regulation) to alleviate such pressures. Nevertheless, different natural features and various governmental modes can generate a number of uncertainties and complexities. Thus, an APRF framework incorporating vulnerability assessment, population–WFE optimization, and scenario analysis can be designed to address the conflict between population development and WFE supply in Beijing city.

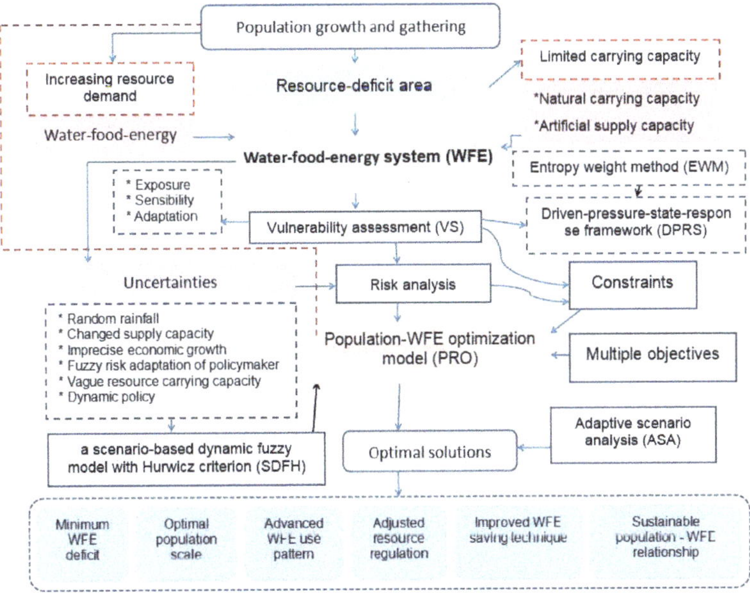

Figure 1. An adaptive population—resources management framework (APRF) under combined policies and its application in Beijing.

2.3. Methodology and Modeling

2.3.1. Method Development

Vulnerability Assessment Based on Driver–Pressure–State–Response (DPSR) Model

In general, the vulnerability of WFE supply can be expressed as a function of exposure (EI) (the proximity of natural resources to pressures and disturbances), sensitivity (SI) (an exposed unit is affected by pressures and disturbances), and adaptability (AI) (the ability of an exposed unit to deal with and recover from adverse effects), which includes various vulnerability indices [33,34]. Since the value of VI incorporates the positive effects from EI and AI, and includes the negative effect from AI, the vulnerability index (VI) of water–food–energy resources can be defined as $\mathrm{VI} = \mathrm{EI} + \mathrm{SI} - \mathrm{AI}$ without any empowerment [35]. Meanwhile, in order to identify the vulnerability of the WFE supply system to population scale or adjustment, an index system named the population-driven WFE vulnerability assessment framework (PWVA) was built (Table 1) [35]. Moreover, a driver–pressure–state–response (DPSR) model was introduced to reflect the cause–effect relationships and interactions between human activities and WFE supply system [36]. Following the principle of comprehensiveness, objectivity, and scientificity, the indicators in PWVA were grouped and assigned to reflect their contributions to population-driven resource vulnerability as follows: population scale, employment rate and corresponding indicators can be deemed as a drivers (denoted as "D") to change the state of economic development and resource utilization (denoted as "S"), which would result in detriments such as water shortage, food crisis, and energy importation (denoted as "P"). Accordingly, the human response or governmental regulation (denoted as "R") can remit the conflict between population development and the WFE supply system. The equations describing the indices based on DPSR are shown in Table 1.

Table 1. Population-driven WFE vulnerability assessment framework.

DPSR Framework	Index	Equation	Vulnerability
Driver	Population scale	Direct data index	Exposure (EI)
	Employment rate	Employment/total labor force	
	The proportion of population for agriculture in industries	Agricultural population/population	
	The proportion of population for industry in industries	Industrial population/population	
	The proportion of population for service industry in industries	Service population/population	
	R&D people	Direct data index	
Pressure	The proportion of agriculture in industries	R&D people in agriculture industry/R&D people	
	The proportion of industry in industries	R&D people in industry/R&D people	
	The proportion of service industry in industries	R&D people in service industry/R&D people	
	Land utilization rate	The area of land developed/total land area	
	Energy yield-to-consumption ratio	Energy output/energy consumption	
	The efficiency of energy utilization	Industrial GDP/energy consumption	
	Total water availability	Direct data index	
	Forest coverage rate	Forest area/total land area	
State	Water resources per capita	Total water resources/population	Sensitivity (SI)
	Forest area per capita	Total forest area/population	
	Energy self-sufficiency gap	Energy consumption − energy output	
	Water shortage rate	(Water consumption − water resource)/ water resource	
	Per capital greening gap	(Per capital green area − standard green value)/standard green value	
	The rate of food self-sufficiency	Grain consumption/grain output	
Response	Ecological environment investment index	The government energy conservation/general budget	Adaptability (AI)
	The government energy conservation	Direct data index	
	Sewage treatment rate	Amount of sewage purification/total sewage	
	Water-saving percentage	Circulating water consumption/water consumption	
	Intensity of soil erosion control	Water and soil loss after treatment/water and soil loss before treatment × 100%	

Then, in order to avoid errors caused by different measured units of indicators and the mutual interference between positive and negative factors, the range method was used to standardize all indicators as follows [37]:

For increasing class indicators,

$$x_{ij} = \frac{x_{ij} - x_{ij}}{x_{max} - x_{min}}; \tag{1a}$$

For decreasing indices,

$$x_{ij} = \frac{x_{max} - x_{ij}}{x_{max} - x_{min}}. \tag{1b}$$

Here, x_{ij} is the j indicator data of the i year, x_{max} is the maximum value of the j indicator, and xmin is the minimum value of the j indicator. Moreover, the entropy value method is introduced as the weight analysis tool, which has the advantages of simple calculation, original data utilization, less information loss, and reduced subjective information [38,39]. The entropy weight calculation formula can be expressed as follows:

$$e_j = -k \sum\nolimits_{i=1}^{m} y_{ij} ln y_{ij}, \tag{1c}$$

$$y_{ij} = x_{ij} / \sum\nolimits_{i=1}^{m} x_{ij}, \tag{1d}$$

$$k = 1/lnm x_{ij}, \tag{1e}$$

$$w_j = 1 - e_j / \sum\nolimits_{j=1}^{n} (1 - e_j), \tag{1f}$$

where e_j is the information entropy of each indicator, m is the number of years, n is the number of indicator items, and x_{ij} is the standardized value of the j indicator in t year.

According to the definition of VI (in Table 1) and entropy weight calculation method (Equations (1a)–(1f)), the exponent of the domain layer and target layer can be calculated by weighted summation as follows:

$$P_{r-vi} = \sum_{n=1}^{m} x_{r-ij} z_j, \tag{1g}$$

where P_{r-vi} represents the r index under the domain layer and the target layer, x_{r-ij} is the specific index (the standardized value of the j index in t year) under the domain layer or the criterion layer, and z_j is the weight of this index calculated according to the entropy value method.

In the survey, since the data used for comprehensive evaluation were all standardized data, this would increase the operative errors. However, standardized data fully retain the information of the original data, which can also reflect the change rate of indicators to a certain extent [40]. Under these situations, the variation trend of different resource data in Beijing results in the intercept term being eliminated after standardization; however, some data still have a time trend. Furthermore, the AIC (Akaike information criterion) was used to determine the optimal lag order for the augmented Dickey–Fuller unit root test (ADF) [41]. Thus, the stationarity of standardized data was analyzed in Table 2. It can be seen that the ADF statistical values of all index data in Beijing were all less than the 1% critical value, revealing a stable trend.

Table 2. Consistency test.

Variable	Test Type (C, T, P)	ADF Statistic	1% Threshold	5% Threshold	10% Threshold	Conclusion
Population scale	(0, 1, 1)	−3.645	−3.75	−3	−2.63	Stable performance
Employment structure	(0, 0, 1)	−3.019	−3.75	−3	−2.63	Stable performance
The proportion of population for agriculture in industries	(0, 1, 2)	−3.104	−3.75	−3	−2.63	Stable performance
The proportion of population for industry in industries	(0, 1, 1)	−3.933	−3.75	−3	−2.63	Stable performance
The proportion of population for service industry in industries	(0, 1, 1)	−3.306	−3.75	−3	−2.63	Stable performance
R&D people	(0, 1, 1)	−4.355	−3.75	−3	−2.63	Stable performance
The proportion of agriculture in industries	(0, 1, 1)	−3.572	−3.75	−3	−2.63	Stable performance
The proportion of industry in industries	(0, 1, 2)	−3.962	−3.75	−3	−2.63	Stable performance
The proportion of service industry in industries	(0, 1, 2)	−3.451	−3.75	−3	−2.63	Stable performance
Land utilization rate	(0, 0, 1)	−3.971	−3.75	−3	−2.63	Stable performance
Energy yield-to-consumption ratio	(0, 0, 1)	−4.946	−3.75	−3	−2.63	Stable performance
The efficiency of energy utilization	(0, 1, 2)	−3.199	−3.75	−3	−2.63	Stable performance
Total water availability	(0, 0, 1)	−4.895	−3.75	−3	−2.63	Stable performance
Forest coverage rate	(0, 1, 0)	−4.619	−3.75	−3	−2.63	Stable performance
Water resources per capita	(0, 0, 2)	−5.544	−3.75	−3	−2.63	Stable performance
Forest area per capita	(0, 1, 0)	−5.388	−3.75	−3	−2.63	Stable performance
Water shortage rate	(0, 0, 2)	−4.514	−3.75	−3	−2.63	Stable performance
Per capita greening gap	(0, 1, 2)	−6.062	−3.75	−3	−2.63	Stable performance
Ecological environment investment index	(0, 1, 1)	−4.927	−3.75	−3	−2.63	Stable performance
The government energy conservation	(0, 0, 1)	−4.273	−3.75	−3	−2.63	Stable performance
Sewage treatment rate	(0, 1, 1)	−3.852	−3.75	−3	−2.63	Stable performance
Intensity of soil erosion control	(0, 0, 2)	−5.984	−3.75	−3	−2.63	Stable performance

Optimization Based on a Scenario-Based Dynamic Fuzzy Model with Hurwicz Criterion (SDFH) Method

Taking into account the vulnerability assessment of WFE supply system, the optimization method should be considered to balance the relationship between population development and WFE management. However, random events (such as random rainfall, sudden agricultural reduction, and dynamic energy plans) impacting the constrained reliability would influence the accuracy of calculation. Therefore, a two-stage stochastic dynamic programming (TSDP) was introduced to build a link between expected targets (i.e., expected resource (WEF) demand based on population scale) and random events (i.e., the reduction in WEF supply capacity caused by drought, plant disease, and energy restriction) as follows [20]:

$$\text{Max } f = uw - \sum_{h=1}^{r} p_h q(v, \delta_h), \tag{2a}$$

subject to

$$R(\delta_h)w + S(\delta_h)v = g(\delta_h), \delta_h \in \Omega, \tag{2b}$$

$$uw - \sum_{h=1}^{r} q(v, \delta_h) \le w, \tag{2c}$$

$$aw \le c, \tag{2d}$$

$$w \ge 0, \tag{2e}$$

$$v \ge 0. \tag{2f}$$

In Model (2a), if the expected WEF demand (initial target or first-stage variable) based on the population scale can be satisfied, the first-stage benefit (i.e., uw) would be obtained; otherwise, a loss or recourse action (i.e., $q(v, \delta_h)$) would be generated. This means that the first benefit can be rectified by the second penalty (e.g., $\sum_{h=1}^{r} p_h q(v, \delta_h)$) when a random event occurs; P_h is the probability of a random event. According to the vulnerability of the WFE supply system driven by population, the possibility of WFE shortage would rectify the expected WEF, as shown in Model (2c) [1,20]. However, data uncertainties due to limited data acquirement (such as error estimation and lost data) cannot be handled by TSDP with probabilistic distributions [25,42–44]. Therefore, a fuzzy credibility constrained programming (FCP) was combined with TSDP as follows:

$$Cr\{aw \le \widetilde{c}\} \ge \alpha. \tag{3a}$$

On the basis of the concept of fuzzy credibility, the credibility measure (Cr) can be expressed as $Cr\{\varsigma \le s\} = \frac{1}{2}(Pos\{\varsigma \le s\} + Nec\{\varsigma \le s\})$ [22]. In general, the credibility level should be greater than 0.5. Thus, Model (3a) can be proven as follows:

$$Cr\{\widetilde{\varsigma} \ge s\} \ge \alpha \Leftrightarrow s \le (2-2\alpha)\varsigma_2 + (2\alpha - 1)\varsigma_1 \Leftrightarrow s \le \varsigma_2 + (1-2\alpha)(\varsigma_2 - \varsigma_1). \tag{3b}$$

In a practical a two-stage stochastic fuzzy credibility programming (TFC) issue, the input of left- or right-hand variables are impacted by various factors, thereby requiring a scenario analysis (SA) [45]. Thus, various adaptive scenarios can be designed to balance the relationship between expected target and actual WFE supply. However, the risk attitudes of policymakers can impact the generation of scenarios due to the experiences and personality traits of policymakers, which cannot be handled by TFC. Therefore, a scenario analysis with Hurwicz criterion was introduced to reflect the vague risk adaptation of policymakers, which is beneficial to obtain compromising alternatives based on eclectic optimistic and pessimistic criteria [46–48]. The scenario-based dynamic fuzzy model with Hurwicz criterion (SDFH) method can be expressed as follows:

$$\text{Max } f_{Hurwicz} = \{\beta * U^{opt} + (1-\beta) * U^{pec}\}, \tag{4a}$$

subject to

$$Cr\left\{[uw - \sum_{h=1}^{r} p_h q(v, \delta_h)] \le U^{opt}\right\} \ge \eta,\ i = 1,2,\ldots,I, \tag{4b}$$

$$Cr\left\{[uw - \sum_{h=1}^{r} p_h q(v,\delta_h)] \geq U^{pec}\right\} \geq \eta,\ i = 1,2,\ldots,I, \tag{4c}$$

$$R(\delta_h)w + S(\delta_h)v = g(\delta_h),\ \delta_h \in \Omega, \tag{4d}$$

$$uw - \sum_{h=1}^{r} q(v,\delta_h) \leq w, \tag{4e}$$

$$aw \leq c_n^2 + (1-2\alpha)(c_n^2 - c_n^1), \tag{4f}$$

$$w \geq 0, \tag{4g}$$

$$v \geq 0. \tag{4h}$$

2.3.2. Modeling Formulation for Practical Application

In a practical framework of APRF, the initial population policy is pre-regulated on the basis of urban planning, which can support economic development, but is restricted by WFE carrying capacity (or natural supply capacity). In general, the initial population policy is calculated as a function of the current population situation (with existing techniques and resource regulation) as a baseline scenario. When the expected demand of WFE based on initial population policy or scale can be satisfied by WEF, a first-stage benefit is generated. However, since the WFE carrying capacity can be influenced by some artificial or objective factors (such as random rainfall or resource regulation change), the shortage of WFE based on the initial population scale can generate an economic loss, which can be regarded as a second-stage recourse action. Under these situations, policymakers would implement a series of policies (including industrial structure adjustment and corresponding employed population change), which can be deemed further second-stage recourse actions to remit the losses of WFE shortage. In a practical APRE, population scale in various industries can be seen as an indicator (i.e., decision variable) to consume WFE resources, which can generate variation in employment structure, industrial layout, and resource use pattern, leading to different WFE stresses. Thus, policymakers can make policy adjustments (such as population adjustment in different industries, resource regulations in WEF supply plans, and technique improvement in supply capacity) to reduce WFE shortages, which can maximize system benefit in a risk-averse and effective manner as follows:

$$\begin{aligned}\text{Max } f_{Hurwicz} &= \{\beta * U^{opt} + (1-\beta) * U^{pec}\} \\ &= \{\beta * > [\text{ICR}_t + \text{BLA}_t + \text{BCT}_t]^{opt} + (1-\beta) * [\text{ICR}_t + \text{BLA}_t + \text{BCT}_t]^{pec}\}\end{aligned} \tag{5a}$$

The notations of objective functions, decision variables, and parameters are shown in the Abbreviations. In Model (5a), $f_{Hurwicz}$ is the total system benefit with the Hurwicz criterion, which includes the income from the current population situation and the corresponding loss from WFE shortage (ICR_t), the benefit and loss from the WFE supply capacity based on population adjustment (BLA_t), and the benefit and loss from the WFE supply capacity based on population adjustment (BLA_t) with consideration of the risk adoption based on the Hurwicz criterion as follows:

(1) Income from current population situation and corresponding loss from WFE shortage (ICR_t):

$$\begin{aligned}&\left(\sum_{t=1}^{3}\sum_{l=1}^{2} IMP_{tl} * \widetilde{BMP}_{tl} - \sum_{l=1}^{2}\sum_{t=1}^{3}\sum_{h=1}^{3} p_{th} * RMP_{tlh} * \widetilde{LMS}_{tl}\right) + \left(\sum_{t=1}^{3}\sum_{j=1}^{1} IAP_{tj} * \widetilde{BAP}_{tj} - \right. \\ &\left.\sum_{j=1}^{1}\sum_{t=1}^{3}\sum_{h=1}^{3} p_{th} * RAP_{tjh} * \widetilde{LAS}_{tj}\right) + \left(\sum_{k=1}^{4}\sum_{t=1}^{3} IEP_{tk} * \widetilde{BIP}_{tk} - \sum_{k=1}^{4}\sum_{t=1}^{3}\sum_{h=1}^{3} p_{th} * REP_{tkh} * \widetilde{LIS}_{tk}\right) \\ &+\left(\sum_{g=1}^{3}\sum_{t=1}^{3} ISP_{tg} * \widetilde{BSP}_{tg} - \sum_{g=1}^{3}\sum_{t=1}^{3}\sum_{h=1}^{3} p_{th} * RSP_{tgh} * \widetilde{LSS}_{tg}\right)\end{aligned} \tag{5b}$$

In Model (5b), three industrial (including agriculture, industry, and service sectors) and one municipal sectors in three periods (t denoted as various periods) are considered. Since WFE consumption patterns vary in different industrial and municipal sectors, two plants

($l = 1$ represents resources for urban residents; $l = 2$ represents resources for rural residents) in the municipal sector, one plant ($j = 1$ denotes irrigation in agriculture) in the agriculture sector, four plants ($k = 1$ and 2 represent high- and medium-consumption industrial plants, $k = 3$ represent other industrial plants, and $k = 4$ represent energy-supply plants) in the industrial sector, and three plants ($g = 1$ represents the traditional service industry; $g = 2$ represents other service industrial plants; $g = 3$ represents green service industrial plants) in the service sector are shown in Model (5b). In Model (5b), the expected population scales (including living population and employed population according to the current industrial structure) regarded as first-stage variables (i.e., IMP_{tl}, IAP_{tj}, IEP_{tk}, ISP_{tg}) can bring about incomes or benefits (i.e., BMP_{tl}, BAP_{tj}, BIP_{tk}, BSP_{tg}), when the resource (including water, coal, and food resources) supply capacity can satisfy or support such population scales. The economic data associated with net system benefit for the population in various sectors (BMP_{tl}, BAP_{tj}, BIP_{tk}, BSP_{tg}) are provided in Table 3. If the WFE resources cannot meet the expected demand based on the population scale, resource-deficient losses are generated, where RMP_{tlh}, RAP_{tjh}, REP_{tkh}, and RSP_{tgh} can be deemed as second-stage variables.

(2) Benefit and loss from WFE supply capacity based on population adjustment (BLA_t):

$$
\begin{aligned}
&\{\sum_{l=1}^{2}\sum_{t=1}^{3} IMP_{lt} * \widetilde{B}MP_{lt} * [wre_{lt} * (1-\eta_{lt}) + cre_{lt} * (1-\delta_{lt})] - \sum_{j=1}^{1}\sum_{i=1}^{3}\sum_{h=1}^{3} p_{th} * AMP_{lth} * \widetilde{L}MP_{lt} * \ [wre_{lt} * (1-\eta_{lt}) \\
&+cre_{lt} * (1-\delta_{lt})]\}\{\sum_{j=1}^{1}\sum_{t=1}^{3} IAP_{tj} * \widetilde{B}AP_{tj} * [wre_{tj} * (1-\eta_{tj}) + cre_{tj} * (1-\delta_{tj})] - \sum_{j=1}^{1}\sum_{i=1}^{3}\sum_{h=1}^{3} p_{th} * AAP_{tjh} * \widetilde{L}AP_{tj} * \\
&[wre_{tj} * (1-\eta_{tj}) + cre_{tj} * (1-\delta_{tj})]\} + \{[\sum_{k=1}^{4}\sum_{t=1}^{3} IEP_{tk} * \widetilde{B}IP_{tk} * [wre_{tk} * (1-\eta_{tk}) + cre_{tk} * (1-\delta_{tk})] - \\
&\sum_{k=1}^{4}\sum_{i=1}^{3}\sum_{h=1}^{3} p_{th} * AEP_{tkh} * \widetilde{L}IP_{tk} * [wre_{tk} * (1-\eta_{tk}) + cre_{tk} * (1-\delta_{tk})]\} + [\sum_{g=1}^{3}\sum_{t=1}^{3} ISP_{tg} * \widetilde{B}S_{tg} * \ [wre_{tg} * (1-\eta_{tg}) \\
&+cre_{tg} * (1-\delta_{tg})] - \sum_{g=1}^{3}\sum_{i=1}^{3}\sum_{h=1}^{3} p_{th} * ASP_{tgh} * \widetilde{L}SP_{tg} * [wre_{tg} * (1-\eta_{tg}) + cre_{tg} * (1-\delta_{tg})]\}
\end{aligned}
\quad . \quad (5c)
$$

Table 3. Policy scenario.

Scenario	Assumption				
	Improvement of Technique Efficiency		Lessen the Limit of Resource Based on Resource Saving		
	Resource Use Efficiency	Retreatment Ratio	Water Resources	Coal Resources	Food Supply
S0	0%	0%	0%	0%	0%
S1	5%	5%	0%	0%	0%
S2	15%	15%	0%	0%	0%
S3	25%	25%	0%	0%	0%
S4	0%	0%	5%	5%	5%
S5	0%	0%	15%	15%	15%
S6	5%	5%	5%	5%	5%
S7	15%	15%	15%	15%	15%

The initial population policy and corresponding population scale can support economic development, albeit while consuming various resources, leading to WFE stresses or losses. Thus, the policies associated with population adjustment (i.e., AMP_{tlh}, AAP_{tjh}, AEP_{tkh}, ASP_{tgh}) would remit the losses of WFE shortage (i.e., LMS_{tl}, LAS_{tj}, LIS_{tk}, LSS_{tg}), but would generate penalties (i.e., LMP_{tl}, LAP_{tj}, LIP_{tk}, LSP_{tg}) due to the shrinking of the economy. The corresponding economic data are provided in Table 4. Furthermore, the policies associated with technique improvement (e.g., the coefficient of water resource consumption or coal use and the coefficient of water-saving or coal-saving) were considered as policy scenarios to remit WFE pressure, as shown in Table 3.

(3) Benefit and cost from technique improvement (BCT_t):

$$\{\sum_{l=1}^{2}\sum_{t=1}^{3} IMP_{lt} * \widetilde{L}MS_{lt} * [ee_{lt} * (1-\phi_{lt}) + ae_{lt} * (1-\mu_{lt})] - \sum_{j=1}^{1}\sum_{i=1}^{3} * RMP_{lth} * CMS_{lt} * [ee_{lt} * (1-\phi_{lt}) + ae_{lt} * (1-\mu_{lt})]\} + \{\sum_{j=1}^{1}\sum_{t=1}^{3} IAP_{tj} * \widetilde{L}AP_{tj} * [ee_{tj} * (1-\phi_{tj}) + ae_{tj} * (1-\mu_{tj})] - \sum_{j=1}^{1}\sum_{i=1}^{3} * RAP_{tjh} * CAS_{tj} * [ee_{tj} * (1-\phi_{tj}) + ae_{tj} * (1-\mu_{tj})]\} + \{[\sum_{k=1}^{4}\sum_{t=1}^{3} IEP_{tk} * \widetilde{L}IP_{tk} * [ee_{tk} * (1-\phi_{tk}) + ae_{tk} * (1-\mu_{tk})] - \sum_{k=1}^{4}\sum_{i=1}^{3} * REP_{tkh} * CIS_{tk} * [ee_{tk} * (1-\phi_{tk}) + ae_{tk} * (1-\mu_{tk})]\} + [\sum_{g=1}^{3}\sum_{t=1}^{3} ISP_{tg} * \widetilde{L}S_{tg} * [ee_{tg} * (1-\phi_{tg}) + ae_{tg} * (1-\mu_{tg})] - \sum_{g=1}^{3}\sum_{i=1}^{3} * RSP_{tgh} * CSS_{tg} * [ee_{tg} * (1-\phi_{tg}) + ae_{tg} * (1-\mu_{tg})]\} \quad (5d)$$

Table 4. Economic data.

		Period 1	Period 2	Period 3
Net System Benefit for Population in Various Sectors (10^3 RMB/Person)				
Municipal sectors	Urban human living	(0.96, 1.02, 1.06)	(0.99, 1.04, 1.08)	(1.02, 1.06, 1.12)
	Rural human living	(0.42, 0.47, 0.50)	(0.45, 0.49, 0.51)	(0.47, 0.51, 0.55)
Agricultural sector	Food resource supply	(2.20, 2.63, 2.88)	(2.32, 2.68, 2.96)	(1.93, 2.36, 2.68)
Industrial sector	Heavy resource-consumption plants	(192.46, 202.32, 206.32)	(178.23, 183.76, 194.13)	(152.36, 148.32, 132.23)
	Medium resource-consumption plants	(12.58, 13.36, 14.58)	(10.82, 11.08, 12.98)	(9.98, 8.25, 7.26)
	Other industrial plants	(29.82, 31.15, 32.87)	(27.32, 29.55, 31.64)	(26.01, 28.25, 29.08)
	Energy-supply plants	(6.21, 7.05, 8.02)	(6.87, 7.21, 8.98)	(7.17, 8.83, 9.76)
Service sector	Traditional service plants	(18.13, 21.43, 25.98)	(20.23, 28.32, 40.12)	(34.32, 45.49, 66.32)
	Other service plants	(21.32, 28.60, 32.32)	(27.32, 32.29, 40.87)	(30.82, 38.41, 47.32)
	Environmentally friendly service plants	(22.32, 26.63, 28.89)	(29.86, 34.81, 38.86)	(36.87, 42.00, 48.82)
Net loss for various sectors (10^3 RMB/person)				
Municipal sectors	Urban human living	(1.90, 1.96, 2.02)	(1.93, 1.99, 2.06)	(1.97, 2.01, 2.08)
	Rural human living	(0.80, 0.85, 0.90)	(0.85, 0.89, 0.93)	(0.89, 0.93, 0.96)
Agricultural sector	Food resource supply	(2.86, 3.18, 3.98)	(2.92, 3.23, 4.02)	(2.76, 2.95, 3.12)
Industrial sector	Heavy resource-consumption plants	(192.46, 202.32, 206.32)	(178.23, 183.76, 194.13)	(152.36, 148.32, 132.23)
	Medium resource-consumption plants	(12.58, 13.36, 14.58)	(10.82, 11.08, 12.98)	(9.98, 8.25, 7.26)
	Other industrial plants	(29.82, 31.15, 32.87)	(27.32, 29.55, 31.64)	(26.01, 28.25, 29.08)
	Energy-supply plants	(7.42, 8.46, 9.02)	(7.82, 8.97, 9.92)	(14.76, 15.40, 16.89)
Service sector	Traditional service plant	(22.32, 23.32, 24.56)	(32.32, 39.24, 42.12)	(40.32, 46.56, 49.12)
	Other service plants	(26.12, 27.12, 28.89)	(30.21, 36.94, 43.01)	(36.01, 41.60, 46.12)
	Environmentally friendly service plants	(27.32, 28.92, 31.21)	(32.72, 38.99, 41.34)	(39.32, 41.60, 44.34)

In Model (5d), the expected population can discharge pollutants, which would disturb the WFE supply capacity to an extent. Therefore, various recycling techniques were considered in the scenario analysis, including the coefficient of recycling by technique improvement, as shown in Table 3. Here, η and δ are the improvement ratios of the resource-saving technique (%), while μ and ϕ are the improvement ratios of the retreatment technique (%). However, the costs of technique improvement (i.e., CMS_{lt}, CAS_{tj}, CIS_{tk}, CSS_{tg}) should be considered.

Moreover, various constraints associated with available resources (WFE), population development, and other economic development scales under various scenarios can be considered as follows:

(1) Constraints of available water resources and corresponding resource regulation:

$$Cr\{\sum_{h=1}^{3}\sum_{t=1}^{3} \widetilde{V}w_{ht} = \sum_{h=1}^{3}\sum_{t=1}^{3}[(\widetilde{R}_{ht} - \widetilde{H}_{ht} - \widetilde{G}_{ht})\} \geq \alpha. \quad (6a)$$

$$Cr\{[wre_{lt}*(1-\eta_{lt})(\sum_{l=1}^{2}\sum_{t=1}^{3}IMP_{lt}-\sum_{j=1}^{1}\sum_{i=1}^{3}\sum_{h=1}^{3}p_{th}*RMP_{lth})]+[wre_{tj}*(1-\eta_{tj})(\sum_{j=1}^{1}\sum_{t=1}^{3}IAP_{tj}- \sum_{j=1}^{1}\sum_{i=1}^{3}\sum_{h=1}^{3}p_{th}*RAP_{tjh})]+[wre_{tj}*(1-\eta_{tj})*(\sum_{k=1}^{4}\sum_{t=1}^{3}IEP_{tk}-\sum_{k=1}^{4}\sum_{i=1}^{3}\sum_{h=1}^{3}p_{th}*REP_{tkh})]+ [wre_{tg}*(1-\eta_{tg})*(\sum_{g=1}^{3}\sum_{t=1}^{3}ISP_{tg}-\sum_{g=1}^{3}\sum_{i=1}^{3}\sum_{h=1}^{3}p_{th}*RSP_{tgh}]\leq(1-\xi_w)*\widetilde{V}w_{ht}\}\geq\alpha\} \quad (6b)$$

Model (6a) shows the constraint of available water resources based on current water resource load, which is equal to the total water availability (including surface and underground water resources) ($\widetilde{R}_{hi}$) minus loss of water ($\widetilde{H}_{hi}$) (including evaporation/infiltration from river) and water requirements of watercourse ($\widetilde{G}_{hi}$). Model (6b) displays that water shortage would occur according to expected demands and available water resources ($\widetilde{V}w_{hj}$), which can be expressed in a credibility-based fuzzy manner due to imprecise information. However, in order to save water, policymakers design resource limit targets (resource regulation scenarios) in planning periods, where ξ_w is the reduced ratio of the resource limit for the resource saving target (%); corresponding policy scenarios associated with resource regulation are displayed in Table 3.

(2) Constraints of available coal resources and corresponding resource regulation:

$$Cr\{[cre_{lt}*(1-\delta_{lt})(\sum_{l=1}^{2}\sum_{t=1}^{3}IMP_{lt}-\sum_{j=1}^{1}\sum_{i=1}^{3}\sum_{h=1}^{3}p_{th}*RMP_{lth})]+[cre_{tj}*(1-\delta_{tj})(\sum_{j=1}^{1}\sum_{t=1}^{3}IAP_{tj}- \sum_{j=1}^{1}\sum_{i=1}^{3}\sum_{h=1}^{3}p_{th}*RAP_{tjh})]+[cre_{tj}*(1-\delta_{tj})*(\sum_{k=1}^{4}\sum_{t=1}^{3}IEP_{tk}-\sum_{k=1}^{4}\sum_{i=1}^{3}\sum_{h=1}^{3}p_{th}*REP_{tkh})]+ [cre_{tg}*(1-\delta_{tg})*(\sum_{g=1}^{3}\sum_{t=1}^{3}ISP_{tg}-\sum_{g=1}^{3}\sum_{i=1}^{3}\sum_{h=1}^{3}p_{th}*RSP_{tgh}]\leq(1-\xi_c)*\widetilde{V}c_{ht}\}\geq\alpha\} \quad (6c)$$

(3) Constraints of available food resources and corresponding resource regulation:

$$Cr\{[[wre_{tj}*(1-\eta_{tj})(\sum_{j=1}^{1}\sum_{t=1}^{3}IAP_{tj}-\sum_{j=1}^{1}\sum_{i=1}^{3}\sum_{h=1}^{3}p_{th}*RAP_{tjh})]*\mathrm{Fre}_{tj}+[wre_{tg}*(1-\eta_{tg})*(\sum_{g=1}^{3}\sum_{t=1}^{3}ISP_{tg} -\sum_{g=1}^{3}\sum_{i=1}^{3}\sum_{h=1}^{3}p_{th}*RSP_{tgh}]*\mathrm{Fre}_{tg}-(\sum_{l=1}^{2}\sum_{t=1}^{3}IMP_{lt}-\sum_{j=1}^{1}\sum_{i=1}^{3}\sum_{h=1}^{3}p_{th}*RMP_{lth})]*\mathrm{Fre}_{lt}+\leq(1-\xi_f)*\widetilde{V}f_{ht}\}\geq\alpha\} \quad (6d)$$

Models (6c) and (6d) present the recourse actions of coal and food shortages to expected demands, which can be restricted by limited available coal and food resources ($\widetilde{V}c_{ht}$ and $\widetilde{V}f_{ht}$) and corresponding resource regulation (ξ_c, ξ_f). In this constraint, policy analysis associated with resource regulation is considered, as shown in Table 3.

(4) Constraints of living population scale for agricultural sector:

$$IAL_{tj}^{\min}\leq\sum_{j=1}^{1}\sum_{t=1}^{3}IAP_{tj}\leq IAL_{tj}^{\max}. \quad (6e)$$

(5) Constraints of employed population scale for industrial sector:

$$IIL_{tk}^{\min}\leq+\sum_{k=1}^{4}\sum_{t=1}^{3}IEP_{tk}\leq IIL_{tk}^{\max}. \quad (6f)$$

(6) Constraints of employed population scale for service sector:

$$ISL_{tg}^{\min}\leq\sum_{g=1}^{3}\sum_{t=1}^{3}ISP_{tg}\leq ISL_{tg}^{\max}. \quad (6g)$$

Models (6e) to (6g) present the scales of agriculture, industry, service, and population development in Beijing city. Here, $IAL_{tj}^{\max}$, $IIL_{tk}^{\min}$, $IIL_{tk}^{\max}$, $ISL_{tg}^{\min}$, and $ISL_{tg}^{\max}$ are the minimum and maximum population scales for agricultural, industrial, and service sectors (person).

(7) Constraints of capacity of water-saving and energy efficiency techniques:

$$\begin{aligned}&\{[\sum_{l=1}^{2}\sum_{t=1}^{3} IMP_{lt} * ee_{lt} * (1-\phi_{lt}) - \sum_{j=1}^{1}\sum_{i=1}^{3}\sum_{h=1}^{3} p_{th} * RMP_{lth} * ee_{lt} * (1-\phi_{lt})] + \{[\sum_{j=1}^{1}\sum_{t=1}^{3} IAP_{tj} * ee_{tj} * (1- \\ &\phi_{tj})] - \sum_{j=1}^{1}\sum_{t=1}^{3}\sum_{h=1}^{3} p_{th} * RAP_{tjh} * ee_{tj} * (1-\phi_{tj})] + [\sum_{k=1}^{4}\sum_{t-1}^{3} IEP_{tk} * ee_{tk} * (1-\phi_{tk}) - [\sum_{k=1}^{4}\sum_{i-1}^{3}\sum_{h-1}^{3} p_{th} * REP_{tkh} * ee_{tk} \\ &*(1-\phi_{tk})] + [\sum_{g=1}^{3}\sum_{t=1}^{3} ISP_{tg} * ee_{tg} * (1-\phi_{tg}) - \sum_{g=1}^{3}\sum_{i=1}^{3}\sum_{h=1}^{3} p_{th} * RSP_{tgh} * ee_{tg} * (1-\phi_{tg})]\} \leq \widetilde{C}w_{ht}\} \geq \alpha\end{aligned} \quad (6h)$$

$$\begin{aligned}&\{\sum_{l=1}^{2}\sum_{t=1}^{3}[IMP_{lt} * ae_{lt} * (1-\mu_{lt}) - \sum_{j=1}^{1}\sum_{i=1}^{3}\sum_{h=1}^{3} p_{th} * RMP_{lth} * ae_{lt} * (1-\mu_{lt})] + [\sum_{j=1}^{1}\sum_{t=1}^{3} IAP_{tj} * + ae_{tj} * (1-\mu_{tj}) \\ &- \sum_{j=1}^{1}\sum_{i=1}^{3}\sum_{h=1}^{3} p_{th} * RAP_{tjh} * ae_{tj} * (1-\mu_{tj})] + [\sum_{k=1}^{4}\sum_{t=1}^{3} IEP_{tk} * ae_{tk} * (1-\mu_{tk}) - [\sum_{k=1}^{4}\sum_{i=1}^{3}\sum_{h=1}^{3} p_{th} * REP_{tkh} * ae_{tk} * (1 \\ &-\mu_{tk})] + [\sum_{g=1}^{3}\sum_{t=1}^{3} ISP_{tg} * ae_{tg} * (1-\mu_{tg}) - \sum_{g=1}^{3}\sum_{i=1}^{3}\sum_{h=1}^{3} p_{th} * RSP_{tgh} * ae_{tg} * (1-\mu_{tg})]\} \leq \widetilde{C}c_{ht}\} \geq \alpha\end{aligned} \quad (6i)$$

Models (6h) and (6i) demonstrate the capacity of the saving technique and recycling technology. Here, Cw_{ht} and Cc_{ht} are the maximal capacity of technique improvement.

(8) Constraints of Hurwicz criterion:

$$Cr\{(\mathrm{ICR}_t + \mathrm{BLA}_t + \mathrm{BCT}_t) \leq U^{opt}\} \geq \alpha. \quad (6j)$$

$$Cr\{(\mathrm{ICR}_t + \mathrm{BLA}_t + \mathrm{BCT}_t) \geq U^{pec}\} \geq \alpha. \quad (6k)$$

(9) Constraints of economic benefit and loss:

$$LAP_{tjh} \leq BAP_{tj}, LIP_{tkh} \leq BEP_{tk}, LSP_{tgh} \leq BSP_{tg}. \quad (6l)$$

(10) Non-negative constraints:

$$LAP_{tjh}, BAP_{tj}, LIP_{tkh}, BEP_{tk}, LSP_{tgh}, BSP_{tg} \geq 0. \quad (6m)$$

$$RAP_{tjh}, BAP_{tj}, RIP_{tkh}, REP_{tk}, RSP_{tgh}, RSP_{tg} \geq 0. \quad (6n)$$

Models (6j) to (6n) present the Hurwicz criterion, as well as various economic benefit or loss and non-negative restrictions.

2.4. Data Acquisition

Table 4 shows the economic data expressed as fuzzy values in three planning periods (5 years for one planning period), which were estimated using the statistical yearbooks of Beijing from 2000 to 2018 with consideration of the economic growth rate [23,24]. Moreover, the random variation in WEF supply capacity could lead to seasonal/uneven distribution of resource availability. For instance, spatial and temporal variation in water availability can result from an uneven distribution of precipitation, which was calculated using previous simulation studies on the annual rainfall of Beijing city (2000 to 2017) [25]. Thus, the probability of low, medium, and high levels of water resources was obtained as 0.2, 0.6, and 0.2 [49,50]. In the study region, the main food production from agriculture can be influenced by rainfall, while the energy supply from hydropower stations can be affected by precipitation under a stable thermal power supply. Thus, the probability of low, medium, and high levels of food and energy resources would be same as the water resources. Furthermore, since the available resources can be influenced by artificial factors (e.g., data deficits and estimation errors), fuzzy programming with a credibility measure was used for the expression of fuzziness.

Table 3 displays the combined policy scenarios of Beijing city, with the aim of reflecting the policy tradeoff among population, technique improvement, and resource regulation. In this study, scenario 0 (S0) represents the basic policy scenario based on the current population—WFE situation, which can generate optimal results according to existent population conditions, economic development, technique levels, and resource regulation. On the basis of S0, three policy scenarios associated with technique improvement (S1 to S3) were designed for three planning periods, using the "empirical method" and "expert consultation method". For instance, the lowest elevated values of resource-use efficiency and retreatment ratio (i.e., 5%) were calculated using previous values (i.e., "empirical method"). With consideration of the speed of technological development in recent years, the highest elevated values of resource-use efficiency and retreatment ratio were estimated by experts (i.e., "expert consultation method"), i.e., 25%. According to the lowest and highest elevated values of resource-use efficiency and retreatment ratio, the medium value (15%) was obtained by the "expert consultation method". Then, two scenarios (S4 and S5) associated with resource regulations were designed according to the same principle, where two levels (5% and 15%) of lower resource limitation were considered. In addition, two policy scenarios (S6 and S7) involving combined policies took technique improvement and resource regulation into account to reflect the tradeoff among various policies.

3. Results and Discussion

3.1. WFE Vulnerability under Basic Policy Scenario (S0)

Figure 2 shows the vulnerability of WFE from 2000 to 2017. The results present that the vulnerability of WFE increased then decreased, with the inflection points occurring in 2012, 2014, and 2016 in line with strategies for population adjustment and technique improvement in Beijing city. In the comparison of exposure (EI), sensitivity (SI), and adaptability (AI) levels, the results reveal that SI levels would be higher than EI levels, demonstrating that population adjustment would influence the WFE system (indicating that WFE system is sensitive to population adjustment). However, the low levels of AI suggest that the effectiveness of government responses to WFE stresses would be relatively low, particularly in 2011.

In general, according to the increase in WFE vulnerability under S0, the failure of the WEF supply system would increase, which would result in a resource shortage. Figure 3 displays the WFE shortages among various sectors under S0 ($\alpha = 0.6$ and $\beta = 0.99$) as follows: (a) water shortages would be mainly influenced by rainfall, presenting a higher value in the dry season and vice versa; (b) the highest coal shortages would occur in industrial plants, particularly in heavy-consumption and other industrial plants (denoted as "HID" and "OID"), representing the main energy-use sectors in Beijing city; (c) food shortages would mainly occur in residential areas due to the large population scale and agriculture transformation to Hebei, which is also influenced by rainfall levels; (d) lower credibility satisfaction levels (α-level) would result in higher resource deficits and vice versa.

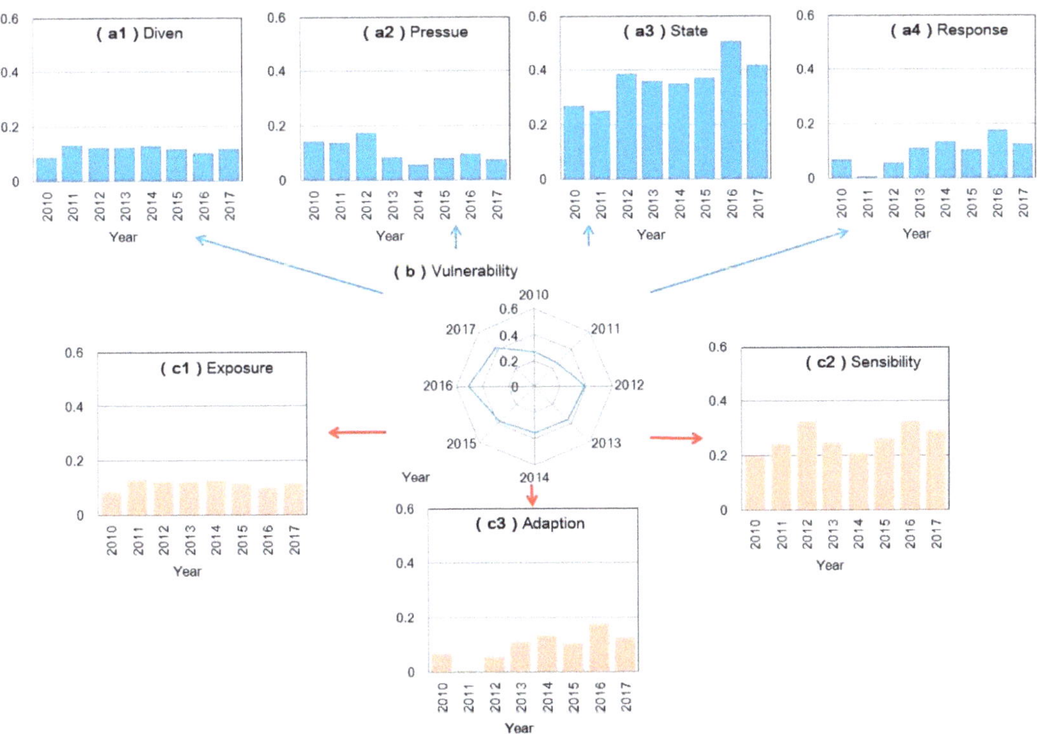

Figure 2. The vulnerability of resources (WEF) from 2000 to 2017 under S0 (without policy adjustment). ((**a1–a4**) are the measure of vulnerability based on Driver–Pressure–State–Response (DPSR) Model; (**b**) is the total vulnerability level; (**c1–c3**) are the exposure, sensitivity, and adaptability levels in vulnerability).

In order to remit resource deficits, the adjustment of population is analyzed in Figure 4, which displays population adjustments among different industrial plants based on water–coal optimization under S0 when $\alpha = 0.99$ and $\beta = 0.6$. The results show that the highest resource-deficient sectors (e.g., HIDs and OIDs) would not lead to the highest population reduction (population adjustment based on industrial transformation), since they had higher incomes or benefits despite great resource stress. On the contrary, the highest population reduction would occur in other service industrial plants (denoted as "OSEs"), due to their lowest benefit per population. Furthermore, the WFE deficit of HIDs and OIDs can be remitted by technique improvement due to their higher-yielding supports; thus, OSEs (deemed as a lower-yielding sector) would adjust the more employed population to relieve WFE pressures and vulnerability.

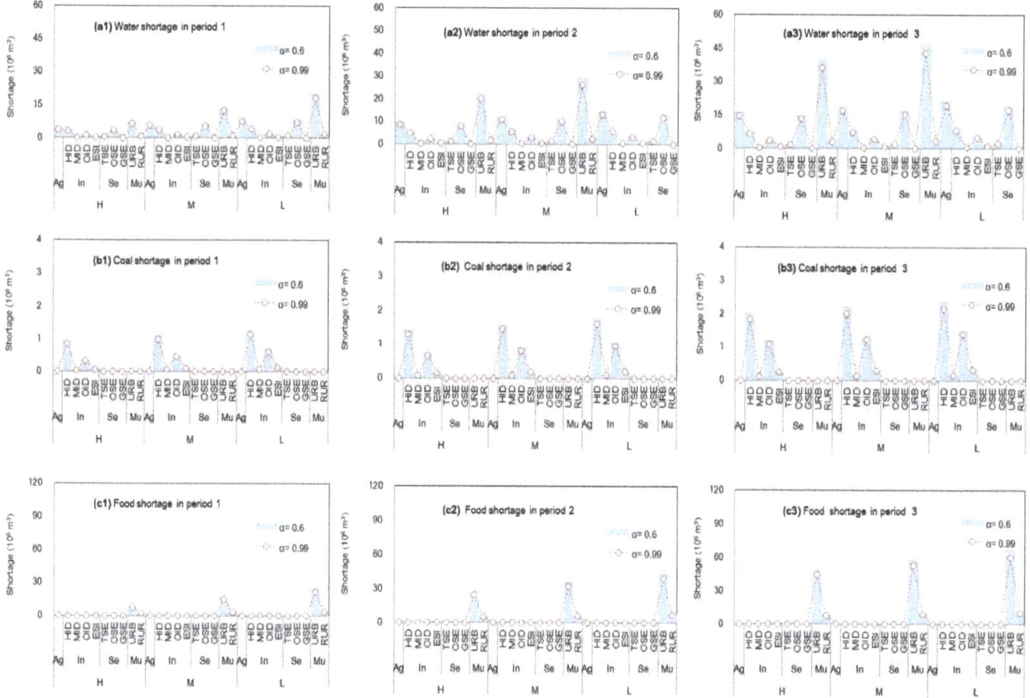

Figure 3. Total water–food–energy shortages among various sectors under S0 (α = 0.6 and β = 0.99) (agricultural sector denoted as "Ag", industrial sector denoted as "In", service sector denoted as "Se", municipal sector denoted as "Mu"; heavy-consumption industry plant denoted as "HID", medium-consumption industry plant denoted as "MID", other industry plant denoted as "OID", energy-supply industry plant denoted as "ESI", traditional service industry plant denoted as "TSE", other service industry plant denoted as "OSE", environmental protection industry plant denoted as "GSE", urban municipal consumption plant denoted as "URB", rural consumption plant denoted as "RUR"; high level of resource availability denoted as "H", medium level of resource availability denoted as "M", low level of resource availability denoted as "L") ((**a1–a3**) are water shortages in various periods; (**b1–b3**) are the coal shortages in various periods; (**c1–c3**) are the food shortages in various periods).

3.2. WFE Shortage and Population Adjustment under Various Policy Scenarios (S1 to S7)

Figure 5 displays the solutions for population adjustments and corresponding WFE shortages in different sectors under various policy scenarios associated with technique improvement (S1 to S3), when α = 0.99 and β = 0.6. The results show that an improvement of resource utilization efficiencies would lead to lower WFE shortages instead of higher population adjustments. Thus, a higher technique improvement level would generate a lower population adjustment. However, technique improvement would generally require financial support; thus, the tradeoff between the income from technique improvement and the cost of technique improvement should be considered.

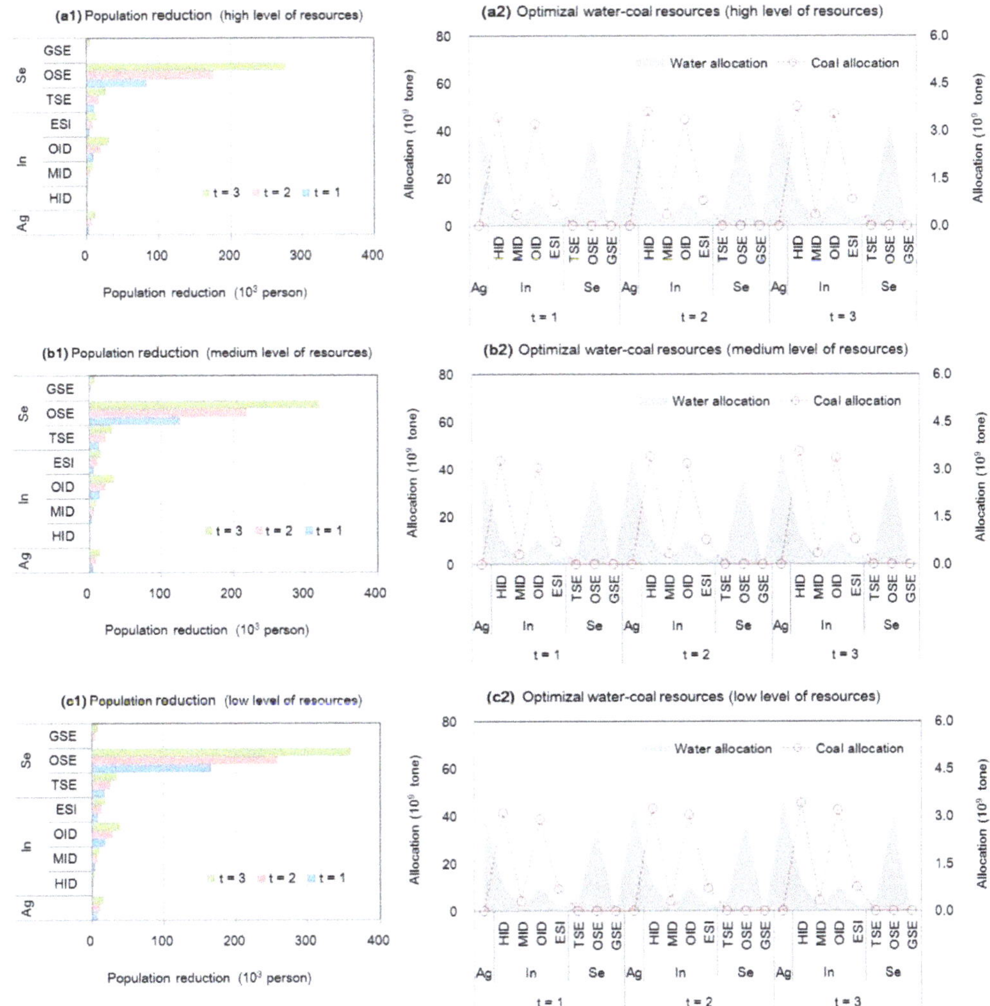

Figure 4. Population adjustments among different industrial plants based on water–coal optimization under S0 when $\alpha = 0.99$ and $\beta = 0.6$ (agricultural sector denoted as "Ag", industrial sector denoted as "In", service sector denoted as "Se", municipal sector denoted as "Mu"; heavy-consumption industry plant denoted as "HID", medium-consumption industry plant denoted as "MID", other industry plant denoted as "OID", energy-supply industry plant denoted as "ESI", traditional service industry plant denoted as "TSE", other service industry plant denoted as "OSE", environmental protection industry plant denoted as "GSE", urban municipal consumption plant denoted as "URB", rural consumption plant denoted as "RUR") ((**a1**,**a2**) are population adjustment and corresponding optimized water-coal resources under high level of resources; (**b1**,**b2**) are population adjustment and corresponding optimized water-coal resources under medium level of resources; (**c1**,**c2**) are population adjustment and corresponding optimized water-coal resources under low level of resources).

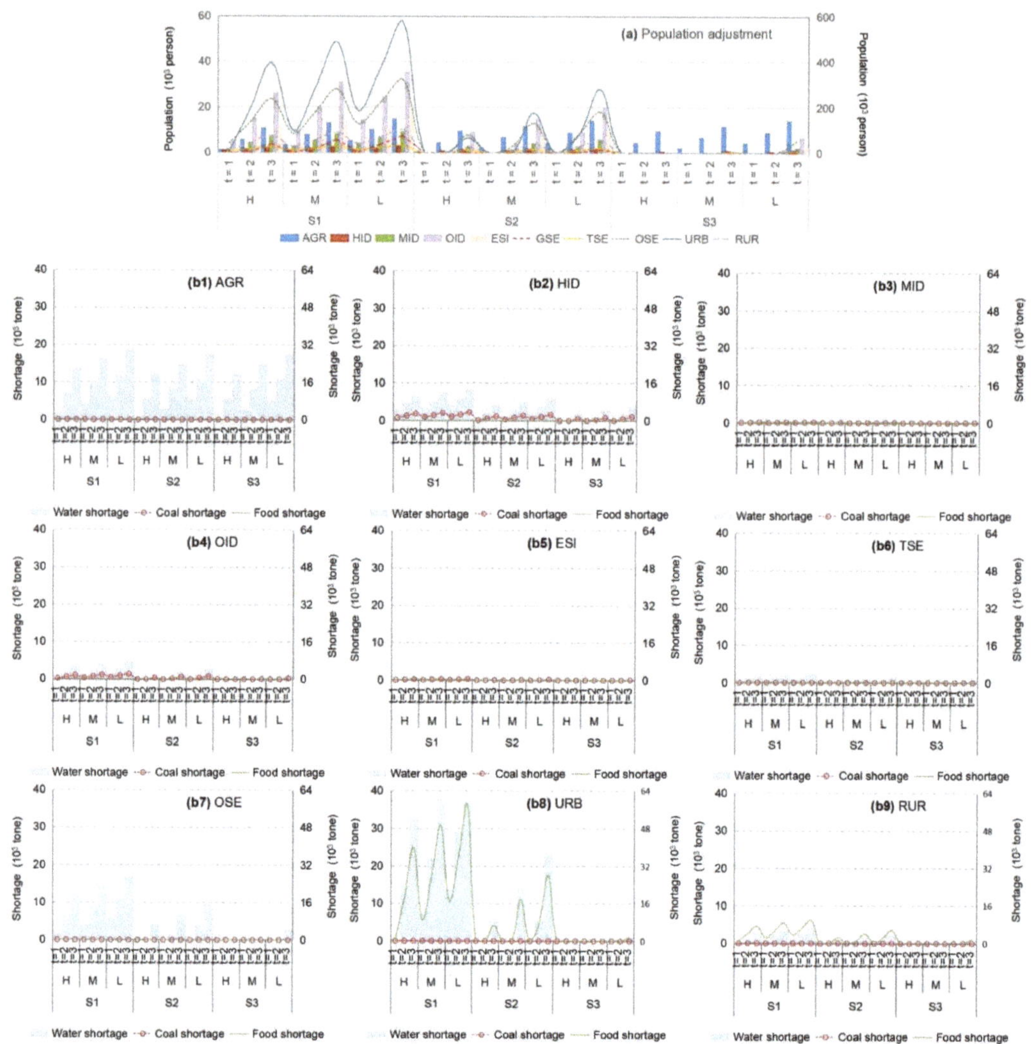

Figure 5. Population adjustments and corresponding water–food–coal shortages in different sectors under S1 to S3 when α = 0.99 and β = 0.6 (agricultural sector denoted as "Ag", industrial sector denoted as "In", service sector denoted as "Se", municipal sector denoted as "Mu"; heavy-consumption industry plant denoted as "HID", medium-consumption industry plant denoted as "MID", other industry plant denoted as "OID", energy-supply industry plant denoted as "ESI", traditional service industry plant denoted as "TSE", other service industry plant denoted as "OSE", environmental protection industry plant denoted as "GSE", urban municipal consumption plant denoted as "URB", rural consumption plant denoted as "RUR"; high level of resource availability denoted as "H", medium level of resource availability denoted as "M", low level of resource availability denoted as "L") ((**a**) is total population adjustment under S1, S2 and S3; (**b1–b9**) are water–food–coal shortages in different sectors under S1 to S3).

Figure 6 presents the satisfaction of WFE targets and corresponding population adjustments under various policy scenarios associated with governmental regulation (S4 to S5), when α = 0.99 and β is 0.6. The results display that the adjustment of population would be reduced by contractible resource regulation. In the comparison of various resources deficits in different sectors, the lowest water satisfaction would take place in the agricultural sector according to lower water resource limits, while the lowest coal satisfaction would occur in heavy-consumption industrial plants and energy-supply plants (e.g., HIDs and ESIs). Although resource regulation can reduce water satisfaction in some plants by reducing direct economic income in the short term, this would compel the companies to save resources to lessen the losses of WFE deficit by pursuing technique improvement, which is more beneficial to technical progress and the water-saving goal achievement in the long term.

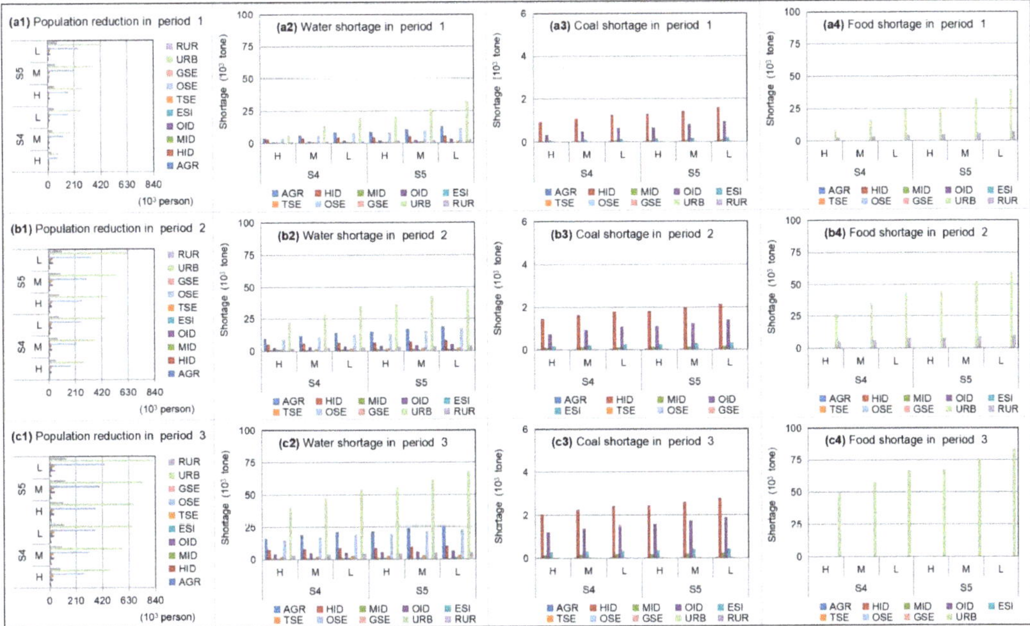

Figure 6. The satisfactions of water–food–energy target and corresponding population adjustments under S4 to S5 when $\alpha = 0.99$ and $\beta = 0.6$ (agricultural sector denoted as "Ag", industrial sector denoted as "In", service sector denoted as "Se", municipal sector denoted as "Mu"; heavy-consumption industry plant denoted as "HID", medium-consumption industry plant denoted as "MID", other industry plant denoted as "OID", energy-supply industry plant denoted as "ESI", traditional service industry plant denoted as "TSE", other service industry plant denoted as "OSE", environmental protection industry plant denoted as "GSE", urban municipal consumption plant denoted as "URB", rural consumption plant denoted as "RUR"; high level of resource availability denoted as "H", medium level of resource availability denoted as "M", low level of resource availability denoted as "L") ((**a1–a4**) are population adjustment and corresponding WEF shortages under S4 and S5 in period 1; (**b1–b4**) are population adjustment and corresponding WEF shortages under S4 and S5 in period 2; (**c1–c4**) are population adjustment and corresponding WEF shortages under S4 and S5 in period 3).

Figure 7 shows the lower population reduction based on combined policies (S1 to S7) compared to S0 ($\alpha = 0.6$ and $\beta = 0.99$). The obtained results demonstrate that an adaptive policy associated with technique improvement can lessen the population reduction in various industrial sectors, whereas a policy associated with resource regulation would increase the population reduction. For example, the lowest population reduction would occur in S3 (the highest technique improvement), while the highest would take place in S5. The combined policies would generate a comprehensive population reduction (such as S6 and S7). The policy tradeoff among population adjustment, resource regulation, and technique improvement can relieve the contradiction between population development and WFE supply capacity.

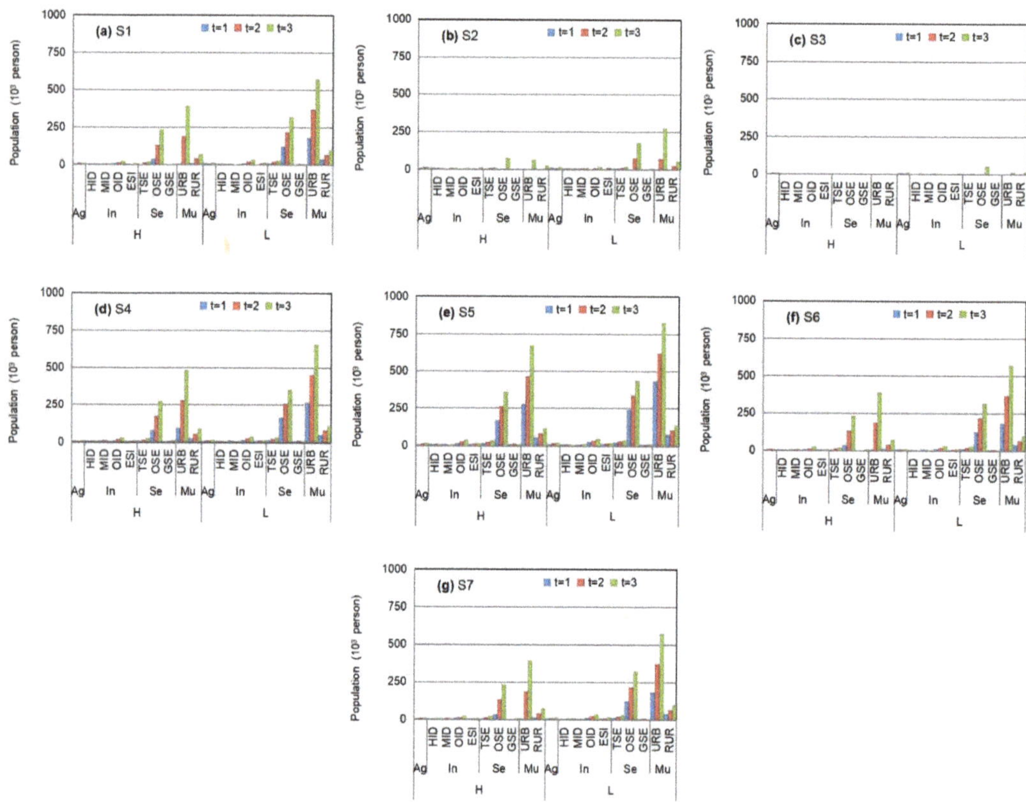

Figure 7. The lower population reduction based on combined policies (S1 to S7) compared to S0 ($\alpha = 0.6$ and $\beta = 0.99$) (agricultural sector denoted as "Ag", industrial sector denoted as "In", service sector denoted as "Se", municipal sector denoted as "Mu"; heavy-consumption industry plant denoted as "HID", medium-consumption industry plant denoted as "MID", other industry plant denoted as "OID", energy-supply industry plant denoted as "ESI", traditional service industry plant denoted as "TSE", other service industry plant denoted as "OSE", environmental protection industry plant denoted as "GSE", urban municipal consumption plant denoted as "URB", rural consumption plant denoted as "RUR"; high level of resource availability denoted as "H", low level of resource availability denoted as "L") ((**a–g**) are the population reduction under S1 to S7).

3.3. System Benefit and Vulnerability Analysis under Various Policy Scenarios (S0 to S7)

Figure 8 presents the vulnerability of resources under S0 to S7, reflecting the variation trend of exposure–sensitivity–adaptability under S0 to S7. With increasing intensity of population adjustment, the exposure of natural resources would decrease accordingly. A conservation population regulation policy would reduce the exposure by about 5.4%; in contrast, an aggressive scenario would reduce the exposure by 30.8%. This indicates that population adjustment, technological progress, and government regulation of resources would not have significant impacts on the exposure level. The adaptability of natural resources in Beijing is influenced by both technological progress and government regulation policies, among which government regulation policies have a greater impact on the adaptability.

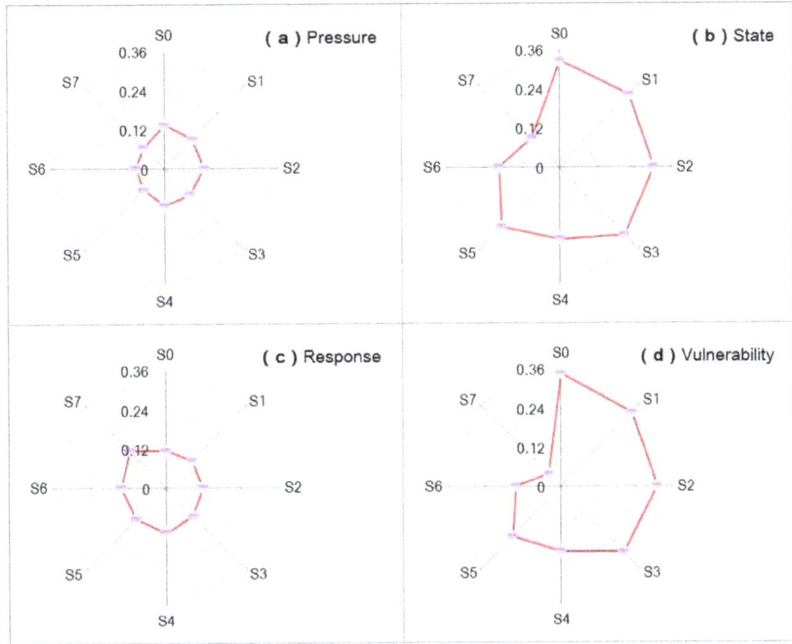

Figure 8. The vulnerability of resources (WEF) under S0 to S7. ((**a**–**c**) are the measure of vulnerability based on Driver–Pressure–State–Response (DPSR) Model; (**d**) is the total vulnerability level).

Figure 9 presents the system benefits and corresponding risk levels under various policy scenarios (from S0 to S7) with the Hurwicz criterion. The obtained results demonstrate the following: (a) the adjustment of population can remit the resource deficit to reduced loss of shortage, which is beneficial for a higher system benefit; (b) although policy scenarios associated with technique improvement can relieve resource stresses, they require financial support, which would generate promotion expenses, leading to decreased benefits (as shown in S1 to S3); (c) policy scenarios associated with resource regulations (e.g., S4 to S5) would generate a reduction in the employed population, leading to lower system benefits; (d) the combined policy scenarios would generate comprehensive benefits (e.g., S6 and S7); (e) a higher α level corresponding to a higher credibility measure (lower violated risk) would generate a lower benefit and vice versa; (f) as the scenario assumption is influenced by the risk preference of policymakers, the Hurwicz criterion (reflected in the β-level) was considered, showing that a higher β level corresponding to a more optimistic attitude can generate a higher benefit, but result in a higher violated risk in the process of decision making and vice versa. Increases in technique requirement and governmental regulation are not suitable for sustainable development of population and WFE management in Beijing. Thus, in general, a combined optimistic–pessimistic result can be obtained when the β level is 0.5.

3.4. Discussion

In this study, population development (such as population scale, growth speed, gathering, and *employment structure*) can be deemed as an important indicator to support economic development and resource consumption, which are pre-regulated at the beginning of planning periods by policymakers for urban planning. Under these situations, population development can be regarded as a driving factor in the vulnerability assessment of WFE (as shown in Table 1), as well as a decision variable in systemic optimization of the P-WEF issue in Beijing city. Population scales and the employed population in various industrial sectors can influence the resource demand and consumption patterns (various consumption coefficients), which would result in differences in WFE vulnerability, resource shortage, optimal population adjustment, and system benefits.

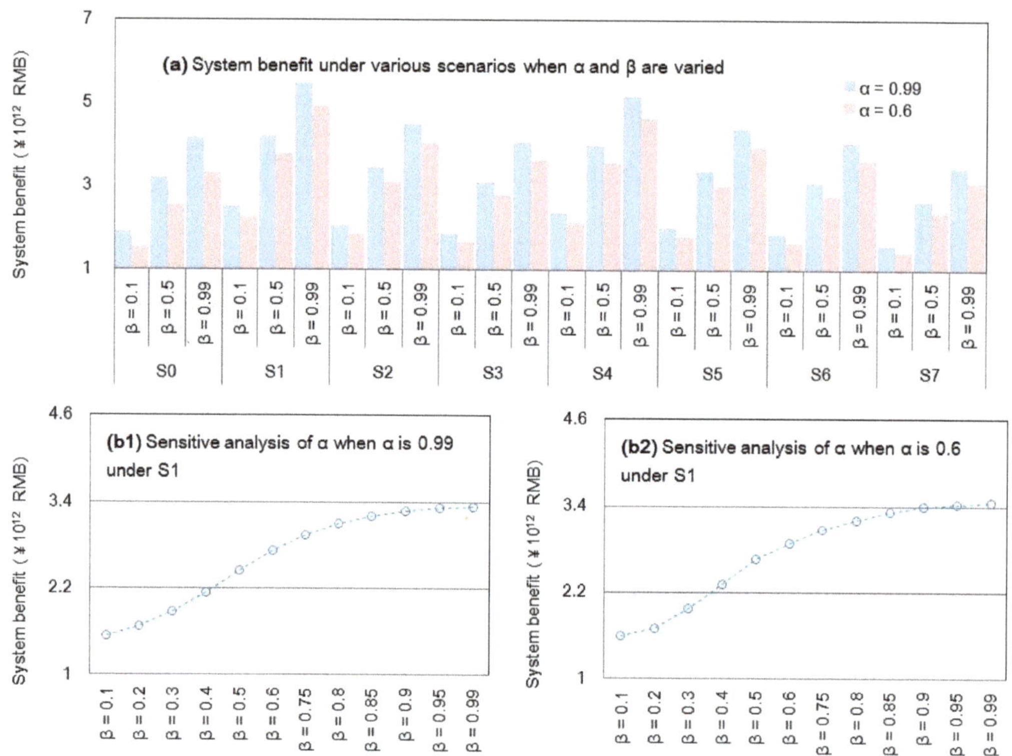

Figure 9. System benefit and risk analysis under S0 to S7.

In a practical P-WEF optimization issue, the implementation of resource regulation and technique improvement can remit resource shortages and vulnerabilities, which would weaken the role of population adjustment. However, a rational design scheme for the initial population policy of the planning period can be considered an effective approach to reduce the losses/penalties of future population adjustment from the long-term perspective of urban development. Thus, various initial population policy scenarios were designed according to the historical situation of population change in Beijing (from 2000 to 2017) and the "14th Five Year Plan" of Beijing as follows: (a) the high and low population development scenarios (S8 and S9) were assumed considering 2.8% and 0.11% growth rates of the population size; (b) three scenarios (S10 to S12) associated with the adjustment of employed population scales were designed according to industrial information and reformed in the "14th Five Year Plan", whereby 1.6%, 4.6%, and 7.1% reductions in the employed population in high-resource-consumption industries (such as agricultural and industrial sectors) were considered. Then, the vulnerability of resources (WEF) under S0 and S8 to S12 were obtained as shown in Figure 10. The results show that both a high growth mode (S8) and a low growth mode (S9) of population size would increase the WFE vulnerability in the short term taking into consideration the current population. Furthermore, it is shown that industrial transformation in Beijing can shrink the employed population scale in high-resource0consumption industries, which would reduce the corresponding WEF vulnerability.

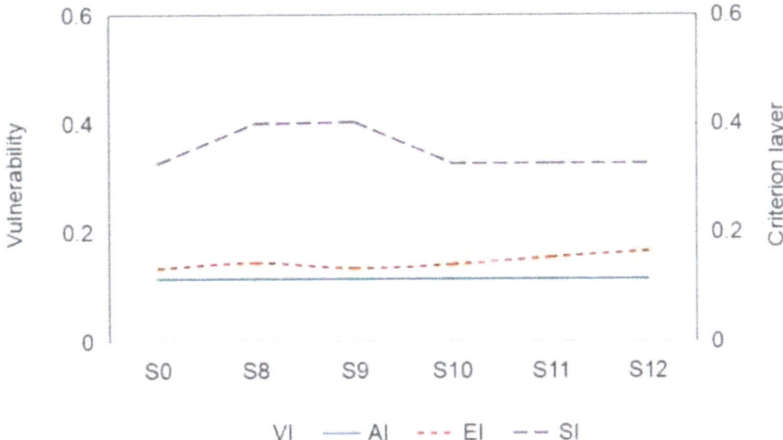

Figure 10. The vulnerability of resources (WEF) under S0 and S8 to S12.

According to the above analysis, the objective of this study was achieved, whereby an adaptive population–resource (including WFE) management framework (APRF) incorporating vulnerability assessment, uncertainty analysis, and systemic optimization methods was developed to optimize the relationship between population development and the water–food–energy (WFE) supply system under combined policies. According to the application of the developed APRF in Beijing, subobjectives were addressed. The vulnerability of resource (including WFE) supply driven by population was calculated and analyzed using the entropy-based driver–pressure–state–response (E-DPSR) model, reflecting the existing WEF vulnerability expressed as exposure, sensitivity, and adaptability based on historical data in a big city. According to the vulnerability assessment, a population–resource (P-WFE) optimization analysis was conducted to identify various policies associated with population adjustment, resource regulation, and technique improvement. A scenario-based dynamic fuzzy model with Hurwicz criterion (SDFH) was developed and embedded into the population–resource (P-WFE) optimization analysis to deal with various types of uncertainties. SDFH can not only build a link between predefined population scale/expected WFE demand and resource shortage penalties due to random water/energy flow, but also handle fuzziness due to data deficiencies. In addition, it is effective in reflecting the risk preferences of policymakers in the process of decision making. Various results associated with WFE shortages, population–economy adjustments, resource vulnerability, and system benefits under various policy scenarios were analyzed, reflecting the tradeoff between population development and WFE management in a risk-averse manner. The above-obtained objectives can facilitate the adjustment of population–resource policies in a big city.

4. Conclusions

With the aid of practical implications of the developed APRF framework in Beijing city, a number of discoveries were made. Firstly, the current population scale and WEF demand cannot accommodate to the regional resource supply capacity in Beijing city, resulting in resource shortages. Secondly, although industrial transformation can prompt the adjustment of the employed population structure to reduce WFE stress and vulnerability, excessive and irrational population policies would damage the smooth operation of the economy in a big city. Thirdly, low efficiencies in the current WFE use pattern and technique level of Beijing enhance the vulnerability of the resource supply system. Although policies associated with technique improvement for resource saving are effective in addressing the conflict among population, WFE, and economy, a higher investment of technology would hinder their application. Thus, how to make a comprehensive policy for balancing the relationships among population adjustment, resource regulation, and technique improvement taking into account the risk preferences of policymaker is an important issue for sustainable development of Beijing. Fourthly, individual polices (such as improvement of resource utilization efficiency, resource regulation, and employed population adjustment based on industrial transformation) have their own advantages in terms of resource shortage reduction, but there are limitations of the high cost of generalization and direct income reduction in the short term. Thus, how to balance the tradeoff between benefit and cost in the long term can be challenging for regional policymakers. Lastly, the differences in the risk attitude of policymakers when confronting uncertain information (e.g., fuzzy resource supply capacity, dynamic expected target, and risk preference) would generate varied policies, which would influence the P-WEF strategy.

Therefore, specific recommendations are proposed. Regional resource carrying capacity should be deemed as an important impact factor in a rational population policy, which could reduce the losses of resources deficits in a big city with a large population scale. Increasing the entry standards for high-consumption enterprises can allow adjusting the employed population structure, which is beneficial for reducing WFE stress/vulnerability and ensuring high-quality economic development. Furthermore, a greener and cleaner production mode should be encouraged to generate a new population employment situation, remitting resource stress from the consumption side. A combination of policies such as market admittance, governmental support, and financial subsidies to support technique improvement should be carried out, which can stimulate resource saving and recycling. Moreover, the government should promote the concept of resource saving to improve the deficiencies of resource consumption from the consumer's perspective. The tradeoff between economic benefits and costs of various policy scenarios should be designed not only in the short term, but also in the long term, which can maximize the positive effects, while minimizing risks to a great extent. Lastly, the risk preferences of policymaker should be considered in comprehensive governance strategies or policies to fortify the robustness of population–resource optimization, thereby achieving sustainable development in Beijing city.

Author Contributions: X.Z. conceptualized the study; H.X. formulated the model; J.L. designed scenarios and analyzed the results; Y.X. (Yong Xue) contributed data to the revised paper; J.Z. revised the paper; Y.X. (Yuqian Xu) edited the format and revised the paper. All authors read and agreed to the published version of the manuscript.

Funding: This research was funded by the National Natural Science Foundation of China (42007412, 51809145).

Institutional Review Board Statement: Not applicable.

Informed Consent Statement: Not applicable.

Data Availability Statement: The data presented in this study are available on request from the corresponding author.

Acknowledgments: The authors are grateful to the editors and the anonymous reviewers for their insightful comments and suggestions.

Conflicts of Interest: The authors declare no conflict of interest.

Abbreviations

Objective fuction	
f	Total system benefit (RMB)
Decision variable	
$IMP_{tl}, IAP_{tj}, IEP_{tk}, ISP_{tg}$	Expected population for municipal, agricultural, industrial, and service sectors in period t (person)
$RMP_{tlh}, RAP_{tjh}, REP_{tkh}, RSP_{tgh}$	The population with resource shortages for municipal, agricultural, industrial, and service sectors in period t (person)
$AMP_{tlh}, AAP_{tjh}, AEP_{tkh}, ASP_{tgh}$	The population adjustment for municipal, agricultural, industrial, and service sectors in period t (person)
Random variable	
$Vw_{ht}, Vc_{ht}, Vf_{ht}$	Available water resources, coal resources, and food in period t (ton)
R_{ht}	Water flow from river of in period t under probability p_{hj} in period t (m^3)
Parameter	
BMP_{tl}	Net benefit of population per volume of resource being satisfied in period t (RMB/person)
$BAP_{tj}, BIP_{tk}, BSP_{tg}$	Net benefit of population per volume of resource being satisfied for agricultural, industrial, and service sectors in period t (RMB/person)
$wre_{tg}, cre_{lt},$	The resource consumption per population for municipal, agricultural, industrial, and service sectors in period t (ton/person)
η, δ	The improvement ratio of resource-saving technique (%)
$LMS_{tl}, LAS_{tj}, LIS_{tk}, LSS_{tg}$	Loss of population with resource shortages for municipal, agricultural, industrial, and service sectors per volume of resources not being satisfied in period t (RMB/person)
$LMP_{tl}, LAP_{tj}, LIP_{tk}, LSP_{tg}$	Loss of population adjustment for municipal, agricultural, industrial, and service sectors per volume of resources not being satisfied in period t (RMB/person)
ee_{lt}, ae_{lt}	The resource consumption per population for municipal, agricultural, industrial, and service sectors with consideration of technique improvement in period t (ton/person)
φ, μ	The improvement ratio of retreatment technique (%)
$CMS_{lt}, CAS_{tj}, CIS_{tk}, CSS_{tg}$	The cost of technique improvement for population in municipal, agricultural, industrial, and service sectors in period t (RMB/person)
H_{ht}	Normal water requirement of watercourse in period t (m^3)
G_{ht}	Evaporation and infiltration loss of water from river in period t (m^3)
ξ_w, ξ_c, ξ_f	The reduced ratio of resource limit for resource-saving target (%)
Cw_{ht}, Cc_{ht}	The maximal capacity of retreatment technology
$IAL_{tj}^{\min}, IAL_{tj}^{\max}$	Minimum and maximum population scale for agricultural sector (person)
$IIL_{tk}^{\min}, IIL_{tk}^{\max}$	Minimum and maximum population scale for industrial sector (person)
$ISL_{tg}^{\min}, ISL_{tg}^{\max}$	Minimum and maximum population scale for service sector (person)

References

1. Zeng, X.; Zhao, J.; Wang, D.; Kong, X.; Huang, G. Scenario analysis of a sustainable water-food nexus optimization with consideration of population-economy regulation in Beijing-Tianjin-Hebei region. *J. Clean. Prod.* **2019**, *228*, 927–940. [CrossRef]
2. Fang, G.; Wang, Q.; Tian, L. Green development of Yangtze River Delta in China under Population-Resources-Environment-Development-Satisfaction perspective. *Sci. Total Environ.* **2020**, *727*, 138710. [CrossRef] [PubMed]
3. Zeng, X.T.; Xiang, H.; Xue, Y.; Su, Y.; Tong, Y.F.; Mao, Z.Y.; Chen, C.; Kong, X.M. A scenario-based optimization frame to adjust current strategy for population-economy-resource-environment harmony in an urban agglomeration, China. *Sustain. Cities Soc.* **2021**, *67*, 102710. [CrossRef]
4. Sun, L.; Pan, B.; Gu, A.; Lu, H.; Wang, W. Energy-water nexus analysis in the Beijing-Tianjin-Hebei region: Case of electricity sector. *Renew. Sustain. Energy Rev.* **2018**, *93*, 27–34. [CrossRef]
5. Chen, C.; Zeng, X.T.; Yu, L.; Huang, G.H.; Li, Y.P. Planning energy-water nexus systems based on a dual risk aversion optimization method under multiple uncertainties. *J. Clean. Prod.* **2020**, *255*, 120100. [CrossRef]
6. Hardin, G. *The Tragedy of the Commons*; H Press: New York, NY, USA, 1968.
7. Horta, I.M.; Camanho, A.S.; Dias, T.G. Residential building resource consumption: A comparison of Portuguese municipalities' performance. *Cities* **2016**, *50*, 54–61. [CrossRef]
8. Tipple, G. Employment from housing: A resource for rapidly growing urban populations. *Cities* **1994**, *6*, 372–376. [CrossRef]
9. Sakanoue, S. A resource-based approach to modelling the dynamics of interacting populations. *Ecol. Model.* **2009**, *220*, 1383–1394. [CrossRef]

10. Bianchi, T.; Jones, C.G.; Shachak, M. Positive feedback of consumer population density on resource supply. *Trends Ecol. Evol.* **1989**, *4*, 234–238. [CrossRef]
11. Gilland, B. World Population and Food Supply: Can Food Production Keep Pace with Population Growth in the Next Half-century? *Food Policy* **2002**, *27*, 47–63. [CrossRef]
12. Al-Shutayri, A.S.; Al-Juaidi, A.E.M. Assessment of future urban water resources supply and demand for Jeddah City based on the WEAP model. *Arab. J. Geosci.* **2019**, *12*, 431. [CrossRef]
13. Vorosmarty, C.; Green, P.; Salisbury, J.; Lammers, R.B. Global water resources: Vulnerability from climate change and population growth. *Science* **2000**, *289*, 284–288. [CrossRef]
14. Nadine, A.; Marshall, R.; Tobin, C.; Paul, A.; Marshall, M.G.; Hobday, A.J. Social vulnerability of marine resource users to extreme weather events. *Ecosystems* **2013**, *16*, 797–809.
15. Yang, X.; Sun, B.Y.; Zhang, J.; Li, Y.Q. Hierarchy evaluation of water resources vulnerability under climate change in Beijing, China. *Nat. Hazards* **2016**, *84*, 63–76. [CrossRef]
16. Chen, J.; Yu, X.; Qiu, L.; Dong, R. Study on vulnerability and coordination of water-energy-food system in Northwest China. *Sustainability* **2018**, *10*, 3712. [CrossRef]
17. Young, O.R. Institutional dynamics: Resilience, vulnerability and adaptation in environmental and resource regimes. *Glob. Environ. Chang.* **2010**, *20*, 21–35. [CrossRef]
18. Li, R.; Fang, S.; Li, J. Regional water use structure optimization under multiple uncertainties based on water resources vulnerability analysis. *Water Resour. Manag.* **2018**, *32*, 1827–1847. [CrossRef]
19. Xiang, X.; Li, Q. Water resources vulnerability assessment and adaptive management based on projection Pursuit model. *J. Coast. Res.* **2020**, *103*, 431. [CrossRef]
20. Huang, G.H.; Loucks, D.P. An inexact two-stage stochastic programming model for water resources management under uncertainty. *Civ. Eng. Environ. Syst.* **2000**, *17*, 95–118. [CrossRef]
21. Guo, P.; Huang, G.H. Two-stage fuzzy chance-constrained programming: Application to water resources management under dual uncertainties. *Stoch. Environ. Res. Risk Assess.* **2009**, *23*, 349–359. [CrossRef]
22. Doolse, G.; Kingwell, R. Robust mathematical programming for natural resource modeling under parametric uncertainty. *Nat. Resour. Model.* **2010**, *23*, 283–302.
23. Inuiguchi, M. Robust optimization by fuzzy linear programming. *Lect. Note Econ. Math. Syst.* **2012**, *658*, 219–239.
24. Ji, X.; Chen, B. Assessing the energy-saving effect of urbanization in China based on stochastic impacts by regression on population, affluence and technology (STIRPAT) model. *J. Clean. Prod.* **2017**, *163*, s306–s314. [CrossRef]
25. Zeng, X.T.; Tong, Y.F.; Cui, L.; Kong, X.M.; Sheng, Y.N.; Chen, L.; Li, Y.P. Population-production-pollution nexus based air pollution management model for alleviating the atmospheric crisis in Beijing, China. *J. Environ. Manag.* **2017**, *197*, 507–521. [CrossRef] [PubMed]
26. Peterson, G.D.; Graeme, S.C.; Stephen, R.C. Scenario Planning: A Tool for Conservation in an Uncertain World. *Conserv. Biol.* **2003**, *17*, 358–366. [CrossRef]
27. Rogers, C.D.F. Engineering future liveable, resilient, sustainable cities using foresight. *Proc. Inst. Civ. Eng.* **2018**, *171*, 3–9. [CrossRef]
28. Salas, J.; Yepes, V. Enhancing sustainability and resilience through multi-level infrastructure planning. *Int. J. Environ. Res. Public Health* **2020**, *17*, 962. [CrossRef]
29. Chen, Z.; Qiu, B. Resilient planning frame for building resilient cities. In Proceedings of the 7th International Association for China Planning Conference, Shanghai, China, 1–29 June 2013.
30. Beijing Water Bureau. *The Water Resources Report of Beijing City (WRB) 2006–2018*; Beijing Water Bureau: Beijing, China, 2019.
31. Beijing Municipal Bureau of Statistics. *The Statistical Yearbook of Beijing City (SYB) 2006–2018*; China Statistics Press: Beijing, China, 2019.
32. Kong, X.M.; Huang, G.H.; Fan, Y.; Li, Y.P.; Zeng, X.; Zhu, Y. Risk analysis for water resources management under dual uncertainties through factorial analysis and fuzzy random value-at-risk. *Stochastic. Environ. Re. Risk. Assess.* **2019**, *31*, 1–16. [CrossRef]
33. Berardy, A.; Chester, M.V. Climate change vulnerability in the food, energy, and water nexus: Concerns for agricultural production in Arizona and its urban export supply. *Environ. Res. Lett.* **2017**, *12*, 035004. [CrossRef]
34. Li, P.Y.; Qian, H.; Wu, J.H. Application of Set Pair Analysis Method Based on Entropy Weight in Groundwater Quality Assessment—A Case Study in Dongsheng City, Northwest China. *J. Chem.* **2011**, *8*, 851–858.
35. Liu, X.; Wang, Y.; Peng, J. Assessing vulnerability to drought based on exposure, sensitivity and adaptive capacity: A case study in middle Inner Mongolia of China. *Chin. Geogr. Sci.* **2013**, *23*, 13–25. [CrossRef]
36. Hem, N.; Peggy, T.B.; Marcia, M.B. Social indicators of vulnerability for fishing communities in the Northern Gulf of California, Mexico: Implications for climate change. *Mar. Policy* **2014**, *45*, 182–193.
37. Ji, Y.; Huang, G.H.; Sun, W. Risk assessment of hydropower stations through an integrated fuzzy entropy-weight multiple criteria decision making method: A case study of the Xiangxi River. *Expert Syst.* **2015**, *42*, 79–86. [CrossRef]
38. He, Y.; Guo, H.; Jin, M.; Ren, P. A linguistic entropy weight method and its application in linguistic multi-attribute group decision making. *Nonlinear Dyn.* **2016**, *84*, 399–404. [CrossRef]
39. Li, L.; Liu, F.; Li, C. Customer satisfaction evaluation method for customized product development using Entropy weight and Analytic Hierarchy Process. *Comput. Ind. Eng.* **2014**, *77*, 80–87. [CrossRef]

40. Halkos, G.E.; Stern, D.I.; Tzeremes, N.G. Economic growth and environmental efficiency: Evidence from US regions. *Econ. Lett.* **2013**, *120*, 48–52. [CrossRef]
41. Coban, V.; Guler, E.; Kılıç, T.; Kandemir, S.Y. Precipitation forecasting in Marmara region of Turkey. *Arab. J. Geosci.* **2021**, *14*, 86. [CrossRef]
42. Krueger, E.; Rao, P.S.C.; Borchardt, D. Quantifying urban water supply security under global change. *Glob. Environ. Chang.* **2019**, *56*, 66–74. [CrossRef]
43. Li, X.; Zhang, Y.; Abbassi, R.; Yang, M.; Zhang, R.; Chen, G. Dynamic probability assessment of urban natural gas pipeline accidents considering integrated external activities. *J. Loss Prev. Process. Ind.* **2021**, *69*, 104388. [CrossRef]
44. Salas, J.; Yepes, V. A discursive, many-objective approach for selecting more-evolved urban vulnerability assessment models. *J. Clean. Prod.* **2018**, *176*, 1231–1244. [CrossRef]
45. Tanaka, H.; Guo, P.; Zimmermann, H.J. Possibility distributions of fuzzy decision variables obtained from possibilistic linear programming problems. *Fuzzy Sets Syst.* **2000**, *13*, 323–332. [CrossRef]
46. Swarta, R.J.; Raskinb, P.; Robinsonc, J. The problem of the future: Sustainability science and scenario analysis. *Glob. Environ. Chang.* **2004**, *14*, 137–146. [CrossRef]
47. Zeng, X.T.; Li, Y.P.; Huang, G.H.; Liu, L. A two-stage interval-stochastic water trading model for allocating water resources of Kaidu-Kongque River in northwestern China. *J. Hydroinform.* **2015**, *17*, 551. [CrossRef]
48. Dong, Y.R.; Li, Y.; Kong, F.L.; Zhang, J.L.; Xi, M. Source structural characteristics and ecological indication of dissolved organic matter extracted from sediments in the primary tributaries of the Dagu River. *Ecol. Indic.* **2020**, *109*, 105576. [CrossRef]
49. Zeng, X.T.; Zhang, S.J.; Feng, J.; Huang, G.H.; Li, Y.P.; Zhang, P.; Chen, J.P.; Li, K.L. A multi-reservoir based water-hydroenergy management model for identifying the risk horizon of regional resources-energy policy under uncertainties. *Energy Convers. Manag.* **2017**, *143*, 66–84. [CrossRef]
50. Kong, X.M.; Huang, G.H.; Li, Y.P.; Fan, Y.R.; Zeng, X.T.; Zhu, Y. Inexact Copula-Based Stochastic Programming Method for Water Resources Management under Multiple Uncertainties. *J. Water Resour. Plan. Manag.* **2018**, *144*, 98. [CrossRef]

International Journal of
Environmental Research and Public Health

MDPI

Article

The Contribution of MCDM to SUMP: The Case of Spanish Cities during 2006–2021

Salvador Garcia-Ayllon [1,*], Eloy Hontoria [2] and Nolberto Munier [3]

1 Department of Civil Engineering, Technical University of Cartagena, Paseo Alfonso XIII, 52, 30203 Cartagena, Spain
2 Department of Business Economics, Technical University of Cartagena, Calle Real, 3, 30201 Cartagena, Spain; eloy.hontoria@upct.es
3 INGENIO (UPV), Sir John A. Mc Donald Boulevard, Kingston, ON KTM 1A3, Canada; nolmunier@yahoo.com
* Correspondence: salvador.ayllon@upct.es

Abstract: Sustainable Urban Mobility Plans (SUMP) are increasingly popular planning tools in cities with environmental issues where numerous actions are usually proposed to reduce pollution from urban transport. However, the diagnosis and implementation of these processes requires broad consensus from all stakeholders and the ability to fit them into urban planning in such a way that it allows the proposals to become realistic actions. In this study, a review of the sustainable urban mobility plans of 47 cities in Spain during the last 15 years has been carried out, analyzing both the diagnosis and proposal of solutions and their subsequent implementation. From the results obtained, a new framework based on a structured hybrid methodology is proposed to aid decision-making for the evaluation of alternatives in the implementation of proposals in SUMP. This hybrid methodology considers experts' and stakeholders' opinion and applies two different multi-criteria decision making (MCDM) methods in different phases to present two rankings of best alternatives. From that experience, an analysis based on the MCDM methods called 'Sequential Interactive Modelling for Urban Systems (SIMUS)' and weighted sum method (WSM) was applied to a case study of the city of Cartagena, a southeastern middle-size city in Spain. This analytic proposal has been transferred to the practical field in the SUMP of Cartagena, the first instrument of this nature developed after COVID-19 in Spain for a relevant city. The results show how this framework, based on a hybrid methodology, allows the development of complex decision mapping processes using these instruments without obviating the need to generate planning tools that can be transferred from the theoretical framework of urban reality.

Keywords: SUMP; urban mobility; city planning; hybrid methodology; multi-criteria decision making (MCDM) methods; SIMUS; weighted sum method (WSM)

Citation: Garcia-Ayllon, S.; Hontoria, E.; Munier, N. The Contribution of MCDM to SUMP: The Case of Spanish Cities during 2006–2021. *Int. J. Environ. Res. Public Health* **2022**, *19*, 294. https://doi.org/10.3390/ijerph19010294

Academic Editor: Paul B. Tchounwou

Received: 25 November 2021
Accepted: 27 December 2021
Published: 28 December 2021

1. Introduction

According to the United Nations, transport is responsible for 25% of greenhouse gases [1], with cars, buses, vans and other actors in the urban mobility of cities responsible for 70% of these emissions [2]. Urban mobility is one of the phenomena that has undergone a process of great transformation in cities in recent decades [3]. Its guiding parameters and derived impacts are among several of the main Sustainable Development Goals (SDGs) of the 2030 Agenda of the United Nations, having a special impact on factors such as air pollution, the rational use of available resources, efficiency and competitiveness of the labor market and the environmental improvement of public space [4]. Mobility indirectly affects SDG 9, which aims to "Build resilient infrastructure, promote inclusive and sustainable industrialization and foster innovation" and SDG 7 "Affordable and clean energy" where sustainable mobility aims at reducing the use of fossil fuels, and more specifically, SDG 11, which seeks to "Make cities and human settlements inclusive, safe, resilient and sustainable" [5,6].

According to the United Nations [7], planning mobility in the cities of the future is a strategic issue for the sustainable development of the planet. For this reason, its delegations are developing projects to use mobility as a tool to achieve the SDGs. As established by [8], sustainable mobility must be equitable, efficient, green and safe. Several authors [9–11] even consider that ensuring mobility in urban areas is essential to achieve at least 7 of the rest of the SDGs: health and well-being (SDG 3), affordable and clean energy (SDG 7), decent work and economic growth (SDG 8), industry, innovation and infrastructure (SDG 9), reduction of inequalities (SDG 10), sustainable cities and communities (SDG 11) and climate action (SDG 13).

A relevant example is the agenda promoted by the Sustainable Transport Division of the Economic Commission for Europe (UNECE, [7]). This comprises of three types of initiatives: analytical, regulatory and capacity building to guide the development of mobility towards the achievement of the SDGs. In its initiatives, the UNECE considers that betting on sustainable mobility as a model of energy transition, the generation of infrastructures for efficient logistics, vehicle regulation, the establishment of investment models in infrastructures, education in road safety and as the axis of urban planning, would have a major impact on up to 13 of the 17 SDGs.

This importance at the international institutional level, however, has not been transmitted sufficiently to the regulatory framework in countries; decision-making in cities being one of the most difficult factors to implement in this matter [8]. The European Union is currently the area in which the most institutional work has been conducted on this matter at the normative field. The regulatory framework for urban mobility began to be developed in 2011 with the publication of the "White Paper on Transport" [12]. In 2013, the so-called "Urban Mobility Package" was approved, which included various initiatives and communications from both the European Parliament and the European Commission [13]. Finally, in 2016 the European strategy for low-emission mobility was published [14]. However, no directive has yet to be approved that specifies urban mobility must be transposed into the national legislation of the different EU countries [8]. Despite this, in the last decade, many cities have developed sustainable urban mobility plans to implement actions and strategies in their urban planning to promote greener mobility and improve urban space by making it more attractive for pedestrians and bicycles.

One of the best-known cases in this field is that of the city of Paris [15], where the philosophy of the so-called "City of 15 minutes" promotes proximity mobility in urban planning more than the traditional solutions of this type of planning (promotion of public transport, improvement of the cycle lane network, modernization of vehicles, etc.) and has been a success that has been exported internationally. This approach goes beyond the traditional philosophy of sustainable mobility based on the execution of infrastructures and the improvement of the public transport service, proposing a sociological change at the urban level that must be implemented through city planning. Through so-called *chrono-urbanism*, changes are sought to traditional zoning paradigms of cities that segregate uses by promoting long-distance displacements through urban proposals that mix uses in neighborhood structures by covering the main demands of any citizen in a temporary radius, so that their main demands can be carried out walking or cycling. Cities such as Melbourne have adopted similar philosophies with their "20-minute city proposal" [16], while other cities such as Milan, through their *Strade Aperte* plan [17,18], have put forward more restrictive proposals to reduce the space for private vehicle circulation in favor of more space for cycling and pedestrian mobility.

However, the urban plans to improve mobility are not a recent invention but are being developed through urban planning with different initiatives in some cities that turned out to be pioneers in this matter. This was the case, for example, of the city of Curitiba, which developed the well-known "Integrated transport network" in the 1980s within the framework of the general urban plan of the city, and which later served as a model for many other cities in Latin America such as Medellin, Bogota, Rio de Janeiro and Santiago de Chile [19].

In Spain, the importance of urban mobility has been reflected in the regulatory framework since 2005, and has been supported by generalist decrees, ministerial orders and calls for aid promoting sustainable mobility [20]. This has led to the development of numerous sustainable urban mobility plans (SUMP) in many cities in Spain during the last 20 years, but has not been supported by a specific technical and regulatory framework. This tool has become quite popular in local planning, sometimes more in the effort to obtain subsidies and European funds for cohesion and development, rather than due to the existence of a true political commitment on the part of local administrations [21]. However, the non-existence of a specific regulatory framework on the matter has given rise to a wide technical heterogeneity, with different approaches both at the diagnostic field and at the level of proposing solutions.

One of the best-known proposals in this context in Spain has been the so-called "Madrid Central" plan approved by the Madrid city council in 2018 [22]. This plan restricted access to the urban center area to the most polluting vehicles, and after various political and social controversies is currently paralyzed due to a change of government in the municipal corporation. However, this is not an isolated case, since there are many urban mobility plans in Spain whose implementation has been defective or even non-existent [23–25]. This situation has occurred either due to the difficulty in generating the necessary consensus with the stakeholders to establish diagnoses and solutions to the current problems, or due to the lack of will of the municipal administrations when implementing the necessary actions [26].

In this context of policy implementation, the multi-criteria decision making (MCDM) methods approaches are currently a widely used tool in the field of multiparametric analysis for the diagnosis and resolution of complex planning problems. A huge amount of successful cases in their application may be found related to complex scenarios' assessment [27] in fields such as logistics [28], environmental management [29], construction industry [30], carbon emissions [31] and even sports tourism [32] and medicine management [33]. Therefore, these tools, because of their characteristics, present a high degree of applicability to complex phenomena of urban planning, such as promoting sustainable mobility.

In this study, an analysis of the implementation of sustainable urban mobility plans in Spain in the last 15 years is carried out for 43 cities. Based on the deficiencies detected, a structured framework based on a MCDM hybrid methodology for SUMP implementation is proposed to address most of the common problems observed. This methodology with two phases, considers in the first phase the experts´ opinion and applies the WSM [34] to obtain an initial ranking of best alternatives for urban mobility. To check the robustness of this initial ranking and to correct some problems detected in the analysis performed in 47 sustainable mobility plans, the stakeholders´ priorities are considered in a second phase of the methodology and an additional ranking through the application of SIMUS [35] is presented. In order to check the robustness of the presented methodology and subsequent framework, both rankings are compared to find discrepancies that could make the authors aware of the reliability of the results.

This methodological proposal has been successfully applied to the case of the sustainable urban mobility plan of the city of Cartagena, the first plan of this category approved after the COVID-19 pandemic in a relevant city in Spain and may be used for other cities that could wish to implement a success sustainable urban mobility plan.

The rest of the paper is structured as follows. Section 2 develops the methodological proposal and describes the case study: Section 2.1 presents how the analysis was performed for 47 SUMPs carried out in Spain; the proposed methodology and framework with SIMUS and the WSM MCDM methods are explained in Section 2.2; and the case study where the proposed methodology is applied (city of Cartagena) is described in Section 2.3. Section 3 provides the most relevant results that are discussed in Section 4. Finally, the conclusions of this research are available in Section 5.

2. Materials and Methods

The research has been carried out in different phases. First, an analysis has been undertaken to assess the status of the Sustainable Urban Mobility plans carried out in Spain. Based on the information about the weaknesses and strengths detected in these urban planning instruments, it is proposed a methodological approach to improve the development of diagnoses and solution proposals. Finally, this framework has been applied in the case of the city of Cartagena.

2.1. Analysis of SUMP Implementation in Spain

For the analysis of the 47 SUMPs evaluated, quantitative and qualitative indicators have been considered to assess the usual profile of the plans. The choice of indicators is aimed at being able to evaluate the statistical correlation between objective parameters of execution and implementation of the plans (quantitative indicators) and parameters of quality and successes of the plans produced (qualitative indicators). The objective is to analyze issues of interest such as, for example, to what extent the existence of a more or less extensive participatory process led to a longer approval time of the SUMP, the relationship between the number of stakeholders involved or the time dedicated to its preparation and the subsequent success of a plan.

The quantitative indicators are the following:

L1. Number of stakeholders involved for the diagnosis. The number of stakeholders referenced in the diagnosis phase of each of the SUMPs has been counted.

L2. Number of actions proposed in the SUMP. To avoid statistical biases due to the heterogeneity of the approaches in the different SUMPs, five categories of actions have been established for each plan, limiting the number of actions accounted for each of them to five. The categories are infrastructural actions (e.g., the construction of cycle lanes), actions to enhance the regulatory framework (e.g., parking regulation in low-emission zones), the active promotion of a more sustainable mobility (e.g., the decarbonization of public transport through the purchase of electric vehicles), actions to improve public space (e.g., the development of superblocks or pedestrianization) and actions of an educational nature (e.g., road safety education plans). These actions are not watertight and there may be actions that can be included in both categories, such as the introduction of a tram that could be understood as the building of infrastructure or the implementation of a less polluting mobility system.

L3. Time necessary for its elaboration (months). To consider time, the official administrative files or the municipalities' own websites have been used, establishing indicative dates through news in the newspapers when specific data were not available from official sources.

On the other hand, the qualitative indicators are the following:

L4. Amplitude of the participatory process of elaboration. Four levels of extent have been established: basic level in the case of SUMPs carried out solely as technical consulting work by an external company; intermediate level for SUMPs carried out at the basic level plus workshops including stakeholder participation; high level for those of intermediate level with administrative procedure of regulated public exposure for presentation of allegations by all citizens; and comprehensive level for those with high level plus implementation of mechanisms for subsequent monitoring of actions by the administration.

L5. Number of compliances, monitoring and verification indicators. Three assessment levels have been established (low, medium and high) depending on the number of monitoring indicators. Those who did not have indicators or made a scant approach in this matter have been established as low, for those who raised it in a purely theoretical way they have been established as medium, and for those who raised a verifiable comprehensive process or who have carried out subsequent revisions and updates of the plan in the following years were considered high.

L6. Proven degree of fulfilment of the SUMP. For the assessment of this indicator, three levels have been established (low, medium, high), taking into account the level of news generated after the approval of the plan. Additionally, we have also considered whether there is any type of monitoring information infrastructure by the municipal administration (low for those who have not generated information after its approval, medium for those who have generated some type of subsequent news flow and high for those who have generated a permanent flow of news through the media or through a specific platform).

2.2. Hybrid MCDM Framework Design

The objective of this methodology is to obtain a reliable ranking of best alternatives in SUMP with which to improve the ease of implementation of SUMPs in cities. To this end, two different MCDM methods will be used: WSM ([36]) and SIMUS ([27]). These two methods will first be briefly described and justified, and then their mixed implementation through a hybrid methodology proposal will be explained.

2.2.1. Weighted Sum Method (WSM)

The WSM is the one most recognizable and simplest of the MCDM methods and its use is sometimes recommended because of the ease of its applicability. WSM is available for a wide spectrum of users including non-technical ones, which is the reason for its choice in this work. This methodological approach is the most common one present in the SUMPs studied when structured proposals for diagnosing problems or strategic selection of alternative solutions are made, since its explanation to city's stakeholders is quite simple. However, it is important to remember that the weights used must be justified based on objective criteria for the methodology to be robust and reliable.

WSM is applicable when evaluating a set of "m" alternatives by means of a set of "n" criteria. To distinguish the importance of a criterion respective to others, weights are considered and the higher the weight is, the higher the importance of the criterion. The importance of alternative A_i denoted as "$Ai^{\,WSM-Score}$" when it is evaluated in terms of criterion Cj and when all criteria are considered simultaneously, is defined as follows:

$$A_i^{WSM-Score} = \sum_{j=1}^{n} W_j A_{ij} \quad for\ i = 1\ to\ m\ \forall i \tag{1}$$

where:
W_j: relative importance of criterion Cj
A_{ij}: score of alternative A_i when evaluated by means of criterion Cj.

2.2.2. Sequential Interactive Modelling for Urban Systems (SIMUS) Method

In urban mobility, it is quite common when modelling city planning to deal with many alternatives and a large set of criteria to assess them. In selecting a MCDM method, the chosen one must be suitable to address this situation in a reasonable computing time and without the need of investing in expensive hardware. The software SIMUS (Munier, Canada) approach is based on Lineal Programming and is able to tackle this scenario.

SIMUS is a MCDM method which is immune to rank reversal (RR, [37,38]). RR can be explained with the following example: given a final ranking of best alternatives is B $\succeq \cap$ A $\succeq \cap$ D $\succeq \cap$ C (symbol '$\succeq \cap$' means that it is preferred or equal to or precede to; therefore, B is preferred to A, which is preferred to D which is preferred to C). This ranking may change when adding a new project or deleting an existing one, i.e., suppose project A is not considered and the MCDM method is run again. The new ranking should be according to common sense: B $\succeq \cap$ D $\succeq \cap$ C, i.e., conserving the ranking precedence in B, D, C. However, it could be that RR appears and the new ranking is D $\succeq \cap$ C $\succeq \cap$ B, which should not happen.

This approach makes the SIMUS approach more attractive for proposing a hybrid framework in the case of urban mobility planning compared to other MCDM methods such as AHP [36], MOORA [39], ELECTRE [40,41], PROMETHEE [42], SAW [41] or TOPSIS [43], for example.

The reason why SIMUS does not suffer RR is mainly based on the rigid mathematical structured followed in the Simplex algorithm which is repeatedly used for each new objective [27], improving the value of the functional 'Z' which is expressed as:

$$Z = \sum_{j=1}^{n} (\alpha_j \; X_j \; \beta_j \; Y_j) \; Maximize/Minimize \tag{2}$$

where:
X and Y are variables or projects
α and β: Scores for projects
j: Number of projects

The Simplex algorithm works using a tableau that has all the data of the problem ordered in a certain manner. In each iteration, the algorithm selects the best project to enter in a new solution by comparing the contribution of all projects (Cj) in improving the solution from the last iteration (Z_j), that is, ($C_j - Z_j$) and obviously chooses that with the greatest difference. Once this selection is made, the Simplex determines the project that must be eliminated from the solution (see [27] for a detailed explanation of parameters). Consequently, if in an existing problem a new project vector that is worse than all the others is added to the system, it will never be considered. By the same token, if a new project is added and is better than another, it will be selected by the same algebraic mechanism.

2.2.3. Hybrid WSM-SIMUS Framework

For the implementation of this methodology, the first step is to analyze the results obtained from the aforementioned indicators of the 47 cities of Spain. The results of this analysis sheds light on the shortcomings and drawbacks when applying theory to real implementations of a SUMP, giving authors guidelines about suitable alternatives to be implemented and the respective criteria to assess them.

For the assessment of alternatives through a framework, a literature review in this field was performed, searching for methods which better match the objectives of this work. Initial output of the proposed hybrid methodology is a first ranking of alternatives obtained through a committee of experts who scored alternatives by means of selected criteria. These criteria were firstly grouped in five clusters, which were weighted according to their importance. To obtain the ranking, a scored matrix was computed by means of the WSM method.

However, reliability of this first ranking had to be checked due to the subjectivity inherent in some stages of the above processes (mainly when weighting the clusters and at a lower level when scoring the alternatives). Based on former appreciations, a second ranking was needed to check the robustness of the initial one.

This second ranking firstly considers stakeholders´ preferences and resources (budget, workforce, time for execution, etc.) and later the assessment of a committee of experts. Both contributions (stakeholders and experts) had to be supported by a solid MCDM method and the principal reason why the Sequential Interactive Modelling for Urban Systems (SIMUS) method was chosen. In the case of Cartagena, 6 experts were contacted, and 35 stakeholders participated on a personal level or as representatives of an institution.

Finally, both rankings were compared searching for high discrepancies that could make the authors aware of shortcomings of the proposed methodology for its application to the city of Cartagena. The process that was followed can be seen in a summarized and schematic way in Figure 1.

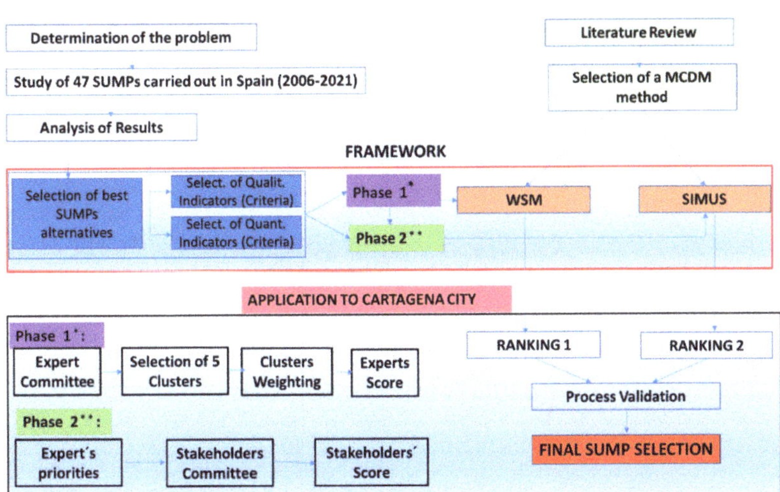

Figure 1. Hybrid WSM–SIMUS methodology proposed for SUMP improvement.

The proposed methodology can be summarized in the following steps:

1. Retrospective study of 47 Sustainable Urban Mobility Plans carried out in Spain in the last 15 years.
2. Diagnosis of existing problems in these planning tools and the implementation of strategic guidelines to aid decision-making based on the results obtained.
3. Selection of the best alternatives for a successful SUMP implementation.
4. Selection of a group of criteria to assess former alternatives.
5. Selection of MCDM parameters of methods based on literature review if needed: The Weight Sum Method (WSM) for Phase 1 and the SIMUS method for Phase 2.

Ranking 1:

6. Based of literature review, 5 clusters were identified (Atmosphere Quality Improvement, Improvement of Healthy Habits, Enhancement of Competitivity, Public Space Improvement and Social Justice) and strategic goals were grouped as criteria in each one of these 5 clusters.
7. Weighting of the clusters: for the application of the WSM, clusters must be weighted according to the literature review carried out by the technical managers of the Sustainable Urban Mobility Plan.
8. Selection of an expert committee to score alternatives by means of the selected criteria.

Based on the 47 SUMPs studied, the authors have analyzed the main reasons why several of the plans were not finally implemented (lack of consensus between stakeholders, lack of budget, discrepancies between stakeholders and municipal technicians or leading experts of the plan, etc.). To avoid these shortcomings in Phase 2, experts´ opinions, which are the first prioritization, are considered first, deleting and generating priority or conditioned relationships for those alternatives that are understood to pose major problems or require prior procedures. This first selection is based on technical, budgetary or administrative reasons. After this first prioritization, the stakeholder´s selection is conducted with these validated alternatives. In this sense, stakeholders follow a half-guided participatory process avoiding focusing their attention on approaches that lead to unrealizable plans. Both selections (stakeholders and experts) are computed in SIMUS, which provides the second ranking in Phase 2.

Ranking 2:

9. Analysis of the experts´ priorities in public workshops or participatory processes.

10. Analysis of available resources (financial, workforce, time, etc.) and preferences based on knowledge of experts and municipal traditional decision makers (political priorities, law limitations, etc.).

Once both rankings have been obtained, a validation process must be carried out contrasting the results obtained for each of the two alternatives proposed to jointly optimize the interests of stakeholders and the limitations imposed by experts or municipal technical managers.

2.3. Application to the Case Study of the City of Cartagena

From the analysis of the statistical correlation between these indicators, the implementation of a structured framework for the city of Cartagena (Figure 2), a south-eastern city of Spain, will be assessed.

Figure 2. Area of study of Cartagena city. Source: Cartagena city council.

The city of Cartagena has been chosen because, from a statistical point of view, it is a case approximately located in the median, based on size and population, of the sample of 47 cities selected to carry out the previous analysis of the sustainable urban mobility plans developed in the last 15 years. The city of Cartagena is a medium-sized city (216,000 inhabitants) located in the southeast of Spain. The city has urban and interurban bus lines, commuter, medium and long-distance rail lines, taxi and VTC services (but no presence of multinational technological platforms such as Uber or Cabify) and has an important urban and periurban network of bike lanes. In its urban area, there are about 600,000 trips every day, of which almost 200,000 are made by car. This case does not have the enormous complexity and interaction of multiple phenomena that can occur in cities such as Madrid (3.2 million) or Barcelona (1.6 million), which may mask the impact of other issues not considered in the analysis such as ride hailing, carpooling, the presence of large transport infrastructures or new mobility forms associated with technological development, the collaborative economy or even the underground economy. Nevertheless, it is large enough to accommodate all the typical casuistry in this type of planning instruments, which is not guaranteed in some of the smaller cities analyzed, such as Ciudad Real (77,000) or Teruel (35,000). The sample of 47 selected SUMPs is, in turn, quite statistically representative of Spain as a country, given that there are 50 provinces in the national territory, with most of their capitals included in the analyzed list.

3. Results

For the presentation of results, the proposed framework was applied. First, a statistical analysis of the diagnostic indicators of the 43 SUMPs analyzed (See Appendix A) was carried out. In a second phase, the results obtained from the analysis of all these plans have served us to apply the proposed methodological framework in the case study of the Sustainable Urban Mobility Plan of Cartagena.

3.1. Analysis of SUMPs in Spain between 2006 and 2021

First, an analysis has been made of the level of presence from a statistical point of view of the different qualitative and quantitative indicators in the selected sample of 47 plans developed in 43 cities in Spain (four of them made second editions to update the previous plan) during the last 15 years. On the other hand, the level of interaction of the results of qualitative and quantitative statistical indicators has been evaluated, by means of a statistical correlation using a linear decision system by least squares (OLS). The results obtained can be seen in Figure 3 and Table 1 for indicators L1 (number of stakeholders), L2 (number of actions in SUMPs), L3 (number of months for implementation), L4 (amplitude of the participatory process of elaboration), L5 (existence of monitoring and verification indicators) and L6 (proven degree of fulfilment of the SUMP).

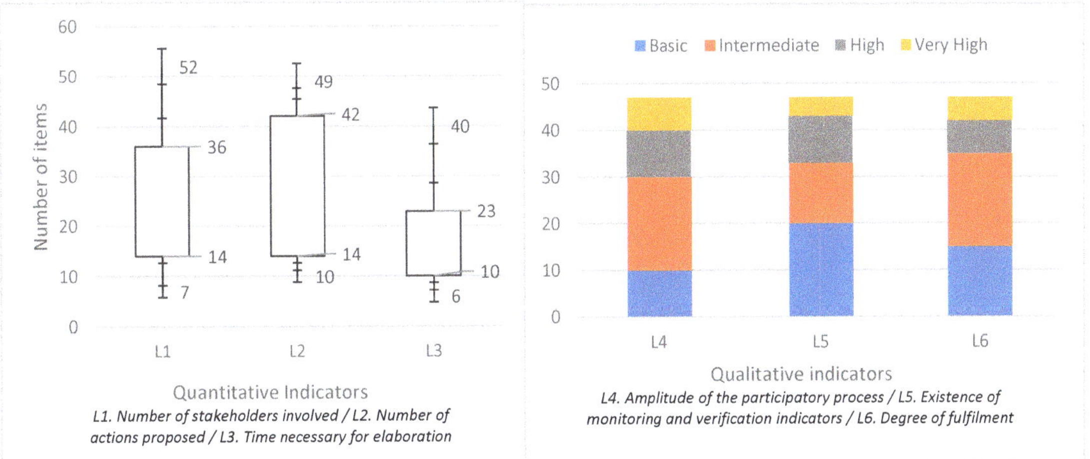

Figure 3. Results obtained for the quantitative and qualitative indicators analyzed in the 43 cities.

If we observe the results obtained, it can be seen that there is a clear statistical correlation between the scarce presence of stakeholders or the absence of a broad participatory process at the SUMP level with the comprehensive proposal of solutions to existing problems. There is also a clear relationship between the absence of follow-up and monitoring mechanisms for these instruments and the failure to achieve their objectives over time. According to what can be observed in the regression coefficients B, the qualitative indicators L4 (amplitude of the participatory process of the SUMP) and L6 (degree of SUMP fulfilment) present the highest level of correlation in general with the quantitative indicators L1 (number of stakeholders involved), L2 (number of actions proposed) and L3 (time necessary for elaboration). On the contrary, the L5 indicator (existence of monitoring and verification indicators) is the one with the weakest regression coefficients, with negative values in some cases (L3 case, close to zero, which rather denotes the absence of correlation). Among the quantitative indicators, the L1 indicator is the one with the most stable behavior, with the L3 being the one with the most heterogeneous values.

In relation to the evaluation of the performance of the model and the relative quality of the statistical model for the given set of data, obtained through the multiple R squared/adjusted R squared values and the Akaike information criterion, respectively, the statement in the previous paragraph is confirmed, with the most robust correlation model being the one that correlates the quantitative indicators with L4, the most robust, and L5, the weakest.

Table 1. Regression analysis using OLS of the existing statistical correlation between quantitative and qualitative indicators.

Indicators	Amplitude of the Participatory Process (L4)				Compliance Indicators (L5)			
	B	Std. Error	*t*	Sign.	B	Std. Error	*t*	Sign.
L1	0.245	0.09	3.384	0.000 *	0.108	0.08	1.992	0.000 *
L2	0.196	0.06	2.767	0.000 *	0.054	0.07	1.682	0.000 *
L3	0.097	0.07	1.810	0.000 *	−0.032	0.09	−3.023	0.000 *
	Akaike's information criterion (AIC): 24,304.5 Multiple R-squared: 0.22 Adjusted R-squared: 0.22 F-statistic: 121.09 Prob (>F) (3,3) degrees of freedom: 0				AIC: 21,009.1 Multiple R-squared: 0.18 Adjusted R-squared: 0.17 F-statistic: 107.34 Prob (>F) (3,3) DF: 0			
Indicators	**Proven degree of fulfilment of the SUMP (L6)**							
	B	**Std. error**	*T*	**Sign.**				
L1	0.219	0.04	2.476	0.000 *				
L2	0.285	0.06	1.232	0.000 *				
L3	0.270	0.05	1.899	0.000 *				
	Akaike´s information criterion (AIC): 22,736.9 Multiple R-squared: 0.22 Adjusted R-squared: 0.21 F-statistic: 143.64 Prob (>F) (3,3) degrees of freedom: 0							

* Significant at 0.01 level.

On the other hand, it is interesting to observe how the majority (62%) of the plans analyzed do not fairly contemplate structured methods of diagnosis and selection of alternatives in decision-making aid. Most of them use a rudimentary or qualitative approach type when this issue is addressed (34%). Only a very exceptional minority (4%) raises structured methods based on objective criteria (usually, type AHP or WSM). Finally, it is interesting to find out that, if applied, the implementation of these improvement mechanisms does not necessarily imply the generation of a longer processing time for the approval of the plan. If we delve into the reasons for the lack of success or the difficulty of approval of the mobility plans studied, we find as the most recurrent causes, the absence of consensus between the stakeholders and the municipal technicians in charge of drafting the plans and the lack of available budget to undertake the actions.

3.2. Application of the Hybrid Framework for the SUMP of the City of Cartagena

Based on the results obtained in the study of SUMPs in Spain, the previously explained methodology has been proposed to improve the implementation process of these planning instruments in cities. The methodological framework described has been applied to the case study of the city of Cartagena, a city that houses the usual characteristics of the statistical sample selected in the previous study.

As a consequence of the application of the proposed framework in its Phase 1, the following 18 × 22 decision matrix (see Table 2) was obtained after the application of the sum weight method using weighting criteria obtained based on a review of scientific literature. The weighting criteria of the different coefficients have been established based on the following documents: general priorities established in the Transport White Paper "Towards a competitive and efficient transport system in the consumption of resources" [12] published in 2011 by the European Commission and the Sustainable Development Goals 2030 of the United Nations organization (Resolution A/RES/70/1 approved by the General Assembly on 25 September 2015 [6]), and specific technical criteria for strategic mobility planning, such as the practical guide for the preparation and implementation of SUMPs published in 2008 by the IDEA foundation [44] and the conclusions established at the Sustainable Urban Mobility Congress held in Bilbao in 2019 [45].

The weighting criteria of the different alternatives evaluated by the expert committee are described below:

- Improvement of environmental quality (IEQ, 20%):
  - Promotion of energy efficiency (EE, quotient × 0.3)
  - Improvement of air quality (AQ, quotient × 0.3)
  - Promotion of noise reduction (NR, quotient × 0.4)
- Promotion of healthy habits (PHH, 20%):
  - Safe and comfortable city for bicycle use (BS, quotient × 0.3)
  - Rationalization of the use of the private car (CUR, quotient × 0.2)
  - Safe and comfortable city for mobility on foot (PS, quotient × 0.3)
  - Promotion of physical exercise (PE, quotient × 0.2)
- Improving competitiveness (IC, 20%):
  - Reduction of travel times (TTR, quotient × 0.2)
  - Infrastructures for more efficient non-motorized mobility (NEVI, quotient × 0.3)
  - Electric vehicle charging infrastructures (EVI, quotient × 0.2)
  - Encouragement of bicycle travel (BUI, quotient × 0.3)
- Improvement of public space (IPS, 20%):
  - Elimination of architectural barriers (SB, quotient × 0.3)
  - Decrease in the occupation of public space by motor vehicles (COR, quotient × 0.2)
  - Promotion of comfortable, inclusive and safe mobility (ISM, quotient × 0.3)
  - Creation of public space for coexistence (CS, quotient × 0.2)
- Social justice (SJ, 20%):
  - Goods accessible to all citizens (HG, quotient × 0.3)
  - Reduction of territory fragmentation and barrier effect (EB, quotient × 0.3)
  - Better quality of life for inhabitants and passers-by (LQ, quotient × 0.4)

In Table 2, scores assigned by WSM to different actions proposed are presented in columns and the five weighted clusters containing respective criteria are shown in rows.

Each cell contains the score given by the expert committee when assessing an alternative regarding a determined criterion, being the maximum score 9 and minimum 1. For this method, Alternative 3 "Building of an integrated and coherent bicycle lane network" is the one out of 26 which gets the highest score as it can be observed in the last row from Table 2 and is consequently the first alternative.

Derived from Table 2, the first ranking of alternatives is obtained and depicted in Table 3 (only 10 most valued alternatives are shown).

Table 2. Decision Matrix Assessment by Expert Committee with Alternatives in columns and Criteria in files.

		Promote Pedestrian Isplacement	Commuter Services	Bike Lanes	Electric MPV	Foster Bike Use	Car Parking	Taxi Fostering	Bus Lines Optimization	Smart Paths	Green Paths	City Center and Suburbs Connections	Superblocks	Road Safety	Pedestrian Center	Intercity Road Safety	Parking Management	Last Mile Logistics	Work Place Transport	IT Transport Manag.	School Paths	Intercity Public Transport	Collab. Public Transport	Traffic Issues Management	20/30 Zones	Intermodality	Cabo Palos Railway
IEQ	EE	1	1	1	9	1	1	1	7	1	1	1	1	1	1	1	1	1	1	7	1	7	7	1	1	7	1
	AQ	9	3	9	8	9	1	1	3	9	9	1	5	1	9	1	1	1	1	1	7	3	3	7	7	3	1
	NR	9	3	9	9	9	5	1	1	9	9	1	5	5	9	5	3	3	3	1	7	1	1	5	5	3	1
PHH	BS	9	1	9	7	9	1	1	1	9	9	1	7	7	9	7	1	1	7	1	9	1	1	3	7	3	1
	PS	9	1	3	3	3	1	1	1	9	9	1	7	5	9	5	1	1	7	1	9	1	1	2	7	1	1
	CUR	9	9	9	9	9	9	3	7	7	9	9	7	1	9	1	3	3	3	1	7	7	7	7	7	8	7
	PE	9	1	9	1	9	1	1	1	9	9	1	5	1	9	1	1	1	7	1	8	1	1	1	1	1	1
IC	TTR	1	7	9	9	1	1	5	9	9	9	9	1	7	1	7	7	7	7	8	7	7	7	8	1	6	1
	NEVI	9	3	9	9	9	1	1	1	9	9	1	7	5	9	5	1	1	6	1	9	1	1	3	1	3	1
	EVI	1	1	1	1	1	1	1	1	1	1	1	1	1	1	1	1	1	1	7	1	1	1	1	1	3	1
	BUI	3	1	9	3	9	1	1	1	8	8	1	7	5	3	5	1	1	6	7	9	1	1	3	7	7	1
IPS	SB	9	7	9	1	1	3	1	1	3	3	1	9	9	9	9	1	1	5	5	7	7	7	3	9	9	1
	COR	9	9	9	9	9	5	1	8	9	9	8	8	1	9	1	8	8	7	1	9	8	8	7	8	7	7
	ISM	9	7	9	9	9	6	7	9	9	9	9	9	9	9	9	4	4	8	7	9	9	9	6	9	9	4
	CS	9	1	9	1	9	9	7	1	9	9	1	9	9	9	9	7	7	7	1	9	1	1	1	8	8	1
SJ	HG	9	7	9	1	1	3	1	1	3	3	1	9	9	9	9	1	1	5	5	7	7	7	3	9	9	1
	EB	9	7	9	1	1	3	1	1	3	3	1	9	9	9	9	1	1	5	5	7	7	7	3	9	9	1
	LQ	9	3	9	9	9	5	1	3	9	9	1	5	5	9	5	3	3	3	1	7	3	7	7	7	3	1
	SCORE	**7.48**	**4.04**	**7.80**	**5.78**	**6.04**	**3.26**	**1.88**	**3.08**	**6.80**	**6.92**	**2.56**	**6.12**	**5.04**	**7.48**	**5.04**	**2.38**	**2.38**	**4.64**	**3.32**	**7.00**	**4.08**	**4.40**	**4.06**	**5.84**	**5.46**	**1.78**

IEQ: Improvement of environmental quality (**EE**: Energy Efficiency; **AQ**: Air Quality; **NR**: Noise Reduction)/**PHH**: Promotion of healthy habits (**BS**: Bike Safety; **PS**: Pedestrian Safety; **CUR**: Car Usage Reduction; **PE**: Physic Exercise)/**IC**: Improving competitiveness (**TTR**: Travel Time Reduction; **NEVI**: Non-Engined Vehicles Infrastructure; **EVI**: Electric Vehicle Infrastructure; **BUI**: Bike Usage Increasement)/**IPS**: Improvement of public space (**SB**: Stop Barriers; **COR**: Car Occupation Reduction; **ISM**: Inclusive/Safety Mob.; **CS**: Coexistence Space)/**SJ**: Social Justice (**HG**: Handy Goods; **EB**: Electric Barriers; **LQ**: Life Quality).

Table 3. Ranking 1 after Phase 1 of the methodology and application of the WSM.

Ranking Position	Alternative Number	Alternative	Score
1	3	Building of an integrated and coherent bicycle lane network	7.8
2	1	Promotion of pedestrian movements	7.48
3	14	Pedestrianization of the Historic Center	7.48
4	20	Generation of safe school itineraries	7
5	10	Greenway Connection	6.92
6	9	Start-up of smart trails	6.80
7	12	Traffic calming through superblocks	6.12
8	5	Recovery from bicycle use	6.04
9	24	Deploy zones 30 and 20 min	5.84
10	4	Implementation of the use of PMV and electric vehicles	5.78

As foreseen in the proposed framework, and aiming to check the robustness of previous Ranking 1, another ranking was obtained after the application of the SIMUS method. For the implementation of this second methodological approach, issues such as the precedence of the actions have been considered for evaluation of alternatives, in order to be able to assess the budgetary needs and technical feasibility of the actions in a combined way when establishing evaluation criteria to the different options. Values obtained with SIMUS algorithm and inputs criteria for scoring the alternatives evaluated can be observed in Table 4.

Table 4 shows the Efficient Result Matrix (ERM) where the 26 alternatives are in columns and the 18 criteria are in rows. The ERM matrix is Pareto Efficient since all scores or results of the different objectives are optimal, that is, they cannot be improved.

Consequently, the final scores for all alternatives are shown in the last row in solid black. According to these scores, Ranking 2 is given and is depicted in Table 5.

Table 4. SIMUS Efficient Result Matrix (ERM) and ranking of alternatives for Cartagena´s City SUMP.

	Promote Pedestrian Displacement	Commuter Services	Bike Lanes	Electric MPV	Foster Bike Use	Car Parking	Taxi Fostering	Bus Lines Optimization	Smart Paths	Green Paths	City Center and Suburbs Connections	Superblocks	Road Safety	Center Pedestrianization.	Intercity Road Safety	Parking Management	Last Mile Logistics	Work Place Transport	IT Transport Manag.	School Paths	Intercity Public Transport	Collab. Public Transport	Traffic Management	20/30 Zones	Intermodality	Cabo Palos Railway
EE		0.06		0.06	0.06	0.06		0.06	0.06	0.06		0.06	0.06		0.06			0.06	0.06		0.06	0.06	0.06	0.06	0.06	
AQ	0.33		0.33											0.33												
NR	0.33		0.33											0.33												
BS	0.25		0.25											0.25						0.25						
PS									0.50	0.50																
CUR	0.33		0.33											0.33												
PE	0.33		0.33											0.33												
TTR									0.50	0.50																
NEVI	0.25		0.25											0.25						0.25						
EVI		0.07		0.07	0.07				0.07	0.07		0.07	0.07		0.07			0.07	0.07		0.07	0.07	0.07	0.07	0.07	
BUI			1.00																							
SB	0.33		0.33											0.33												
COR	0.25		0.25											0.25						0.25						
ISM		0.03		0.09	0.09				0.09	0.09		0.09	0.09		0.09			0.03	0.03		0.03	0.03	0.03	0.09	0.09	
CS	0.25		0.25											0.25						0.25						
HG	0.33		0.33											0.33												
EB	0.33		0.33											0.33												
LQ	0.33		0.33											0.33												
SOC		0.06		0.06	0.06	0.06		0.06	0.06	0.06		0.06	0.06		0.06			0.06	0.06		0.06	0.06	0.06	0.06	0.06	
PF	3.67	0.16	4.67	0.22	0.22	0.06	0.00	0.06	1.22	1.22	0.00	0.22	0.22	3.67	0.22	0.00	0.00	0.16	0.16	1.00	0.16	0.16	0.16	0.22	0.22	0.00
NPF	12	3	13	3	3	1	0	1	5	5	0	3	3	12	3	0	0	3	3	4	3	3	3	3	3	0
RESULT	0.67	0.17	0.72	0.17	0.17	0.06	0.00	0.06	0.28	0.28	0.00	0.17	0.17	0.67	0.17	0.00	0.00	0.17	0.17	0.22	0.17	0.17	0.17	0.17	0.17	0.00

EE: Energy Efficiency; **AQ**: Air Quality; **NR**: Noise Reduction; **BS**: **Bike** Safety; **PS**: Pedestrian Safety; **CUR**: Car Usage Reduction; **PE**: Physic Exercise; **TTR**: Travel Time Reduction;
NEVI: Non-Engined Vehicles Infrastructure; **EVI**: Electric Vehicle Infrastructure; **BUI**: Bike Usage Increasement; **SB**: Stop Barriers; **COR**: Car Occupation Reduction;
ISM: Inclusive/Safety Mob.; **CS**: Coexistence Space; **HG**: Handy Goods; **EB**: Electric Barriers; **LQ**: Life Quality. **SIMUS OUTPUTS** (IN GREY COLOUR): **SOC** = SUM OF COLUMNS;
PF = PARTICIPATION FACTOR; **NPF** = NORMALIZED PARTICIPATION FACTOR. **RANKING: Alt.3—Alt.1—Alt.14—Alt.9—Alt.10—Alt.20—Alt.4—Alt.5—Alt.12—Alt.13—Alt.15—Alt.24—Alt.25—Alt.2—Alt.18—Alt.19—Alt.21—Alt.22—Alt.23—Alt.6—Alt.8—Alt.7—Alt.11—Alt.16—Alt.17—Alt.26.**

Table 5. Ranking 2 after Phase 2 of the methodology and the application of SIMUS method (only 10 most valued alternatives are shown).

Ranking Position	Alternative Number	Alternative	Score
1	3	Building of an integrated and coherent bicycle lane network	0.72
2	1	Promotion of pedestrian movements	0.67
3	14	Pedestrianization of the Historic Center	0.67
4	9	Start-up of smart trails	0.28
5	10	Greenway Connection	0.28
6	20	Generation of safe school itineraries	0.22
7	4	Implementation of the use of VMP and electric vehicles	0.17
8	5	Recovery from bicycle use	0.17
9	12	Traffic calming through superblocks	0.17
10	13	Road safety improvements	0.17

For a better comparison of Ranking 1 and Ranking 2, both are shown vis-a-vis in Table 6 to check differences. Horizontal arrows indicate exact correspondence between the two rankings. Obliquus arrows show the difference in positions between scores of the two rankings.

Table 6. Comparison of Ranking 1 and Ranking 2 to validate the proposed methodology.

Alternatives	Ranking 1 Best Alternatives	Ranking 2 Best Alternatives	Scores WSM (Ranking 1)	Scores SIMUS (Ranking 2)
Building a bicycle lane network	3	3	7.8	0.72
Promotion of pedestrian movements	1	1	7.48	0.67
Pedestrianization of the Historic Center	14	14	7.48	0.67
Generation of safe school itineraries	20	9	7	0.28
Greenway Connection	10	10	6.92	0.28
Start-up of smart trails	9	20	6.80	0.22
Traffic calming through superblocks	12	4	6.12	0.17
Recovery from bicycle use	5	5	6.04	0.17
Deploy zones 30 and 20 min	24	12	5.84	0.17
Implementation of the use of PMV and electric vehicles	4	13	5.78	0.17

4. Discussion

The analysis carried out for the last 15 years in Spain has been quite revealing about a problem of urban planning in cities that, due to its fairly recent nature, has had little attention paid to it in the current scientific literature [46–49]. One of the issues found during the bibliographic review of the 47 urban mobility plans analyzed, was the existence of a certain heterogeneity in their approach.

However, we can distinguish chronologically two distinct groups. In this sense, we can talk about a first-generation urban mobility plan from the first proposals from 2006 to approximately 2013, and a second-generation plan from around 2014 to the present. The date of passage from one period to another does not respond to any specific regulatory or technical milestone, but rather to the fact that by that time almost all the main capital cities of the country had already developed an instrument of this category [26].

Regarding these first-generation plans, it should be noted that it is from the point of view of their approach to fairly heterogeneous proposals, as there was no regulatory pattern beyond certain recommendations made by some institutions [44]. Although from the point of view of its results, we have very unequal cases, at the level of supplementation, it should be noted that in many cases these were proposals rather voluntaristic, and subsequently, there was scarce later implementation.

Sometimes, the lack of success of this type of plan responds to the absence of a political commitment beyond obtaining the subsidies that the approval of these instruments granted local corporations at that time. However, in other cases, the problems for their implementation stemmed from technical deficiencies in the plans themselves, because of the lack of verification and monitoring tools or due to the absence of a realistic approach in relation to their objectives.

In the case of the so-called second-generation plans, we find a greater homogeneity of approaches, although uneven results remained. In this case, it should be noted that this period includes plans developed for the first time and others from cities that have proceeded to review their sustainable urban mobility plan drawn up during the previous stage to develop a more up-to-date one.

The greater homogeneity of approaches responds to the existence of a growing technical literature and greater experience of local administrations in this matter, although, there is still no regulatory technical framework available, as it happens at present in other countries [50–53]. Regarding its results, although the time frame to observe them in this case is much shorter, we found an absence of real commitments by local authorities. In addition, technical or budgetary problems, regarding the difficulty of making realistic approaches and reaching the necessary consensus with stakeholders for the implementation of actions persist.

If we make a comparison with the performance at an international level, we can find similar phenomena (see for example active travel policies in Italy [54] or regulatory shortcomings for defining SUMPs in Portugal or the Czech Republic [55]) and issues with different problems (large cities casuistry with societies highly aware of sustainable mobility in Europe [56] or problems in Latin America linked to poor urban planning, see [57]). However, the need to implement new analysis methodologies that evaluate the operation and design of SUMPs and help improve their capacity for success continues to be a constant that continues to be raised from different approaches at the international level [58–60].

In this context, the framework proposal resulting from this work can be of great help in improving the implementation of new tools, both in the diagnostic and in the solution proposal phase of SUMPs. It is a common situation when dealing with a large set of alternatives to be implemented in a planning tool such as SUMP that some of them may precede others. For example, to implement solar panels to recharge bikes in a bike station, it is first necessary to build bike stations, as well as and perhaps before that stage is the need to build bike lanes in specific areas of a city.

In the case of decisions based on consensus with stakeholders, this precedence from some alternatives to others is important for several reasons, like the available time to finish the complete project and/or the available budget each year. In the case study of Cartagena, this key factor was performed by computing stakeholders´ precedence by the SIMUS algorithm, the output of which may be observed when comparing alternative 4 'Implementation of the use of PMV and electric vehicles' in both rankings, where a discrepancy of 3 positions is found. As can be seen in Table 6, there have been significant changes in the ranking of preferences for alternatives, the consideration of which, thanks to this hybrid process, avoids several of the problems detected in the study of the 47 SUMPs.

Hired experts and local government officials are well acquainted with these technical, legal or budgetary issues that usually condition the viability of the implementation of these urban planning tools [61]. However, in the current situation, in which social participation has become a backbone of any urban planning proposal, no SUMP can be realistically implemented without the endorsement of numerous stakeholders such as neighborhood associations, users of bicycles and personal mobility vehicles, representatives of affected institutions, merchants, etc. [62].

Therefore, it is necessary to develop more sophisticated and objective tools that can combine these two approaches in a realistic way to avoid the problems detected in the 47 SUMPs analyzed. Regarding the difference obtained in the results between both analysis methodologies, it should be noted that no drastic changes should be expected (which would not make sense), but relevant changes whose introduction optimized using SIMUS and its precedence parameters possibly helps to avoid problems later during the implementation of the SUMP.

In this context, the proposed working methodological framework covers many of the deficiencies detected with the implementation of a hybrid methodology that combines the simplicity of application of an objective WSM methodology based on weighted evaluation indicators and the capability to implement factors of precedence provided by SIMUS. Although the results obtained for the case study of Cartagena do not show a radical difference in the results between the two approaches, they do allow the adjustment of some parameters of prioritization and validation of actions. These improvements may avoid leaving some loose ends, which in many cases may later entail a delay, or even blockage, of the SUMP start-up for budgetary, political or administrative problems.

The proposed methodological framework has limitations since it implies knowing the internal functioning of the precedence systems of the diagnostic phases and the proposed proposals so that it really allows an effective optimization of the SUMP. This approach has been effective in a medium-sized city such as Cartagena, where the number of variables and their interactions is reasonably manageable. However, it would be interesting for future lines of research to deepen the effectiveness of this methodology by evaluating its ability to optimize processes for proposals for solutions in urban mobility planning instruments in larger and more complex cities such as Madrid or Barcelona.

5. Conclusions

The development of Sustainable Urban Mobility Plans in numerous cities in Spain during the last 15 years without the existence of a specific regulatory framework has given rise to a varied catalogue of actions with various problems. In this study, 47 plans of this nature developed in different Spanish cities have been analyzed, observing how many of these planning instruments have had difficulties both in diagnosing problems and in implementing solutions. By means of a statistical analysis, it has been contrasted how there is a clear correlation between the implementation of these instruments with rudimentary participatory processes or the scarcity of indicators for subsequent monitoring with the failure of these strategies to improve urban mobility in cities.

Based on this diagnosis, a structured hybrid MCDM framework based on WSM and SIMUS methods has been proposed. The results obtained with the application of this methodology for the implementation of the SUMP of the city of Cartagena, show how the implementation of analytic mechanisms in the SUMPs of middle-sized cities such as Cartagena can facilitate the achievement of their objectives.

Author Contributions: Conceptualization, S.G.-A. and E.H.; methodology S.G.-A., E.H. and N.M.; formal analysis, S.G.-A. and E.H.; investigation, S.G.-A. and E.H.; resources, S.G.-A.; data curation S.G.-A.; supervision, N.M.; project administration, S.G.-A., E.H. and N.M.; funding acquisition, E.H. validation, S.G.-A., E.H. and N.M. All authors have read and agreed to the published version of the manuscript.

Funding: Ministry of Science, Innovation and Universities of Spain (MICINN) Project RTI2018-099139-B-C21 with ERDF for financial support.

Institutional Review Board Statement: Not applicable.

Informed Consent Statement: Not applicable.

Acknowledgments: Authors acknowledge the data and institutional support given by the local authorities of the city of Cartagena to carry out this research.

Conflicts of Interest: The authors declare no conflict of interest.

Appendix A

The cities where SUMPs have been analyzed for the study are as follows:

Table A1. List of cities in Spain with SUMP analyzed.

City	Start Date	Approval Date	Source
Madrid	March 2013	December 2014	madrid.es
Barcelona	September 2012 December 2020	March 2015 In process	barcelona.cat barcelona.cat
Vitoria-Gasteiz	2005 September 2019	October 2007 In process	vitoria-gasteiz.org vitoria-gasteiz.org
Sevilla	2019	May 2021	sevilla.org
Málaga	2015	July 2020	movilidad.malaga.eu
Valencia	2011	December 2013	www.upv.es
Murcia	2009	May 2013	www.murcia.es
Alicante	2011	December 2013	www.alicante.es
Bilbao	2007 2016	2011 May 2018	pmus.bilbao.eus pmus.bilbao.eus
Vigo	2011 2021	June 2014 In process	hoxe.vigo.org hoxe.vigo.org
Burgos	2005 November 2018	2006 In process	aytoburgos.es aytoburgos.es
Terrasa	No data 2014	2002 May 2016	terrassa.cat terrassa.cat
Santander	2008	February 2010	santander.es
Albacete	January 2007	July 2010	albacete.es
Badalona	2009	June 2015	badalona.cat
Elche	2013	2015	elche.es
Castellon	2007	2009	castello.es
Ponferrada	November 2007	June 2014	ponferrada.org
Leganes	2008	July 2010	leganes.org
Fuenlabrada	2006 2016	September 2008 November 2019	ayto-fuenlabrada.es ayto-fuenlabrada.es
San Vicente de Raspeig	2006 2021	2008 In process	raspeig.es raspeig.es
Torrejón de Ardoz	2019	May 2021	ayto-torrejon.es
Reus	2010	March 2012	reus.cat
S. Fernando de Henares	2007	2009	ayto-sanfernando.com
Palma de Mallorca	2012	October 2014	mobipalma.mobi
Ourense	2011	May 2012	ourense.gal
San Sebastian	2006	September 2008	donostiafutura.com
Tarragona	2010	September 2012	tarragona.cat
Gerona	October 2012	December 2014	web.girona.cat
Lleida	2008	November 2011	mobilitat.paeria.cat
Zaragoza	2006 October 2016	2008 March 2019	zaragoza.es zaragoza.es

Table A1. *Cont.*

City	Start Date	Approval Date	Source
Valladolid	2005 2019	November 2007 April 2021	pimussva.es pimussva.es
Córdoba	April 2011	October 2013	pmus.cordoba.es
Jaen	March 2021	In process	pmusjaen.com
Granada	2010	February 2013	movilidadgranada.com
Ciudad Real	2010	March 2012	ciudadreal.es
Cádiz	2012	June 2013	institucional.cadiz.es
Salamanca	2011	Julio 2013	ingenieriacivil. aytosalamanca.es
Logroño	2011	November 2013	logro~no.es
Teruel	2010	Junio 2012	urbanteruel.es
La Coruña	2011	December 2013	coruna.gal
Pontevedra	2016	April 2020	aestrada.gal
Lorca	2016	May 2017	movilidad.lorca.es
Cartagena	October 2020	September 2021	cartagena.es

Technical note: the number of SUMPs analyzed to prepare the statistical analysis with indicators is 47 SUMPs from 43 cities in Spain. This is because there are 4 cities that have two versions of their SUMP on this question, it must also be specified that in the list in the appendix there are 10 cities in this situation; however, 6 of them have recently finished their second version of their SUMP or they still have it in process, so for the purposes of statistical analysis, they have not been able to be computed since they do not have enough background to analyze the compliance indicator L6. City 44 is Cartagena, the city of the case study, which, like the one mentioned, approved its SUMP recently, so it has not been computed in the statistical analysis of Phase 1.

References

1. Deenapanray, P.N.K.; Khadun, N.A. Land transport greenhouse gas mitigation scenarios for Mauritius based on modelling transport demand. *Transp. Res. Interdiscip. Perspect.* **2021**, *9*, 100299. [CrossRef]
2. Gustafsson, M.; Svensson, N.; Eklund, M.; Dahl Öberg, J.; Vehabovic, A. Well-to-wheel greenhouse gas emissions of heavy-duty transports: Influence of electricity carbon intensity. *Transp. Res. Part D Transp. Environ.* **2021**, *93*, 102757. [CrossRef]
3. Fan, C.; Jiang, X.; Mostafavi, A. Evaluating crisis perturbations on urban mobility using adaptive reinforcement learning. *Sustain. Cities Soc.* **2021**, *75*, 103367. [CrossRef]
4. United Nations Department of Economic and Social Affairs. *The Sustainable Development Goals Report 2019*; United Nations Department of Economic and Social Affairs: New York, NY, USA, 2019. [CrossRef]
5. United Nations Transport and the Sustainable Development Goals. Available online: https://unece.org/transport-and-sustainable-development-goals (accessed on 25 November 2021).
6. Macmillan, A.; Smith, M.; Witten, K.; Woodward, A.; Hosking, J.; Wild, K.; Field, A. Suburb-level changes for active transport to meet the SDGs: Causal theory and a New Zealand case study. *Sci. Total Environ.* **2020**, *714*, 136678. [CrossRef] [PubMed]
7. United Nations Economic Commission for Europe. *Urban Mobility and Public Transport Situation in UNECE Capitals*; United Nations Economic Commission for Europe: Geneva, Switzerland, 2016.
8. Miskolczi, M.; Földes, D.; Munkácsy, A.; Jászberényi, M. Urban mobility scenarios until the 2030s. *Sustain. Cities Soc.* **2021**, *72*, 103029. [CrossRef]
9. Tiwari, G.; Mohan, D. Challenges of localizing mobility SDGs in Indian cities. *IATSS Res.* **2021**, *45*, 1–2. [CrossRef]
10. Arfanuzzaman, M. Harnessing artificial intelligence and big data for SDGs and prosperous urban future in South Asia. *Environ. Sustain. Indic.* **2021**, *11*, 100127. [CrossRef]
11. Pamucar, D.; Deveci, M.; Gokasar, I.; Işık, M.; Zizovic, M. Circular economy concepts in urban mobility alternatives using integrated DIBR method and fuzzy Dombi CoCoSo model. *J. Clean. Prod.* **2021**, *323*, 129096. [CrossRef]
12. European Environment Agency Transport White Paper. Available online: https://www.eea.europa.eu/policy-documents/transport-white-paper-2011 (accessed on 25 November 2021).
13. European Commission Sustainable Transport—New Urban Mobility Framework. Available online: https://ec.europa.eu/info/law/better-regulation/have-your-say/initiatives/12916-Transporte-sostenible-nuevo-marco-de-movilidad-urbana_es (accessed on 25 November 2021).
14. European Commission A European Strategy for Low-Emission Mobility. Available online: https://www.eea.europa.eu/policy-documents/a-european-strategy-for-low (accessed on 25 November 2021).
15. Paris City Hall Paris Ville du Quart D'heure, ou le Pari de la proximité. Available online: https://www.paris.fr/dossiers/paris-ville-du-quart-d-heure-ou-le-pari-de-la-proximite-37 (accessed on 25 November 2021).

16. Lara Allende, A.; Stephan, A. Life cycle embodied, operational and mobility-related energy and greenhouse gas emissions analysis of a green development in Melbourne, Australia. *Appl. Energy* **2022**, *305*, 117886. [CrossRef]
17. Pucci, P. Spatial dimensions of electric mobility—Scenarios for efficient and fair diffusion of electric vehicles in the Milan Urban Region. *Cities* **2021**, *110*, 103069. [CrossRef]
18. Pucci, P. Mobility behaviours in peri-urban areas. The Milan Urban Region case study. *Transp. Res. Procedia* **2017**, *25*, 4229–4244. [CrossRef]
19. Guzman, L.A.; Arellana, J.; Oviedo, D.; Moncada Aristizábal, C.A. COVID-19, activity and mobility patterns in Bogotá. Are we ready for a '15-minute city'? *Travel Behav. Soc.* **2021**, *24*, 245–256. [CrossRef]
20. Gonzalez, J.N.; Perez-Doval, J.; Gomez, J.; Vassallo, J.M. What impact do private vehicle restrictions in urban areas have on car ownership? Empirical evidence from the city of Madrid. *Cities* **2021**, *116*, 103301. [CrossRef]
21. Muñiz, I.; Sánchez, V. Urban Spatial form and Structure and Greenhouse-gas Emissions from Commuting in the Metropolitan Zone of Mexico Valley. *Ecol. Econ.* **2018**, *147*, 353–364. [CrossRef]
22. Romero, F.; Gomez, J.; Rangel, T.; Vassallo, J.M. Impact of restrictions to tackle high pollution episodes in Madrid: Modal share change in commuting corridors. *Transp. Res. Part D Transp. Environ.* **2019**, *77*, 77–91. [CrossRef]
23. Lopez-Carreiro, I.; Monzon, A. Evaluating sustainability and innovation of mobility patterns in Spanish cities. Analysis by size and urban typology. *Sustain. Cities Soc.* **2018**, *38*, 684–696. [CrossRef]
24. Santana-Santana, S.B.; Peña-Alonso, C.; Pérez-Chacón Espino, E. Assessing physical accessibility conditions to tourist attractions. The case of Maspalomas Costa Canaria urban area (Gran Canaria, Spain). *Appl. Geogr.* **2020**, *125*, 102327. [CrossRef]
25. Serrano-López, R.; Linares-Unamunzaga, A.; Muñoz San Emeterio, C. Urban sustainable mobility and planning policies. A Spanish mid-sized city case. *Cities* **2019**, *95*, 102356. [CrossRef]
26. Mozos-Blanco, M.Á.; Pozo-Menéndez, E.; Arce-Ruiz, R.; Baucells-Aletà, N. The way to sustainable mobility. A comparative analysis of sustainable mobility plans in Spain. *Transp. Policy* **2018**, *72*, 45–54. [CrossRef]
27. Munier, N.; Hontoria, E.; Jiménez-Saez, F. Strategic Approach in Multi-Criteria Decision Making—A Practical Guide for Complex Scenarios. In *International Series in Operations Research & Management Science*; Springer Nature Switzerland AG: Cham, Switzerland, 2019; Volume 275, pp. 3–13. ISBN 9783030027254. [CrossRef]
28. Wen, Z.; Liao, H.; Ren, R.; Bai, C.; Zavadskas, E.K.; Antucheviciene, J.; Al-Barakati, A. Cold chain logistics management of medicine with an integrated multi-criteria decision-making method. *Int. J. Environ. Res. Public Health* **2019**, *16*, 4843. [CrossRef]
29. García-Ayllón, S. New strategies to improve co-management in enclosed coastal seas and wetlands subjected to complex environments: Socio-economic analysis applied to an international recovery success case study after an environmental crisis. *Sustainability* **2019**, *11*, 1039. [CrossRef]
30. Heidary Dahooie, J.; Vanaki, A.S.; Firoozfar, H.R.; Zavadskas, E.K.; Čereška, A. An Extension of the Failure Mode and Effect Analysis with Hesitant Fuzzy Sets to Assess the Occupational Hazards in the Construction Industry. *Int. J. Environ. Res. Public Health* **2020**, *17*, 1442. [CrossRef]
31. Wang, C.; Yang, Q.; Dai, S. Supplier selection and order allocation under a carbon emission trading scheme: A case study from China. *Int. J. Environ. Res. Public Health* **2020**, *17*, 111. [CrossRef]
32. Yang, J.-J.; Chuang, Y.-C.; Lo, H.-W.; Lee, T.-I. A two-stage MCDM model for exploring the influential relationships of sustainable sports tourism criteria in Taichung City. *Int. J. Environ. Res. Public Health* **2020**, *17*, 2319. [CrossRef]
33. Huang, C.-Y.; Hsieh, H.-L.; Chen, H. Evaluating the investment projects of spinal medical device firms using the real option and DANP-mV based MCDM methods. *Int. J. Environ. Res. Public Health* **2020**, *172*, 3335. [CrossRef] [PubMed]
34. Mitropoulos, L.K.; Prevedouros, P.D. Urban transportation vehicle sustainability assessment with a comparative study of weighted sum and fuzzy methods. *J. Urban Plan. Dev.* **2016**, *142*, 04016013. [CrossRef]
35. Munier, N. Methodology to select a set of urban sustainability indicators to measure the state of the city, and performance assessment. *Ecol. Indic.* **2011**, *11*, 1020–1026. [CrossRef]
36. Chourabi, Z.; Khedher, F.; Babay, A.; Cheikhrouhou, M. Multi-criteria decision making in workforce choice using AHP, WSM and WPM. *J. Text. Inst.* **2019**, *110*, 1092–1101. [CrossRef]
37. Aires, R.F.F.; Ferreira, L. The rank reversal problem in multi-criteria decision making: A literature review. *Pesqui. Oper.* **2018**, *38*, 331–362. [CrossRef]
38. Maleki, H.; Zahir, S. A comprehensive literature review of the rank reversal phenomenon in the analytic hierarchy process. *J. Multi-Criteria Decis. Anal.* **2013**, *20*, 141–155. [CrossRef]
39. Agrawal, R. Sustainable material selection for additive manufacturing technologies: A critical analysis of rank reversal approach. *J. Clean. Prod.* **2021**, *296*, 126500. [CrossRef]
40. Figueira, J.R.; Greco, S.; Roy, B.; Słowiński, R. An Overview of ELECTRE Methods and their Recent Extensions. *J. Multi-Criteria Decis. Anal.* **2013**, *20*, 61–85. [CrossRef]
41. Wang, Y.-M.; Luo, Y. On rank reversal in decision analysis. *Math. Comput. Model.* **2009**, *49*, 1221–1229. [CrossRef]
42. Verly, C.; De Smet, Y. Some results about rank reversal instances in the PROMETHEE methods. *Int. J. Multicriteria Decis. Mak.* **2013**, *3*, 325–345. [CrossRef]
43. García-Cascales, M.S.; Lamata, M.T. On rank reversal and TOPSIS method. *Math. Comput. Model.* **2012**, *56*, 123–132. [CrossRef]

44. Instituto para la Diversificación y Ahorro de la Energía Guía Práctica para la Elaboración e Implantación de Planes de Movilidad Urbana Sostenible. Available online: https://www.idae.es/publicaciones/pmus-guia-practica-para-la-elaboracion-e-implantacion-de-planes-de-movilidad-urbana (accessed on 25 November 2021).
45. Sustainable Urban Mobility Congress. Available online: https://sumbilbao19.com/documentos/ (accessed on 25 November 2021).
46. Melkonyan, A.; Koch, J.; Lohmar, F.; Kamath, V.; Munteanu, V.; Alexander Schmidt, J.; Bleischwitz, R. Integrated urban mobility policies in metropolitan areas: A system dynamics approach for the Rhine-Ruhr metropolitan region in Germany. *Sustain. Cities Soc.* **2020**, *61*, 102358. [CrossRef]
47. Russo, F.; Rindone, C. Regional Transport Plans: From Direction Role Denied to Common Rules Identified. *Sustainability* **2021**, *13*, 9052. [CrossRef]
48. Ali, M.; Kamal, M.D.; Tahir, A.; Atif, S. Fuel Consumption Monitoring through COPERT Model—A Case Study for Urban Sustainability. *Sustainability* **2021**, *13*, 11614. [CrossRef]
49. Spadaro, I.; Pirlone, F. Sustainable Urban Mobility Plan and Health Security. *Sustainability* **2021**, *13*, 4403. [CrossRef]
50. Ortega, J.; Tóth, J.; Péter, T. A Comprehensive Model to Study the Dynamic Accessibility of the Park & Ride System. *Sustainability* **2021**, *13*, 4064.
51. Vujadinović, R.; Jovanović, J.Š.; Plevnik, A.; Mladenovič, L.; Rye, T. Key Challenges in the Status Analysis for the Sustainable Urban Mobility Plan in Podgorica, Montenegro. *Sustainability* **2021**, *13*, 1037. [CrossRef]
52. Morfoulaki, M.; Papathanasiou, J. Use of PROMETHEE MCDA Method for Ranking Alternative Measures of Sustainable Urban Mobility Planning. *Mathematics* **2021**, *9*, 602. [CrossRef]
53. Conticelli, E.; Gobbi, G.; Saavedra Rosas, P.I.; Tondelli, S. Assessing the Performance of Modal Interchange for Ensuring Seamless and Sustainable Mobility in European Cities. *Sustainability* **2021**, *13*, 1001. [CrossRef]
54. Maltese, I.; Gatta, V.; Marcucci, E. Active Travel in Sustainable Urban Mobility Plans. An Italian overview. *Res. Transp. Bus. Manag.* **2021**, *40*, 100621. [CrossRef]
55. Klímová, A.; Pinho, P. National policies and municipal practices: A comparative study of Czech and Portuguese urban mobility plans. *Case Stud. Transp. Policy* **2020**, *8*, 1247–1255. [CrossRef]
56. Lindenau, M.; Böhler-Baedeker, S. Citizen and Stakeholder Involvement: A Precondition for Sustainable Urban Mobility. *Transp. Res. Procedia* **2014**, *4*, 347–360. [CrossRef]
57. Barandier, J.R., Jr. Niterói's central area urban redevelopment project: Planning to achieve sustainable mobility in metropolitan area of Rio de Janeiro. *Transp. Res. Procedia* **2017**, *25*, 3116–3128. [CrossRef]
58. May, A.D. Encouraging good practice in the development of Sustainable Urban Mobility Plans. *Case Stud. Transp. Policy* **2015**, *3*, 3–11. [CrossRef]
59. Arsenio, E.; Martens, K.; Di Ciommo, F. Sustainable urban mobility plans: Bridging climate change and equity targets? *Res. Transp. Econ.* **2016**, *55*, 30–39. [CrossRef]
60. Sampaio, C.; Macedo, E.; Coelho, M.C.; Bandeira, J.M. Economic and environmental analysis of measures from a Sustainability Urban Mobility Plan—Application to a small sized city. *Transp. Res. Procedia* **2020**, *48*, 2580–2588. [CrossRef]
61. Agnoletti, M. The degradation of traditional landscape in a mountain area of Tuscany during the 19th and 20th centuries: Implications for biodiversity and sustainable management. *For. Ecol. Manag.* **2007**, *249*, 5–17. [CrossRef]
62. Jiménez, P.; María-Dolores, D.; Beltrán, S. An Integrative and Sustainable Workplace Mobility Plan: The Case Study of Navantia-Cartagena (Spain). *Sustainability* **2020**, *12*, 10301. [CrossRef]

International Journal of
Environmental Research and Public Health

MDPI

Article

The Impact of Environmental Uncertainty on Corporate Innovation: Empirical Evidence from an Emerging Economy

Jinyong Chen [1,*], Weijia Shu [1], Xiaochi Wang [1], Muhammad Safdar Sial [2], Mariana Sehleanu [3] and Daniel Badulescu [3]

1 Business School, Hubei University, Wuhan 430000, China; junkswv@163.com (W.S.); wxchubukj@163.com (X.W.)
2 Department of Management Sciences, COMSATS University Islamabad (C.U.I.), Islamabad 44000, Pakistan; safdarsial@comsats.edu.pk
3 Department of Economics and Business, Faculty of Economic Sciences, University of Oradea, 410087 Oradea, Romania; mavancea@uoradea.ro (M.S.); dbadulescu@uoradea.ro (D.B.)
* Correspondence: chenjinyong@hubu.edu.cn

Abstract: The paper analyzes the effect of environmental uncertainty on corporate technological innovation from the perspective of an innovation value chain under the institutional background of China. This paper not only discusses the intermediary effect of agency problems on environmental uncertainty and corporate technological innovation but also deeply explores the influence of information transparency, government subsidies, and other mechanisms to alleviate agency problems on environmental uncertainty and corporate technological innovation. We use the data of listed companies in China from 2008 to 2019 as the research sample, and the results show that, in general, environmental uncertainty has a negative effect on both input and output of technological innovation, and the negative effect can last for two years. Further research shows that the agency problem has an intermediary effect on the environmental uncertainty and corporate technology innovation, and the environmental uncertainty aggravates the agency problem, which hinders the input and output of corporate technology innovation. As an important mechanism to alleviate the agency problems, information transparency and government subsidies can effectively alleviate the agency conflict, thus reducing the inhibition of environmental uncertainty on the input and output of technological innovation. Our findings contribute to the discussion of driving factors for technological innovation in the context of China's system. Our results provide useful insights into the link between environmental uncertainty and corporate innovation for economic academics and practitioners alike.

Keywords: environmental uncertainty; information transparency; enterprise technology innovation; government subsidies

Citation: Chen, J.; Shu, W.; Wang, X.; Sial, M.S.; Sehleanu, M.; Badulescu, D. The Impact of Environmental Uncertainty on Corporate Innovation: Empirical Evidence from an Emerging Economy. *Int. J. Environ. Res. Public Health* **2022**, *19*, 334. https://doi.org/10.3390/ijerph19010334

Academic Editors: Salvador García-Ayllón Veintimilla and Antonio Espín Tomás

Received: 15 November 2021
Accepted: 24 December 2021
Published: 29 December 2021

1. Introduction

As an emerging economy in the world, China's economy has entered a new stage, which is different from the rapid growth period of the past 30 years, and it is in urgent need of innovation to drive high-quality economic development. As the micro-subject of economic development, the innovation activities of corporations, especially technological innovation, are not only directly related to their long-term survival and sustainable development in an increasingly fierce competitive environment but also related to the core competitiveness and competitive position of industries, regions and countries. However, corporate technological innovation faces the influence of environmental uncertainty such as trade friction, policy change, and market demand change. Environmental uncertainty is an unpredictable change, which is risky and ambiguous [1]. In addition, the agency problem is common in corporate governance in East Asian countries, which will also affect corporate technological innovation. Therefore, the paper explores the influence of environmental uncertainty on technological innovation under the institutional background of agency conflict.

The existing literature includes a significant amount of research on the motivation of corporate technological innovation. Focusing on the external influencing factors of corporate technological innovation, such as financial policy, industrial policy and industry characteristics [2–4], and internal influencing factors, such as corporate governance structure, managers' shareholding ratio, executive incentive, corporation size, and nature of equity [5–7], a great deal of results have been achieved. However, most of the existing literature has not considered the risks faced by corporations and environmental uncertainty. The innovation investment decisions made by management are influenced by multiple factors, and any environmental uncertainty has risks, which in turn affects management's innovation investment decisions [8]. There are two important points in the literature about the influence of environmental uncertainty on technological innovation. One is the "opportunity-oriented effect" of environmental uncertainty on corporate technological innovation [9,10], and the other is the "risk avoidance effect" [11,12].

However, the current research mainly focuses on whether the relationship between them is linear or non-linear. On the one hand, it does not deeply explore the path and mechanism of "opportunity-oriented" or "risk-avoiding" caused by environmental uncertainty to corporate technological innovation. On the other hand, the existing literature mainly focuses on the situation of developed countries in Europe and the United States. The research on the environmental uncertainty of corporate technological innovation in emerging economies in the world needs to be enriched, and its institutional background should also be considered. Based on the above analysis, the paper integrates innovation theory, information asymmetry theory and principal-agent theory. Under the institutional background of China and from the perspective of an innovation value chain, the OLS model, Poisson model and ZIP model are used to analyze the effect of environmental uncertainty on corporate technological innovation. This paper not only discusses the intermediary effect of the agency problem on environmental uncertainty and corporate technological innovation but also deeply explores the influence of information transparency, government subsidies and other mechanisms to alleviate the agency problem on environmental uncertainty and corporate technological innovation.

The contribution of the paper is reflected in three aspects: (1) On the basis of the existing literature, this paper further explores the path and mechanism of environmental uncertainty on corporate technological innovation, especially whether there is "opportunity-oriented effect" or "risk-avoiding effect" in emerging countries, which enriches the related literature on technological innovation motivation and economic consequences of environmental uncertainty; (2) combined with China's institutional background and the prevalence of agency problems, the paper further discusses whether the environmental uncertainty affects corporate technological innovation through the agency problem, which makes up for the lack of relevant research on its influencing mechanism; (3) this paper attempts to alleviate the agency conflict from the perspective of internal information transparency and external government subsidies and integrate it into the research framework of environmental uncertainty on corporate technological innovation, so as to provide a theoretical and empirical basis for corporate technological innovation development and government departments to formulate relevant policies.

The paper is organized as follows: Section 2 provides the theoretical background and research hypothesis. Section 3 shows our methodology. Section 4 presents our results and discussion, and Section 5 shows conclusions, policy implications, limitations and future prospects.

2. Theoretical Background and Research Hypothesis

Innovation activity is a long-term investment activity with long cycles and high risk [13]. When the external environment situation is serious, enterprises, in order to deal with emergencies that may result from environmental uncertainty, based on preventive motivation, usually tend to choose conservative investment strategies, reduce the capital of innovative investment, and maintain a high free cash flow [8] to deal with market shocks and fierce market competition and ease the pressure of survival. When the external environment is more uncertain, the less accurate management is in judging the merits of innovative investment projects, preferring to delay innovation investment or reduce capital investment, and only when the environment stabilizes and uncertainty is manageable or disappears will companies put innovative investment projects on the agenda [14].

In the research on the motivation and economic consequences of enterprise technological innovation, many documents are often confined to a single stage of innovation activities or generally study them as a whole while ignoring the value chain of innovation activities. Innovation activities should be subdivided into different stages, and domestic scholars divided innovation activities into two stages: technology research and development and achievement transformation. The innovation capabilities of enterprises at different stages are affected by different factors, and the various stages are interrelated and interact with each other. In the technology development stage, companies will organize R&D personnel to use existing resources to develop new technologies and products by investing time and money, etc. This stage is usually manifested as a large amount of investment in R&D expenses. The greater the environmental uncertainty, the riskier the market [15]. Managers will adopt a conservative attitude in operating the enterprise, thereby reducing the enterprise's investment in innovation.

Hypothesis 1a (H1a). *In the technological development stage, environmental uncertainty has an inhibitory effect on technological innovation.*

In the achievement transformation stage, the investment in the research and development stage is transformed into technological achievements or new products produced, and the sales of products create income for the enterprise and bring about the improvement of the economic efficiency of the enterprise. When the uncertainty of the external environment rises, on the one hand, environmental uncertainty increases the difficulty of evaluating management's business decisions, causing management to make decisions more cautiously and adopt a conservative or herd decision strategy in order to avoid making mistakes and damaging interests, which makes the investment level of innovation activities restricted [16]; on the other hand, increased environmental uncertainty may have a greater impact on the company and even destroy the company's existing innovation potential, leading to short-sighted management and not accepting innovation activities that could obtain potentially high-return through taking risks to avoid risk, which hinders the technological innovation of enterprises. Accordingly, the following hypotheses are proposed:

Hypothesis 1b (H1b). *At the stage of achievement transformation, environmental uncertainty has an inhibitory effect on technological innovation.*

Agency problems are common in modern enterprises, and information asymmetry is the main reason for increasing agency costs. Large shareholders hold a high proportion of shares and participate in the business decision-making of enterprises [17]. Based on the high cost and uncertainty of R&D projects, large shareholders may tend to avoid risks [18]. The greater the environmental uncertainty, the more difficult it is for the external supervisory authority or the media to supervise the major shareholders, which covers up the responsibility of the executives for investment failures [19], and the more difficult it is for the major shareholders' infringement to be found, which increases the executive's personal interest motives, and the major shareholders' payment cost has been reduced. Major shareholders are the decision makers of major issues of the company, and technological innovation is a project that requires long-term investment and has a relatively high cost, so major shareholders may abandon R&D projects out of consideration of risks and returns [20]. Major shareholders with a high proportion of shares hold an evasive attitude towards high-risk R&D projects, and the probability of abandoning R&D increases [21]. The management mechanism and governance level of an enterprise affect the investment of innovative activities, and environmental uncertainty increases the degree of information asymmetry, creating conditions for large shareholders to encroach on the interests of small shareholders, which will exacerbate the problem of agency conflicts, resulting in a lack of the driving force of sustained high-level innovation investment by controlling shareholders. Therefore, environmental uncertainty exacerbates the emergence of agency problems, thereby further inhibiting the innovation activities of enterprises. Based on the above analysis, the following hypotheses are proposed:

Hypothesis 2 (H2). *Under the circumstance that other conditions remain unchanged, environmental uncertainty exacerbates the second type of agency problem, which has a restraining effect on enterprise technological innovation activities.*

According to the theory of information asymmetry, information asymmetry is common among investors [22]. When a company is in a high degree of environmental uncertainty, the asymmetry of information intensifies, and the decision-making of the company's management is affected, causing R&D activities to face greater threats and weaken the enthusiasm of management for innovation [23]. The innovation activities of enterprises need financial support, so when there is a "funding gap" in the investment innovation activities of enterprises, especially small or new enterprises are vulnerable to insufficient investment caused by external factors, and there are also reasons for insufficient internal investment, which leads to an increase in the cost of enterprise innovation. It is difficult to sustain innovation activities [24]. The problem of insufficient capital for corporate innovation investment has two aspects. On the one hand, it is due to external financing constraints, and on the other hand, it is due to internal management incentives [25].

Corporate transparency reduces the sensitivity of management turnover to poor innovative output. It also increases innovative efficiency through its governance role in facilitating efficient allocation of R&D capital. These findings illuminate the unique roles and mechanisms of transparency in promoting innovation incentives and outcomes. Motivating innovation is important in many incentive problems. The optimal innovation-motivating incentive scheme exhibits substantial tolerance (or even reward), so it regulates the resistance of corporate executives to innovative activities and reduces the professional risks of executives. [26–28]. The increase in information transparency creates an atmosphere of tolerance for failure, thereby reducing the risks faced by managers, stimulating R&D motivation and promoting the output of results. Based on the above analysis, Hypothesis 3 is proposed:

Hypothesis 3 (H3). *As long as other conditions remain unchanged, improving information transparency can help alleviate the inhibitory effect of environmental uncertainty on enterprise technological innovation.*

The production and operation activities carried out by enterprises have their own specific market environment. When the environmental uncertainty changes, corporate investment will change with it [29]. When the market situation is good and the environmental uncertainty it faces is low, corporate management can predict and supervise technological changes and other changes in a timely and accurate manner, so as to make correct business decisions. The government provides free subsidies to provide companies with capital turnover opportunities, which help promote enterprises to carry out innovative activities. The severe market situation has increased the uncertainty in the external environment of the company, and various risks have also arisen, which has increased the degree of information asymmetry, and the management lacks sufficient information, so it is difficult to estimate the benefits and costs, and difficult to accurately assess the risk of decision-making. Internally, they will face greater operational and financial risks, and external investors cannot easily invest, which makes companies prone to greater financing constraints and increases the pressure on companies to survive.

Direct government transfer payments or indirect tax reductions provide companies with net cash flow, reduce the capital cost of R&D activities, reduce the uncertainty and risk of innovation, and help stimulate companies to invest in innovative projects [30]. Government subsidies transmit a positive signal to the outside when the uncertainty of the external environment increases, and major government financial incentives were positively influential to innovative economic performance of firms, alleviate the external moral hazard of enterprises, provide enterprises with invisible guarantees, and bring innovative resources [31].

When the external environment is uncertain, the impact of government subsidies on enterprises becomes more and more important. Government subsidies have a significant crowding-out influence on enterprises' R&D investment behavior, and the influence is further moderated by the attributes of enterprise ownership [32,33]. The state encourages enterprises to innovate and conditionally provides government subsidies to attract enterprises to carry out technological innovation. Enterprises can take advantage of the opportunities brought about by environmental changes and the direct or indirect support of the government to increase innovation activities and gain core advantages [34]. Therefore, this paper proposes Hypothesis 4:

Hypothesis 4 (H4). *Under the circumstance that other conditions remain unchanged, government subsidies can help alleviate the inhibitory effect of environmental uncertainty on enterprise technological innovation.*

The research idea of this article is shown in Figure 1.

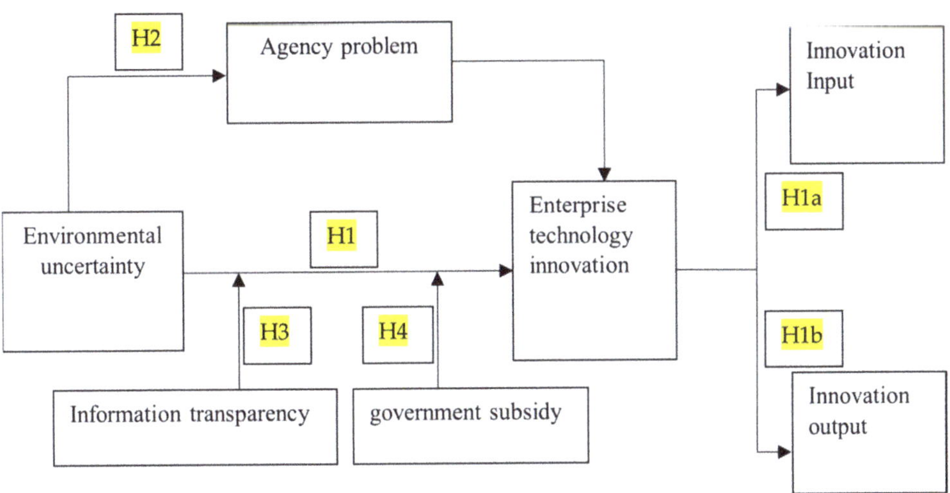

Figure 1. Research idea map.

3. Methodology

3.1. Research Samples and Data Sources

This paper obtains the financial data of the domestic listed A-share companies from 2008 to 2019 from the CSMAR database and makes full use of the data processing software Stata15 and Excel(Microsoft, Washington, WA, U.S.A) to sort and analyze. The initial research sample excluded ST listed companies, financial and insurance listed companies, samples with missing data on R&D investment, innovation output and important variables, and 10,323 samples were obtained. We winsorized the upper and lower 1% quantiles for continuous variables to reduce the deviation of extreme values from the empirical results.

3.2. Definition of Main Variables

3.2.1. Explained Variable

This article studies innovation activities from the two stages of the innovation value chain, namely technology research and development and achievement transformation. In the technology research and development stage, the method of Hoang Luong et al. [35] is used to measure the R&D expenditures by taking the natural logarithm. In the stage of achievement transformation, the methods of Suk Bong Choi et al. [36] and Jing Chi et al. [37] are used to measure the total number of patent applications (Patents), including invention patents, exterior design, and utility models. Appendix A Table A1 shows the definitions and calculation methods of specific variables.

3.2.2. Explanatory Variable

Environmental uncertainty is an explanatory variable. Environmental uncertainty has been characterized in terms of dynamism. Dynamism refers to the environmental instability that makes it difficult to predict changes and affects the volatility that a business unit faces. Typically, the volatility of industry sales and income is used to proxy dynamism [38]. In addition, the company's operating income data from the past 5 years is used to calculate the standard deviation of abnormal sales income in the past 5 years to measure the fluctuation of its income. Then, it is adjusted in consideration of industry standards, and the industry-adjusted value is calculated as the environmental uncertainty. See Equation (1) for details.

$$\text{Sale} = \varphi_0 + \varphi_1 Year + \varepsilon \tag{1}$$

Among them, Sale is operating income, and Year is the annual variable

3.2.3. Moderator: Information Transparency

Information transparency is reflected by accrued surplus. The larger the value, the opaquer the information. Learning from the methods of scholars Hutton, et al. [39], the accrued surplus calculated by the revised Jones model measures the severity of the white-washing of corporate profits. The larger the Trans obtained by Equation (4), the larger the accrued surplus and the lower the information transparency. In order to facilitate the analysis, 1-Tran is used to forward the indicator.

$$\begin{aligned} \text{TA}_{i,t}/\text{Asset}_{i,t-1} = {} & \alpha_1(1/\text{Asset}_{i,t-1}) + \alpha_2(\Delta\text{REV}_{i,t}/\text{Asset}_{i,t-1}) \\ & +\alpha_3(\Delta PPE_{i,t}/\text{Asset}_{i,t-1}) + \varepsilon_{i,t} \end{aligned} \tag{2}$$

$$\text{DA}_{i,t} = \text{TA}_{i,t} - \left\{ \begin{array}{c} \alpha_1(1/\text{Asset}_{i,t-1}) + \alpha_2(\Delta\text{REV}_{i,t} - \Delta\text{REC}_{i,t})/\text{Asset}_{i,t-1} \\ +\alpha_3(\Delta PPE_{i,t}/\text{Asset}_{i,t-1}) \end{array} \right\} \tag{3}$$

$$\text{Tran}_{i,t} = \{Abs(DA_{i,t-1}) + Abs(DA_{i,t-2}) + Abs(DA_{i,t-3})\}/3 \tag{4}$$

$$\text{Trans}_{i,t} = 1 - \text{Tran}_{i,t} \tag{5}$$

Among them, TA is the total accrued surplus, Asset is the total assets of the lagging period, ΔREV is the ratio of the change in sales income to the total assets of the previous year, ΔPPE is the change in the original value of fixed assets, and ΔREC is the change in accounts receivable.

3.2.4. Government Subsidy

The government subsidy in this article is the government subsidy under the non-operating income item in the notes of the financial statements of the CSMAR Database. We learned from scholars Martin, Hud et al. [40], etc., to divide this value by total assets to measure government subsidy.

3.2.5. Mediator

The second type of agency problem is the mediating variable, which draws on the methods of domestic scholars and is measured by the ratio of other receivables to total assets. The ratio is expressed by agency.

Using the mediating effect test model of Andrew Ferguson, Amanda, J., Fairchild and Mahesh, Srinivasan [41–43], we explored the impact of environmental uncertainty on two-stage innovation activities and the ways in which it acts, as shown in Figure 2.

$$\mathrm{Y} = \beta_0 + \beta_1 \mathrm{EU} + \beta_2 \mathrm{Controls} + \beta_3 \sum Ind + \beta_4 \sum Year + \varepsilon \quad (6)$$

$$Agency = \gamma_0 + \gamma_1 EU + \gamma_2 \mathrm{Controls} + \gamma_3 \sum Ind + \gamma_4 \sum Year + \varepsilon \quad (7)$$

$$\mathrm{Y} = \theta_0 + \theta_1 Agency + \theta_2 \mathrm{Controls} + \theta_3 \sum Ind + \theta_4 \sum Year + \varepsilon \quad (8)$$

$$\mathrm{Y} = \lambda_0 + \lambda_1 EU + \lambda_2 \mathrm{Agency} + \lambda_3 \mathrm{Controls} + \lambda_4 \sum Ind + \lambda_5 \sum Year + \varepsilon \quad (9)$$

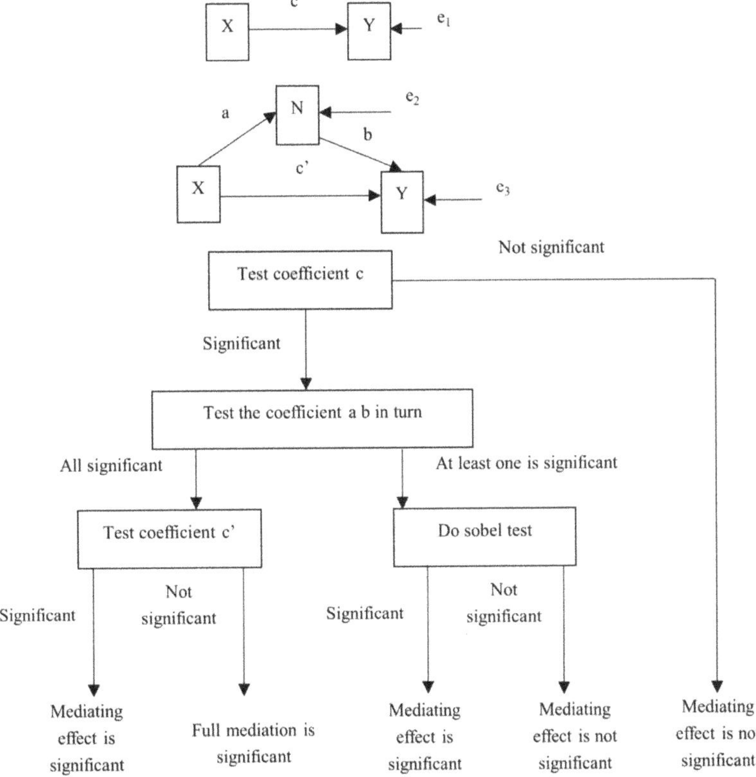

Figure 2. Mediating effect test procedure.

Among them, in the inspection technology research and development stage, Y is the innovation input R&D; in the inspection achievement transformation stage, Y is the innovation output expressed by the number of patent applications (Patents) and the natural logarithm of the number of patent applications (LPatents). If β_1 is significantly negative, assume that H1 is verified, θ_1 is significantly positive, and γ_1, λ_1, and λ_2 are significantly negative, then the agency problem has a significant mediating effect, and Hypothesis 2 is verified.

In order to verify the moderating effect of a moderator on environmental uncertainty and technological innovation, use the model of Mohit, Srivastava et al. [44] as reference to test innovation input; use the model of Jing, Chi et al. [37] as reference to test innovation output.

$$\begin{aligned} \text{R\&D} = \; & \chi_0 + \chi_1 \text{EU} + \chi_2 M + \chi_3 EU \times M + \chi_4 \text{ROA} + \chi_5 \text{SIZE} + \chi_6 \text{LEV} + \chi_7 \text{TOB} + \\ & \chi_8 \text{JYYZ} + \chi_9 \text{XJBL} + \chi_{10} Cash + \chi_{11} \text{Dud} + \chi_{12} \text{CGB} + \chi_{13} \text{Dual} + \chi_{14} \text{LDBL} + \\ & \chi_{15} \text{QYCS} + \chi_{16} \textstyle\sum Ind + \chi_{17} \sum Year + \varepsilon \end{aligned} \tag{10}$$

$$\begin{aligned} \text{Patens} = \; & \delta_0 + \delta_1 \text{EU} + \delta_2 \text{W} + \delta_3 EU \times \text{W} + \delta_4 \text{ROA} + \delta_5 \text{SIZE} + \delta_6 \text{LEV} + \delta_7 \text{TOB} + \\ & \delta_8 \text{JYYZ} + \delta_9 \text{XJBL} + \delta_{10} Cash + \delta_{11} \text{Dud} + \delta_{12} \text{CGB} + \delta_{13} \text{Dual} + \delta_{14} \text{LDBL} + \\ & \delta_{15} \text{QYCS} + \delta_{16} \textstyle\sum Ind + \delta_{17} \sum Year + \varepsilon \end{aligned} \tag{11}$$

Model (10) draws on scholars Daniela and Coluccia, who used an OLS model [45]. For Equation (11), the dependent variable INNOVATION is the number of enterprise patent rights, so we used the Poisson model. Because the rate of sample observations of zero is 20.98%, with the phenomenon of zero-inflated, using the general Poisson model may cause the problem of biased model estimation coefficients. Therefore, this paper refers to the Zero-inflated Poisson (ZIP) model proposed by Lambert (1992), which was first proposed by Lambert (1992) and applied to the study of the damage rate of manufacturing products. Therefore, this paper uses the Zero-inflated Poisson model to conduct empirical analysis [46].

When examining the moderating effect of information transparency on environmental uncertainty and technological innovation, M, W is Trans, and introducing the interaction term between environmental uncertainty EU and information transparency Trans; when examining the moderating effect of government subsidies on environmental uncertainty and technological innovation, M, W is SUB, introducing the interaction term between EU and government subsidy SUB with environmental uncertainty.

The definitions of main variables are described in Appendix A Table A1, and we performed a Panel Unit Root Test on the variable EU to test whether the data are stable and regress the unbalanced panel data. Due to space constraints, we placed the test and regression results in Appendix A Table A2.

4. Results and Discussion

4.1. Descriptive Statistical Analysis

Table 1 shows the descriptive statistical results of each variable. It can be seen from the table that the average value of environmental uncertainty (EU) is 0.135, the minimum value is 0.013, and the maximum value is 0.961, indicating that the uncertainties faced by enterprises have differences. The median of investment in technological innovation (R&D) is 17.999, and the average is 17.950, indicating that the overall innovation investment level of listed companies is acceptable. The minimum value of innovation output, the number of patents (Patents), is 0, the average value is 75.409, and the maximum value is 4282. The gap is large, indicating that the results formed by different enterprises after investing in innovation capital are quite different and input does not necessarily bring output. As a result, the effect of inter-enterprise achievement transformation needs to be improved. The median of government subsidy (SUB) is 16.616, the maximum is 22.172, and the minimum is 0, indicating that different companies receive a large difference in the intensity of government subsidies. The minimum value of information transparency (Trans) is 0.613, the maximum value is 1.430, and the standard deviation is 0.0687. There are differences in information transparency between different companies, but the difference is not significant.

Table 1. Descriptive statistics of each variable.

Main Variable	Sample Size	Mean	Median	Minimum	Maximum	Standard Deviation
LPatents	2645	2.8497	3.1781	0.0000	8.3624	1.8491
Patents	2645	75.4091	23.0000	0.0000	4282	260.8831
R&D	10,323	17.9500	17.9999	13.7090	21.4619	1.3312
EU	10,323	0.1346	0.1005	0.0133	0.9607	0.1151
SUB	10,323	16.5139	16.6161	0.0000	22.1719	1.8988
Trans	10,323	0.9940	0.9946	0.6125	1.4299	0.0687
Agency	10,323	0.0175	0.0100	0.0002	0.2446	0.0231
ROA	10,323	0.0392	0.0346	−0.2245	0.2080	0.0475
SIZE	10,323	22.2836	22.1819	19.5409	25.8882	1.0481
LEV	10,323	0.4254	0.4205	0.0491	0.9343	0.1825
TOB	10,323	2.1097	1.7587	0.8888	8.1375	1.1429
Dual	10,323	0.7575	1.0000	0.0000	1.0000	0.4286

4.2. Correlation Analysis

Table 2 is the regression table of the correlation between the main variables. The correlation coefficients of environmental uncertainty (EU), patents (Patents) and innovation input (R&D) are −0.008 and −0.058, respectively, indicating that environmental uncertainty (EU) has a negative impact on the innovation activities of the enterprise. The correlation coefficients between government subsidies (SUB), patents (Patents) and innovation input (R&D) are 0.182 and 0.358, respectively, indicating that government subsidy (SUB) has a positive impact on the innovation activities of enterprises. The correlation coefficients between agency, patents (Patents) and innovation input (R&D) are −0.055 and −0.117, respectively, indicating that the second type of agency problem between enterprises negatively affects the technological innovation of enterprises. It can be seen from Table 2 that the correlation coefficients between the variables are all lower than 0.5. Since the VIF value of all variables is less than 10, there is no multicollinearity problem among the variables.

Table 2. Correlation analysis of the main variables.

Variable	Patents	R&D	EU	SUB	Trans	Agency	ROA	SIZE
Patents	1							
R&D	0.322 ***	1						
EU	−0.008	−0.058 ***	1					
SUB	0.182 ***	0.358 ***	−0.01	1				
Trans	0.005	−0.059 ***	−0.011	0.004	1			
Agency	−0.055 ***	−0.117 ***	0.054 ***	−0.011	0.015	1		
ROA	0.086 ***	0.172 ***	−0.045 ***	0.049 ***	−0.303 ***	−0.132 ***	1.000	
SIZE	0.255 ***	0.473 ***	0.055 ***	0.389 ***	−0.067 ***	−0.005	0.036 ***	1.000

Note: *** represent significance at the levels of 1%, respectively.

4.3. Regression Analysis

4.3.1. Regression Analysis of Environmental Uncertainty and Enterprise Technological Innovation

(1) Innovation input in technology research and development stage

In order to explore the relationship between the two in depth, model (6) is used to test Hypothesis H1a to verify the relationship between environmental uncertainty and technological innovation investment. Column (1) of Table 3 shows environmental uncertainty (EU) and technological innovation regression results in the technological development stage, and the regression coefficient between EU and technological innovation input (R&D) is −0.521, which is significantly negatively correlated at the 1% confidence level, indicating that EU has an inhibitory effect on innovation input (R&D). It can be seen that the greater the environmental uncertainty, the more significant the inhibitory effect on enterprises' investment in technological innovation, which verifies Hypothesis 1a.

Table 3. Regression analysis of environmental uncertainty and enterprise technological innovation.

Variable	(1) R&D	(2) LPatents	(3) Patents
EU	−0.521 ***	−0.612 *	−0.344 ***
	(−6.07)	(−1.89)	(−14.70)
ROA	2.791 ***	0.332	1.035 ***
	(10.86)	(0.36)	(15.42)
SIZE	0.789 ***	0.659 ***	0.976 ***
	(58.52)	(13.72)	(333.03)
LEV	0.666 ***	1.516 ***	2.696 ***
	(5.28)	(3.22)	(57.65)
TOB	0.061 ***	0.119 ***	0.247 ***
	(5.32)	(3.08)	(94.37)
Dual	−0.070 ***	−0.310 ***	0.045 ***
	(−2.97)	(−3.82)	(8.07)
Constant	−0.970 ***	−12.975 ***	−21.093 ***
	(−3.04)	(−11.52)	(−228.64)
Industry, Year		Control	
Observation	1023	2645	2645
Adjusted R^2	0.450	0.137	-
F	223.0	12.39	-

Note: * and *** represent significant at the 10%, and 1%, confidence levels, respectively. The t value is in parentheses.

Among the control variables, it can be seen from column (1) of Table 3 that the regression coefficients of enterprise size (SIZE) and profitability index (ROA) are 0.789 and 2.791, respectively, which are both significantly positive at the 1% level, indicating that the larger the scale, the better the profitability. The better innovation foundation and innovation resources companies have, the more they invest in innovation activities. The regression coefficient of investment opportunity (TOB) is positive, indicating that the more investment opportunities a company has, the more likely it is to choose an innovative project to invest in, thereby promoting the company's input in technological innovation. The regression coefficient of the net business cycle (JYYZ) is significantly negative, which shows that the longer the net business cycle, the more unfavorable the company's innovation investment.

(2) Innovative output at the stage of achievement transformation

In order to explore the relationship between the two in depth, model (6) is used to test Hypothesis 1b to verify the impact of environmental uncertainty on the innovation output of the enterprise in the transformation stage of the results. The results of multiple regression are shown in Table 3. Column (2) of Table 3 shows the regression results of environmental uncertainty (EU) and technological innovation in the achievement transformation stage, and innovation output (LPatents) is the natural logarithm of the number of patent applications, which is a continuous variable, using the OLS model test Hypothesis 1b. In column (2) of Table 3, the regression coefficient between EU and innovation output (LPatents) is −0.612, which is significantly negative at the 10% confidence level, indicating that EU has an inhibitory effect on innovation output (LPatents). It can be seen that the greater the environmental uncertainty, the more unfavorable the technological innovation output of the enterprise, which verifies Hypothesis 1b.

Taking into account the data characteristics of the number of patent applications in China, the explained variables of the model are converted from continuous variables to the number of patent applications to explain. As patent data have many values of 0, the Poisson model is more consistent with the number features, and multiple regression models make the regression results more robust. From column (3) of Table 3, it can be seen that the regression coefficient between environmental uncertainty (EU) and technological innovation output (Patents) is −0.344, which is a significant negative correlation, indicating that EU's influence on innovation output (Patents) has an inhibitory effect. It can be concluded that the greater the environmental uncertainty, the more unfavorable the technological innovation output of the enterprise, which verifies Hypothesis 1.

4.3.2. Regression Analysis of Environmental Uncertainty, the Second Type of Agency Problem and Enterprise Technological Innovation

In the technology research and development stage, models (6)–(9) were used to test Hypothesis 2 to verify the relationship between environmental uncertainty and technological innovation input and the mediating effect of the second type of agency problem on environmental uncertainty and technological innovation input. From column (1) of Table 4, it can be seen that the regression coefficient between environmental uncertainty EU and technological innovation input (R&D) is −0.521, which is significantly negatively correlated at the 1% confidence level, indicating that EU's contribution to innovation input (R&D) has an inhibitory effect. Further testing the mediation effect, from column (2) of Table 4, it can be seen that the correlation coefficient between agency and innovation input (R&D) is −2.693, which is significantly negative at the 1% confidence level, indicating that the second type of agency problem is common among enterprises and affects enterprises' input in technological innovation and inhibits innovation. From column (3) of Table 4, it can be seen that the regression coefficient between agency and the environmental uncertainty EU is 0.006, which is significantly positive at the 1% level, that is, the higher the environmental uncertainty, the more serious the enterprise's second type of agency problem. From column (4) of Table 4, we can see that in model (9), the coefficients of agency, Environmental Uncertainty (EU) and the company's technological innovation input (R&D) are −2.606 and −0.505, respectively, which are both significantly positive at the 1% level. Combining the regression results of the previous models, we show that the greater the environmental uncertainty, the more significant the inhibitory effect on the enterprise's technological innovation investment. In addition, it exacerbates the second type of agency problem, which has a negative impact on enterprises' investment in technological innovation indirectly. This verifies Hypothesis 2.

Table 4. Regression analysis of environmental uncertainty, the second type of agency problem and innovation input.

Variable	(1) R&D	(2) R&D	(3) Agency	(4) R&D
EU	−0.521 ***		0.006 ***	−0.505 ***
	(-6.07)		(3.35)	(−5.88)
Agency		−2.693 ***		−2.606 ***
		(−6.02)		(−5.83)
ROA	2.791 ***	2.771 ***	−0.010 *	2.764 ***
	(10.86)	(10.78)	(−1.84)	(10.77)
SIZE	0.789 ***	0.781 ***	−0.003 ***	0.782 ***
	(58.52)	(57.67)	(−9.22)	(57.85)
LEV	0.666 ***	0.697 ***	0.013 ***	0.701 ***
	(5.28)	(5.52)	(4.80)	(5.56)
TOB	0.061 ***	0.065 ***	0.001 ***	0.064 ***
	(5.32)	(5.63)	(4.31)	(5.57)
Dual	−0.070 ***	−0.062 ***	0.002 ***	−0.066 ***
	(-2.97)	(−2.65)	(3.03)	(−2.80)
Constant	−0.970 ***	−0.820 **	0.078 ***	−0.766 **
	(−3.04)	(−2.55)	(11.14)	(−2.39)
Industry, Year		Control		
Observation	10,323	10,323	10,323	10,323
Adjusted R^2	0.450	0.450	0.112	0.451
F	223.0	222.9	35.22	218.8

Note: *, **, and *** represent significance at the 10%, 5%, and 1% confidence levels, respectively. The t value is in parentheses.

In the achievement transformation stage, we used models (6)–(9) to test Hypothesis 2 to verify the effect of environmental uncertainty on firms' innovation output and the mediating effect of the second type of agency problem between environmental uncertainty and technological innovation output. The results of the multiple regressions are shown in Table 5. Column (1) of Table 5 shows the regression results of environmental uncertainty (EU) and technological innovation in the achievement transformation stage. Innovation output (LPatents), which is the natural logarithm of the number of patent applications, is a continuous variable, and the OLS model was used to test Hypothesis 1. In column (1) of Table 5, the regression coefficient of EU and innovation output (LPatents) is −0.612, which is significantly negative at the 10% confidence level, indicating that EU has an inhibitory effect on innovation output (LPatents). Further testing the mediating effect, in column (2) of Table 5, the correlation coefficient between agency and innovation output (LPatents) is −3.934, which is significantly negative at the 5% confidence level, indicating that the second type of agency problem is prevalent among firms, which affects their technological innovation output and plays an inhibitory role on innovation. As can be seen in column (3) of Table 5, the regression coefficient between Agency and EU is 0.006, which is significantly positive at the 1% level, that is, the higher the environmental uncertainty, the more serious the second type of agency problem of firms. In column (4) of Table 5, it can be seen that in model (9), the coefficient of agency with firms' technological innovation output (LPatents) is −3.804, which is significantly negative, and the coefficient of environmental uncertainty (EU) with firms' technological innovation output (LPatents) is -0.586, which is significantly negative at the 10% level. Combining the regression results of the previous models, it can be concluded that the greater the environmental uncertainty, the more unfavorable the technological innovation output of the firm, and also, it indirectly has a negative impact on the technological innovation output of the firm by exacerbating the second type of agency problem. This verifies Hypothesis 2.

Table 5. Regression analysis of environmental uncertainty, the second type of agency problem and innovation output 1.

Variable	(1) LPatents	(2) LPatents	(3) Agency	(4) LPatents
EU	−0.612 *		0.006 ***	−0.586 *
	(−1.89)		(3.35)	(−1.81)
Agency		−3.934 **		−3.804 **
		(−2.14)		(−2.07)
ROA	0.332	0.335	−0.010 *	0.280
	(0.36)	(0.37)	(−1.84)	(0.31)
SIZE	0.659 ***	0.646 ***	−0.003 ***	0.647 ***
	(13.72)	(13.36)	(−9.22)	(13.39)
LEV	1.516 ***	1.581 ***	0.013 ***	1.617 ***
	(3.22)	(3.35)	(4.80)	(3.42)
TOB	0.119 ***	0.122 ***	0.001 ***	0.121 ***
	(3.08)	(3.15)	(4.31)	(3.13)
Dual	−0.310 ***	−0.301 ***	0.002 ***	−0.297 ***
	(−3.82)	(−3.71)	(3.03)	(−3.65)
Constant	−12.975 ***	−12.726 ***	0.078 ***	−12.705 ***
	(−11.52)	(−11.23)	(11.14)	(−11.21)
Industry, Year	Control			
Observation	2,645	2,645	10,323	2,645
Adjusted R^2	0.137	0.138	0.112	0.139
F	12.39	12.42	35.22	12.19

Note: *, **, and *** represent significance at the 10%, 5%, and 1% confidence levels, respectively. The t value is in parentheses.

Table 6 shows the regression results of the Poisson model. It can be seen from column (1) of Table 6 that the regression coefficient between environmental uncertainty (EU) and technological innovation output (Patents) is −0.344, which is significantly negative, indicating that EU's impact on innovation output (Patents) has an inhibitory effect. To further test the mediation effect, in the column (2) of Table 6, the correlation coefficient between agency and innovation output (Patents) is −4.682, which is significantly negative, indicating that the second type of agency problem generally exists between different companies and affects the company's output of technological innovation, inhibiting innovation. In column (3) of Table 6, the regression coefficient between agency and EU is 0.006, which is significantly positive, that is, the higher the environmental uncertainty, the more serious the second type of agency problem of the enterprise. In column (4) of Table 6, using model (9) to test, the coefficient of agency and firm's technological innovation output (Patents) is −4.603, which is significantly negative. At the 1% confidence level, the environmental uncertainty EU and the technological innovation output (Patents) of the enterprise are significantly negative with a regression coefficient of −0.321. Combining the regression results of the previous models and using the principle of mediating the effect test, it is found that the greater the environmental uncertainty, the more unfavorable the technological innovation output of the enterprise. In addition, it also indirectly negatively affects the technological innovation output of enterprises by exacerbating the second type of agency problem. This verifies Hypothesis 2.

Table 6. Regression analysis of environmental uncertainty, the second type of agency problem and innovation output 2.

Variable	(1) Patents	(2) Patents	(3) Agency	(4) Patents
EU	−0.344 *** (−14.70)		0.006 *** (3.35)	−0.321 *** (−13.66)
Agency		−4.682 *** (−26.93)		−4.603 *** (−26.43)
ROA	1.035 *** (15.42)	0.992 *** (14.85)	−0.010 * (−1.84)	0.992 *** (14.79)
SIZE	0.976 *** (333.03)	0.968 *** (329.01)	−0.003 *** (−9.22)	0.969 *** (328.79)
LEV	2.696 *** (57.65)	2.723 *** (58.54)	0.013 *** (4.80)	2.757 *** (59.11)
TOB	0.247 *** (94.37)	0.248 *** (94.81)	0.001 *** (4.31)	0.246 *** (93.90)
Dual	0.045 *** (8.07)	0.053 *** (9.61)	0.002 *** (3.03)	0.051 *** (9.22)
Constant	−21.093 *** (−228.64)	−20.892 *** (−225.78)	0.078 *** (11.14)	−20.881 *** (−225.60)
Industry, Year		Control		
Observation	2645	2645	10,323	2645

Note: * and *** represent significance at the 10%, 5%, and 1% confidence levels, respectively. The t value is in parentheses.

4.3.3. Regression Analysis of Environmental Uncertainty, Information Transparency and Enterprise Technological Innovation

We used models (10) and (11) to test Hypothesis 3 to verify the moderating effect of information transparency on environmental uncertainty and technological innovation. The regression results are listed in Table 7. In the technology research and development stage, in column (1) of Table 7, at a confidence level of 1%, environmental uncertainty (EU) is significantly negatively correlated with the technological innovation input (R&D), and its regression coefficient is −2.489, indicating that EU has a restraining effect on innovation investment (R&D). After introducing the interaction terms of environmental uncertainty and information transparency, EU × Trans is significantly positively correlated at the level of 5%, and its regression coefficient is 1.982, indicating that corporate information transparency can help alleviate the inhibitory effect of the environmental uncertainty on enterprises' investment in technological innovation.

Table 7. The moderating effect of information transparency on environmental uncertainty and enterprise technological innovation.

Variable	(1) R&D	(2) Patents	(3) Patents
EU	−2.489 ***	−3.225 ***	−4.649 ***
	(−2.67)	(−10.17)	(−15.05)
Trans	0.550 **	2.684 ***	2.665 ***
	(2.45)	(43.38)	(44.01)
EU × Trans	1.982 **	2.803 ***	4.525 ***
	(2.12)	(8.83)	(14.59)
ROA	3.368 ***	4.852 ***	5.860 ***
	(13.62)	(77.33)	(89.87)
SIZE	0.794 ***	0.993 ***	0.962 ***
	(58.86)	(336.52)	(320.35)
LEV	0.657 ***	2.504 ***	1.570 ***
	(5.22)	(53.37)	(33.74)
TOB	0.061 ***	0.246 ***	0.220 ***
	(5.31)	(93.33)	(84.09)
Dual	−0.070 ***	0.050 ***	0.114 ***
	(−2.99)	(9.06)	(20.72)
Constant	−1.618 ***	−23.954 ***	−23.009 ***
	(−4.04)	(−208.74)	(−200.87)
Industry, Year		Control	
Observation	10,323	2645	2645
Adjusted R^2	0.451	-	-
F	218.8	-	-

Note: ** and *** represent significance at the 5% and 10%, confidence levels, respectively. The t value is in parentheses.

In the stage of achievement transformation, considering the data characteristics of the number of patent applications in China, the explained variables of the model are converted from continuous variables to the number of patent applications to explain. Since patent data have many phenomena with a value of 0, the POISSON model and ZIP model test are more in line with the quantitative characteristics, and multiple regression models make the regression results more robust. Column (2) of Table 7 is the regression result of the POISSON model. It can be seen that EU and technological innovation output (Patents) are significantly negatively correlated at the 1% confidence level, and the regression coefficient is −3.225, indicating that EU has an inhibitory impact on innovation output. After introducing the interaction term of environmental uncertainty and information transparency, EU × Trans is significantly positively correlated at the 1% level, and the regression coefficient is 2.803, indicating that corporate information transparency can help alleviate the inhibitory impact of environmental uncertainty on corporate technological innovation output. Combining information transparency can alleviate the inhibitory effect of enterprises on innovation input and innovation output when facing environmental uncertainty, Hypothesis 3 in this article has been verified.

Among the control variables, it can be seen from columns (1) and (2) of Table 7 that the regression coefficients of enterprise size SIZE and profitability index ROA are both significantly positive at the 1% level, indicating that the larger the scale and the greater the profitability of the enterprise, the more it can promote the technological innovation of the enterprise. The regression coefficient of investment opportunity TOB is positive, indicating that the more investment opportunities a company has, the greater the possibility of choosing innovative projects for investment, which is conducive to promoting the technological innovation of the company. The regression coefficient of the net business cycle JYYZ is significantly negative, indicating that the longer the net business cycle, the more unfavorable to the innovation of the enterprise.

Column (3) of Table 7 is the regression result of the ZIP model. It can be seen that whether the EU × Trans (the interaction term of environmental uncertainty and information transparency) or the control variables, The results are consistent with the regression results of the OLS model and Poisson model. At the 1% confidence level, EU has a significantly negative correlation with technological innovation output (Patents), and the regression coefficient is −4.649, indicating that EU has an inhibitory effect on innovation output of patents. After introducing the interaction terms of environmental uncertainty and information transparency, EU × Trans is significantly positively correlated at the 1% level, and the regression coefficient is 4.525, indicating that corporate information transparency can help alleviate the inhibitory impact of environmental uncertainty on corporate technological innovation output. Combining information transparency can alleviate the inhibitory effect of enterprises on innovation input and innovation output when facing environmental uncertainty, Hypothesis 3 in this article has been verified.

4.3.4. Regression Analysis of Environmental Uncertainty, Government Subsidies and Enterprise Technological Innovation

We used models (10) and (11) to test Hypothesis 4 to verify the moderating effect of government subsidies on environmental uncertainty and technological innovation. The regression results are listed in Table 8. In column (1) of Table 8, EU has a significant negative correlation with technological innovation input (R&D) at the 1% confidence level, and the regression coefficient is −2.650, indicating that EU has an inhibitory effect on innovation input (R&D). After introducing the interaction term between environmental uncertainty and government subsidies, EU × SUB is significantly positive at the 1% level, and the regression coefficient is 0.129, indicating that corporate government subsidies can help alleviate the inhibitory effect of EU on corporate technological innovation investment.

Table 8. The moderating effect of government subsidies on environmental uncertainty and enterprise technological innovation.

Variable	(1) RD	(2) Patents	(3) Patents
EU	−2.650 ***	−1.395 ***	−6.460 ***
	(−3.24)	(−4.59)	(−18.95)
SUB	0.083 ***	0.270 ***	0.212 ***
	(9.96)	(80.83)	(62.95)
EU × SUB	0.129 ***	0.064 ***	0.365 ***
	(2.62)	(3.70)	(18.82)
ROA	2.947 ***	3.812 ***	4.480 ***
	(12.50)	(59.99)	(69.26)
SIZE	0.702 ***	0.725 ***	0.702 ***
	(49.51)	(197.69)	(186.77)
LEV	0.607 ***	1.887 ***	1.144 ***
	(4.88)	(41.18)	(25.57)
TOB	0.055 ***	0.212 ***	0.189 ***
	(4.87)	(82.56)	(73.37)
Dual	−0.062 ***	0.037 ***	0.130 ***
	(−2.68)	(6.69)	(23.50)
Constant	−0.415	−19.985 ***	−18.276 ***
	(−1.25)	(−198.85)	(−176.29)
Industry, Year		Control	
Observation	10,323	2,645	2,645
Adjusted R^2	0.465	-	-
F	231.4	-	-

Note: *** represent significant at the 1% confidence levels, respectively. The t value is in parentheses.

Table 8 shows the regression results of the POISSON model and the ZIP model. Column (2) of Table 8 is the regression result of the POISSON model. It can be seen that at the 1% confidence level, EU and technological innovation output patents are significantly negatively correlated, and the regression coefficient is −1.395, indicating that EU has an inhibitory effect on innovation output. After introducing the interaction term between environmental uncertainty and government subsidies, EU × SUB is significantly positively correlated at the 1% level with a regression coefficient of 0.064, indicating that corporate government subsidies can help alleviate EU's inhibitory effect on corporate technological innovation output. Combining government subsidies can alleviate the inhibitory effect on innovation input and innovation output when facing environmental uncertainty, Hypothesis 4 in this paper is verified.

Among the control variables, in columns (1) and (2) of Table 8, the regression coefficients of enterprise size (SIZE) and profitability index (ROA) are both significantly positive at the 1% level, indicating that the larger the scale and the greater profitability of the enterprise, the more it can promote technological innovation. The regression coefficient of investment opportunity (TOB) is significantly positive, indicating that the more investment opportunities a company has, the more likely it is to choose an innovative project to invest, which is conducive to promoting the technological innovation of the company. The regression coefficient of the net business cycle (JYYZ) is significantly negative, indicating that the longer the net business cycle, the more adverse the effect will be on the innovation activities of the enterprise.

Column (3) of Table 8 is the regression result of the ZIP model. It can be seen that whether it is the interaction term EU × SUB between environmental uncertainty and government subsidies or the control variables, they are consistent with the regression results of the previous model. At the 1% confidence level, EU has a significant negative correlation with technological innovation output patents, and the regression coefficient is −6.46, indicating that EU has an inhibitory effect on innovation output patents. After introducing the interaction term between environmental uncertainty and government subsidies, EU × SUB is significantly positively correlated at the 1% level, with a regression coefficient of 0.365, indicating that corporate government subsidies can help alleviate EU's inhibitory effect on corporate technological innovation output. Combining government subsidies can alleviate the inhibitory effect on innovation input and innovation output when facing environmental uncertainty, Hypothesis 4 in this paper is verified.

In addition, we have also conducted panel data regression on the main models, and the research conclusions are still consistent. See Appendix A Table A3 and Appendix A Table A4 for the specific regression results.

5. Conclusions, Policy Implications, Limitations and Future Prospects

5.1. Conclusions

Based on the existing literature research, we analyze the effects and paths of environmental uncertainty on corporate technological innovation using data from Chinese listed companies in Shanghai and Shenzhen A markets from 2008–2019. This paper not only discusses the intermediary effect of the agency problem on environmental uncertainty and corporate technological innovation but also deeply explores the influence of information transparency, government subsidies and other mechanisms to alleviate the agency problem on environmental uncertainty and corporate technological innovation, and draws the following conclusions.

First, the "risk-avoiding" effect of environmental uncertainty on corporate technology innovation is greater than the "opportunity-oriented" effect. The greater the environmental uncertainty faced by corporations, the more obvious the inhibition effect on corporate technology innovation. Specifically, the stronger the environmental uncertainty is, the less the investment in corporate innovation will be reduced, and the effect of corporations' innovation output will be reduced in the stage of achievement transformation. Further exploration reveals that the inhibition effect of environmental uncertainty on corporate technological innovation will last for at least 2 years.

Second, the agency problem is widespread in Chinese corporations, especially the second kind of agency problem between controlling shareholders and minority shareholders. The results show that the more severe the second kind of agency problem is, the more unfavorable it is to technological innovation in listed corporations, significantly reducing corporations' incentives to invest in sustained innovation and inhibiting innovation inputs and innovation outputs. Moreover, environmental uncertainty directly aggravates the second kind of agency problem and affects corporate technological innovation activities through agency problems, reducing corporate technological innovation inputs and innovation outputs. Further research shows that the environmental uncertainty faced by corporations can aggravate the second kind of agency problem, which will have a restraining effect on technological innovation of corporations for two years.

Third, both internal governance mechanisms (information transparency) and external governance mechanisms (government subsidies), which alleviate the agency problem, have a moderating effect on environmental uncertainty and corporate technological innovation. The improvement of information transparency and government subsidies is helpful in alleviating the inhibition effect of environmental uncertainty on corporate technological innovation inputs and outputs. Further research shows that when corporations face environmental uncertainty, they can transform the ongoing two-year negative effect of environmental uncertainty on their technological innovation by improving the transparency of corporate information.

5.2. Policy Implications

Based on the research conclusions, we put forward the following suggestions:

First, based on the two-stage innovation activities, in the technology R&D stage, corporations need to establish an innovation risk control mechanism oriented to environmental uncertainty, encourage corporations to optimize the allocation of corporate technological innovation resources, and maintain the stability and sustainability of R&D investment. In the achievement transformation stage, it is necessary to avoid the disconnection between R&D and transformation. Corporations should improve the output and quality of technological innovation, which will bring good revenue, thus in turn supporting technological R&D and achieving the circular development of innovation activities.

Second, corporations should continuously improve internal supervision mechanisms, improve their governance capacity and optimize corporate governance structure. Corporations are facing increasingly fierce environmental changes, while the external market changes are random and unpredictable. Corporations should improve their ability to prevent and control uncertainty, enhance their own internal governance capacity and incentive systems, and reduce moral hazards. It is necessary to give full play to the function of equity checks and balances. Corporations should improve the degree of equity checks and balances, reduce the second kind of agency problems between major shareholders and minority shareholders, and reduce the agency conflicts caused by agency problems.

Third, corporations should improve the transparency of corporate information and information disclosure mechanisms. The government should increase support for technological innovation, while strengthening the management and supervision of government subsidy funds. Both approaches are helpful to alleviate the inhibition effect of the agency problem on environmental uncertainty and corporate technological innovation. Improving the transparency of corporate information is conducive to alleviating information asymmetry. The government should give full play to the incentive role of financial subsidies, increase support for corporations eligible for subsidies, and promote the effective implementation of innovative activities. Both approaches reduce the negative impact of uncertainty and agency problems on innovation activities.

5.3. Limitations and Future Prospects

On the one hand, the measurement of environmental uncertainty deserves further exploration. At present, most scholars use the fluctuation of operating income to measure it, instead of measuring it from multiple dimensions that affect environmental uncertainty. On the other hand, restricted by the information disclosure and databases of Chinese corporations, technological innovation is measured only by the quantity of R&D investment and patent rights output, but quality indicators such as patent citations are not adopted. In addition, although part of the patent datum is collected manually, there are limitations due to the high number of missing values of patent data in CSMAR and other databases.

This paper explores the effect of environmental uncertainty on the technological innovation activities of corporations in two stages and its path and mechanism based on theoretical and empirical analysis research from the perspective of the innovation value chain. We provide some ideas for the study of the economic consequences caused by environmental uncertainty and also expand the research on the motivation of corporate technological innovation activities. However, the research of environmental uncertainty can be further refined in the future to explore the mechanism of its impact on technological innovation from different dimensions. On the other hand, whether there are other paths and mechanisms of environmental uncertainty affecting technological innovation can be further explored in the future.

Author Contributions: All of the authors contributed to conceptualization, formal analysis, investigation, methodology, and writing and editing of the original draft. All authors have read and agreed to the published version of the manuscript.

Funding: This research received no external funding.

Institutional Review Board Statement: Not applicable.

Informed Consent Statement: Not applicable.

Data Availability Statement: The data will be made available on request from the corresponding author.

Conflicts of Interest: The authors declare no conflict of interest.

Appendix A

Table A1. Definition of the main variables.

Variable Type	Variable Name	Variable Symbol	Variable Definition
Explained Variable	Innovation investment	R&D	Natural logarithm of R&D expenses
	Innovation output	Patents	The number of patent applications, including the total number of applications for invention patents, designs and utility models
		LPatents	Natural logarithm of the number of patent applications
Explanatory Variable	Environmental uncertainty	LPatents	The industry-adjusted ratio of the standard deviation of sales revenue in the past 5 years to the average value of sales revenue in the past 5 years
Moderator	government subsidy	SUB	Government subsidy/total assets
	Information transparency	Trans	Accrued surplus calculated by the revised Jones model
Mediator	Second type of agency problem	Agency	Ratio of other receivables to total assets

Table A1. *Cont.*

Variable Type	Variable Name	Variable Symbol	Variable Definition
Controlled Variable	Profitability	ROA	Net asset interest rate = Net profit/average total assets
	Company size	SIZE	Natural logarithm of total assets at the end of the year
	Investment opportunities	TOB	Tobin's Q value
	Asset-liability ratio	LEV	Liabilities/assets of the balance sheet disclosed at the end of the year
	Two jobs in one	Dual	whether the general manager concurrently serve as the chairman of the board
	Industry	Ind	Ind is an industry dummy variable
	Year	Year	Year is the annual dummy variable

Table A2. Panel unit root test results.

Method	Statistic	Prob. **	Cross-Section	Obs
Levin, Lin and Chu t	−903.337	0.0000	1266	5951
Im, Pesaran and Shin W-stat	−60.8075	0.0000	921	4916
ADF-Fisher Chi-square	3506.88	0.0000	1266	5951
PP-Fisher Chi-	4333.32	0.0000	1266	5951

Note: ** Probabilities for Fisher tests are computed using an asymptotic Chi-square distribution. All other tests assume asymptotic normality.

Table A3. OLS Regression (panel data).

Variables	RD	LPatents
EU	−0.238 ***	−0.905 **
	(−3.48)	(−2.21)
ROA	0.482 ***	0.199
	(2.71)	(0.19)
SIZE	0.946 ***	0.659 ***
	(65.85)	(8.13)
LEV	−0.345 ***	2.521 ***
	(−3.09)	(3.82)
TOB	0.004	0.074 *
	(0.59)	(1.94)
JYYZ	0.000	−0.000
	(1.31)	(−0.39)
XJBL	−0.070 ***	−0.052
	(−4.35)	(−0.58)
Cash	0.321 ***	−0.160
	(2.75)	(−0.24)
Dud	−0.205	−0.938
	(−1.02)	(−0.79)
CGB	0.002	0.020 **
	(1.35)	(2.52)
Dual	−0.016	−0.098
	(−0.73)	(−0.78)
LDBL	0.005	0.100
	(0.48)	(1.50)
QYCS	−0.042 ***	−0.210 **
	(−3.03)	(−2.41)

Table A3. *Cont.*

Variables	RD	LPatents
Constant	−2.835 ***	−12.325 ***
	(−8.53)	(−6.62)
Observations	8,442	2,131
R-squared	0.434	0.077
Number of firms	1,879	891
Adjusted R^2	0.270	−0.602
F	386.0	7.875

Note: *, ** and *** represent significant at the 10%, 5%, and 1% confidence levels, respectively. The t value is in parentheses.

Table A4. Poisson regression (panel data).

Variables	Patents
EU	−0.197 ***
	(−7.85)
ROA	2.959 ***
	(39.21)
SIZE	0.916 ***
	(281.78)
LEV	2.379 ***
	(49.34)
TOB	0.220 ***
	(75.99)
JYYZ	−0.004 ***
	(−37.31)
XJBL	−0.148 ***
	(−20.66)
Cash	2.198 ***
	(41.19)
Dud	1.872 ***
	(39.59)
CGB	0.015 ***
	(66.57)
Dual	−0.052 ***
	(−8.28)
LDBL	0.057 ***
	(15.24)
QYCS	−0.464 ***
	(−55.65)
Industry	Control
Year	Control
Constant	−19.266 ***
	(−192.73)
Pseudo R^2	0.4059

Note: *** represents significant at the 1% confidence level. The t value is in parentheses.

References

1. Priem, R.L.; Love, L.G.; Shaffer, M.A. Executives' Perceptions of uncertainty sources: A numerical tax-onomy and underlying dimensions. *J. Manag.* **2002**, *6*, 725–746.
2. Lv, M.; Bai, M. Evaluation of China's carbon emission trading policy from corporate innovation. *Financ. Res. Lett.* **2021**, *39*, 101565. [CrossRef]
3. Jiang, Z.; Wang, Z.; Lan, X. How environmental regulations affect corporate innovation? The coupling mechanism of mandatory rules and voluntary management. *Technol. Soc.* **2021**, *65*, 101575. [CrossRef]
4. Xing, C.; Zhang, Y.; Tripe, D. Green credit policy and corporate access to bank loans in China: The role of environmental disclosure and green innovation. *Int. Rev. Financ. Anal.* **2021**, *77*, 101838. [CrossRef]
5. Ellis, J.; Smith, J.; White, R. Corruption and corporate innovation. *Int. Rev. Financ. Anal.* **2020**, *7*, 2124–2149. [CrossRef]
6. Pastor, L.; Veronesi, P. Uncertainty about government policy and stock prices. *J. Finance* **2012**, *4*, 1219–1264. [CrossRef]

7. Tan, Y.; Tian, X.; Zhang, X. The real effect of partial privatization on corporate innovation: Evidence from China's split share structure reform. *J. Corp. Financ.* **2020**, *64*, 101661. [CrossRef]
8. Almeida, H.; Campello, M. Financial constraints, asset tangibility and cor-porate investment. *Rev. Financ. Stud.* **2007**, *5*, 1429–1460. [CrossRef]
9. Gong, L.; Liu, Z.; Rong, Y. Inclusive leadership, ambidextrous innovation and organizational performance: The moderating role of environment uncertainty. *Leadersh. Organ. Dev. J.* **2021**, *5*, 783–801. [CrossRef]
10. Guan, J.; Xu, H.; Huo, D. Economic policy uncertainty and corporate innovation: Evidence from China. *Pacific Basin Finance J.* **2021**, *67*, 101542. [CrossRef]
11. Hirshleifer, D.; Welch, I. An Economic Approach to the psychology of change: Amnesia, inertia, and impulsiveness. *J. Econ. Manag. Strategy* **2004**, *3*, 379–421.
12. Wang, C.; Qureshi, I.; Guo, F. Corporate Social Responsibility and Disruptive Innovation: The moderating effects of environmental turbulence. *J. Bus. Res.* **2022**, *139*, 1435–1450. [CrossRef]
13. Yiwei, D.; Zhaoxia, X. Bank lending and corporate innovation: Evidence from SFAS 166/167. *Contemp. Account. Res.* **2021**, *4*, 3017–3052.
14. Bernanke, B.S. Irreversibility, uncertainty, and cyclical investment. *NB-ER Working Papers* **1983**, *1*, 85–106.
15. Verdu, A.J.; Tamayo, I.; Ruiz-Moreno, A. The moderating effect of environmental uncertainty on the relationship between real options and technological innovation in high-tech firms. *Technovation* **2012**, *9–10*, 579–590. [CrossRef]
16. Baum, C.F.; Caglayan, M.; Ozkan, N. The impact of macroeconomic uncertainty on non-financial firms' demand for liquidity. *J. Financ. Econ.* **2006**, *4*, 289–349. [CrossRef]
17. Dai, L.; Shen, R.; Zhang, B. Does the media spotlight burn or spur innovation? *Rev. Financ. Stud.* **2021**, *26*, 343–390. [CrossRef]
18. Zhong, R.I. Transparency and firm innovation. *J. Account. Econ.* **2018**, *1*, 67–93. [CrossRef]
19. Guan, Y.; Zhang, L.; Zheng, L. Managerial liability and corporate innovation: Evidence from a legal shock. *J. Corp. Finance* **2021**, *69*, 102022. [CrossRef]
20. Dreyer, C.; Schulz, O. Investor horizons and corporate policies under uncertainty. *Rev. Finance Econ.* **2021**. [CrossRef]
21. Armstrong, G.S.; Guay, W.R.; Wber, J.P. The Role of information and financ-ial reporting in corporate governanceand debt contrancting. *J. Account. Econ.* **2010**, *50*, 179–234. [CrossRef]
22. Lee, W.-J.; Pittman, J.; Saffar, W. Political uncertainty and cost stickiness: Evidence from national elections around the world. *Contemp. Account. Res.* **2019**, *2*, 1107–1139.
23. Dechow, P.M.; Skinner, D. Earnings management: Reconciling the views of accounting academics, practitioners and regulators. *J. Account. Horiz.* **2000**, *14*, 235–250. [CrossRef]
24. Healy, P.M.; Palepu, K.G. Information asymmetry, corporate disclosure, and the capital markets: A review of the empirical disclosure literature. *J. Account. Econ.* **2001**, *3*, 405–440. [CrossRef]
25. Ferreira, D.; Manso, G.; Silva, A.C. Incentives to innovate and the decision to go public or private. *Rev. Financ. Stud.* **2014**, *1*, 256–300. [CrossRef]
26. Manso, G. Motivating Innovation. *J. Finance* **2011**, *5*, 1823–1860. [CrossRef]
27. Tsanga, A.; Wang, K.T.; Liu, S. Integrating corporate social responsibility criteria into executive compensation and firm innovation: International evidence. *J. Corp. Financ.* **2021**, *70*, 102070. [CrossRef]
28. Kong, D.; Zhao, Y.; Liu, S. Trust and innovation: Evidence from CEOs' early-life experience. *J. Corp. Finance* **2021**, *69*, 101984. [CrossRef]
29. Li, K.; Xia, B.; Chen, Y. Environmental uncertainty, financing constraints and corporate investment: Evidence from China. *Pacific Basin Finance J.* **2021**, *70*, 101665. [CrossRef]
30. Hall, B.H.; Lerner, J. Chapter 14-The financing of R&D and innovation. *Hand. Econ. Innov. North Holland* **2010**, *1*, 609–639.
31. Shu, C.; Wang, Q.; Gao, S. Firm patenting, innovations, and government institutional support as a double-edged sword. *J. Prod. Innov. Manag.* **2015**, *2*, 290–305. [CrossRef]
32. Yu, F.; Guo, Y.; Le-Nguyen, K. The impact of government subs-idies and enterprises' R&D investment: A panel data study from renewableenergy in China. *Energy Policy* **2016**, *89*, 106–113.
33. Zhou, Z. Government ownership and exposure to political uncertainty:Evidence from China. *J. Bank. Financ.* **2017**, *84*, 152–165. [CrossRef]
34. Ng, J.; Saffar, W.; Zhang, J.J. Policy uncertainty and loan loss provisions in the banking industry. *Rev. Account. Stud.* **2020**, *25*, 726–777. [CrossRef]
35. Luong, H.; Moshirian, F.; Nguyen, L. How do foreign institutional investors enhance firm innovation? *J. Financ. Quant. Anal.* **2017**, *4*, 1449–1490. [CrossRef]
36. Choi, S.B.; Lee, S.H.; Williams, C. Ownership and firm innovation in a transition economy: Evidence from China. *Res. Policy* **2011**, *3*, 441–452. [CrossRef]
37. Chi, J.; Liao, J.; Yang, J. Institutional stock ownership and firm innovation: Evidence from China. *J. Multinatl. Financ. Manag.* **2019**, *50*, 44–57. [CrossRef]
38. Xue, L.; Ray, G.; Sambamurthy, V. Efficiency or innovation: How do industry environments moderate the effects of firms' IT asset portfolios? *MIS QUART* **2012**, *2*, 509–528. [CrossRef]
39. Hutton, A.P.; Marcus, A.J.; Tehranian, H. Opaque financial reports, R2, and crash risk. *J. Financ. Econ.* **2009**, *1*, 67–86. [CrossRef]

40. Hud, M.; Hussinger, K. The impact of R&D subsidies during the crisis. *Res. Policy* **2015**, *10*, 1844–1855.
41. Srinivasan, M.; Mukherjee, D.; Gaur, A.S. Buyer–supplier partnership quality and supply chain performance: Moderating role of risks, and environmental uncertainty. *Eur. Manag. J.* **2011**, *4*, 260–271. [CrossRef]
42. Ferguson, A.; Lam, P. Government policy uncertainty and stock prices: The case of Australia's uranium industry. *Energy Econ.* **2012**, *4*, 1219–1264. [CrossRef]
43. Fairchild, A.J.; MacKinnon, D.P. A General model for testing mediation and moderation effects. *Prev. Sci.* **2009**, *10*, 87–99. [CrossRef]
44. Srivastava, M.; Moser, R.; Hartmann, E. The networking behavior of Indian executives under environmental uncertainty abroad: An exploratory analysis. *J. Bus. Res.* **2018**, *82*, 230–245. [CrossRef]
45. Coluccia, D.; Dabić, M.; Giudice, M.D. R&D innovation indicator and its effects on the market. An empirical assessment from a financial perspective. *J. Bus. Res.* **2020**, *119*, 259–271.
46. Lambert, D. Zero-inflated poisson regression, with an application to defects in manufacturing. *Technometrics* **1992**, *1*, 1–14. [CrossRef]

International Journal of
Environmental Research and Public Health

MDPI

Article

Promoting Inclusive Outdoor Recreation in National Park Governance: A Comparative Perspective from Canada and Spain

Maria José Aguilar-Carrasco [1,*], Eric Gielen [1], Maria Vallés-Planells [2], Francisco Galiana [2], Mercedes Almenar-Muñoz [3] and Cecil Konijnendijk [4]

1 Department of Urbanism, School of Civil Engineering, Universitat Politècnica de València, 46022 Valencia, Spain; egielen@urb.upv.es
2 Department of Agrifood and Rural Engineering, School of Agricultural Engineering and Environment, Universitat Politècnica de València, 46022 Valencia, Spain; convalpl@agf.upv.es (M.V.-P.); fgaliana@agf.upv.es (F.G.)
3 Department of Urbanism, School of Architecture, Universitat Politècnica de València, 46022 Valencia, Spain; meralmuo@urb.upv.es
4 Department of Forest Resources Management, Faculty of Forestry, University of British Columbia, Vancouver, BC V6T 1Z4, Canada; cecil.konijnendijk@ubc.ca
* Correspondence: maagcar@aaa.upv.es

Abstract: While national parks (NPs) have for a long time made substantial contributions to visitor well-being, many spaces remain out of reach of people with disabilities (PwDs). This is partly due to a lack of policies that take accessibility for broader intersectional audiences into consideration. This paper evaluates governance and legal frameworks in NPs in both Canada and Spain. A decision-making framework based on intersectionality realities is proposed to assess current conditions of environmental good governance using a set of descriptors created to scrutinize laws and technical documents that can promote equitable access to NPs. To validate results derived from the regulatory evaluation, semistructured interviews with park managers were carried out. Results revealed the importance of incorporating equity discourses into policies that regulate NP networks to guarantee that all the intersectional realities for park uses are considered in their management. Furthermore, when a country develops a well-structured federal framework under which the rights of PwDs are ensured, it transcends other fields of law. Differences between the Canadian and the Spanish situation are highlighted, as well as the need for links between higher-level policies and laws and on-the-ground implementation, with NP management plans playing an important role.

Keywords: accessibility; environmental equity; legislation; people with disabilities; public use; good governance; intersectionality; stewardship

Citation: Aguilar-Carrasco, M.J.; Gielen, E.; Vallés-Planells, M.; Galiana, F.; Almenar-Muñoz, M.; Konijnendijk, C. Promoting Inclusive Outdoor Recreation in National Park Governance: A Comparative Perspective from Canada and Spain. *Int. J. Environ. Res. Public Health* **2022**, *19*, 2566. https://doi.org/10.3390/ijerph19052566

Academic Editors: Salvador García-Ayllón Veintimilla and Antonio Espín Tomás

Received: 29 December 2021
Accepted: 17 February 2022
Published: 23 February 2022

Publisher's Note: MDPI stays neutral with regard to jurisdictional claims in published maps and institutional affiliations.

1. Introduction

Direct contact with nature provides benefits to human health [1,2]. Natural protected areas (NPAs) are spaces of special interest due to the quality of their ecosystems, which has made them desirable places to visit [3]. According to data from the International Union for Conservation of Nature (IUCN) published in the Protected Planet Report 2012 [4], 12.7% of the world's land and 1.6% of the world's ocean are recognized as NPAs. Among these, national parks (NPs) are eagerly anticipated by society and have the objective of protecting large-scale ecological processes to maintain their ecosystem services and functions, which sustain human life as we know it. In addition, educational purposes are a main objective and are focused on enjoying the outdoor opportunities they provide [5–10].

Despite the fact that NPs have become a tourist attraction due to the increase in the number of visitors reported annually [3], there are still some places that exclude a significant percentage of society, such as people with disabilities (PwDs), more specifically those who have mobility/motor disabilities (PwMDs), who historically have been cut off from leisure activity in natural areas. The lack of data on PwDs in NP visits in statistics published by them demonstrates, at the very least, that they are not being considered. The fact that PwMDs enjoy less of NPs than those who are not disabled [11,12] may be for various reasons: the individual's own decision, a lack of infrastructure and equipment facilitating their access, or not having enough information about these places [13–16].

Disabilities is an umbrella term, covering problems in body function or structure, or activity limitations, such as difficulties encountered by an individual in executing a task or action, or a restriction on participation understood as a problem experienced in involvement in life situations [17]. The main purpose of the United Nations' 1993 Convention on the Rights of Persons with Disabilities (CRPD) was to change attitudes towards and approach to PwDs so that they are treated as "subjects" with rights capable of claiming them and making their own decisions, as well as being active members of society [17,18].The Standard Rules on the Equalization of Opportunities for Persons with Disabilities was adopted in the 85th plenary meeting on 20 December 1993, which concluded in the resolution A/RES/48/96, 4 March 1994 [19], also in the constitution at least of Western nations. Subsequently, policies of each country must regulate all concerns to guarantee PwD rights (work, education, health, leisure, etc.) so that they may subsequently be taken into consideration and applied to any other field, which may directly or indirectly affect their needs, assessing the scope of their integration thanks to active participation in decision making, so that their inclusion will be effective and efficient [14,18,20,21]. These considerations must always be subject to the primary objective of NPs, which is the protection of ecological integrity [5,6]. In this way, it would seem advisable to build frameworks from a transformative intersectional approach. This paradigm allows for a more critical focus, developing a multidirectional crossroads of interconnections of power, identity, and discrimination. Gender and other human social subjectivities, such as abilities or disabilities, are also considered [22–24]. In the case of PwDs, not only the lack of the environments' accessibility, but also their participation in decision making on issues that directly affect them has not been taken into consideration [25–28].

Several authors have analyzed issues that may affect PwDs' enjoyment of natural environments. Some are related, whether they are NPs or an urban green environment [11,29–31]. Newman and Park (n/i) focused on more specific aspects of the public use of NPs, introducing the outdoor recreation access route (ORAR) based on accessibility standards for outdoor activities, which have been developed from the most advanced policies in this area, the US and the UK legislation [32].

A path's accessibility indicator has been configured through geospatial analysis [15] where wheelability and walkability are the main goal to advance in accessible configurations of wild or urban parks for PwMDs throughout their integration into outdoor activities [32]. Highlighting NPs' educational purposes, Sugerman [33] suggested the implementation of an inclusive education model in outdoor adventure activities in parks. The involvement of actors with disabilities in decision making on NP accessibility was also recently addressed by Groulx et al. [25], who agree about the obviousness of involving a population affected by the lack of accessibility in any research and decision making on matters that directly affect them. Finally, one criterion that defines and characterizes the sustainability of natural area tourism, according to sustainable development goals (SDGs), is active outdoor participation for PwDs [10,26,27].

To date, causes that can constrain PwMD experiences when visiting green wild areas have been studied in a sectoral way. However, how these issues can be overcome has not been systematically evaluated from the point of view of whether policies enable or disable their inclusive and equitable experience to go into the woods. The existence of regulations that address inclusion (international, national, and regional) applied carefully to urban environments [34] suggests the need to address the issue systematically. Whether these regulations and their management instruments adequately integrate from the start their needs for using and enjoying NPs needs to be studied.

Therefore, it is a matter of verifying whether it is regulatory frameworks themselves, both regarding NPAs and PwDs, that are not allowing them to relate to natural environments and identifying parameters from an intersectional point of view that are conditioning them in their enjoyment of outdoor activities. This could be undertaken using a decision-making framework that must be based on their needs to predict to what extent the regulatory framework has gaps in it that impede the development of a governance that allows PwDs to enjoy nature.

The central question of this paper is how intersectionality in legal frameworks can contribute to dealing with previously neglected inequalities and thus to promoting a more inclusive and more equal enjoyment of NPs and their natural values by PwMDs. International frameworks and key references, such as [8,35–37], provide the basis for this evaluation. Equity inputs related to the governance, legislation, and management of NPs are examined to promote accessible environment outputs.

2. Materials and Methods

A decision-making framework is proposed to assess how adapted governance [35] in NP management is to the requirements established by intersectionality in terms of inclusive public use [23]. The object of this analysis are the international and national norms and the management instruments that regulate NP conservation activities at different levels of decision making [6], including the international treaty for the conservation of nature [38] and the plan management of the public use of each NP.

To validate good governance [36,39] in NP legislation in a systematic way, the framework includes a set of descriptors based on standards that include the limiting needs of PwDs in terms of accessibility to NPs. These descriptors are applied in two different countries to show similarities and divergences between their legislations. Finally, two case studies are analyzed, with data being collected through semistructured interviews [40] with park managers, to validate the results derived from the evaluation of the regulations and to explore to what extent the legislation's philosophy is being transferred to the parks' planning instruments.

2.1. Ambit of Study

The study is focused on the Canadian and Spanish legal frameworks. Both countries have signed the Convention on the Rights of Persons with Disabilities (CRPD) of the United Nations (1993), as well as the Convention on Biological Diversity (CBD) [38], which is the basis for incorporating intersectionality into regulatory frameworks. The Standard Rules on the Equalization of Opportunities for Persons with Disabilities was adopted in the 85th plenary meeting on 20 December 1993, which concluded in the resolution A/RES/48/96, 4 March 1994. It advises that states should ensure that all systems of society and the environment related to services, activities, information, and documentation are made available to all, particularly to PwDs [19]. Canada signed the CRPD on 30 March 2007, and its ratification was on 11 March 2010 [41]. Canada also ratified the Convention on Biological Diversity (CBD) on 4 December 1992 and became a party on 29 December 1993 [42]. Spain signed CRPD on 27 September 2007, ratifying it on 24 September 2009 [43], while the CBD was ratified on 21 December 1993, followed by becoming a party on 21 March 1994 [44]. It is assumed that if treaties have been ratified, federal policies and legal frameworks will be revised accordingly if needed. The legal system in Canada is based on common law, which is a system based on jurisprudence or case law, since its main source is judicial decision. Its main characteristics are: (i) there is not always a written constitution or codified laws, (ii) and judicial decisions are binding [45].

In Canada, the rights of PwDs were recognized in the Charter of Rights and Freedoms (1981), under point 15 Equality Rights, and regulated in the Canadian Human Rights Act of 1985 (CHR, 1985). The purpose of this act was to combat discrimination (Arts. 5 to 14) and to ensure a management plan with inclusive outdoor recreation at regulated NPs (Arts. 17 to 24) [46,47]. Recently, the act to ensure a barrier-free Canada, S.C. 2019, c.10, commonly known as Accessible Canada Act (ACA), which came into force in July 2019, recommends removing existing disability barriers faced by PwDs in matters coming under Canadian federal jurisdiction. In addition, the creation of new barriers should be foreseen, providing a structure that complies with accessibility standards [48].

Initially, NPs in Canada were considered places for recreation and tourism rather than for ecological protection, which later became the main objective for parks under the Canada National Parks Act of 1979 (CNPA) [49]. In 1985, the amended CNPA emphasized that public use should be regulated through each NP's management plan according to the NP's zoning. This will ensure the ecological integrity of natural resources, which is the most important objective to protect some areas under the CNPA [50]. NPs were considered a special type of public land administered by the federal government under the provisions of the CNPA. Parks Canada is the federal agency (within the Department of Canadian Heritage) that manages the whole Canadian system of protected natural and cultural heritage made up of NPs, national marine conservation areas, national urban parks, national historic sites, heritage buildings and townsite communities in NPs, heritage railway stations, and so forth [51,52]. To complete the NP network, each natural area should be represented by a new park. Policy frameworks were developed as the National Parks System Plan (1997). Each new amended park management plan must be approved by the federal minister in charge of NPs and tabled in parliament. According to the Parks Act Regulation, it is a constitutional requirement that NP lands be federal government property through an official agreement. Where lands are subject to a comprehensive land claim settlement by aboriginal communities (First Nations), Parks Canada will work closely with them throughout the process of founding the new park. Although the federal government regulates NPs, they all must have a management plan, which should be developed by regions and local communities in a participatory manner. Management plans have a 15-year lifespan, with reviews every 5. Each park management plan provides a park-specific roadmap for delivering the core elements of Parks Canada's mandate, namely, visitor experience, public understanding and awareness, and heritage resource protection. Its network is represented by 39 natural areas from the Atlantic to the Pacific Ocean and Arctic coast [49,53].

Spanish legislation is based on Roman law, unlike Canadian law. These are legal norms that are written out, which means everyone has the opportunity to know them. It is a right based on sectoral law codes that enshrine basic rights and obligations [45].

The Spanish Constitution (1978), in Articles 14 and 49, promotes the absence of discriminatory practices and establishes the mandatory role of public authorities in ensuring the integration of people with disabilities. Furthermore, Spain regulates more specific PwD rights through the Royal Degree-Law 1/2013 of November 29, a general act on the rights of people with disabilities and their social inclusion [54,55].

The first Spanish NP legal framework was in 1916 with the National Park Act and the official founding of the first two NPs: Picos de Europa and Ordesa y Monte Perdido—then called Covadonga Mountain—and Ordesa Valley and Ara River National Parks. The Spanish initiative was inspired by the example of the United States' NPs, following a romantic vision that prevailed over scientific criteria related to landscape or biodiversity [56]. This perspective also characterized Spanish NP governance as formalized in the laws of 1954, 1975, and 1989. Law 4/1989, 27 March, on the Conservation of Natural Areas, Flora, and Fauna, introduced technical planning of natural resources and transferred some of the authority to the regions [57]. This process was ratified by several decisions of the Spanish Supreme Court (102/1995, 194/2004, 101/2005, and 99/2013). According to the latter, and as confirmed in today's National Parks Act 30/2014 of 3 December, the state has the authority to manage the National Park Master Network Plan. The Spanish regions are in charge of managing and governing NPs through their mandatory plan without any interference from the state [58], unlike in Canadian park governance. The Spanish NP network will be complete when it includes 40 representative natural systems according to the Spanish Natural Systems Annex to the law. As shown above, Spain has two important governance documents: first, the National Park Master Network Plan [59], which is the most important tool for planning and regulating NPs for a maximum of 10 years, and second, the regions in which an NP is located must develop the Public Use Plan and Management and Use Guidelines. The latter comprise the NPs' regular planning instrument with a set of general guidelines and rules for park use and management, also with a minimum effective term of 10 years (Arts. 19 and 20) [60].

In Table 1, a brief description of both countries' NP network is provided. It has been constructed using data from the Canada National Parks Act (S.C. 2000, c. 32) (2000) [52], Canada's National Park System Plan (1997) [53], Parks Canada Guiding Principles and Operational Policies [61], Law 30/2014 of 3 December 2014 of Spanish National Parks [60], and Royal Decree 389/2016 of 22 October 2016, which approves the master plan of the Spanish National Parks Network [59].

Table 1. National parks network characteristics in Canada and Spain defined by their respective regulatory frameworks. Canada classifies parks ecology in natural regions between terrestrial (TNR) and marine (MNR); Spain defines it as natural systems (NS). Governance models of NPs have been typified by Dudley (2008) as: (A): governance by government; (B): shared governance; (C): private governance, and (D): governance by indigenous people and local communities.

Dimensions/Concepts	Canada	Spain
National Park Network	47	16
Total land covered by national park protection type in km^2	450,000.0	3845.9
Percentage land of the country occupied by NP law	2.25	0.76
Percentage land occupied by NPs (national goal) upon completion of the network	3	-
Park network goal according to the ecology classification criteria	39 TNR 29 MNR	40 NS
NP system network achievement as of 2020	28 TNR 5 MNR	12 NS
NP land ownership	100% public (federal government after an official agreement is signed between First Nations and Canadian Government)	82% public (municipality, 45%; State, 20%; and regions, 17%) 18% private
NPs' governance classification by Dudley (2008)	A (government)	A (government)
NPs' zoning and uses without accessibility possibilities	Special preservation Wilderness	Reserve Restricted
NPs' zoning and uses with accessibility possibilities [1,2]	Natural environment Outdoor recreation Park services	Moderate Special use Traditional settlements

[1] In Canada, *Natural environment* is the park area where outdoor recreation activities are permitted to raise awareness of the cultural and natural values of the park. *Outdoor recreation* is an area with essential services and facilities whose defining feature is direct access by motorized vehicle. *Park services* is the park area that contains a concentration of visitor services and support facilities [49,50,53]. [2] In Spain, *Moderate zone* is the park area where going into the wilderness is permitted and there are low-impact infrastructures because that area combines conservation with cultural values, such as traditional agricultural uses and forestry (communal forest). *Special* is the area where major buildings, facilities, and infrastructures tend to be located within the park if deemed necessary. *Traditional settlements* are an exceptional circumstance where there are populated areas. To ensure citizens' basic rights, these areas have been established as a one-area land with various uses [59,60].

Despite their differences regarding legislation and regulations for NPs and their management, both countries have signed relevant international treaties. Thus, in principle they have incorporated the philosophies of these treaties into their federal legal framework. Additionally, both countries have a consolidated NP network with similarities and differences, which are described above. Analysis of the legal framework of both countries allows us to detect dysfunctions or gaps in the standard and make proposals for regulatory improvement.

For the local-level study, Yoho National Park in British Columbia, Canada, and Aigüestortes i Estany de Sant Maurici (AESM) National Park in Catalonia, Spain, were analyzed. Both NPs cover on-land areas and have a management plan that manages the areas intended for public use. Yoho NP's dates from 2003, and the new one is currently undergoing its approval process, while AESM NP adopted its current plan in 2010. Both NPs have bio-geo-cultural similarities, such as being part of an alpine mountain system (the Canadian Rocky Mountains and Spanish Pyrenees) and important heritage values on their lands [62–64]. Their mountain landscapes also represent a challenge in terms of recreational accessibility and are very popular between nature and adventure tourists [65,66].

2.2. Governance Analysis of NP Management

Based on governance theory and the policy arrangement approach [28,30,35,36], our methodology was developed to decide whether it is possible to reconcile the uses of NPs, considering intersectional realities, with their ecological preservation through good governance and responsible stewardship [67,68].

The framework is based on the tetrahedron model proposed by Liefferink et al. (2006) and the key principles of governance proposed by FAO (2011). The tetrahedron model proposed by Liefferink et al., (2006) as part of the policy arrangement approach, was applied as the analytical framework to undertake an in-depth evaluation guided by the inclusive NP outdoor recreation principle. It proposes evaluating policy practices according to four dimensions: discourses, rules, actors, and resources [39]. In our case, we apply those as:

- Discourses: international treaties and their philosophies;
- Rules: laws and plan management regulations;
- Actors: people or organizations with the capacity to impact decisions about the management plan and those who administrate resources, including PwDs;
- Resources: planning tools and budgets of stewardship.

These dimensions are connected to the key principles proposed by FAO (2011) to assess and monitor forest good governance [36]—effectiveness, participation, accountability, and equity/fairness.

Data from relevant literature and reports were obtained from administration home pages in both countries, and collection was completed in June 2021. The Spanish literature was complemented with directives from the European Union law from the EUR-Lex website [69]. Relevant international treaties were also downloaded from their official websites. Further information was obtained by a thematic search of online scientific databases (Web of Science, Scopus, and Google Scholar). The keywords used for this wider-scoped search included: inclusive public use, national parks, governance, stewardship, intersectionality, facilities, outdoor recreation, people with disabilities, equity, and accessibility. A wide range of publications was considered, including peer-reviewed journal articles, books, reports, conference proceedings, and policy and guideline documents.

Next, a comparative table was built with a set of 14 descriptors structured according to [39] the dimensions (Table 2). Each descriptor is also related to the FAO key principles defined above. Descriptors were constructed inspired by Director's Order #42, accessibility for visitors with disabilities in US National Park Service programs and services [70], the UK Countryside for All Good Practice Guide [71], the EUROPARC-España (2005) manual of public use concepts [72], the EUROPARC-España (2007) catalog of best practices regarding accessibility in protected natural areas [73], the US guide to managing the NP System Plan [74], and the catalogue of good accessibility practices in natural protected areas, and taking into account the author's experience as a PwD enjoying parks.

Table 2. Definition and objective of each descriptor to evaluate NPs' legal framework, related governance dimensions or policy practices in the tetrahedron model proposed by Liefferink et al. (2006) [33], and some of the FAO key principles (2011) [32].

Dimension	Descriptor [1]	Definition	Principle
Discourses	Nature for all objective (NAO)	The extent to which universal accessibility is included as a planning objective.	Equity/fairness
	Removing barriers program (REB)	The extent to which detection, elimination, and preventing barriers is included as a planning objective.	
Rules	Inclusive legal framework (ILF)	The inclusion of accessibility standards in the federal regulatory frameworks.	Effectiveness
	Active public participation (APP)	An intersectional and active institutional participation in policymaking with powers and responsibilities defined by NPs' legal framework.	Participation
	Technical accessibility standard (TAS)	The inclusion of accessibility standards in NP management plan.	Effectiveness
	Outdoor wilderness experience (OWE)	Active involvement in outdoor activities.	
	Periodic evaluation program (PEP)	Periodic evaluations to detect issues in accessibility services.	Accountability
Actors	Justice equity participation (JEP)	The extent to which the legal framework provides opportunities to minorities to become involved in the decision-making process voluntarily or mandatorily.	Participation
	Park interpreter staff (PIS)	Staff is required to have accessibility skills or provided with training to improve theirs.	Effectiveness
Resources	Inclusive outdoor walkability (IOW)	Amenities specified by law to improve outdoor accessibility.	Effectiveness
	Website disability information (WDI)	The extent to which outdoor accessibility is specified on park's official digital platforms.	Accountability
	Educational and informational programs (EIP)	Educational programs based on inclusiveness.	Effectiveness
	Access to significant natural features (ANA)	Access for all people to significant natural and cultural park points.	Effectiveness
	Transportation systems (TRS)	Accessible land and water transport within the parks' perimeters.	Effectiveness

[1] Descriptors have been defined to closely assess the legal framework of national parks. Each descriptor is designed to study all relevant aspects for full inclusion of PwMDs.

Data were compiled for both countries considering federal and regional NP regulations and planning tools: documents that are binding or mandatory and others that are not. All these were scrutinized using the accessibility descriptors to see whether the FAO principles were taken into consideration in the explanatory memorandum, chapters, and articles (the body or content) of the consulted documents. Voluntary or mandatory mentions of each descriptor were checked. Each indicator was then qualitatively typified according to this scale:

- A: provisions in the law, guides, or management plan statements;
- B: no specific provisions but stipulated in the law;
- C: no specific regulation;
- NA: not applicable, because not focused on wilderness experience.

The results from the legal framework and management instruments of the NP network are organized in four tables, one for each dimension [39]. The regulations and planning tools for each country are typified according to the aforementioned dimension and the scale on which they operate.

The study was then completed with a local level analysis of both countries. The objective was to evaluate how the philosophy of the federal regulation influenced park policies' applicability in favor of improving their accessibility for PwMDs.

The park management plan provides a sound basis for studying the equity of NPs' recreational uses, where amenities must be developed to make it possible for PwDs to enjoy nature. NP management plans were analyzed through semistructured interviews with managers from both NPs selected [40]. This part of the study was approved by the Behavioural Research Ethics Board of the University of British Columbia (Approval Certificate Number: H20-00582) and Parks Canada Agency Research and Collection Permit (Permit Number: YNP-2020-35637). The Yoho NP questionnaires and full data table are included in Appendix A. The AESM questionnaire and full data table are in Appendix B.

3. Results

3.1. Legal Framework Assessment

Ten legal documents for Canada and seven for Spain, which correspond to the documents in force on which NPs' network management is based, were assessed. A major difference between the two countries was detected between the two park agencies. The Canadian Parks Agency is accountable of NPs and other types of heritage that are under its jurisdiction, such as national historic sites, federal heritage buildings, and townsite communities within NPs [51,52], while in Spain heritage sites are managed by other bodies. Therefore, specific acts referring to Canadian heritage managed by the Canadian Parks Agency are also included for the Canada parks legal framework.

Tables 3–6 show the comparative results of governance for the discourses, rules, actors, and resource dimensions, respectively. Laws and regulations and guides and technical documents are separated and hierarchically organized from the national to the local level. The laws and regulations evaluated for Canada were: the Canada National Parks Act (S.C. 2000, c.32), Canada National Marine Conservation Areas Act (S.C. 2002, c,18), Parks Canada Agency Act (S.C. 1998, c. 31), Historic Sites and Monuments Act (R.S.C. 1985, c.H-4), NP General Regulations (SOR/78-213), and Heritage Railway Stations Protection Act (R.S.C. 1985, c.52, 4th Supp.) [49,75–79]. In terms of guidelines and management documents, the NP System Plan (1997), Parks Canada Guiding Principles and Operational Policies (PCGPOP) (2019), Parks Canada Agency Departmental Plan 2019–2020 and 2020–2021, and Yoho NP of Canada Management Plan (2010) [51–53,61,80] should be mentioned. For Spain, the laws and regulation analyzed were: the Spanish National Parks Act 30/2014, Natural and Biodiversity Heritage Law (42/2007, December 13, last amended 15 September 2015), and Law 7/1988 of 30 March on the reclassification of the AESM NP. Moreover, the NP Master Plan (Royal Decree 389/2016 of 22 October), Uses and Management Master Plan of the AESM Park (Decree 39/2003 of 4 February), and AESM Park's Public Use Plan were evaluated as guidelines and management documents [59,60,81–84].

The information provided in Table 3 shows that:

1. The nature for all objective (NAO) lacks specific regulation in the Canadian laws and regulations (C). Although NAO is still not properly integrated, the PCGPOP stipulates this descriptor even though there are no specific provisions. The 2019–2020 departmental plan (with an A score) proposes a more determined commitment to access for PwDs to parks. In the Spanish case, the NP act (with an A) regulates in its fifth article the right for all to access nature, but the decree creating AESM does not include any regulation regarding this concern. Of note are the results regarding guidelines (A qualification) and management documents, which are clearly committed to favoring PwDs' access to nature.
2. The removing barriers program (REB) objective has not been considered in the Canadian legal framework and management guides (C score). In the Spanish case, the law dictates that all barriers should be removed to improve PwDs' access to parks (A qualification). This mandate has been transposed to the management tools; however, mandatory stipulations do not appear (B qualification).

Table 3. Comparison table of the governance discourse dimension for the nature for all (NAO) and removing barriers program (REB) objectives.

Descriptor	NAO	REB
FAO Principle	Equity/Fairness	Equity/Fairness
Canada: Law and regulation [1]		
Canada National Parks Act (CNPA)	C	C
Canada National Marine Conservation Areas Act	C	C
Parks Canada Agency Act	NA	NA
Historic Sites and Monuments Act	NA	C
NPs' General Regulations	C	C
Heritage Railway Stations Protection Act	NA	C
Canada: Guidelines and management documents [1]		
Parks Canada Guiding Principles and Operational Policies (PCGPOP)	B	C
NP System Plan (1997)	C	C
Parks Canada Agency Departmental Plan 2019–2020	A	C
Yoho NP of Canada Management Plan (2010)	C	C
Spain: Law and regulation [1]		
Spanish National Parks Act (SNPA)	A	A
Natural and Biodiversity Heritage Law	NA	NA
AESM national park creation decree, 21 October 1955	C	C
Law 7/1988 of AESM NP reclassification	NA	NA
Spain: Guidelines and management documents [1]		
NP Master Plan	A	B
Uses and Management Master Plan	A	B
AESM NP Public Use Plan	A	B

[1] Typified scale: A: provisions in the law; B: no specific provision but stipulated in the law; C: no specific regulation; NA: not applicable, because not focused on the wilderness experience.

Table 4. Comparison table for the governance rules dimension, focusing on inclusive legal framework (ILF), active public participation (APP), technical accessibility standard (TAS), outdoor wilderness experience (WEX), and periodic evaluation program (PEP).

Descriptor	ILF	APP	TAS	OWE	PEP
FAO Principle	Effectiveness	Participation	Effectiveness	Effectiveness	Accountability
Canada: Law and regulation [1]					
Canada National Parks Act (CNPA)	C	A	C	C	A
Canada National Marine Conservation Areas Act	C	A	C	C	A
Parks Canada Agency Act	NA	NA	NA	NA	A
Historic Sites and Monuments Act	C	NA	C	NA	A
NP General Regulations	C	NA	C	C	C
Heritage Railway Stations Protection Act	C	NA	C	NA	C
Canada: Guidelines and management documents [1]					
Parks Canada Guiding Principles and Operational Policies (PCGPOP)	B	A	C	B	A
NP System Plan (1997)	NA	A	NA	NA	A
Parks Canada Agency Departmental Plan 2019–2020	A	B	C	B	A
Yoho NP of Canada Management Plan (2010)	C	C	C	C	A
Spain: Law and regulation [1]					
Spanish National Parks Act (SNPA)	A	A	B	B	A
Natural and Biodiversity Heritage Law	B	A	NA	NA	A
AESM national park creation decree	C	C	C	C	C
Law 7/1988 of AESM NP reclassification	NA	C	NA	NA	NA
Spain: Guidelines and management documents [1]					
NP Master Plan	A	A	B	B	A
Uses and Management Master Plan	A	A	C	B	A
AESM NP Public Use Plan	A	A	B	A	A

[1] Typified scale: A: provisions in the law; B: no specific provision but stipulated in the law; C: no specific regulation; NA: not applicable, because not focused on the wilderness experience.

Table 5. Comparison table for the actor dimension in relation to justice equity participation (JEP) and park interpreters staff (PIS).

Descriptors	JEP	PIS
FAO Principle	**Participation**	**Effectiveness**
Canada: Law and regulation [1]		
Canada National Parks Act (CNPA)	A	C
Canada National Marine Conservation Areas Act	C	C
Parks Canada Agency Act	NA	B
Historic Sites and Monuments Act	NA	C
NP General Regulations	NA	NA
Heritage Railway Stations Protection Act	C	C
Canada: Guidelines and management documents [1]		
Parks Canada Guiding Principles and Operational Policies (PCGPOP)	B	C
NP System Plan (1997)	A	NA
Parks Canada Agency Departmental Plan 2019–2020	B	C
Yoho NP of Canada Management Plan (2010)	B	C
Spain: Law and regulation [1]		
Spanish National Parks Act (SNPA)	A	C
Natural and Biodiversity Heritage Law	A	NA
AESM national park creation decree	C	C
Law 7/1988 of AESM NP reclassification	NA	NA
Spain: Guidelines and management documents [1]		
NP Master Plan	A	C
Uses and Management Master Plan	A	C
AESM NP Public Use Plan	A	C

[1] Typified scale: provisions in the law; B: no specific provisions but stipulated in the law; C: no specific regulation; NA: not applicable, because not focused on wilderness experience.

Table 6. Comparison table for the resources dimension in relation to inclusive outdoor walkability (IOW), website disability information (WDI), educational and informational programs (EIP), access to significant natural features (ANA), and transportation systems (TRS).

Descriptor	IOW	WDI	EIP	ANA	TRS
FAO Principle	**Effectiveness**	**Accountability**	**Effectiveness**	**Effectiveness**	**Effectiveness**
Canada: Law and regulation [1]					
Canada National Parks Act (CNPA)	C	C	C	C	C
Canada National Marine Conservation Areas Act	C	C	C	C	C
Parks Canada Agency Act	C	C	C	C	C
Historic Sites and Monuments Act	C	C	C	C	C
NP General Regulations	C	C	C	C	C
Heritage Railway Stations Protection Act	NA	C	C	NA	C
Canada: Guidelines and management documents [1]					
Parks Canada Guiding Principles and Operational Policies (PCGPOP)	C	C	B	B	C
NP System Plan (1997)	NA	NA	C	NA	NA
Parks Canada Agency Departmental Plan 2019–2020	C	C	C	B	C
Yoho NP of Canada Management Plan (2010)	C	C	C	C	C
Spain: Law and regulation [1]					
Spanish National Parks Act (SNPA)	C	C	C	B	C
Natural and Biodiversity Heritage Law	NA	NA	C	NA	NA
AESM NP creation decree	C	C	C	C	C
AESM NP reclassification law	C	C	C	C	C
Spain: Guidelines and management documents [1]					
NP Master Plan	A	C	B	A	C
Uses and Management Master Plan	B	C	A	A	B
NP Public Use Plan	B	C	A	A	B

[1] Typified scale: A: provisions in the law; B: no specific provisions but stipulated in the law; C: no specific regulation; NA: not applicable, because of missing to focus on wilderness experience.

In Table 4, descriptors for governance rules dimension show:

1. The inclusive legal framework (ILF) applied to Canada clearly demonstrates the lack of provisions in laws and regulations studied (C qualification) as no reference was found on accessibility for PwDs. Despite the fact that no specific regulations or stipulations are in place in the acts, their management tools, such as PCGPOP and Parks Canada Agency Departmental Plan 2019–2020, seem to correct this aspect. The first highlights the idea that the information must be accessible to all visitors, and the second considers accessibility for PwDs as a main goal to be able to foster equitable tourism in parks. No other specific regulations were found (C score) because there is no mention of how to undertake accessibility activities in the park. The analysis for Spain confirms that the legal framework is based on the accessibility statement: inclusive public use of parks is a main objective in the NP act, which is subjected on its ecology integrity. This philosophy has been transposed into the technical documents, so Spain receives an A qualification.
2. Active public participation (APP) evaluated in the Canadian case through laws and regulations shows a good result in terms of participation (A qualification). First Nations and local communities are considered from the start when selecting the potential NP area, so they have the opportunity to be integrated into the whole management decision process. In Spain, public participation is considered in the SNPA. Article 2 promotes collaboration with the public authority and the active participation of individuals or associations. In this line, NP authorities must particularly encourage the involvement of private or public stakeholders. All this is developed further in Articles 35 and 38, with the public participation process and how NP information should be conveyed to citizens. The NP Master Plan also includes a provision on the first point where the park authorities are urged to encourage community collaboration and active participation.
3. Technical accessibility standard (TAS) obtained a C for Canada both within laws and regulations and in guidance and management documents. In the Spanish case, although SNPA indicates accessibility as a requirement in the public use of parks, applying accessibility standards is not required (B score). As the technical documents are inspired by this law, there is once again a lack of information about this point.
4. Outdoor wilderness experience (OWE) is not included in any provisions in Canada's legal framework for NPs. Only the Parks Canada Agency Departmental Plan 2019–2020 earns a B qualification because accessibility to parks is stipulated as a future improvement. Spanish legal analysis shows no specific provision for WEX, but it is stipulated in the law (B score) because both the SNPA and the network master management plan indicate the importance of promoting accessibility, although without regulating it. A similar situation can be seen in guidelines and management documents: the public use plan does not specify how to achieve an inclusive wilderness experience outdoors.
5. Periodic evaluation program (PEP) is the descriptor most covered in both legal frameworks. The Canadian legal framework earned an A because stewardship reinforces the idea of running periodic evaluations. The Spanish case also includes provisions for periodic evaluations of parks and management plans.

Descriptors for the actors' dimension as shown in Table 5:

1. Justice equity participation (JEP) is contemplated in the CNPA, point 12(1), with a public consultation requirement to involve citizens and stakeholders, including PwD, in the decision-making process (A qualification). This point develops how the minister responsible for the NP must provide opportunities for public participation at every territorial level, including the participation of aboriginal organizations and representatives from park communities. In guidelines and management documents, this aspect is reinforced in the NP System Plan (1997) (A qualification), although the departmental plan and Yoho NP management plan are not sufficiently specific in terms of inclusiveness for PwDs (B qualification). In Spain, the NP law includes a provision that forces the implementation of park planning instruments to guarantee

public participation and accessibility, which is even more fully developed in the NP Master Plan.

2. The park interpreters staff (PIS) descriptor is not considered in any specific regulation (C qualification) in either country. There are no provisions about park staff skills in communication with PwDs in laws or regulations or in guidelines and management documents.

Analysis of the governance resources dimension summarized in Table 6 shows:

1. Inclusive outdoor walkability (IOW) results obtain a C qualification for the Canadian legal framework as none of the provisions have been properly developed. In the case of Spain, there is a similar result for laws and regulations, although the later NP Master Plan states that visitor services must be designed and developed, taking into account universal accessibility. Even so, the study of the park management plans revealed that no specific actions have been developed (B qualification).
2. The website disability information (WDI) indicator obtains a C qualification in both countries as no provisions were identified in the NPs' legal frameworks on how parks should manage their websites in terms of information for PwDs planning a visit to the park.
3. Educational and informational programs' (EIP) results do not score better than IOW or WDI in Canadian regulation. There is only one citation in this regard in the Parks Canada Guiding Principles and Operational Policies (PCGPOP). A few more were identified in the Spanish case, resulting in an A and B qualification for the guidance and management documents. These documents develop and highlight the importance of addressing all intersectional realities in the parks' educational programs.
4. Access to significant natural features (ANA) is not represented in the Canadian legal framework; only the PCGPOP and departmental plan consider this question as a challenge. On the contrary, the Spanish A score for guidance and management documents shows a certain commitment to PwDs' access to the most significant landscapes in the parks.
5. The transportation systems (TRS) indicator is underdeveloped in Canadian law and regulations, and not specifically for Yoho NP despite its important heritage related to transportation systems. In Spain, legal and regulation items are no better, although management documents received a B score as accessibility is a main objective.

3.2. *Case Study*

Managers of Yoho NP consider inclusiveness to be covered in the NP act because the law indicates that "NPs are dedicated to all Canadians." However, they recognized that accessibility could be improved with further action. This statement is also valid for the Parks Canada Guiding Principles and Operational Policies and the National Park System Plan. There is, however, no specific framework for including accessibility factors in the management plan actions. The Yoho NP manager admitted that together with reconciliation with First Nations, greening, and sustainability, accessibility will be one of the future core practices. Currently, the NP management plan is being renewed with the intention of including some specifications regarding accessibility. PwD associations, such as the Rocky Mountain Adaptive Association, have been actively involved in the new plan in an effort to take into consideration the realities and opinions of the entire community and add some ideas to incorporate accessibility amenities to the park. In addition, a satisfaction survey was published on the NP's official website, giving all people the opportunity to participate, but unfortunately, no data for PwDs were collected. The NP manager also commented that the park administration would be interested in exploring new amenities to offer new accessible paths, which can provide new opportunities for people with mobility impairments to explore Yoho landscapes.

Since 1994, the AESM NP has undertaken actions to improve accessibility. This was first done by creating the "Breaking Barriers" program in 1993, and then by building accessibility infrastructure in 1998. These actions were undertaken before the NP management plan was published. AESM NP even received two awards from the Spanish association of people with visual impairments (ONCE). In 2005, a new NP strategic plan was developed, which included new regulations and projects to improve accessibility for PwDs at visitor centers. The park has been in permanent contact with ONCE, looking for new accessibility strategies for the park. Moreover, the Breaking Barriers program undertakes educational actions designed for scholars with special learning needs. According to the manager, the involvement of PwD organizations has benefited the development of the NP management plan. NP managers have data about visitors with disabilities, but it should be noted that people do not always go to the park through the visitor center, making it difficult to fully track them. The NP manager stated in the interview that one of the park's new objectives would be to develop new paths to create more opportunities for PwDs to explore other landscapes.

In Appendices A and B, detailed tables for the Yoho (A1) and AESM (A2) NPs list the past and future actions that the park managers have taken and will take concerning accessibility.

4. Discussion

We analyzed the governance and legal situation in Canada and Spain through 14 descriptors, applying the framework of policy arrangements structured in dimensions proposed by Liefferink et al. (2006) [39] to answer the main question of how intersectional legal frameworks can contribute to promoting a more inclusive NP for citizens with disabilities.

4.1. Dimensions

The discourse dimension shows that principles of equity or fairness are not yet well established in the Canadian system according to results obtained from descriptors' applicability, in line with Kovacs Bruns and London's conclusions (2010) [20]. The accessibility topic was not introduced into governance discourse until 2019 when the PCGPOP [61] came into force and in the last two departmental plan reports (2019–2020) [51,52]. In Spain, those principles have been taken into account in its respective legal framework as shown by the nature for all objective (NAO) and removing barriers program (REB) (Table 3), which have been incorporated into the SNPA through the discourse by the allocation of resources and their benefits as a main objective in the act (Art. 5) [60]. Moreover, EUROPARC-España [72] has provided a framework to plan new accessibility plan challenges as a result of an assessment of the Spanish park network using accessibility criteria, reporting the state of the entire park services, indoor and outdoor.

Regarding the NP case studies, the equity and fairness discourse seems to be working on the amendments of national policies and regulations when they are incorporated into the NP plan. Both NPs have been fostering new equity perspectives in guidance and management documents with a clear commitment to favoring access to nature for PwDs. In Canada, the proposal aims to develop guides that provide equal opportunities, integrating informational techniques to make park themes accessible to all visitors. In addition, the Park Plan Management Draft for Yoho expresses a new equity vision in key strategy 2, "True to Place Experiences" (points 4.1 and 4.2) [85]. AESM is pioneering new avenues related to accessibility for PwDs within the park. Its manager aims to explore new options to provide more accessibility paths to points in the park.

Overall, Canada shows a lower level of effectiveness than Spain in terms of rules. In the Canadian legal framework, references to accessibility are missing, when CNPA [49] could have incorporated an intersectional vision to public use of parks, which has been developed in Art. 38 (14.3, c). This is confirmed by a case study of Yoho NP, where insignificant accessibility measures were implemented to improve the wilderness experience through outdoor opportunities. Although in Parks Canada Guiding Principles and Operational Policies (2019) brief references are found on accessibility for PwDs, no technical accessibility standards (TAS) have been included to provide parks amenities like other countries, such as the US and the UK, have [31,70–72,85]. Recent management tools, such as the Parks Canada Agency Departmental Plan 2019–2020 and 2020–2021 [55,56], could signify a changing trend.

In the Spanish legal framework, there are specific references to the inclusive public use of national parks (Art. 5) [60]. This approach also influences the NPs' planning tools, such as the Network Master Plan, where conservation is intersected with inclusive uses, providing them as amenities required by PwDs (Section 1.3-b, Art. 12) [59,84]. Moreover, this park's legal framework's effectiveness can be supported by PwD rights regulated beyond just the Spanish Constitution [54], since it has been supplemented with a specific legal PwD framework [21,55]. In addition, as a member of the European Union, Spain must transpose EU directives into its legal system to guarantee equity in active lives for PwDs in society. AESM as a case study seems to be even more effective than the federal rules themselves. The "Breaking Barriers" program, developed in 1993 when the park was designated, has carried out indoor–outdoor accessibility actions, highlighting its pioneering intersectional educational program in line with Sugerman's [33] suggestions on how to integrate PwDs into outdoor educational activities in parks.

However, neither the Canadian nor the Spanish legal frameworks and technical documents include standard accessibility specifications about how to achieve accessibility communication skills for park staff (PIS) or transportation systems (TRS) for PwDs. Only educational and informational programs (EIP) and access to significant natural features (ANA) show a higher level of effectiveness in Canada and Spain since guidelines and management documents specifically highlight the importance of accessibility referred to in these two items. Specifically, inclusive outdoor wilderness experience (OWE) has been only developed in Spain at the regional–local level through AESM NP public plan uses [84].

The participation principle is guaranteed in terms of rules for stewardship in both countries. Comparative analysis shows several processes for the participation of citizens and stakeholders in decision making either directly or through legitimate intermediaries representing their interests in both countries. However, there is no equitable participation that considers PwDs. In the case of Canada, a good stewardship example is that of First Nations' involvement in the parks' decision making [51,52]. However, justice equity participation (JEP) of PwDs is not clearly defined by the park frameworks as it does not regulate which amenities must be included to provide equal participation opportunities for PwDs in those processes [49,53,61]. For instance, Yoho's management plan renewal is based on a park visitor satisfaction survey, which provides a road map to deal with new park challenges, but no data have been obtained from PwDs, either because there are no visits from this segment of the population or because they did not participate in it. Aware of this lack of information, Yoho's manager is in touch with a PwD association. In Spain, the extent to which PwD engagement in decision making is made possible is not adequately specified. At the federal level, Law 27/2006, which provided the active public participation framework at the federal level, did not include any representative PwD organizations in its environmental advisory council (Title III, Art. 19) [86]. Locally, AESM has a board of trustees composed of different administrations and entities, which currently does not include any organizations of PwDs, although they might be added in the future. [87].

National park governance accountability seems to have reached similar positive results in both countries. Canada's NP network has a consistent and well-structured monitoring system through its recurring reporting on park status. Spanish national parks stipulate in SNPA (Art. 21) [60] the parks' ongoing assessment through the autonomous region's agency. The Spanish Parks Master Plan mandates an annual statement and a triannual report submitted to the Spanish senate and published publicly [88]. Yoho's management renovation plan has been successful, and the draft is now on the official park website. Likewise, SNPA states that a use and management master plan, and a public use plan, valid for 10 years, has to be developed after a new NP is sanctioned [60]. When plan revisions take longer than the recommended deadline, the periodic evaluation program descriptor (PEP) should be revised as we found in the AESM management plan dated from 2003 [84]. Disability information on the public website is an essential park qualification tool for people planning visits to NPs [89]. However, the mechanism to provide information through official channels (WDI) [25] has not been addressed in the Canadian or the Spanish NPs' legal frameworks. Despite the fact that website disability information (WDI) is rarely included in park rules [25], it has been confirmed that the Canadian draft management park plan suggests that WDI is a major challenge. It should provide visitors on the trip cycle with the correct expectations to facilitate their real park experiences [89].

4.2. Integrated Analysis of Dimensions

Analysis through the set of proposed dimension descriptors confirms the relevance of developing a specific legal framework to protect PwDs' rights at the federal level, as stated by Kovacs Bruns and London [20] and Groulx et al. [25]. However, the rights of PwDs to access parks must also be regulated in the NP law itself, considering all intersectional realities in those areas where public use of the park is allowed and regulated, to eliminate barriers and prevent inequalities. This is confirmed by the Spanish parks framework. Even though these rights have been established since 2013 with the latest amended law on PwD rights [55], it does not seem that it has been enough to establish equitable and inclusive outdoor activities in NPs [21].

International treaties offer good guidance for incorporating intersectional realities into NP use, but first, federal legal frameworks and park agencies must integrate them and incorporate them into their governance to address systematic barriers. Canada is on the path to comply with the international treaties on the rights of PwDs, while Spain has 40 years of experience regulating the paradigm of social inclusion [23]. A new promising set of regulations, such as the British Columbia Accessibility Act (2018) [90] and the Accessible Canada Act (ACA) in 2019 [48,91], have provided new opportunities to promote sustainable park tourism, accessible to all able and non-able people [10,25–27]. Furthermore, the PCGPOP could incorporate some input from standards published in Canada, such as the recommendation from the BC Building Code [92] or from other countries such as the US or the UK, which have developed accessibility standards for parks [31]. In the Spanish case, there are accessibility standards in other fields, such as those described in the basic document (SUA), specifically chapter 9 on safety of use and accessibility, which could be applied to plan indoor and outdoor infrastructures [93].

Both countries have shown some gaps in parks' legal frameworks in terms of providing an inclusive environment for PwDs. Good governance is based on cooperative and mutually supportive relationships among all state stakeholders (government, private, and public entities). Moreover, the right to enjoy a full life in society of those sectors of society who have been prevented access to nature and ignored throughout history should not be neglected to comply with ratified international treaties [22–24]. This must be taken into consideration as common goals for park agencies to move towards inclusive environments in outdoor park activities in line with sustainable tourism [16,26,27]. In addition, to carry out these types of policies, park rules must be supplemented, incorporating inclusive public park use standards [31]. To meet the needs of PwDs, they must play a participatory role in decision making focused on improving outdoor accessibility activities, contributing to their experience [14,25]. Those decisions should be binding and incorporated into future amendments to park management plans.

4.3. Study Limitations

Our analysis of PwDs and NP legal frameworks does not take into account potential additional aspects that could be indirectly linked to the study. We consider that future research should address the information and communication technologies (ICTs) applied to protected natural spaces as a main research topic in both regulation lines and their applicability in a study case. Moreover, to promote an inclusive ICT platform, special mention should be made on both the content and the performance of web-based information to design it to be accessible to all people, incorporating intersectional perspectives. There also exist structural factors of importance, such as environmental, geographical, distance, and population distribution, which can influence NP accessibility. The distance from the most populated areas to the Canadian NPs is much greater than that in Spain, which can be a relevant factor in terms of park visitors.

The objective of analyzing both management documents from two parks is to verify the positive or negative impact of federal level policies on PwDs' lives and highlight the importance of legislating from the perspective of intersectional realities and needs, since if this is not the case, as has been shown, at lower levels such as the local, the actions carried out may not satisfy the needs of all citizens. Obviously, the NPs chosen are not necessarily representative of other NPs in their countries. Perhaps more case studies could provide a better overview of the park network to evaluate how the application of the legal framework can influence park facilities for each management plan.

In addition, the applicability of parks' accessibility standards must be adapted to each park's characteristics. In line with this, future research should undertake intersectional public park use aptitude tests.

5. Conclusions

The promotion of inclusive outdoor recreational activities in NPs requires specific policies to promote accessibility for PwDs in federal legislation, which can inspire incorporating inclusiveness input into the regulation of NP laws, where equity should be designated as a main objective.

A comparison of inclusive tools in parks' legal frameworks in Canada and Spain shows some advances in terms of improving park accessibility. Canada is going to be able to meet the equity challenge in the near future, but it should first consider amending the parks act, emphasizing accessibility in its section on park facilities. Spain has the framework to keep moving forward in equity, but more tools should be added to its parks act.

The dimensions of policy arrangements and FAO principles shed light on the good governance framework. A set of proposed descriptors have offered a holistic way to evaluate the integration of the equity and fairness philosophy into parks' legal framework. To demonstrate whether there is a framework that has been based on intersectionality, the application of descriptors has started in the field of discourses, then checking that it has permeated into rules, resources, and actors. This has allowed us to verify the status of the issue in both countries.

Improvements based on the descriptors proposed can help to achieve the dimensions of good governance in terms of meeting the principles of effectiveness, participation, and accountability to support sustainable tourism. Consequently, the inclusion of PwDs in park activities implies a standardization of the norms through park accessibility standards at the federal and regional level.

The institutional framework may allow or limit the equity of NP use. Administrations who have the responsibility to correct this lack of equity in the use of parks must endeavor to include PwDs (individuals or through their respective representatives) in participation bodies. For this reason, a report with recommendations will be drawn up for the competent body since the Advisory Council of Environmental Parks (Spain) and the AESM Board of Trustees do not have any PwD association among their members.

Future research should undertake a comprehensive evaluation of information and communication technologies (ICTs) in two research lines: (i) regulations to update legislation in accordance with current social trends and (ii) evaluating content and format from an intersectionality perspective. It is assumed that public use must be inclusive, and it cannot generate park conservation problems. Considering its implementation in the NPs, both NP agencies should undertake to establish an adequate regulatory framework for correct ICT implementation.

Author Contributions: All of the authors contributed to conceptualization, formal analysis, investigation, methodology, and writing and editing of the original draft. All authors have contributed read and agreed to the published version of the manuscript.

Funding: This research was funded by [Funding for open access charge: Universitat Politècnica de València] grant number [PAID-12-21].

Institutional Review Board Statement: The study was conducted in accordance with the Declaration of Helsinki, and approved by the Behavioural Research Ethics Board of the University of British Columbia (Approval Certificate Number: H20-00582) and Parks Canada Agency Research and Collection Permit (Permit Number: YNP-2020-35637).

Informed Consent Statement: Informed consent was obtained from all subjects involved in the study.

Data Availability Statement: The data will be made available on request from the corresponding author.

Acknowledgments: The authors would like to gratefully acknowledge the contributions of the visitor experience manager for the Lake Louise, Yoho, and Kootenay Field Unit and Mercé Aniz, Aigüestortes i Estany de Sant Maurici National Park manager.

Conflicts of Interest: The authors declare no conflict of interest.

Appendix A. Yoho National Park Questionnaire (Interview Was Undertaken in a Local Language, English)

1. Is inclusive use reflected in the park's governance, legislative frameworks, and management plans? If so, please explain.
   - National Parks Act (1979). Last amended on 12 December 2017. Could you briefly explain your response?
     - Yes
     - No
   - National Parks General Regulations (SOR/78-213) last amended on 23 November 2018. Could you briefly explain your response?
     - Yes
     - No
   - National Park System Plan (1997). Could you briefly explain your response?
     - Yes
     - No
   - Parks Canada Guiding Principles and Operational Policies. Could you briefly explain your response?
     - Yes
     - No
   - Park-specific management plan (Yoho National Park, 2010). Could you briefly explain your response?
     - Yes
     - No

   When responding to questions 2, 3, and 4, please use Table A1.

2. Which, if any, accessibility improvements have you observed since your park's management plan was approved in 2010?
3. Have there been investments in accessible infrastructure since the management plan was approved? If so, using Table A1 below, indicate where the areas and programs highlighted in the 2010 management plan are.
4. In your opinion, has the implementation of accessibility improvements been successful in promoting universal use? In what ways? What do you think could be improved?
5. Have you worked with associations or groups that represent the interests of people with mobility impairments when making the improvements outlined above?

6. Are you monitoring park visits? If so, are visitors registered in terms of visitor profile, and specifically, do you keep track of visitors with mobility impairments? Can you share those visitor statistics with us? Do you have any other information available on park use by visitors with mobility impairments?
7. Finally, as an expert on the Yoho National Park, would you be interested in exploring the possibility of assessing and designing new accessible trails to facilitate access to the NP's landscapes/resources for wheelchair users?

Questions 8 and 9 only refer to the Yoho National Park when park management is actually involved in a renewal process (table is available to help you answer the following questions)

8. The new park management plan is in the process of being renewed; which associations are participating in this development? Are those data available?
9. In your state of the park assessment (SOPA), you concluded that the results have been "good" in increasing the number of park visitors and their enjoyment of the park. The next question is to know if you included visitors with disability (and how many of them there were) in this satisfaction survey. Do you believe that this evaluation should include people with disability to produce a realistic map of society?

Table A1. Yoho National Park data table results based on the accessibility actions highlighted in the 2010 management plan.

Management Plan Actions	Accessibility Improvement	Investment or Budget	Funding Type (Pub/Private)	Funding's Organization	Start Date	Finish Date	Outcome	Challenges
Hiking programs, park exhibits, and electronic media	Interpretive panels in the town field with braille for visually impaired	Unknown, project is +12 years old	Government	Parks Canada			Positive	Keeping offer updated and ensuring products met the needs of user groups
Reception and orientation at both west and east entrances	Accessible washrooms at east entrance	CAD 360,000	Government	Parks Canada	2013	2017	Positive	
Expand range of recreational, leisure, and learning opportunities	None							
Infrastructure at Kicking Horse and Monarch Campgrounds	New in 2020, accessible washroom and shower building	CAD 1,615,000		Parks Canada	2017	2019	Positive	
Winter recreation opportunities	None							
Kicking Horse Pass to the Last Spike Cultural Landscape		Unknown		Parks Canada				
Target youth, urban Canadians, and new Canadians: Kicking Horse Pass, the Spiral Tunnels, and the Burgess	None							
Shale materials to communicate the significance of cultural resources and the World Heritage Site	Accessible digital interpretation for remote access	CAD 1000		Parks Canada	2020	2020	Positive	Satellite/cell phone reception, reaching broader audiences
Special events and new recreational activities that promote public understanding and appreciation of the park	None			Parks Canada				

Appendix B. Aigüestortes i Estany de Sant Maurici National Park Questionnaire (Interview was Undertaken in a Local Language, Spanish)

1. As NP director, could you please provide your perspective, interpretation, and opinion regarding compliance with the accessibility objectives of the following documents:
   - Law 30/2014. Could you briefly explain your answer?
     - Yes
     - No

- Decree 39/2003 of 4 February approving the NP's Guiding Use and Management Plan (PRUG). Could you briefly explain your answer?
  - Yes
  - No
- Public Use Plan (PUP). Could you briefly explain your answer?
  - Yes
  - No

Please use Table A2 when answering the following questions about the actions listed in Article 12 of the Public Use Plan.

2. What improvements in terms of accessibility have been made since the approval of the national park's PRUG and PUP? Could you provide current data on what kinds of improvements have been made in the following items (PUP Article 12)?
3. Have there been any investments in infrastructure and equipment to improve accessibility since the PRUG (2003) and the PUP were approved? Could you provide current data on investments made in the following items (PUP Article 12)?
4. In your opinion, has the implementation of accessibility improvements been successful in promoting inclusive use of the national park? How have you seen that effect to be positive or negative from a social and environmental point of view?
5. When implementing accessibility improvements and, at the time, drawing up the PUP, was the opinion/collaboration of people, associations, and/or representatives that defend the interests of persons with disabilities taken into account? If so, could you please name the organizations that participated in the process?
6. Is there a record of visits to the national park? If so, are visitors being registered with additional personal information, including a record of visitors with disabilities using any sort of classification? If so, could you please share with our research team the statistical data? Do you have any other information about visitors with reduced mobility?
7. Finally, and knowing that there are already some accessible routes, we would like to know from your technical perspective and as an expert on the natural resources of the national park if you think it would be feasible to explore new entrances and routes to expand the network in terms of accessibility and to offer new landscapes/natural resources in the park to people with motor disabilities.

Table A2. Aigüestortes i Estany de Sant Maurici data table results based on the accessibility actions highlighted in the 2010 management plan.

PUP Actions	Accessibility Improvement	Investment and/or Budget	Funding Type (Public/Private)	Funding Organization	Start Date	Finish Date	Outcome (Positive and Negative)	Challenges
Education and environmental interpretation programs	"Breaking Barriers" program: specifically dedicated to people with disabilities	EUR 12,000/year + NPs workers	Public	NP authority and Regional Government of Catalonia	1993	-	Positive: 2000 students/year, 1000 of them local, as well as around 2000 visitors enjoyed the environmental educational activities. Additionally, activities organized with the ASPID foundation, which are focused on special needs' educational centers.	To promote guided visitations. To provide visitors and give students opportunities to visit villages inside the park area. Accessibility program evaluation with ONCE. Searching for new funding to develop a new park accessibility plan.
Enjoyment of nature: new options offered	Accessibility materials (tactile/transportable models)	±EUR 5000/year +NP personal staff	Public	NP authority and Catalonia Administration	2005	-	Positive: Huge diversity of indoor and outdoor nature activities (e.g., Starlight).	Deseasonalize tourism with a more well-rounded offering of activities.
Visitor centers	Periodic accessibility evaluation of visitor centers to improve it	EUR 50,000/year	Public	NP authority and Catalonia Administration		-	Positive: Visitor centers receive and inform visitors and have free Wi-Fi, permanent exhibitions, and audiovisual materials in many languages.	Adapt/promote heritage buildings, such as visitor centers, especially those located at the main entrances.
Paths	There are 6 paths with available length to people with disabilities	Since 1993, the park has made paths more accessible. In 2008. Obra Social la Caixa funded building two accessible catwalks in the park	Public	NP Catalonia and local Administrations	1993	-	Positive: Acceptable pathway network (200 km), 8–10 km available to people with disabilities located in 6 different and significant landscapes, including high mountains.	Strike a balance between high mountains and visitor security. Be able to provide updated information on the trails (especially in winter). Analyzing new proposals to expand accessible path network.

Table A2. *Cont.*

PUP Actions	Accessibility Improvement	Investment and/or Budget	Funding Type (Public/Private)	Funding Organization	Start Date	Finish Date	Outcome (Positive and Negative)	Challenges
Other services and/or public use equipment	Access to the parking lot at the "San Mauricio" lake and the "Planell" area through standard disability permits		Public	NP authority and Catalonia Administration		-	Positive: Carros de Foc path tourism program, all mountain shelters, along the path, work together promoting local products. Los Pastores ecomuseum with livestock activity promotion. Toirigo Environment Informational Center and Campsite. Sale of informational material at visitor centers. Negative: The park has detected that some people park their cars in disabled parking-only spots. Unfortunately, the park cannot punish infractors because it does not have the power to do so.	
Signage	Signs in braille	EUR 10,000/year	Public	NP authority, Catalonia and local administrations	2000	-	Positive: Good signage even in remote and high mountain areas. Billboards and maps in the main park entrances and villages. Negative: Problem with signage maintenance in winter.	Improvement of signage system taking into consideration distances and unevenness. Currently only time it takes is indicated.
Public transport	Not available		Public	Catalonia Administration		-	Positive: Access by taxi available to most emblematic park points. In summer, visitors can travel by cable car located in the south or take a bus around the park. Only available in summer.	Ensure entire loop of bus all year round (or most of the year).

References

1. Lackey, N.Q.; Tysor, D.A.; McNay, G.D.; Joyner, L.; Baker, K.H.; Hodge, C. Mental health benefits of nature-based recreation: A systematic review. *Ann. Leis. Res.* **2019**, *4*, 379–391. [CrossRef]
2. Zhang, G.; Poulsen, D.V.; Lygum, V.L.; Corazon, S.S.; Gramkow, M.C.; Stigsdotter, U.K. Health-promoting nature access for people with mobility impairments: A systematic review. *Int. J. Environ. Res. Public Health* **2017**, *14*, 703. [CrossRef] [PubMed]
3. Siikamäki, P.; Kangas, K.; Paasivaara, A. Biodiversity attracts visitors to national parks. *Biodivers. Conserv.* **2015**, *24*, 2521–2534. [CrossRef]
4. Bertzky, B.; Corrigan, C.; Kemsey, J.; Kenney, S.; Ravilious, C.; Besançon, C.; Burgess, N. *Protected Planet Report 2012: Tracking Progress towards Global Targets for Protected Areas (2012)*; IUCN: Gland, Switzerland; UNEP-WCMC: Cambridge, UK, 2012; Available online: https://www.unep-wcmc.org/system/dataset_file_fields/files/000/000/001/original/PPR2012_en.pdf?1395065803 (accessed on 1 January 2021).
5. Dudley, N. (Ed.) *Guidelines for Applying Protected Area Management Categories*; IUCN: Gland, Switzerland, 2013; 86p, Available online: https://portals.iucn.org/library/sites/library/files/documents/PAG-021.pdf (accessed on 28 December 2021).
6. Lausche, B. *Guidelines for Protected Areas Legislation*; IUCN: Gland, Switzerland, 2011; 370p.
7. Thede, A.K.; Haider, W.; Rutherford, M.B. Zoning in national parks: Are Canadian zoning practices outdated? *J. Sustain. Tour.* **2014**, *22*, 626–645. [CrossRef]
8. Leung, Y.-F.; Spenceley, A.; Hvenegaard, G.; Buckley, R. (Eds.) *Tourism and Visitor Management in Protected Areas: Guidelines for Sustainability. Best Practice Protected Area Guidelines*; Series No. 27; IUCN: Gland, Switzerland, 2018; 120p.
9. Muñoz, L.; Hausner, V.H. What do the IUCN categories really protect? A case study of the alpine regions in Spain. *Sustainability* **2013**, *5*, 2367–2388. [CrossRef]
10. Pérez-Calderón, E.; Prieto-Ballester, J.M.; Miguel-Barrado, V.; Milanés-Montero, P. Perception of sustainability of Spanish national parks: Public use, tourism and rural development. *Sustainability* **2020**, *12*, 1333. [CrossRef]
11. Devine, M.A. "Being a 'doer' instead of a 'viewer'": The role of inclusive leisure contexts in determining social acceptance for people with disabilities. *J. Leis. Res.* **2004**, *36*, 137–159. [CrossRef]
12. Shen, X.J.; Huang, R.H.; Zhang, J.S. U.S. national parks accessibility and visitation. *J. Mt. Sci.* **2019**, *16*, 2894–2906. [CrossRef]
13. Corazon, S.S.; Christoffersen Gramkow, M.; Varning Poulsen, D.; Linn Lygum, V.; Zhang, G.; Karlsson Stigsdotter, U. I would really like to visit the forest, but it is just too difficult: A qualitative study on mobility disability and green spaces. *Scand. J. Disabil. Res.* **2019**, *21*, 1–13. [CrossRef]
14. Farrington, J.H. The new narrative of accessibility: Its potential contribution to discourses in (transport) geography. *J. Transp. Geogr.* **2007**, *15*, 319–330. [CrossRef]
15. Setola, N.; Marzi, L.; Torricelli, C. Accessibility indicator for a trails network in a Nature Park as part of the environmental assessment framework. *Environ. Impact Assess. Rev.* **2018**, *69*, 1–15. [CrossRef]
16. Bianchi, P.; Cappelletti, G.M.; Mafrolla, E.; Edgardo Sica, E.; Sisto, R. Accessible tourism in natural park areas: A social network analysis to discard barriers and provide information for people with disabilities. *Sustainability* **2020**, *12*, 9915. [CrossRef]
17. Iwarsson, S.; Sthal, A. Accessibility, usability and universal design—Positioning and definition of concepts describing person-environment relationships. *Disabil. Rehabil.* **2003**, *25*, 57–66. [CrossRef]
18. United Nations. Universal Declaration of Human Rights 1948. Available online: https://www.un.org/en/about-us/universal-declaration-of-human-rights (accessed on 10 January 2019).
19. United Nations. Department of Economic and Social Affairs Disability. *The Standard Rules on the Equalization of Opportunities for Persons with Disabilities (A/RES/48/96)*. Available online: https://www.un.org/development/desa/disabilities/standard-rules-on-the-equalization-of-opportunities-for-persons-with-disabilities.html (accessed on 23 June 2020).
20. Kovacs Bruns, K.; London, G.L. Analyzing the impact of disability legislation in Canada and the United States. *J. Disabil. Policy Stud.* **2010**, *20*, 205–218. [CrossRef]
21. Citarella, A.; Sanchez Iglesias, A.I.; González Ballester, S.; Gentil Gutiérrez, A.N.; Maldonado Briegas, J.J.; Vicente Castro, F.; Gozález Bernal, J. Being disabled persons in Spain: Policies, stakeholders and services. *Int. J. Dev. Educ. Psychol. INFAD Rev. Psicol.* **2020**, *1*, 507–516. [CrossRef]
22. Lombardo, E.; Verloo, M. Institutionalizing Intersectionality in the European Union? *Int. Fem. J. Polit.* **2009**, *11*, 478–495. [CrossRef]
23. Watson, B.; Scraton, S.J. Leisure studies and intersectionality. *Leis. Stud.* **2013**, *32*, 35–47. [CrossRef]
24. Mumbi Maina-Okori, N.; Renee Koushik, J.; Wilson, A. Reimagining intersectionality in environmental and sustainability education: A critical literature review. *J. Environ. Educ.* **2018**, *49*, 286–296. [CrossRef]
25. Groulx, M.; Lemieux, C.; Freeman, S.; Cameron, J.; Wright, P.A.; Healy, T. Participatory planning for the future of accessible nature. *Local Environ.* **2021**, *26*, 808–824. [CrossRef]
26. Sisto, R.; Cappelletti, G.M.; Bianchi, P.; Sica, E. Sustainable and accessible tourism in natural areas: A participatory approach. *Curr. Issues Tour.* **2021**, *2021*, 1–18. [CrossRef]
27. Sica, E.; Sisto, R.; Bianchi, P.; Giulio Cappelletti, G. Inclusivity and responsible tourism: Designing a trademark for a national park area. *Sustainability* **2021**, *13*, 13. [CrossRef]
28. Wirtz, Z.; Hagerman, S.; Hauer, R.J.; Konijnendijk van den Bosch, C. What makes urban forest governance successful?—A study among Canadian experts. *Urban For. Urban Green.* **2021**, *58*, 126901. [CrossRef]

29. Perry, M.; Cotes, L.; Horton, B.; Kunac, R.; Snell, I.; Taylor, B.; Wright, A.; Devan, H. Enticing" but not necessarily a "space designed for me": Experiences of urban park use by older adults with disability. *Int. J. Environ. Res. Public Health* **2021**, *18*, 552. [CrossRef] [PubMed]
30. Lawrence, A.; De Vreese, R.; Johnston, M.; Konijnendijk van den Bosch, C.; Sanesi, G. Urban forest governance: Towards a framework for comparing approaches. *Urban For. Urban Green.* **2013**, *12*, 464–473. [CrossRef]
31. Newman & Park (n/i). Accessibility Standards in Outdoor Environments. Available online: https://citeseerx.ist.psu.edu/viewdoc/download?doi=10.1.1.626.4095&rep=rep1&type=pdf (accessed on 17 January 2020).
32. Menzies, A.; Mazan, C.; Borisoff, J.F.; Mattie, J.L.; Ben Mortenson, W.B. Outdoor recreation among wheeled mobility users: Perceived barriers and facilitators. *Disabil. Rehabil. Assist. Technol.* **2021**, *16*, 384–390. [CrossRef]
33. Sugerman, D. Inclusive outdoor education: Facilitating groups that include people with disabilities. *J. Exp. Educ. Winter* **2001**, *24*, 166–172. [CrossRef]
34. Pinna, F.; Garau, C.; Annunziata, A. A literature review on urban usability and accessibility to investigate the related criteria for equality in the city. In *Computational Science and Its Applications—ICCSA 2021*; Lecture Notes in Computer Science; Gervasi, O., Ed.; Springer: Cham, Switzerland, 2021; Volume 12958. [CrossRef]
35. Arts, B.; Leroy, P. *Institutional Dynamics in Environmental Governance*; Springer: Dordrecht, The Netherlands, 2006; pp. 1–21.
36. FAO. *Framework for Assessing and Monitoring Forest Governance. The Program on Forests Food and Agriculture Organization (PROFOR)*; United Nations: Rome, Italy, 2011; Available online: https://www.fao.org/climatechange/27526-0cc61ecc084048c7a9425f64942df70a8.pdf (accessed on 8 January 2021).
37. Campese, J.; Nakangu, B.; Silverman, A.; Springer, J. IUCN Commission on Environmental, Economic and Social Policy 2016. Natural Resource Governance Framework Assessment Guide. Available online: https://www.iucn.org/commissions/commission-environmental-economic-and-social-policy/our-work/knowledge-baskets/natural-resource-governance (accessed on 8 September 2020).
38. Convection on Biological Diversity. List of the Parties. 1992. Available online: https://www.cbd.int/information/parties.shtml (accessed on 8 January 2021).
39. Liefferink, D. The dynamics of policy arrangements: Turning round the tetrahedron. In *Institutional Dynamics in Environmental Governance*; Arts, B., Leroy, P., Eds.; Springer: Dordrecht, The Netherlands, 2006; Volume 45, pp. 45–69.
40. Creswell, J.W. *Qualitative Inquiry & Research Design: Choosing Among Five Approaches*, 4th ed.; SAGE Publications, Inc.: Los Angeles, CA, USA, 2018; ISBN 978-1-5063-3020-4.
41. Government of Canada. The Government of Canada Tables the Optional Protocol to the United Nations Convention on the Rights of Persons with Disabilities. Available online: https://www.canada.ca/en/employment-social-development/news/2017/11/the_government_ofcanadatablestheoptionalprotocoltotheunitednatio.html (accessed on 18 August 2019).
42. Government of Canada. Ratification of the Convention on Biological Diversity (CBD). 1993. Available online: https://www.canada.ca/en/environment-climate-change/corporate/international-affairs/partnerships-organizations/biological-diversity-convention.html (accessed on 18 August 2019).
43. Gobierno de España. Ministerio de la Presidencia, Relación con las Cortes y Memoria Democrática. Agencia Estatal Boletín Oficial del Estado. Instrumento de Ratificación de la Convección sobre los Derechos de las Personas con Discapacidad, Hecho en Nueva York el 13 de Diciembre de 2006. BOE» núm. 96, de 21 de Abril de 2008, Páginas 20648 a 20659 (12p). Available online: https://www.boe.es/eli/es/ai/2006/12/13/(1) (accessed on 18 August 2019).
44. Gobierno de España. Ministerio para la transición ecológica y reto demográfico. Convenio sobre la Diversidad Biológica. 1992. Available online: https://www.miteco.gob.es/es/biodiversidad/temas/conservacion-de-la-biodiversidad/conservacion-de-la-biodiversidad-en-el-mundo/cb_mundo_convenio_diversidad_biologica.aspx (accessed on 18 August 2019).
45. López, H. Derecho Romano y el Sistema Jurídico Anglosajón. 2014. Available online: https://derechoromanounivia.wordpress.com/2014/08/07/el-derecho-romano-y-el-sistema-juridico-anglosajon/ (accessed on 28 December 2021).
46. Government of Canada. Canadian Charter of Rights and Freedoms. 1981. Available online: https://open.canada.ca/data/en/dataset/06d31e10-a2a8-4d53-9ff3-567714a0a9f3 (accessed on 1 August 2019).
47. Justice Laws Website. Canadian Human Rights Act (R.S.C., 1985, c. H-6). Available online: https://laws.justice.gc.ca/eng/acts/H-6/rpdc.html (accessed on 1 August 2019).
48. Jacobs, L.; Anderson, M.; Rohr, R.; Perry, T. The Annotated Accessible Canada Act—Complete Text. 2021. Available online: https://scholar.uwindsor.ca/lawpub/12 (accessed on 30 May 2021).
49. Government of Canada. Justice Laws Website. Canada National Parks Act (S.C. 2000, c. 32). Available online: https://laws-lois.justice.gc.ca/eng/acts/N-14.01/ (accessed on 20 January 2019).
50. Needham, M.D.; Dearden, P.; Rollins, R.; McNamee, K. *Parks and Protected Areas in Canada. Planning and Management*, 4th ed.; Oxford University Press Canada: North York, ON, Canada, 2016; pp. 3–38.
51. Government of Canada. Parks Canada. 2019–20 Departmental Plan. 2019. Available online: https://www.pc.gc.ca/en/agence-agency/bib-lib/plans/dp/dp2019-20/index (accessed on 20 February 2021).
52. Government of Canada. Parks Canada. 2020–21 Departmental Plan. 2020. Available online: https://www.pc.gc.ca/en/agence-agency/bib-lib/plans/dp/dp2020-21/index (accessed on 20 February 2021).
53. Government of Canada. Parks Canada. National Park System Plan. Canadian Heritage Parks Canada, Canada, 1997; pp. 2–13, 20–21. Available online: http://parkscanadahistory.com/publications/system-plan-eng-1.pdf (accessed on 2 September 2019).

54. Gobierno de España. Ministerio de la Presidencia, Relación con las Cortes y Memoria Democrática. Agencia Estatal Boletín Oficial del Estado. Constitución Española, 1978. BOE» núm. 311, de 29 December 1978. Available online: https://www.boe.es/eli/es/c/1978/12/27/(1)/con (accessed on 1 August 2019).
55. Gobierno de España. Ministerio de la Presidencia. Relación con las Cortes y Memoria Democrática. Agencia Estatal Boletín Oficial del Estado. Real Decreto Legislativo 1/2013, de 29 de Noviembre, por el que se Aprueba el Texto Refundido de la Ley General de las Personas con Discapacidad y de su Inclusión Social. BOE» núm. 289, de 3 de Diciembre de 2013, Páginas 95635 a 95673 (39p). Available online: https://www.boe.es/eli/es/rdlg/2013/11/29/1 (accessed on 1 August 2019).
56. Santamarina Campos, B. El inicio de la protección de la naturaleza en España. Orígenes y balance de la conservación. *Rev. Española Investig. Sociol.* **2019**, *168*, 55–72. [CrossRef]
57. Almenar-Muñoz, M. El reciente avance en la protección de las zonas húmedas en la Comunidad Valenciana. *Actual. Jurídica Ambiental.* **2016**, *63*, 1–10.
58. López-Ramón, F. Trayectoria del régimen jurídico de los Parques Nacional en España. *Ambient. Rev. Minist. Medio Ambiente* **2014**, *106*, 82–89. Available online: https://www.miteco.gob.es/ministerio/pags/Biblioteca/Revistas/pdf_AM%2FAmbienta_2014_106_82_89.pdf (accessed on 20 February 2021).
59. Gobierno de España. Ministerio de la Presidencia, Relación con las Cortes y Memoria Democrática. Agencia Estatal Boletín Oficial del Estado. Real Decreto 389/2016 de 22 de Octubre por el que se Aprueba el Plan Director de la Red de Parques Nacionales. «BOE» núm. 257, de 24 de Octubre de 2016, Páginas 74051 a 74076 (26p). Departamento: Ministerio de Agricultura, Alimentación y Medio Ambiente. Ref.: BOE-A-2016-9690. Available online: https://www.boe.es/eli/es/rd/2016/10/22/389 (accessed on 20 September 2019).
60. Gobierno de España. Ministerio de la Presidencia, Relación con las Cortes y Memoria Democrática. Agencia Estatal Boletín Oficial del Estado. Ley 30/2014 de 3 de Diciembre de Parques Nacionales. BOE» núm. 293, de 4 de Diciembre de 2014, Páginas 99762 a 99792 (31 págs.). Available online: https://www.boe.es/eli/es/l/2014/12/03/30 (accessed on 20 January 2019).
61. Government of Canada. Parks Canada. Parks Canada Guiding Principles and Operational Policies. 2018. Available online: https://www.pc.gc.ca/en/docs/pc/poli/princip (accessed on 8 July 2020).
62. Government of Canada. Yoho National Park. Nature, Science and Culture. Available online: https://www.pc.gc.ca/en/pn-np/bc/yoho/culture (accessed on 20 January 2019).
63. Government of Canada. The Recognition and Reconciliation of Rights Policy for Treaty Negotiations in British Columbia. Available online: https://www.rcaanc-cirnac.gc.ca/eng/1567636002269/1567636037453 (accessed on 2 September 2019).
64. Gobierno de España. Ministerio Para la Transición Ecológica y reto Demográfico. Aigüestortes i Estany de Sant Maurici: Valores Culturales. Available online: https://www.miteco.gob.es/es/red-parques-nacionales/nuestros-parques/aiguestortes/valores-naturales/default.aspx (accessed on 19 December 2019).
65. Statista Research Department. National Park Visitors in Canada 2011–2020. 2021. Available online: https://www.statista.com/statistics/501269/national-park-visitors-in-canada/#:~{}:text=National%20park%20visitors%20in%20Canada%202011%2D2020&text=In%20the%20fiscal%20year%202019,15.9%20million%20the%20previous%20year.&text=The%20national%20parks%20of%20Canada%20are%20managed%20by%20Parks%20Canada%20Agency. (accessed on 28 December 2021).
66. Gobierno de España. Ministerio para la Transición Ecológica y el Reto Demográfico. Red de Parques Nacionales: Visitantes. Available online: https://www.miteco.gob.es/es/red-parques-nacionales/la-red/gestion/visitasppnn_tcm30-67283.pdf (accessed on 28 December 2021).
67. Worrell, R.; Appleby, M.C. Stewardship of natural resources: Definition, ethical and practical aspects. *J. Agric. Environ. Ethics* **1999**, *12*, 263–277. [CrossRef]
68. Saner, M.; Wilson, J. Stewardship, Good Governance and Ethics (1 December 2003). Institute On Governance Policy Brief No. 19. Available online: https://papers.ssrn.com/sol3/papers.cfm?abstract_id=1555815 (accessed on 19 December 2019).
69. EUR-Lex. Access to European Union Law. Available online: https://eur-lex.europa.eu/homepage.html?locale=en (accessed on 25 January 2020).
70. United States Department of the Interior. National Park Service. Director's Order No. 42: Accessibility for Visitors with Disabilities. 2000. Available online: https://www.nps.gov/subjects/policy/upload/DO_42_11-3-2000.pdf (accessed on 15 February 2019).
71. Fieldfare Trust. *Countryside for All Good Practice Guide. A Guide to Disabled People's Access in the Countryside*; The Fieldfare Trust, Ltd.: London, UK, 2005; Available online: https://www.pathsforall.org.uk/mediaLibrary/other/english/countryside-for-all-guide.pdf (accessed on 25 January 2020).
72. EUROPARC-España. *Manual Sobre Conceptos de uso Público en los Espacios Naturales Protegidos*; Fundación Fernando González Bernáldez: Madrid, Spain, 2005; 94p.
73. EUROPARC-España. *Catálogo de Buenas Prácticas en Materia de Accesibilidad en Espacios Naturales Protegidos*; Fundación Fernando González Bernáldez: Madrid, Spain, 2007; 343p.
74. U.S. Department of the Interior. National Park Service. Management Policies. The Guide to Managing the National Park System. 2006. Available online: https://www.nps.gov/policy/mp/policies.html#_Toc157232893 (accessed on 21 November 2019).
75. Government of Canada. Justice Laws Website. Canada National Marine Conservation Areas Act (S.C.2002, c. 18). Available online: https://laws-lois.justice.gc.ca/eng/acts/c-7.3/ (accessed on 20 February 2021).
76. Government of Canada. Justice Laws Website. Parks Canada Agency Act (S.C. 1998, c.31). Available online: https://laws-lois.justice.gc.ca/eng/acts/P-0.4/ (accessed on 20 February 2021).

77. Government of Canada. Justice Law Website. Historic Sites and Monuments Act, R.S.C., 1985, c. H-4 (2013). Available online: https://laws-lois.justice.gc.ca/eng/acts/H-4/page-1.html (accessed on 20 September 2019).
78. Government of Canada. Justice Laws Website. National Parks General Regulations (SOR/78-213). Available online: https://laws-lois.justice.gc.ca/eng/regulations/sor-78-213/index.html (accessed on 12 July 2019).
79. Government of Canada. Justice Laws Website. Heritage Railway Stations Protection Act, R.S.C.1985, c.52 (4th Supp.). Available online: https://laws-lois.justice.gc.ca/eng/acts/h-3.5/ (accessed on 12 July 2019).
80. Government of Canada. Parks Canada. Yoho National Park. Park Management Plan. 2010. Available online: https://www.pc.gc.ca/en/pn-np/bc/yoho/gestion-mgmt/plandirecteur-mgmtplan (accessed on 12 July 2019).
81. Gobierno de España. Ministerio de la Presidencia, Relaciones con las Cortes y Memoria Democrática. Agencia Estatal del Boletín Oficial del Estado. Ley 42/2007, de 13 de Diciembre, del Patrimonio Natural y de la Biodiversidad. Available online: https://www.boe.es/eli/es/l/2007/12/13/42/con (accessed on 20 July 2019).
82. Gobierno de España. Ministerio de la Presidencia, Relaciones con las Cortes y Memoria Democrática. Agencia Estatal del Boletín Oficial del Estado. Ley 7/1988, de 30 de Marzo, de Reclasificación del Parque Nacional de Aigüestortes y Lago de Sant Maurici. Available online: https://www.boe.es/eli/es-ct/l/1988/03/30/7 (accessed on 14 July 2019).
83. Gobierno de España. Ministerio para la Transición Ecológica y Reto Demográfico. Decreto 39/2003 de 4 de Febrero de 2003 del Plan Rector de Uso y Gestión de Aigüestortes i l´Estany de Sant Maurici. DOGC nº 3825. Available online: https://www.miteco.gob.es/es/red-parques-nacionales/la-red/gestion/prug_aigues_tcm30-59508.pdf (accessed on 14 July 2019).
84. Gencat. Parcs Naturals de Catalunya. Plà d´Us Públic del Parc Nacional de Aigüestortes I Estany de Sant Maurici. Available online: http://parcsnaturals.gencat.cat/web/.content/Xarxa-de-parcs/aiguestortes/Coneix-nostra-feina/instruments-planificacio/plans-programes-especifics/pla-us-public/PDF/43_109693.pdf (accessed on 14 July 2019).
85. Government of Canada. Parks Canada. Yoho National Park of Canada Draft Management Plan. 2021. Available online: https://www.pc.gc.ca/en/pn-np/bc/yoho/gestion-mgmt/plandirecteur-mgmtplan/ebauche-draft (accessed on 1 June 2021).
86. Gobierno de España. Ministerio de la Presidencia, Relaciones con las Cortes y Memoria Democrática. Ley 27/2006, de 18 de Julio, por la que se Regulan los Derechos de Acceso a la Información, de Participación Pública y de Acceso a la Justicia en Materia de Medio Ambiente (Incorpora las Directivas 2003/4/CE y 2003/35/CE). BOE» núm. 171, de 19/07/2006. Available online: https://www.boe.es/eli/es/l/2006/07/18/27/con (accessed on 10 February 2021).
87. Gencat. Parc Nacionla d´Aigüestortes i Estany de Sant Maurici. Òrgan Rector. Available online: http://parcsnaturals.gencat.cat/ca/xarxa-de-parcs/aiguestortes/coneix-la-nostra-feina/qui-som/organ-rector/ (accessed on 10 February 2021).
88. Gobierno de España. Vicepresidencia Tercera del Gobierno. Ministerio para la Transición Ecológica y el Reto Demográfico. Plan de Seguimiento y Evaluación de la Red de Parques Nacionales, 2016; p. 72. Available online: https://www.miteco.gob.es/es/red-parques-nacionales/plan-seguimiento-evaluacion/seguimiento.aspx (accessed on 10 February 2021).
89. Buhalis, D.; Michopoulou, E. Information-enabled tourism destination marketing: Addressing the accessibility market. *Curr. Issues Tour.* **2011**, *14*, 145–168. [CrossRef]
90. British Columbia Government Laws. Bill M 219. 2018 British Columbia Accessibility Act. 2018. Available online: https://www.bclaws.gov.bc.ca/civix/document/id/lc/billsprevious/3rd41st:m219-1 (accessed on 1 August 2019).
91. Government of Canada. Justice Laws Website. Accessible Canada Act (S.C. 2019, c. 10). Available online: https://laws-lois.justice.gc.ca/eng/acts/A-0.6/ (accessed on 1 June 2021).
92. British Columbia. Technical Bulletins. Information Bulletin Building and Safety Standards Branch. Accessibility in the 2018 British Columbia Building Code of 24 August 2018. Available online: https://www2.gov.bc.ca/gov/content/industry/construction-industry/building-codes-standards/bc-codes/technical-bulletins (accessed on 1 June 2021).
93. Gobierno de España. Ministerio de la Presidencia, Relaciones con las Cortes y Memoria Democrática. Agencia Estatal Boletín Oficial del Estado. Real Decreto 732/2019, de 20 de Diciembre, por el que se Modifica el Código Técnico de la Edificación, Aprobado por el Real Decreto 314/2006, de 17 de Marzo. Documento Básico SUA, Seguridad de Utilización y Accesibilidad de 20 de Diciembre de 2019. Available online: https://www.boe.es/diario_boe/txt.php?id=BOE-A-2019-18528 (accessed on 1 June 2021).

International Journal of
Environmental Research and Public Health

Article

Coastal Monitoring Using Unmanned Aerial Vehicles (UAVs) for the Management of the Spanish Mediterranean Coast: The Case of Almenara-Sagunto

Vicent Esteban Chapapría *, José Serra Peris and José A. González-Escrivá

Ports & Coastal Engineering, Universitat Politècnica Valencia, 46021 Valencia, Spain; jserra@upv.es (J.S.P.); jgonzale@upv.es (J.A.G.-E.)
* Correspondence: vesteban@upv.es

Abstract: The concentration of the world's population in coastal areas means an increase in pressure on the environment and coastal ecosystems. The impacts of climate change affect natural biophysical and ecological systems and human health. Research has been developed to create coastal monitoring with Unmanned Aerial Vehicles (UAVs) that allow data to be obtained and methodologies that integrate computer vision algorithms for 3D and image processing techniques for analysis, combined with maritime information. The Valencian oval is located on the Spanish Mediterranean coast and registers significant coastal erosion. It is a densely populated area, with high economic relevance and tourist activity. The main goals of the developed research in this coastal area include creating a methodology of data collection that identifies environmental indicators significant to community health and uses in the coastal areas, to test progression of interventions and to assess coastal erosion detection and monitoring. The final objective is to aid in decision-making and coastal management. Sediment characterization was obtained, and continuous maritime information was collected. The dynamic evolution of coastal areas was researched by using UAVs on the Spanish Mediterranean coast. This technique is suitable for measuring medium to small coastal changes. Flight planning was carried out using the grid mode and adapted to areas in order to obtain a homogeneous pixel size and precision. This monitoring program takes advantage of technological development with very low economic costs and is a good tool for making decisions that must be based on scientific information. With the monitoring work, an annual erosion between 12 and 6 m was detected. The monitoring program has evidenced the shoreline trend as a result of the impact of rigid structures, mainly ports and groins, in promoting down-drift erosion processes in the area.

Keywords: population; monitoring; coastal erosion; integrated coastal zone management; Unmanned Aerial Vehicles (UAVs)

Citation: Chapapría, V.E.; Peris, J.S.; González-Escrivá, J.A. Coastal Monitoring Using Unmanned Aerial Vehicles (UAVs) for the Management of the Spanish Mediterranean Coast: The Case of Almenara-Sagunto. *Int. J. Environ. Res. Public Health* **2022**, *19*, 5457. https://doi.org/10.3390/ijerph19095457

Academic Editors: Salvador García-Ayllón Veintimilla, Antonio Espín Tomás and Paul B. Tchounwou

Received: 7 March 2022
Accepted: 25 April 2022
Published: 29 April 2022

1. Introduction

The impacts of climate change affect natural biophysical and ecological systems and human health [1–4]. Coastal erosion is affecting the environment and economic activities globally [5]. A wide spectrum of different processes is involved [6] at different temporal and spatial scales, including hydrodynamic and other processes. The most at-risk populations are those on the coast, due to their large size and high exposure to rising sea levels, storm surges, coastal erosion and river flooding, which is increased by human-induced pressures on coastal areas [7,8].

Since the 1950s, coastal areas have steadily increased in population [9]. This is a generalized process throughout the world. In Spain, one in three people live in a town on the coast [10]. The concentration of the world's population in coastal areas means an increase in pressure on the environment and coastal ecosystems. Projections indicate that by 2050, 70% of the world's population will live in urban areas and agglomerations [11]. This is reflected in the characteristics of concentration and rapid and sustained growth. In turn, however, there are two other characteristics to consider: dispersion of the population and conurbation of two or more cities. Therefore, the study of coastal urban development on a global scale is of great interest for the integrated management of coastal areas.

Research on the evolution of cities and coastal agglomerations between 1945 and 2012 established that the number and size of cities, and the populations living in them, are continuously increasing. As a result, the concept of city and coastal agglomeration (CCA) is considered as referring to those with more than 100,000 inhabitants located less than 100 km from the coastline. In turn, however, there are two other characteristics to consider: dispersion of the population and conurbation of two or more cities. Mechanisms to determine the relationship between various ecological, geographic and socioeconomic factors and urban development on the coast have been analyzed [9]. An investigation was carried out using UN population census data with the use of Google Earth, which allowed for the observation of cities and their development, as well as associated infrastructure, especially ports and roads, and their relationship with the area. There are currently more than 2100 CCAs with more than 100,000 inhabitants in the world, where almost 1500 million people live [9]. In addition to this population, there are also other smaller population centers and those of coastal rural areas. There are areas throughout all Mediterranean countries where this is so. Many of these coastal areas are tourist destinations and tourism economic impact is relevant. CCAs refer to spaces that relate directly, fluidly, intensely and in a double direction to the terrestrial social environment and marine coastal ecosystems. Consequently, there is convergence between the integrated coastal zone management (ICZM) and urban management. As a result, they are of extraordinary importance for the integrated management of coastal areas [12–16].

Coastal erosion is a process that degrades coastal profiles and mainly occurs due to natural factors (e.g., related to climate change) and overcrowding (e.g., urbanization and massive tourism) [17]. While the first cause can be considered as a slow process, the growing presence of humans is leading to the rapid aging of coasts. Mediterranean Sea authorities are focusing on the necessity for a systematic and comprehensive approach to the management of littoral areas. Coastal areas are at risk due to rising sea levels associated with climate change. Therefore, the stability of these areas is becoming increasingly important, and sedimentary materials play an increasingly important role in determining the amount of damage from flooding and erosion [18]. The fundamentals of sedimentary transport have been studied [19–22]. Some research on used management approaches in the studied area demonstrate the inefficiency of hard measures to stop erosion and their negative impact on environmental quality [23]. Scale and resolution of the images used in some investigations which provide very good information on the evolution of the coast is large [24]. Due to varying processes and their different scales, data acquisition and the analysis of coastal evolution and its impacts have not been periodically systematized in coastal management practices [25]. Coastal evolution and erosion have been monitored in a variety of ways [12,26–28]. It has been promoted at regional and municipal levels [7,29–32]. It has been executed on dunes, estuaries, port environments, cliffs, etc., and coastal observatories have been created in recent years [25,29,33–35]. Research must explain and properly combine time and space scales associated with sediment transport [6,19].

Management strategies must be based on scientific information. Monitoring data provide adequate tools to understand coastal processes and the analysis of what is happening during each period. Remote sensing techniques, which use satellite imagery to measure the morphology and scale of sandbars, have been introduced. Tracking techniques based on Landsat and Sentinel 2 images with the use of geographic information systems (GIS) and terrestrial surveys can be achieved to obtain indicators on coastal areas by using landmarks, global positioning system (GPS) light detection and ranging (LiDAR) and tri-dimensional (3D) scanners. However, the shoreline detection problem has still not been solved, since there are no algorithms that can be used without taking into account the objectives and characteristics of remote sensing images [36]. The resolution of the images has increased markedly. However, although the resolution has increased significantly, it is still not suitable for measuring small-scale dimensional changes. [31].

More recently, coastal zone surveys, including foreshore, backshore and dunes, have acquired and processed high-resolution data obtained by vertical take-off and landing (VTOL) drones, also known as UAV with cameras [37–39]. Research has been developed for monitoring with UAVs and it has allowed to obtain data and apply methodologies to integrate artificial vision algorithms for 3D and image processing.

The objectives and goals of the developed research for the region are to assess the state of coastal erosion by using integrated systems to provide a comprehensive and complete model for coastal erosion detection and monitoring and to promote a clear pathway that will serve for decision-making and coastal management.

2. Study Area

The Valencian oval is located on the Spanish Mediterranean coast (Figure 1) and registers significant coastal erosion. Due to coastal erosion, environmental problems and difficulties for the recreational and tourist uses of the coast have increased in recent decades. The Valencian coast has a total length of 470 km. Its central area is 256 km in length and extends from Ebro Delta to Cape San Antonio. The coastline is generally in the direction of S60° E.

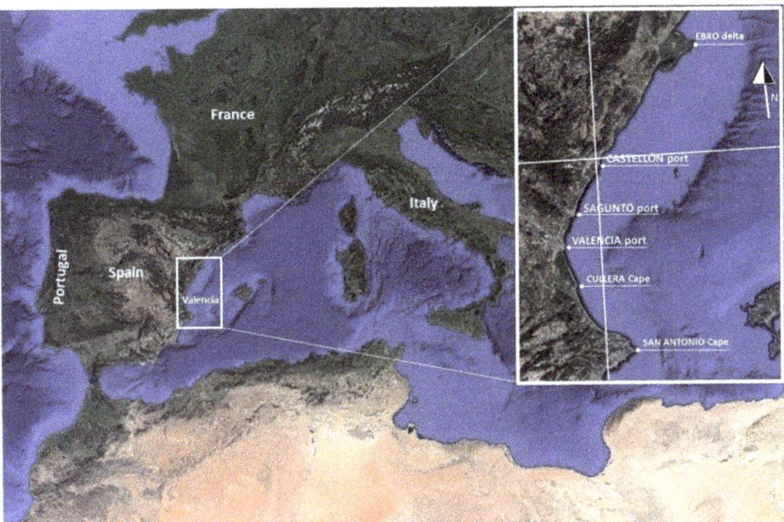

Figure 1. Valencian oval location (prepared by the authors based on Google Earth).

The total population of the Valencian coast is 4.64 million inhabitants, and it has increased by almost 2 million in the last 50 years. The central region of Valencia extends to the north and south of the city and has a continuous conurbation of more than 2.9 million inhabitants. Currently, the central coastal area has a population density of 1238 inhabitants per square kilometer. It is an area with high tourist use and very important economic activity. Valencia's port specializes in container traffic and has become a Mediterranean leader in this field, with the metropolitan area being a relevant logistic center. The effects of occupation and increased use, urbanization, degradation and destruction of coastal spaces, construction of port infrastructure and other infrastructure for land transport and hydraulic regulation have produced important sedimentary imbalances and an alteration of coastal dynamics with the consequent modification of coastal morphologies (e.g., beaches, deltas, dunes, mouths, etc.); these accumulations are the result of sedimentation and changes in profiles. Erosion in many coastal areas has increased due to the complex effects of climate change. Rising sea levels and changes in weather patterns have intensified the frequency of extreme events occurring on our planet [1,3,34–36].

An evaluation of the situation of the sedimentary balance, determining zones in accretion and others in erosion throughout the last 15 years, was obtained for the stretch of coast between the south port of Castellón and the port of Sagunto. Recently, to define interventions in areas with coastal erosion, topographical and bathymetric surveys have been developed in Almenara and the Sagunto beaches. This is a coastal area with extensive sandy beaches and dunes and is only interrupted by ports and small river mouths. Figure 2 shows the location of the coastal front where the monitoring tasks were developed. From north to south are: (a) Malvarrosa beach, between Queralt and Quartell, with a total length of 1200 m; (b) Corinto beach, south of Quartell, with a length of 1300 m; and (c) to the south, Almardà beach, with a coast length of 2030 m. The total length of the coast is 4530 m.

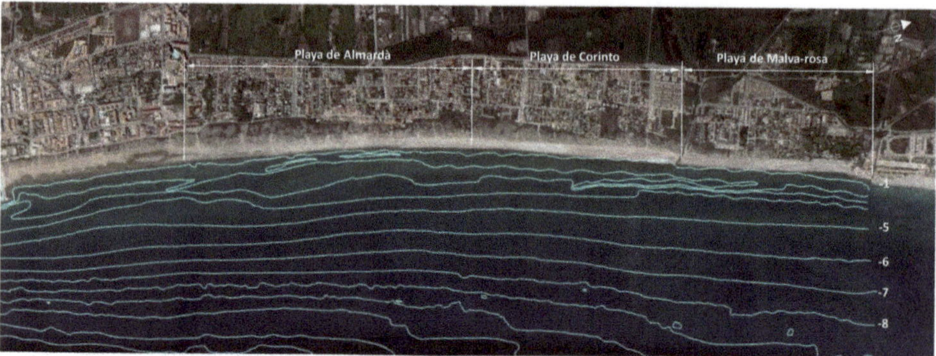

Figure 2. Bathymetry in the Almenara–Sagunto coastal area (prepared by the authors based on Google Earth).

In September 2021, the stabilization works of the La Llosa and Almenara coastline began; these are based on a construction project that aims to stabilize the coastline in this area. The actions planned in this project focus on the beaches of La Llosa, Almenara and to the north of Sagunto, undertaking, among other things, the transfer of winnowed gravel from the Malvarrosa beach to the south of the Gola de Queralt, as well as the extension of jetties. From the point of view of fauna in the dune coastal area, it is worth mentioning the importance of the nesting *Charadrius alexandrinus*, a bird included in some catalogues of endangered fauna species in the vulnerable category. The coastal regeneration works have had to consider the nesting period in a very special way. For this reason, this project must stop during the spring period.

In order to assess negative effects on beaches, especially those located in the northern area of Almardà, Corinto and Malvarrosa beaches, the monitoring of coastal areas needs to consider whether or not there is coastal erosion as a result of these regeneration works on Almenara beach. The follow-up will focus on analyzing the evolution of the Sagunto beaches and coastal area. This monitoring was extended for three months, from November 2021 to January 2022.

3. Materials and Methods

This study aims to utilize a multidisciplinary approach by combining different analyses of maritime information, coastal processes and monitoring technologies to establish the causes of coastal erosion. An integrated analysis offers advantages in obtaining comparisons and simulations of dynamic situations in order to explore possible strategies to overcome erosion and, consequently, sustain public health and uses, economic growth, minimize population risk and maintain biodiversity in coastal ecosystems. The methodology in this study aims to achieve coastal erosion management and coastal erosion research, especially on vulnerable areas, which could be based on a risk assessment [14,34,37,38].

There is no periodic information on the geometry of the coast and its temporal evolution available. There are some assessments of advances and retreats of the coastline, but they are not systematic, nor do they take into account the position of the sea level and its effects; they are also not correlated with the maritime climate between each observation period. For this reason, it is necessary for the purposes of this study to correctly carry out these controls of the resulting geometry. To this end, determinations were made in the field through the use of surveys, which are described in detail later. The sources of information on coastal geometry and coastal sediments that were used have the methodological purpose of allowing comparisons to be made between different periods of time, so that the behavior and evolution of the beaches and the coastal area can be deduced, and sedimentary movements quantified.

To monitor the coastal–littoral area under investigation, it is necessary to develop a complete methodology [19,35] that allows for the analysis of, at least, the following information, so as to determine the following characteristics and their evolution over time:

- Sedimentary materials present on the beaches and contributions and movements made;
- Geometry of backshore, beach face and surf zone profile;
- Sea level in the period of observation and survey;
- Maritime climate data, especially having regard to wind and wave regimes in the monitoring period;
- Analysis of storms and their effects in the monitoring period;
- Anthropic actions taken, such as works, occupations, etc.

In addition, it was necessary to carry out work and analyze the granulometric characteristics of coastal materials, both native and those that may have been contributed. The collection of information in this regard must be as broad as possible, especially with regard to the implementation period. Surveys were performed over three months. For the purposes of monitoring and controlling the effects to be analyzed, continuous maritime information is being collected regarding the following:

- Sea level, through the analysis of tides provided by the Network of Tide Gauges (Red de Mareógrafos, REDMAR);
- Storm surges, based on prediction data, real-time records and an analysis of historical data from the State Ports Public Agency (Ente Público Puertos del Estado, EPPE) program through the link https://www.puertos.es/oceanografia/Paginas/portus.aspx (accessed on 2 December 2021).

3.1. Sediment Characteristics and Maritime Conditions

To obtain sediment characterization, which allows for the establishment of potential sediment sources to characterize coastal conditions in the study area, sediment samples were taken and analyzed in the laboratory. Sediment sampling was performed at a central point of each of the three beaches: Malvarrosa, Corinto and Almardá beaches (Figure 2). Granulometric tests of the sampled sediments were developed once a sample was washed and organic matter was removed using ASTM sieves. The carbonate contents of the samples were determined by removing it with acid. The mineralogical analysis of the samples was carried out by visual observation and counting of fractions of ASTM series by means of a microscope. The existence of blue quartz in the fractions was also obtained since this mineral is trace in that area of the origin of the sediments.

The sedimentary transport by waves and storm surges was estimated. The wave climate in the area was established [40–44] using different data sources. Tide data were obtained from Valencia tide gauge (REDMAR network) managed by Organismo Público Puertos del Estado (OPPE), which is an hourly data series provided by the Maritime Climate Program including data collected since 1992. To determine storm surge conditions, GOS 2.1 (Global Ocean Surges) reanalysis database [31] was used. The numerical model used in the GOS reanalysis is the ROMS (Regional Ocean Modeling System), a three-dimensional model developed by Rutgers' Ocean Modeling Group (http://marine.rutgers.edu/po/index.php?model=roms, accessed on 9 November 2021). Relationships between the different sea reference levels were considered. Wind conditions (mean and extreme annual conditions) were determined, obtained through dynamic downscaling using the WRF-ARW 3.1.1 model (weather research and forecasting and advanced research dynamical solver) from the ERA-interim atmospheric reanalysis [40], developed by the European Center for Medium-Range Weather Forecasts (ECMWF). Wave data at deep water conditions were obtained, which are also from the GOW 2.1 (Global Ocean Waves 2.1 [44]). This reanalysis includes data collected since 1989 and has an hourly temporal resolution and a spatial resolution of 0.125° along the Mediterranean. The numerical model used for the generation of the reanalysis was the Wave Watch III model developed by the USA's National Oceanic and Atmospheric Administration&National Centers for Environmental Prediction, NOAA/NCEP. The SWAN-OLUCA mixed numerical model for the wave propagation to the coastline was applied [44] to obtain the breaking currents.

Aerial and satellite views were also analyzed [43,45,46]. The longshore transport rate was estimated [47,48] using the following equation:

$$Q\,(m^3/year) = 2027 \times 10^6 \times H_0{}^{5/2} \times sen(2\alpha_0) \times cos(\alpha_0)^{1/4} \times K_p \times K_g$$

where longshore transport rate, Q, is the amount of littoral drift to the right or to the left, past a point on the shoreline in an annual period. Wave conditions are computed with H_0, significant wave height in deep water depths. Angle between wave crest and shoreline is α_0. Coefficients, K_p, K_g, correct the probability of the wave height and direction, and the geometric configuration of the sector with the capacity to generate transport.

3.2. UAV System Data Collection and Zero Line

To monitor the geometry of the coast, the first task carried out was to establish control areas and points: stretches of coastline and profiles. Then, the downscaling of the information obtained from the Sentinel Hub satellite images, which is a cloud-based GIS platform for the distribution, management and analysis of satellite data, was addressed. Annual orthophoto images of the study area were obtained from the regional government of Valencia through the Institut Cartogràfic Valencià link, https://icv.gva.es/es/cartografia-tematica (accessed on 5 November 2021). All this allowed us to start the monitoring works.

For precision monitoring, field work was launched with the aim of carrying out monthly large-scale geometric surveys using drone systems. The processing of different images allows for remote sensing on shoreline detection and for the monitoring of coastal areas [36,37,39]. Applications of UAV technology can contribute to integrated coastal zone management (ICZM) and achieve the UN sustainable development goals (SDGs) [49]. The dynamic evolution of coastal areas in different areas has been researched by using UAVs [50–53]. In this study, a combination of different data is used to obtain an interdisciplinary approach.

For photogrammetric flights for geometric purposes, a global navigation satellite system (GNSS) base belonging to the air system was used to provide precise georeferencing in real time kinematic (RTK) to the images captured from the air. Flight planning was carried out in grid mode and adapted to the existing terrain in order to obtain homogeneous pixel size and precision.

The GNSS base of the aerial system is located in topographical bases. Precise XYZ coordinates were previously given to these bases, relying on the reference framework that defines the Spanish official reference system (which is the ETRS89 in UTM projection).

These coordinates were measured using a dual-frequency GNSS system with an accuracy of less than 1 cm.

In each flight, a series of support and control points (SACP) were also mediated with the dual-frequency GNSS system in order to provide precision guarantees. A precision report was delivered in each of the photogrammetric flights for the geometric analysis of the areas it has flown over. Flights were carried out with a professional drone (DJI Phantom 4 RTK and DRTK Base) and the GNSS system used to measure the SACP was a multi-constellation Spectra with a Trimble TDC600 controller (Figure 3). Once the areas to be flown over have been defined and located, flight planning was carried out for each area. This is performed using the drone's own native software.

Figure 3. The drone and GNSS base used for photogrammetric flights (provided by General Drones, SL).

The planning parameters are to fly at a height relative to the ground of 36 m to obtain a pixel size or ground sample distance (GSD) of 1 cm/pixel, with the drone's camera oriented completely from above so that the photo capture overlaps 80% longitudinally and 90% horizontally and transversely. Subsequently, the images captured by the drone are downloaded and processed using the drone's native software to enable precise georeferencing with respect to the data measured by the drone's GNSS system. This was performed post-process, whereby the position given by the GNSS base in the fraction of a second that the camera captured each of the aerial images was synchronized.

Next, Pix4D software is used to process all the georeferenced images in order to obtain the necessary metric products. These are the 3D Point Cloud, orthophotography and digital terrain model (DTM). From these products, other cartographic products can be obtained ex post in any format, including both raster and vector.

Finally, the zero line (ZL) is defined along the entire coastline using GPS methodology. The study area does not have a significant tide. Small time scale sea level changes are mainly due to wind set-up, storm surges and wave conditions. The swash zone forms the land–ocean boundary at the landward edge of the surf zone, where waves runup and rundown the beach face. The ZL is defined by runup and rundown interpolation in the swash zone. This method allows us to precisely define the edge of the beach. It can be complemented with the correction of the sea level by astronomical or meteorological tides and wave oscillation. The data collection is extended up to the −1.00 level in profiles taken between 25 m. Consequently, detailed information can be obtained from the most exposed and dynamic sections of the submerged beach profile. The total length of the stretch of coast to be monitored with the ZL is 4530 m long. All the obtained information must be analyzed throughout the monitoring time with information on the maritime climate.

4. Results

The northern part of the Spanish Mediterranean coastline is generally in the direction of S60°E; it is fairly straight and low-lying without natural indentations. There are a large number of port infrastructure and coastal protection systems, including jetties, dykes and breakwaters. It suffers considerable erosion. The coast, especially in the Valencian oval, registers significant erosion due to the north–south movement of materials and the absence of sediments from the north, which is mainly due to the presence of docking installations and other construction. Beaches are mostly composed of D_{50} = 0.25 mm sand and occasionally gravel close to river mouths. Many bathymetric studies have detected submarine materials, mainly composed of fine sand of D_{50} = 0.11–0.15 mm [49].

Problems of erosion have been occurring in recent decades. The effects of occupation and increased use, urbanization, degradation and destruction of coastal spaces, construction of port infrastructure and others for land transport and hydraulic regulation, etc., have produced significant sedimentary imbalances. The Valencian oval coast, particularly between the port of Castellón and the port of Valencia, has intensively recorded these effects. Along this coast, longitudinal protection, groins, detached breakwaters and many artificial beach nourishments have been built. All these structures, especially the ports and groins, have produced significant effects on littoral transport rates.

On the northern beaches, especially in Almenara and the Malvarrosa beaches, the fine sands have disappeared due to the action of littoral transport, which makes bathing on this beach difficult during the tourist period, especially for elderly people. Sediments in the surf zone comprise 51.3% gravel. On the beaches of Corinto and Almardá, the gravel percentages vary between 5.3 and 22.1%. The D_{50} of sediments on these beaches varies between 0.19 and 0.29 mm (Figure 4). For comparison, results of the mean sediment sizes on these beaches in 2007 and 2008 can be seen in Table 1.

Figure 4. Section and point sediments sampling in Malvarrosa beach (prepared by the authors based on Google Earth).

Table 1. Granulometric D_{50} size results in November 2007 (A) and September 2008 (B) [41].

Almardá Beach			Corinto Beach		
Height/Depth	A D_{50} (mm)	B D_{50} (mm)	Height/Depth	A D_{50} (mm)	B D_{50} (mm)
2	0.21	0.22	2	0.30	0.24
1	0.50	0.25	1	>2	>2
0	1.80	0.30	0	>2	>2
−1	0.14	0.18	−1	0.19	0.20
−2	0.19	0.16	−2	0.13	0.17
−3	0.12	0.14	−3	0.11	0.14
−4	0.12	0.13	−4	0.17	0.13
−5	0.18	0.12	−5	0.13	0.15
−6	0.16	0.12	−6	0.11	0.16
−7	0.19	0.11	−7	0.11	0.15
−8	0.11	0.11	−8	0.18	0.15

In September 2021, sediment samples were taken in two sections of Malvarrosa beach, 1 and 2, (Figure 4). In each section, samples were obtained on the coastline (1a and 2b), at 1 m above sea level on the beach (1c and 2d), and at 2 m above sea level on the beach (1e and 2f). Results of the granulometric analysis of these samples can be seen in Table 2, where D_{50} size, percentages and types of different sediments can be seen.

Table 2. Granulometric analysis, D50 size, % and type of sediments in Malavarrosa beach September 2021Malvarrosa beach.

ASTM nr	Sieving Size (mm)	Height (in Meters) of the Beach Sample Collection					
		Shoreline		+1		+2	
		1a	2b	1c	2d	1e	2f
		% Retained		% Retained		% Retained	
2 1/2	63	0	0	0	0	0	0
1	10	0	0	0	0	0	0
3	6.73	0	0	0	0	0	0
5	4	0.3	4.6	3.8	7.3	0.3	0.2
10	2	6.4	17.2	3.5	8.5	0.1	1.0
18	1	12.1	20.4	4.3	12.3	0.2	5.8
25	0.717	9.7	9.8	6.3	6.7	2.8	11.0
40	0.425	20.2	19.3	25.9	16.3	16.4	15.0
80	0.180	25.5	16.7	31.7	27.7	41.2	27.1
120	0.125	20.9	11.1	22.2	20.4	34.6	13.0
200	0.075	4.9	0.9	2.1	0.7	4.0	2.5
>200	<0.075	0	0	0.2	0.1	0.4	0.1
D_{50} (in mm)		0.38	1.04	0.25	0.34	0.27	0.32
% gravel		6.7	21.8	7.3	15.8	0.4	1.2
% sand		93.3	78.2	92.5	84.1	99.2	98.6
% silt and clays		0	0	0.2	0.1	0.4	0.2
Beach material classification		sand	sand and gravel	sand	sand	sand	sand

Research has been carried out based on aerial photos, which determined the volumes of additions and losses of sediments on the coastline. Figure 5 shows statistics maritime conditions considered to estimate longshore transport rates. Black line in Figure 5 represents registered conditions. Figure 6 shows the evolution of the coastline, calculated by photogrammetry, between Almenara and Sagunto. Evolution of the coastline is represented in different periods with reference to the original line, the coastline, in 1947. The Malvarrosa beach shows greater erosion, which decreases towards the south in the Corinto and Almardá beaches. The Almardà beach registers accumulations. The coastline of the Almardá beach was extended between 1994 and 2000.

At Corinto beach, in one 1.4 km stretch a volume of 820,152 m^3 accumulated between 1957 and 1965, and there was an erosion of 1340,584 m^3; therefore, it can be concluded that the transport capacity was approximately 65,000 m^3/year. During the analyzed period, the detected trend is the recession in the whole section. There are more pronounced variations at the Malvarrosa and Corinto beaches, while at Almardà, the variation is very uniform with a tendency for accretion or apparent stability.

Figure 7 represents the rate of evolution. This rate has an average annual value of 33 cm (−0.33 m) for the entire coastal area. The beaches of Malvarrosa and Corinto La Costa suffer from erosion, while Almardà shows a slight tendency towards accumulation. All this is clearly due to the north–south direction of net littoral solid transport.

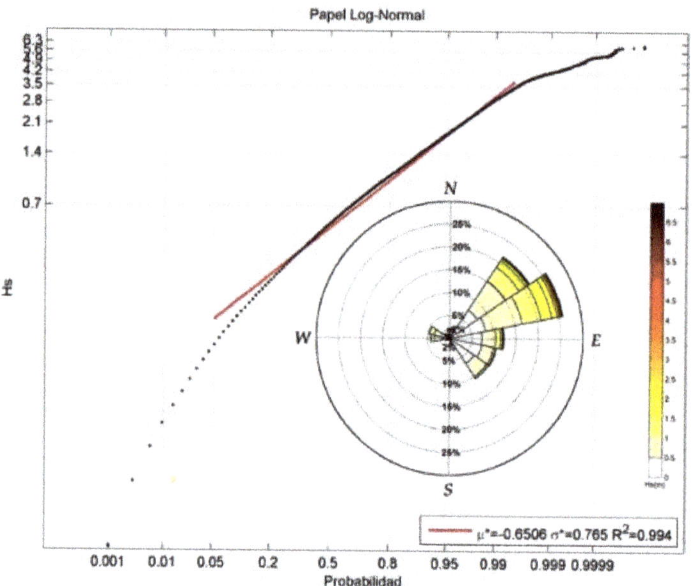

Figure 5. Mean scalar regime and significant waves in deep water coastal area of Almenara–Sagunto, showing relation between significant wave height in deep water depths, Hs, and probability of non-exceedance in x-axis (Instituto de Hidráulica Ambiental de Cantabria).

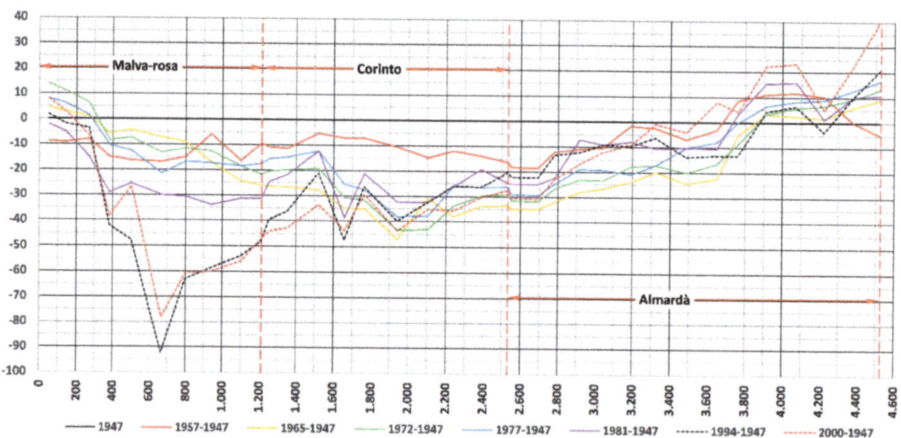

Figure 6. Shoreline changes in meters (y-axis) and coastal evolution in Malvarrosa, Corinto and Almardá beaches (x-axis, distance from the north end of the section studied to the south) (prepared by the authors).

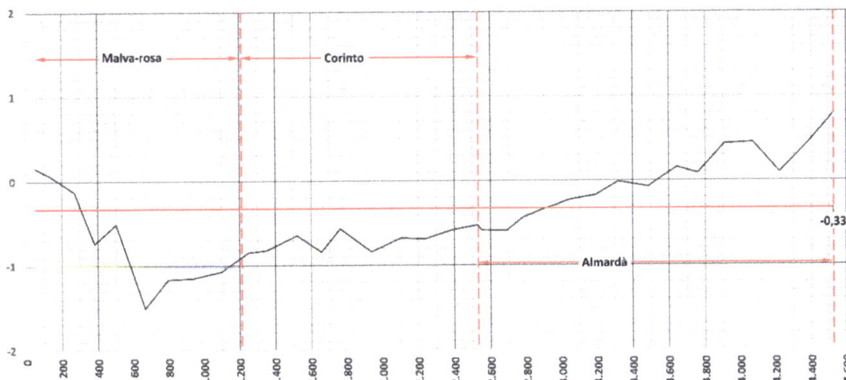

Figure 7. Coastal evolution rates at Malvarrosa, Corinto and Almardá beaches referring to 1947 situation (x-axis, distance from the north end of the section studied to the south) (prepared by the authors).

To monitor the geometric changes in the coast, three campaigns with UAV were carried out, one per month, between November and December 2021 and January 2022. Figure 8 shows the areas where campaigns were developed. Figure 9 shows some UAV image results in these areas and campaigns of the total monitored stretch of coast. In these areas there are no significant changes in sea level due to tides. The changes recorded in the campaign period involved variations in sea level with a maximum amplitude of 12 cm due to weather conditions. Consequently, the influence of this amplitude of sea level changes in the width of the beaches is not relevant. The slope of the swash zone has an average slope of 34%. Consequently, variations in sea level of 12 cm imply maximum width changes of 35 cm.

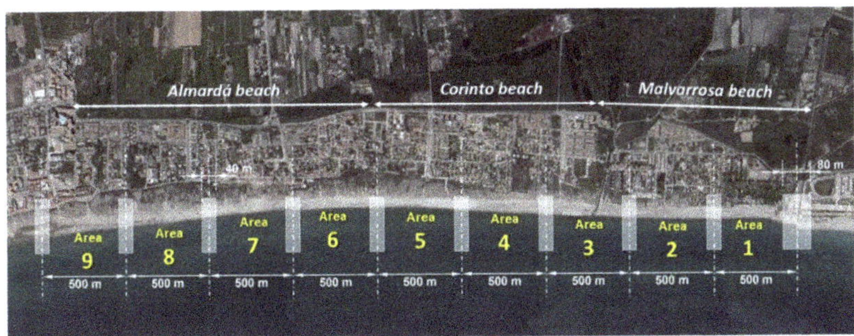

Figure 8. Different coastal areas considered for monitoring works at Malvarrosa, Corinto and Almardá beaches (prepared by the authors).

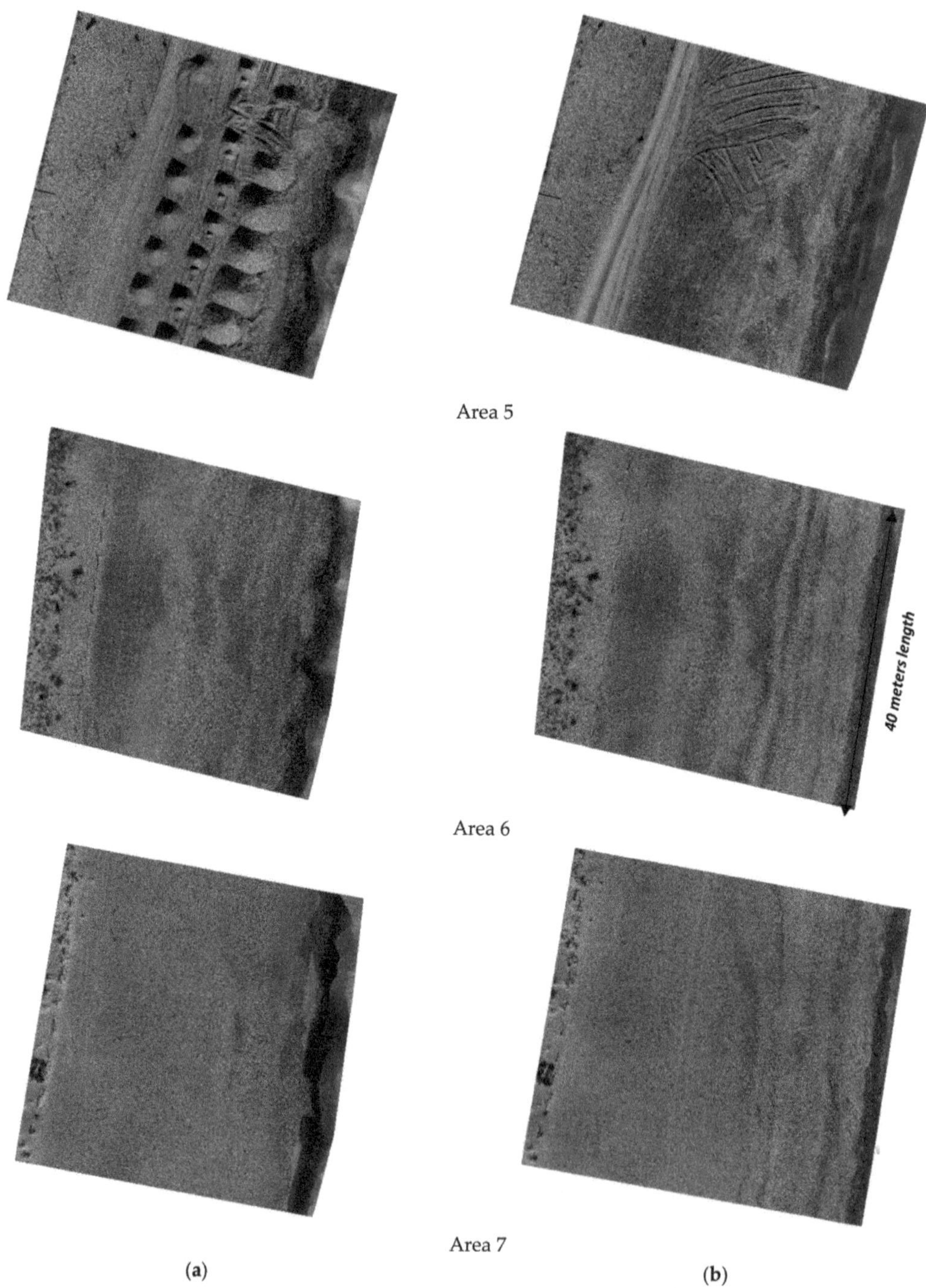

Figure 9. Different coastal area images in different campaigns showing changes and short-term coastal evolution at Malvarrosa, Corinto and Almardá beaches: (**a**) November 2021; (**b**) January 2022. Images are rotated to show the north facing upwards (prepared by the authors).

In the study area, it was possible to detect through monitoring an annual erosion of 12 m on Malvarrosa (Figure 10) beach and 6 m on Corinto beach. This evolution of the beaches can be defined as Class 6-High Erosion [7]. This definition of the beach evolution classes establishes seven classes, from Class 1-Very High Accretion to Class 7-Very High Erosion. The monitoring program has evidenced the shoreline trend as a result of the impact of rigid structures, mainly ports and groins, in promoting down-drift erosion processes in the area.

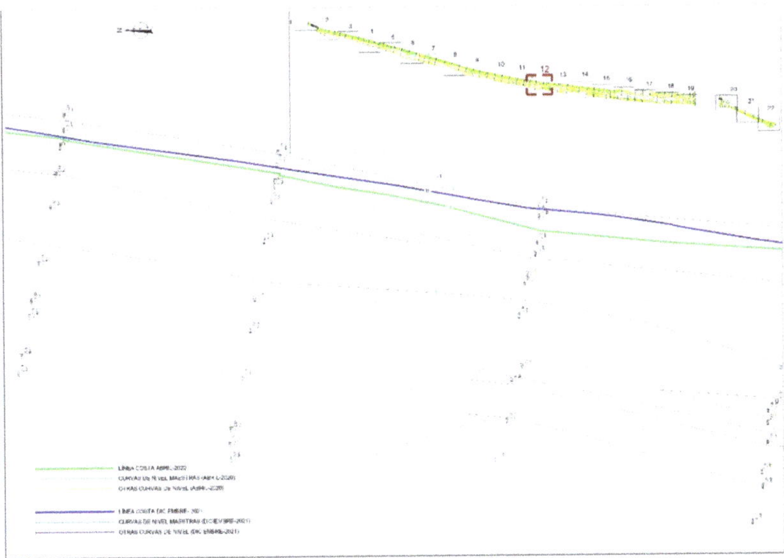

Figure 10. Evolution of coastline in Malvarrosa beach since April 2021 (green line), to December 2022 (blue line) (prepared by the authors).

5. Discussion

Cities and urban areas expand their spaces, and the services they manage converge with the management of coastal spaces and necessary preservation. This has produced the alteration of coastal dynamics with the consequent modification of coastal morphologies (e.g., beaches, deltas, dunes, mouths, etc.), erosion, accumulations due to sedimentation, changes in profiles, etc. The vulnerability of the coast is rapidly increasing due to climate change. Vulnerability is manifested via increases in flood levels, beach profile changes, erosion, increases and changes in coastal morphology, variations in littoral transport rates and coastal erosion. Ports can create coastal erosion by altering wave patterns. The research that was carried out attempts to clarify environmental effects in managing port-induced coastal erosion occurring at beaches, which are extensively used by the population.

Changes in weather patterns due to climate change produce an intensification of the frequency of extreme sea events. Beaches in this area are showing coastal sedimentary movements that lead to the appearance of gravel and its propagation towards the south in very significant magnitudes. This has been very noticeable after recent extraordinary storms in September 2019 and the Gloria storm in January 2020.

On the Spanish Mediterranean coast, more than 90% of the sand and gravel comes from rivers, torrents and streams, while the rest comes from erosion materials from cliffs, in addition to vegetable matter. Granulometric results show that sediment beaches along the Sagunto and Almenara coast are a mix of sand and gravel. Until a few years ago, Almenara and Sagunto beaches maintained a sand cover, giving the impression of sandy beaches.

Recently, erosive processes in this coastal area have displaced the sand and exposed the gravel (Figure 11).

Figure 11. December 2021, image of Malvarrosa beach facing northward. Coastal erosion has caused the disappearance of the sand cover. Coarse sediments, such as gravel, remain on the beach.

The obtained shoreline change rates show that problems in the coast between the ports of Castellón and Sagunto are the result of general degradation of natural conditions, resulting in a fragmented coast with sedimentary imbalances generated by ports and other infrastructure, human use and invasive urbanization of coastal spaces. River regulation and sediment reduction supplies to the coast and the effects of climate change have increased erosion. At the same time, to correct the situation, there have been no investments made according to environmental, natural and socioeconomic values. ICZM under the principle of sustainability is needed. The current result is a degraded and fragmented coast, which requires management and intervention through coordinated action and sustainability.

Over the years, the imbalance in sediment and transport, which is strongly influenced by port constructions and groins, has been sought to be mitigated through the building of various defense works. The entire area under investigation here has been altered from its natural initial dynamic. The areas with the greatest problems are those in which strong transport and urban areas coincide. The current situation is attributable to successive transformations that have taken place over time in an initially continuous coast, with an area of lagoons and incipient dunes. The construction of the ports of Sagunto in 1902, Castellón in 1915 and Burriana in 1932 reduced littoral transport rates, and a significant volume of sediments were retained upstream of these ports. The urbanism developed over the years, which was more intense from the 1960s/1970s and affected natural conditions, especially in dune areas. Solutions given to these problems were local, with coastal defenses mostly developed in front of urban centers, which gave rise to even greater sedimentary imbalances, forming important concavities downstream (Figure 12).

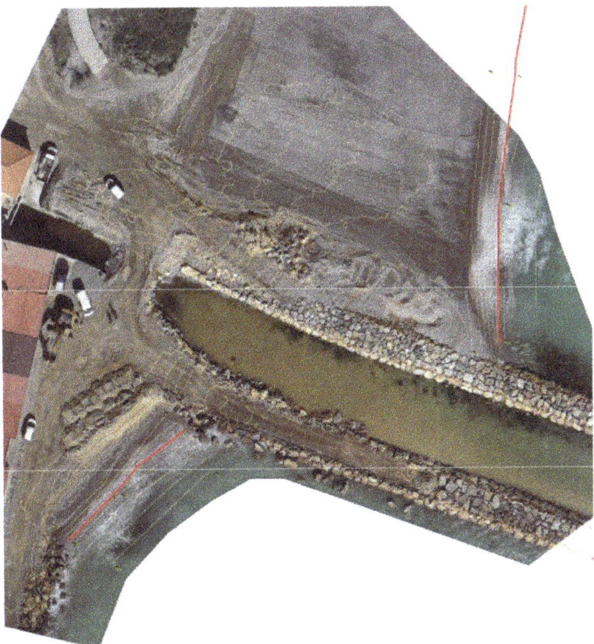

Figure 12. Orthophotography of defense coastal erosion works at Almenara and Sagunto beaches; ZL (red lines) and DTM (yellow lines) from UAV.

The diagnosis of sections between the port of Burriana and the port of Sagunto recommend a reanalysis study on all the currently existing defense works, taking into account climate change and extending to critical points with serious current erosion. With the start-up of the stabilization works on the coast of Almenara and Sagunto, it was possible to develop a monitoring system that provides data on coastal evolution on an appropriate time and space scale. Dunes and beach changes were evaluated. Coastal management requires us to define its characteristics and evolution. The task of continuously collecting information is a basic necessity for proper management and, in addition, today there are increasingly powerful and accessible technological means for it.

ICZM is a key element for the sustainable development of the coastal areas. One of the consequences of climate change is the increase in global coastal erosion and the loss of valuable coastal areas. A large number of development activities, such as construction of ports, urbanization, massive tourism, overexploitation of aquifers and the presence of dams, are increasing the vulnerability of coastal areas. Due to extraordinary storms on the coast, there is a great urgency for ICZM. Action must be addressed with a robust system based on coastal monitoring and the evaluation of results. Recent trends in coastal erosion mitigation are addressing towards soft and innovative methods, since hard methods have important impacts such as high cost and down-drift erosion. The debate on coastal sustainability must consider the effectiveness and adequacy of necessary investments, consolidating them into conservation plans and ICZM.

6. Conclusions

To act against coastal erosion and correct the current situation affecting dunes and beaches, it is necessary to promote soft and innovative solutions such as sand bypassing, dune rehabilitation and dune vegetation. These solutions will also work to prevent coastal erosion due to stronger storms caused by climate change. During the stabilization works of a coastal area on the Spanish Mediterranean coast, a monitoring investigation was developed based on the use of UAVs and maritime and sediment information. Coastal monitoring provided appropriate information because of its spatial and temporal dimensions and data quality to strengthen scientific knowledge and sustainability.

This monitoring program made it possible to manage very precise information, and high-resolution images can be processed to detect changes, both from the works in progress and from natural changes due to the action of winds, waves and tides. This monitoring program is a good tool for making decisions that must be based on scientific information and takes advantage of technological development with very low economic costs.

Author Contributions: Conceptualization, methodology, validation, formal analysis and investigation, V.E.C., J.S.P. and J.A.G.-E.; writing—original draft preparation, writing—review and editing and supervision, V.E.C. All authors have read and agreed to the published version of the manuscript.

Funding: This research was partially funded by Sagunto's municipality, grant number 882472F.

Institutional Review Board Statement: Not applicable.

Conflicts of Interest: The authors declare no conflict of interest.

References

1. Bell, C.; Masys, A.J. Climate Change, Extreme Weather Events and Global Health Security a Lens into Vulnerabilities. In *Advanced Sciences and Technologies for Security Applications*; Springer: Cham, Switzerland, 2020.
2. Ford, J.D.; Bell, T.; Couture, N.J. Perspectives on Canada's North Coast Region. In *Canada's Marine Coasts in a Changing Climate*; Government of Canada: Ottawa, ON, Canada, 2016.
3. Islam, M.M.; Shahin, M.; Miraj, M.; Ghosh, S.; Islam, M.N.; Islam, I. Climate Change Impacts and Mitigation Approach: Coastal Landscape, Transport, and Health Aspects. In *Springer Climate*; Springer: Cham, Switzerland, 2021.
4. Lukoseviciute, G.; Panagopoulos, T. Management Priorities from Tourists' Perspectives and Beach Quality Assessment as Tools to Support Sustainable Coastal Tourism. *Ocean Coast. Manag.* **2021**, *208*, 105646. [CrossRef]
5. Saadon, M.S.I.; Ab Wahida, N.S.; Othman, M.R.; Nor, D.A.M.; Mokhtar, F.S.; Nordin, N.; Kowang, T.O.; Nordin, L. An Evaluation of the Impact of Coastal Erosion to the Environment and Economic Activities at Mengabang Telipot, Terengganu. *J. Crit. Rev.* **2020**, *7*, 1132–1136. [CrossRef]
6. Gracia, F.J.; Anfuso, G.; Benavente, J.; del Río, L.; Domínguez, L.; Martínez, J.A. Monitoring Coastal Erosion at Different Temporal Scales on Sandy Beaches: Application to the Spanish Gulf of Cadiz Coast. *J. Coast. Res.* **2005**, *49*, 22–27.
7. Molina, R.; Anfuso, G.; Manno, G.; Prieto, F.J.G. The Mediterranean Coast of Andalusia (Spain): Medium-Term Evolution and Impacts of Coastal Structures. *Sustainability* **2019**, *11*, 3539. [CrossRef]
8. Sumner, P.D.; Rughooputh, S.D.D.V.; Boojhawon, R.; Dhurmea, K.; Hedding, D.W.; le Roux, J.; Pasnin, O.; Tatayah, V.; Zaynab, A.; Nel, W. Erosion Studies on Mauritius: Overview and Research Opportunities. *South Afr. Geogr. J.* **2021**, *103*, 65–81. [CrossRef]
9. Barragán, J.M.; de Andrés, M. Analysis and Trends of the World's Coastal Cities and Agglomerations. *Ocean Coast. Manag.* **2015**, *114*, 11–20. [CrossRef]
10. De Andres, M.; Manuel Barragan, J. Urban-Coastal Development. Study Method for Quantifying in a Global Scale. *Rev. Estud. Andal.* **2016**, *33*, 11–14.
11. UN-Habitat. *Planning Sustainable Cities Global Report on Human Settlements 2009*; UN-Habitat: Nairobi, Kenya, 2016.
12. Williams, A.T.; Rangel-Buitrago, N.; Pranzini, E.; Anfuso, G. The Management of Coastal Erosion. *Ocean Coast. Manag.* **2018**, *156*, 4–20. [CrossRef]
13. Jonah, F.E.; Mensah, E.A.; Edziyie, R.E.; Agbo, N.W.; Adjei-Boateng, D. Coastal Erosion in Ghana: Causes, Policies, and Management. *Coast. Manag.* **2016**, *44*, 116–130. [CrossRef]
14. Rangel-Buitrago, N.; Neal, W.J.; de Jonge, V.N. Risk Assessment as Tool for Coastal Erosion Management. *Ocean Coast. Manag.* **2020**, *186*, 105099. [CrossRef]
15. Rangel-Buitrago, N.; de Jonge, V.N.; Neal, W. How to Make Integrated Coastal Erosion Management a Reality. *Ocean Coast. Manag.* **2018**, *156*, 290–299. [CrossRef]
16. Martínez, C.; Contreras-López, M.; Winckler, P.; Hidalgo, H.; Godoy, E.; Agredano, R. Coastal Erosion in Central Chile: A New Hazard? *Ocean Coast. Manag.* **2018**, *156*, 141–155. [CrossRef]
17. Girau, R.; Anedda, M.; Fadda, M.; Farina, M.; Floris, A.; Sole, M.; Giusto, D. Coastal Monitoring System Based on Social Internet of Things Platform. *IEEE Internet Things J.* **2020**, *7*, 1260–1272. [CrossRef]
18. Filho, W.L.; Hunt, J.; Lingos, A.; Platje, J.; Vieira, L.W.; Will, M.; Gavriletea, M.D. The Unsustainable Use of Sand: Reporting on a Global Problem. *Sustainability* **2021**, *13*, 3356. [CrossRef]
19. Ouillon, S. Why and How Do We Study Sediment Transport? Focus on Coastal Zones and Ongoing Methods. *Water* **2018**, *10*, 390. [CrossRef]
20. Esteban, V.; Serra, J.; Diez, J.J. Beach Erosion in the South End of the "Ovalo Valenciano". In Proceedings of the International Conference on Coastal and Port Engineering in Developing Countries, Colombo, Sri Lanka, 20–26 March 1983; Volume 1.

21. Diez, J.J.; Arenillas, M.; Esteban, V. The Performance of Different Shore Protection Systems in the East-Spanish Mediterranean Coasts. 1966. Available online: https://orcid.org/0000-0001-7007-9357 (accessed on 10 November 2021).
22. Lopez, M.; Lopez, I.; Aragones, L.; Serra, J.C.; Esteban, V. The Erosion on the East Coast of Spain: Wear of Particles, Mineral Composition, Carbonates and Posidonia Oceanica. *Sci. Total Environ.* **2016**, *572*, 487–497. [CrossRef]
23. Semeoshenkova, V.; Newton, A. Overview of Erosion and Beach Quality Issues in Three Southern European Countries: Portugal, Spain and Italy. *Ocean Coast. Manag.* **2015**, *118*, 12–21. [CrossRef]
24. Cabezas-Rabadán, C.; Pardo-Pascual, J.E.; Almonacid-Caballer, J.; Rodilla, M. Detecting Problematic Beach Widths for the Recreational Function along the Gulf of Valencia (Spain) from Landsat 8 Subpixel Shorelines. *Appl. Geogr.* **2019**, *110*, 102047. [CrossRef]
25. Laporte-Fauret, Q.; Marieu, V.; Castelle, B.; Michalet, R.; Bujan, S.; Rosebery, D. Low-Cost UAV for High-Resolution and Large-Scale Coastal Dune Change Monitoring Using Photogrammetry. *J. Mar. Sci. Eng.* **2019**, *7*, 63. [CrossRef]
26. Jena, B.K.; Sathish Kumar, D.; Kintada, K. Operational Strategy to Monitor Coastal Erosion in Tropical Areas. *Int. J. Ocean. Clim. Syst.* **2017**, *8*, 135–143. [CrossRef]
27. Vousdoukas, M.I.; Ranasinghe, R.; Mentaschi, L.; Plomaritis, T.A.; Athanasiou, P.; Luijendijk, A.; Feyen, L. Sandy Coastlines under Threat of Erosion. *Nat. Clim. Chang.* **2020**, *10*, 260–263. [CrossRef]
28. Li, R.; Liu, J.K.; Felus, Y. Spatial Modeling and Analysis for Shoreline Change Detection and Coastal Erosion Monitoring. *Mar. Geod.* **2001**, *24*, 1–12. [CrossRef]
29. Kerguillec, R.; Audère, M.; Baltzer, A.; Debaine, F.; Fattal, P.; Juigner, M.; Launeau, P.; le Mauff, B.; Luquet, F.; Maanan, M.; et al. Monitoring and Management of Coastal Hazards: Creation of a Regional Observatory of Coastal Erosion and Storm Surges in the Pays de La Loire Region (Atlantic Coast, France). *Ocean Coast. Manag.* **2019**, *181*, 104904. [CrossRef]
30. Palutikof, J.P.; Rissik, D.; Webb, S.; Tonmoy, F.N.; Boulter, S.L.; Leitch, A.M.; Perez Vidaurre, A.C.; Campbell, M.J. CoastAdapt: An Adaptation Decision Support Framework for Australia's Coastal Managers. *Clim. Chang.* **2019**, *153*, 491–507. [CrossRef]
31. Leatherman, S.P. Coastal Erosion and the United States National Flood Insurance Program. *Ocean Coast. Manag.* **2018**, *156*, 35–42. [CrossRef]
32. Randazzo, G.; Lanza, S. Regional Plan against Coastal Erosion: A Conceptual Model for Sicily. *Land* **2020**, *9*, 307. [CrossRef]
33. Bagot, P.; Huybrechts, N.; Sergent, P. Satellite-Derived Topography and Morphological Evolution around Authie Macrotidal Estuary (France). *J. Mar. Sci. Eng.* **2021**, *9*, 1354. [CrossRef]
34. Silva, R.; Martínez, M.L.; Hesp, P.A.; Catalan, P.; Osorio, A.F.; Martell, R.; Fossati, M.; da Silva, G.M.; Mariño-Tapia, I.; Pereira, P.; et al. Present and Future Challenges of Coastal Erosion in Latin America. *J. Coast. Res.* **2014**, *71*, 1–16. [CrossRef]
35. Bio, A.; Bastos, L.; Granja, H.; Pinho, J.L.S.; Goncalves, J.A.; Henriques, R.; Madeira, S.; Magalhaes, A.; Rodrigues, D. Methods for Coastal Monitoring and Erosion Risk Assessment: Two Portuguese Case Studies. *J. Integr. Coast. Zone Manag.* **2015**, *15*, 47–53. [CrossRef]
36. Toure, S.; Diop, O.; Kpalma, K.; Maiga, A.S. Shoreline Detection Using Optical Remote Sensing: A Review. *ISPRS Int. J. Geo-Inf.* **2019**, *8*, 75. [CrossRef]
37. Papakonstantinou, A.; Topouzelis, K.; Pavlogeorgatos, G. Coastline Zones Identification and 3D Coastal Mapping Using UAV Spatial Data. *ISPRS Int. J. Geo-Inf.* **2016**, *5*, 75. [CrossRef]
38. Topouzelis, K.; Papakonstantinou, A.; Doukari, M. Coastline Change Detection Using Unmanned Aerial Vehicles and Image Processing Techniques. *Accept. Fresenius Environ. Bull.* **2017**, *26*, 5564–5571.
39. Nikolakopoulos, K.G.; Sardelianos, D.; Fakiris, E.; Papatheodorou, G. New Perspectives in Coastal Monitoring. *Earth Resour. Environ. Remote Sens. GIS Appl. X* **2019**, *11156*, 1115602.
40. Lim, C.; Kim, T.; Lee, S.; Yeon, Y.J.; Lee, J.L. Quantitative Interpretation of Risk Potential of Beach Erosion Due to Coastal Zone Development. *Nat. Hazards Earth Syst. Sci. Discuss.* **2021**, 180. [CrossRef]
41. Javier, D.J.; Veiga, E.M.; Esteban, V. *Effects of Climate Change on Coastal Evolution*; Springer: Berlin/Heidelberg, Germany, 2015; ISBN 9783319093000/9783319092997.
42. Fitton, J.M.; Hansom, J.D.; Rennie, A.F. A Method for Modelling Coastal Erosion Risk: The Example of Scotland. *Nat. Hazards* **2018**, *91*, 931–961. [CrossRef]
43. Fraser, C.; Bernatchez, P.; Dugas, S. Development of a GIS Coastal Land-Use Planning Tool for Coastal Erosion Adaptation Based on the Exposure of Buildings and Infrastructure to Coastal Erosion, Québec, Canada. *Geomat. Nat. Hazards Risk* **2017**, *8*, 1103–1125. [CrossRef]
44. Instituto de Hidráulica Ambiental de la Universidad de Cantabria. *Elaboración de La Metodología y Bases de Datos Para La Proyección de Impactos Del Cambio Climático a Lo Largo de La Costa Española*; Ministerio para la Transición Ecológica: Madrid, Spain, 2019.
45. Dee, D.P.; Uppala, S.M.; Simmons, A.J.; Berrisford, P.; Poli, P.; Kobayashi, S.; Andrae, U.; Balmaseda, M.A.; Balsamo, G.; Bauer, P.; et al. The ERA-Interim Reanalysis: Configuration and Performance of the Data Assimilation System. *Q. J. R. Meteorol. Soc.* **2011**, *137*, 553–597. [CrossRef]
46. Mentaschi, L.; Vousdoukas, M.I.; Pekel, J.F.; Voukouvalas, E.; Feyen, L. Global Long-Term Observations of Coastal Erosion and Accretion. *Sci. Rep.* **2018**, *8*, 12876. [CrossRef]
47. U.S. Army Corps of Engineers. *Coastal Engineering Manual*; U.S. Army Corps of Engineers: Washington, DC, USA, 2002.
48. Coastal Engineering Research Center (U.S.). *Shore Protection Manual: Volume I and II*; Coastal Engineering Research Center (U.S.): Vicksburg, MS, USA, 1984. [CrossRef]

49. Kandrot, S.; Hayes, S.; Holloway, P. Applications of Uncrewed Aerial Vehicles (UAV) Technology to Support Integrated Coastal Zone Management and the UN Sustainable Development Goals at the Coast. *Estuaries Coasts* **2021**, 1–20. [CrossRef]
50. Callaghan, K.; Engelbrecht, J.; Kemp, J. The Use of Landsat and Aerial Photography for the Assessment of Coastal Erosion and Erosion Susceptibility in False Bay, South Africa. *S. Afr. J. Geomat.* **2015**, *4*, 65. [CrossRef]
51. Yoo, C.I.; Oh, T.S. Beach Volume Change Using Uav Photogrammetry Songjung Beach, Korea. *Int. Arch. Photogramm. Remote Sens. Spat. Inf. Sci.* **2016**, *XLI-B8*, 1201–1205. [CrossRef]
52. Chang, Y.S.; Jin, J.Y.; Jeong, W.M.; Kim, C.H.; Do, J.D. Video Monitoring of Shoreline Positions in Hujeong Beach, Korea. *Appl. Sci.* **2019**, *9*, 4984. [CrossRef]
53. Lu, C.H.; Chyi, S.J. Using UAV-SfM to Monitor the Dynamic Evolution of a Beach on Penghu Islands. *Terr. Atmos. Ocean. Sci.* **2020**, *31*, 283–293. [CrossRef]

MDPI
St. Alban-Anlage 66
4052 Basel
Switzerland
Tel. +41 61 683 77 34
Fax +41 61 302 89 18
www.mdpi.com

International Journal of Environmental Research and Public Health Editorial Office
E-mail: ijerph@mdpi.com
www.mdpi.com/journal/ijerph

www.ingramcontent.com/pod-product-compliance
Lightning Source LLC
LaVergne TN
LVHW071611170726
843515LV00009B/2370

* 9 7 8 3 0 3 6 5 7 3 8 0 9 *